# Vitamin D and Human Health

# Vitamin D and Human Health

Special Issue Editor

**Michal Zmijewski**

MDPI • Basel • Beijing • Wuhan • Barcelona • Belgrade

*Special Issue Editor*
Michal Zmijewski
Gdanski Uniwersytet Medyczny
Poland

*Editorial Office*
MDPI
St. Alban-Anlage 66
4052 Basel, Switzerland

This is a reprint of articles from the Special Issue published online in the open access journal *International Journal of Molecular Sciences* (ISSN 1422-0067) from 2018 to 2019 (available at: https://www.mdpi.com/journal/ijms/special_issues/vitamin_D_human_health)

For citation purposes, cite each article independently as indicated on the article page online and as indicated below:

LastName, A.A.; LastName, B.B.; LastName, C.C. Article Title. *Journal Name* **Year**, *Article Number*, Page Range.

**ISBN 978-3-03897-586-1 (Pbk)**
**ISBN 978-3-03897-587-8 (PDF)**

# Contents

# About the Special Issue Editor

**Michal Zmijewski** received a doctoral degree (2003) from the University of Gdansk, Poland (Department of Biochemistry). During postdoctoral training, he worked in the laboratory of Prof. Andrzej Slominski (2004–2008, Center of Cancer Research, UTHSC, Memphis, TN, USA). After returning to Poland, he defended his habilitation thesis (2010) and currently he works at the Department of Histology, Medical University of Gdansk (MUG), Gdansk, Poland. In 2016, the title of Professor of Medical Sciences was granted to Michal Zmijewski by the President of Poland. Since 2012 (two terms: 2012–16 and 2016–20) he has served as Vice-Dean of the Faculty of Medicine, MUG. He is also the member of several scientific organizations and the Vice-President of Polish Histochemical and Cytochemical Society. Prof. Zmijewski is a well-established specialist in dermato-endocrinology, a relatively new discipline combining dermatology and neuroendocrinology. Human skin is not only subjected to neuroendocrine regulation but also actively synthesizes a number of neuropeptides; steroid hormones; vitamin D; and other biologically active molecules. Their local production is essential for the integrity of the epidermal barrier, protection against harmful factors, and the regulation of inflammatory responses. Prof. Zmijewski mainly focuses on multidimensional studies on the role of vitamin D in the pathogenesis, prevention, and treatment of civilization diseases. He also specializes in the discovery and testing of new anticancer drugs. In his laboratory, he is investigating several vitamin D derivatives and their classic metabolites as potential therapeutic agents. He is also interested in the skin's response to stress, with a particular emphasis on changes in the expression of the elements of cutaneous hypothalamic–hypothalamic–adrenal axis (HPA) axis during the differentiation of keratinocytes or the course of skin diseases, including psoriasis, atopic dermatitis, or mastocytosis. He is conducting several projects in collaboration with scientists from Poland, US, Germany, UK, Australia, and Italy. Prof. Zmijewski is the author of 80 scientific publications, seven book chapters, and more than 100 abstracts and invited presentations. He was and continues to be a primary investigator (PI) or co-investigator on several grants (NCN, MNiSW, Program Strategmed, Poland; and NIH, USA). His cumulative impact factor is IF<280 ($h$-index 32).

International Journal of
*Molecular Sciences*

MDPI

*Editorial*

# Vitamin D and Human Health

**Michal A. Zmijewski**

Department of Histology, Medical University of Gdańsk, 80-211 Gdańsk, Poland; mzmijewski@gumed.edu.pl

Received: 28 December 2018; Accepted: 29 December 2018; Published: 3 January 2019

**Abstract:** Vitamin D is currently one of the hottest topics in research and clinics, as well as in everyday life. Over the past decades, scientists gathered overwhelming evidence indicating that the observed global vitamin D deficiency not only has a negative impact on human skeletal system, but also facilitates development and progression of multiple disease of civilization, including cardiovascular diseases, diabetes, autoimmune disease, and cancer. This Special Issue, entitled "Vitamin D and Human Health", summarizes recent advances in our understanding of pleiotropic activity of vitamin D in the form of eight comprehensive reviews. Furthermore, eight research papers provide new insight into vitamin D research and highlight new directions.

**Keywords:** vitamin D; analogs of vitamin D; vitamin D deficiency; supplementation; vitamin D activity and metabolism; extra-skeletal effects of vitamin D; therapy and prevention

---

## 1. Introduction

The active form of vitamin D (1,25(OH)2D3, calcitriol) regulates calcium–phosphate homeostasis through the interaction with vitamin D receptor (VDR). It also has a huge impact on the proper functioning of musculoskeletal, immune, nervous, and cardiovascular systems. It is well known that despite huge progress, the technical revolution caused substantial changes in the environment and a human life. An introduction of diets based on highly processed food, an indoor lifestyle, and sun avoidance greatly contributed to the development of the global vitamin D deficiency. A low level of vitamin D is strongly correlated with a decreased calcium level, which in turn leads to inadequate mineralization of bones with subsequent development of rickets in children or osteoporosis in adults. It results not only in bone deformation, but also in high susceptibility of falls and bone fractures. Thus, proper vitamin D supplementation according to recent standards is essential for maintenance of the body homeostasis [1–4]. In spite of tremendous efforts and accumulating data concerning the impact of vitamin D on human life, there is still the need for extensive studies on molecular mechanisms activated by vitamin D, which would underline potential benefits of this pleiotropic hormone. On the other hand, clinical significance of vitamin D needs to verified through a series of large, randomized, controlled long-term trials based on comparison of serum levels of 25(OH)D3 rather than doses of supplementations.

## 2. Vitamin D and Health

Vitamin D deficiency is inseparably connected to demineralization of bones, which results in an increased susceptibility to fractures. Atteritano and coworkers presented a case study showing the relation between low vitamin D levels and susceptibility to bone fragility fractures in HIV-positive patients [5]. De Luca et al. showed that the presence of specific alleles of *Fok*I single nucleotide polymorphism (SNP) in the vitamin D receptor (*VDR*) gene affects cell proliferation and inflammatory response. The study was contacted on intervertebral disc cells derived from patients with discopathy or spondylolisthesis. Interestingly, it seems that the *Ff* variant of *Fok*I genotype of the *VDR* gene is more responsive to the anti-inflammatory effects of vitamin D and could be used as a diagnostic factor in discopathies [6]. Furthermore, the effects of the *VDR* SNPs were also investigated in acute pancreatitis (AP). It was found that a particular SNP (allele T in *Taq*1) is almost three times more frequent in AP patents in comparison with alcohol-abuse controls [7]. Thus, it seems that alteration in vitamin D signaling through the presence of unique SNPs in the *VDR* gene may be associated with predisposition to certain pathologies.

The involvement of vitamin D in the regulation of the functions of the cardiovascular system and its pleasurable impact on hypertension is currently under debate and intensive investigation. Legarts et al. [8]. summarized our current knowledge concerning the role of vitamin D in regulation of blood pressure and development of hypertension. It has to be underlined that multiple animal studies involving VDR-null mutants have shown that vitamin D has a direct impact on regulation of the renin–angiotensin–aldosterone axis and *VDR* mutations, or a low level of vitamin D results in an increase in the blood pressure. However, human trials or retrospective studies have not found a clear association between vitamin D level or its supplementation with hypertension. It could have been the problem with suboptimal study design and hopefully ongoing large scale, randomized studies will validate potential benefits of vitamin D in the treatment of hypertension.

Vitamin D deficiency is often association with several neurological diseases, as the vitamin D receptor is expressed in several brain structures including the hippocampus, hypothalamus, substantia nigra, and thalamus. Most importantly, vitamin D regulates the expression of neurotrophins, including neural growth factor (NGF) and neurotransmitters (acetylcholine, dopamine, and gamma-aminobutyric acid). In the current issue of *IJMS*, Morretti and coworkers [9] summarized our current knowledge concerning the role of vitamin D in the prevention and treatment of neurological disorders, focusing on multiple sclerosis, stroke, and Alzheimer's and Parkinson's diseases.

### 3. Vitamin D and Immune Response

It is well established that vitamin D inhibits proliferation and induces differentiation of the cells of different linages and is essential for regeneration of the epithelial barrier, as well as maturation of immune cells. For example, lymphocytes, neutrophils, monocytes, and dendritic cells not only express VDR and are direct targets for 1,25(OH)2D3, but also activate circulating 25(OH)D3 through hydroxylation by CYP27B1 [10]. The immunomodulatory effects of 1,25(OH)2D3 include switching between cell-mediated response (Th1) and humoral immunity (Th2). Vitamin D activates macrophages and production of antimicrobial peptides by epithelial and immune cells, which could be essential in the eradication of bacterial or viral infections. It is not surprising that an occurrence of the sessional infections, such as influenza, is often linked to vitamin D deficiency. Keeping in mind the various effects of vitamin D on immune response, Gruber-Bzura [11] discussed the potential role of vitamin D in influenza prevention and treatment. It has to be underlined that an impact of vitamin D on the immune system is usually cell type, tissue, or organ dependent. For instance, it was recently suggested that vitamin D could be useful in the prevention and treatment of autoimmune diseases such as multiple sclerosis, type 1 diabetes mellitus, rheumatoid arthritis, or systemic lupus erythematous (SLR). The consequence of vitamin D deficiency in the lupus development and progression was reviewed by Mak [12]. Strikingly, exposure to UV light is a major contributor to SLR flare up, thus the sun avoidance behavior only aggravates vitamin D deficiency in patients with lupus. On the other hand, a few recent clinical studies suggested not only a correlation of vitamin D deficiency with the severity of lupus, but also that proper supplementation may inhibit the production of autoantibodies, decrease the Th1/Th17 and memory B cells fractions, and reduce fatigue [12]. Furthermore, an increased activity of the immune system, including production of specific antibodies, is also the most important cause of graft-versus-host disease in recipients of allogeneic hematopoietic stem cell transplantation. Thus, the modulatory role of vitamin D may decrease adverse effects of graft-versus-host disease [10].

### 4. Vitamin D and Cancer

It is well established that the low level of vitamin D is associated with an increased risk of any type of cancer and a decrease survival rate, mainly because of an increased severity of the symptoms and metastatic potential of malignancies [13]. Very promising clinical studies analyzed by Medrano [10] suggested that vitamin D supplementation is significantly associated with an increase in overall survival and lower risk of relapse of myeloid, but not lymphoid malignancies in transplant recipients. The possible link between vitamin D and an immune regulation of the tumor microenvironment was also discussed by Liu et al. [13]. It is well established that vitamin D modulates an immune response through the inactivation with the NFκB pathway. In the tumor stroma, secretion of cytokines and prostaglandins is essential for the propagation of cancer cells, but vitamin D, through the downregulation of NFκB and cyclooxygenase 2 (COX-2), can attenuate their secretion. On the other hand, Pawlik and coworkers [14] observed that vitamin D and its analogs (PRI-2191 and PRI-2205) modulate the prevalence of a certain fraction of lymphocytes (an increase number of T helper lymphocytes (Th2), regulatory T (Treg), granulocytes, and B lymphocytes), but reduce the fraction of TCD4+, TCD4+CD25+, and TCD8+ cells in the 4T1 mouse mammary gland cancer model. It was accompanied by the modulation of the level of pro-tumorogenic cytokines in the serum. It seems that the modulatory effects of vitamin D in a cancer treatment may also include the adverse effects, which should be considered.

Cancer metastasis is the most important problem in the treatment of any type of cancer. For instance, in melanoma, metastasis dramatically decreases the survival rate of patients [15]. Many studies have shown recently that vitamin D and its analogs can be used in adjuvant radio-therapy (see recent review [16]). In the current issue of *IJMS*, Podgórska et al. [17] documented that treatment with either 1,25(OH)2D3 or 25(OH)D3 sensitized human (SKMEL-188) and Bomirski's hamster melanoma cells to low doses of proton beam radiation. Interestingly, vitamin D is also considered in the treatment of benign tumors such as uterine fibroids, derived from smooth muscle cells of the uterus. As reviewed by Ciebiera and coworkers [18], a few clinical studies have shown that low serum levels of 25(OH)D3 or the presence of specific SNPs of the genes related to vitamin D metabolism or activity correlate with the occurrence of uterine fibroids. Thus, keeping in mind antiproliferative and antifibrotic properties of vitamin D, authors suggested its potential beneficial effects not only in prevention, but also in the treatment of uterine fibroids [18].

## 5. Vitamin D Analogs

For many years, both supplementation and clinical uses of vitamin D were limited because of the potential occurrence of hypercalcemia. Thus, many laboratories around the world have investigated vitamin D analogs, which do not affect calcium level, but still possess antiproliferative and immunomodulatory properties of the active form of vitamin D. Hundreds of synthetic analogs have been investigated so far, with some pleasurable effects. Interestingly, resent studies have shown that not only does 1,25(OH)2D3 possesses biological activity, but also its precursor, 25(OH)D3, could effectively inhibit proliferation of melanoma cells [19] or be used as a radio-sensitizing agent in the melanoma treatment [17]. It is also well established that the cholesterol side-chain cleavage enzyme P450scc (*CYP11A1*) could catalase the synthesis of several vitamin D hydroxyderivatives. Those compounds were shown to be the potent inhibitors of a cell proliferation with immunomodulatory properties (see recent review [15]). Slominski and coworkers [20] have demonstrated that one of the products of CYP450ssc enzymatic activity, 20,23(OH)2D3, and the active form of vitamin D (1,25(OH)2D3) share similar, but also activate unique genomic targets. This observation could at least partially explain the decreased impact of 20,23(OH)2D3 on the serum level of calcium in comparison with calcitriol. In the current issue of *IJMS*, Wasiewicz et al. [19] compared the antiproliferative activity of 1,25(OH)2D3, synthetic calcipotriol, and a short side-chain vitamin D analog 21-hydoxypregnacalciferol (21(OH)pD) on three melanoma cell lines. Interestingly, it was shown that the antiproliferative activity of 21(OH)pD was not fullly dependent on the expression of VDR. This particular observation could be of great importance because in melanoma (like in many other cancers), a decreased level of VDR correlates with disease progression [15]. Finally, diverse effects of two vitamin D analogs, (24*R*)-1,24-dihydroxyvitamin D3 (PRI-2191) and 5,6-*trans* isomer of calcipotriol (PRI-2205), on the tumor microenvironment and metastasis of 4T1 mouse mammary gland cancer were studied by Prof. Wietrzyk's group ([14] and see discussion above).

## 6. New Cellular Targets for Vitamin D and Its Analogs

It is well established that the active form of vitamin D (1,25(OH)2D3) binding to the VDR–RXR complex and its subsequent translocation to the nucleus activates the classic genomic pathway. However, the existence of a fast nongenomic vitamin D response with the involvement of cell membrane bond VDR and/or protein disulfide isomerase PDIA3 was also postulated [2]. The modulation of immune response by nongenomic pathway was also discussed by Medrano [10]. Interestingly, a recent study from Prof Slominski's group [20] suggested that an aryl hydrocarbon receptor (AhR) is a new unique target for 20,23(OH)2D3. This unexpected observation has opened new therapeutic options for this unique vitamin D analog.

Recently, mitochondria have been recognized as a potential target for the action of vitamin D. Ricca and colleagues [21] showed that VDR plays a crucial role in the regulation of mitochondrial respiration and protects cells from an excessive production of reactive oxygen species (ROS) and subsequent cell damage. This is in line with our recent observation that vitamin D and its analogs modulate mitochondrial membrane potential; production of reactive oxygen species (ROS); and expression of ROS-associated genes, including catalase and superoxide dismutases (SOD1 and SOD2) [22]. On the other hand, Abu el Maaty and coworkers have recently investigated the potential targeting of thioredoxin-interacting protein (TXNIP) by vitamin D [23]. TXNIP is known to play a pivotal role in the regulation of glucose and redox homeostasis and its expression was shown to be modulated by vitamin D. However, the current study [23] postulated more complex interactions between vitamin D and TXNIP. The effects of vitamin D on TXNIP expression were shown to be cancer cell line specific and glucose dependent. Furthermore, there are also indications that vitamin D affects TXNIP protein stability during prolongated incubation.

Finally, our recent studies showed that VDR is not fully required for antiproliferative activities of short side chained analogs of vitamin D such as 21-hydroxypregnacaliferol (21(OH) pD [19,24]. However, the potential intracellular pathways activated by these vitamin D analogs still remain to be discovered.

## 7. Conclusions

This Special Issue gives insight into the evolving field of vitamin D regarding its mechanisms of action, deficiency, supplementation, health benefits, and clinical applications.

There is ongoing debate as to whether vitamin D should be treated only as a supplement, eventually used in prophylactics, or if it could be also considered in the therapy of multiple disorders. Having in mind pleiotropic, modulatory effects of vitamin D, the serum level of 25-OH D3 should be always considered as an important diagnostic factor, especially in the case of vitamin D deficiency. Multiple clinical trials also showed positive effects of vitamin D supplementation on overall human health, and suggested its possible use in the treatment of several diseases, including cancer. However, further large studies are still required in order to validate the potential benefits and safety of vitamin D in clinics. On the other hand, low calcemic analogs are a very promising alternative for calcitriol, and new pathways activated by vitamin D and its analogs broadened our knowledge concerning the role of vitamin D in human health and disease.

**Conflicts of Interest:** The authors declare no conflict of interest.

## References

1. Płudowski, P.; Karczmarewicz, E.; Bayer, M.; Carter, G.; Chlebna-Sokół, D.; Czech-Kowalska, J.; Dębski, R.; Decsi, T.; Dobrzańska, A.; Franek, E.; et al. Practical guidelines for the supplementation of vitamin D and the

treatment of deficits in Central Europe—Recommended vitamin D intakes in the general population and groups at risk of vitamin D deficiency. *Endokrynol. Polska* **2013**, *64*, 319–327. [CrossRef]

2. Piotrowska, A.; Wierzbicka, J.; Żmijewski, M.A. Vitamin D in the skin physiology and pathology. *Acta Biochim. Pol.* **2016**, *63*, 17–29. [CrossRef] [PubMed]
3. Holick, M.F. The vitamin D deficiency pandemic and consequences for nonskeletal health: Mechanisms of action. *Mol. Asp. Med.* **2008**, *29*, 361–368. [CrossRef] [PubMed]
4. Pludowski, P.; Holick, M.F.; Grant, W.B.; Bischoff-Ferrari, H.A.; Gordon, C.M.; Hanley, D.A.; Heaney, R.P.; Murad, M.; Weaver, C. Vitamin D supplementation guidelines. *J. Steroid Biochem. Mol. Biol.* **2018**, *175*, 125–135. [CrossRef] [PubMed]
5. Atteritano, M.; Mirarchi, L.; Venanzi-Rullo, E.; Santoro, D.; Iaria, C.; Catalano, A.; Lasco, A.; Arcoraci, V.; Lo Gullo, A.; Bitto, A.; et al. Vitamin D Status and the Relationship with Bone Fragility Fractures in HIV-Infected Patients: A Case Control Study. *Int. J. Mol. Sci.* **2018**, *19*, 119. [CrossRef] [PubMed]
6. De Luca, P.; de Girolamo, L.; Perucca Orfei, C.; Viganò, M.; Cecchinato, R.; Brayda-Bruno, M.; Colombini, A. Vitamin D's Effect on the Proliferation and Inflammation of Human Intervertebral Disc Cells in Relation to the Functional Vitamin D Receptor Gene FokI Polymorphism. *Int. J. Mol. Sci.* **2018**, *19*, 2002. [CrossRef]
7. Cieślińska, A.; Kostyra, E.; Fiedorowicz, E.; Snarska, J.; Kordulewska, N.; Kiper, K.; Savelkoul, H.F.J. Single Nucleotide Polymorphisms in the Vitamin D Receptor Gene. *Int. J. Mol. Sci.* **2018**, *19*, 1919. [CrossRef] [PubMed]
8. Legarth, C.; Grimm, D.; Wehland, M.; Bauer, J.; Krüger, M. The Impact of Vitamin D in the Treatment of Essential Hypertension. *Int. J. Mol. Sci.* **2018**, *19*, 455. [CrossRef] [PubMed]
9. Moretti, R.; Morelli, M.E.; Caruso, P. Vitamin D in Neurological Diseases: A Rationale for a Pathogenic Impact. *Int. J. Mol. Sci.* **2018**, *19*, 2245. [CrossRef]
10. Medrano, M.; Carrillo-Cruz, E.; Montero, I.; Perez-Simon, J.A. Vitamin D: Effect on Haematopoiesis and Immune System and Clinical Applications. *Int. J. Mol. Sci.* **2018**, *19*, 2663. [CrossRef] [PubMed]
11. Gruber-Bzura, B.M. Vitamin D and Influenza-Prevention or Therapy? *Int. J. Mol. Sci.* **2018**, *19*, 2419. [CrossRef] [PubMed]
12. Mak, A. The Impact of Vitamin D on the Immunopathophysiology, Disease Activity, and Extra-Musculoskeletal Manifestations of Systemic Lupus Erythematosus. *Int. J. Mol. Sci.* **2018**, *19*, 2355. [CrossRef] [PubMed]
13. Liu, W.; Zhang, L.; Xu, H.-J.; Li, Y.; Hu, C.-M.; Yang, J.-Y.; Sun, M.-Y. The Anti-Inflammatory Effects of Vitamin D in Tumorigenesis. *Int. J. Mol. Sci.* **2018**, *19*, 2736. [CrossRef] [PubMed]
14. Pawlik, A.; Anisiewicz, A.; Filip-Psurska, B.; Nowak, M.; Turlej, E.; Trynda, J.; Banach, J.; Gretkierewicz, P.; Wietrzyk, J. Calcitriol and Its Analogs Establish the Immunosuppressive Microenvironment That Drives Metastasis in 4T1 Mouse Mammary Gland Cancer. *Int. J. Mol. Sci.* **2018**, *19*, 2116. [CrossRef] [PubMed]
15. Slominski, A.T.; Brożyna, A.A.; Skobowiat, C.; Zmijewski, M.A.; Kim, T.K.; Janjetovic, Z.; Oak, J.; Jozwicki, Z.; Jetten, A.; Mason, Z.; et al. On the role of classical and novel forms of vitamin D in melanoma progression and management. *J. Steroid Biochem. Mol. Biol.* **2018**, *177*, 159–170. [CrossRef] [PubMed]
16. Śniegocka, M.; Podgórska, E.; Płonka, P.M.; Elas, M.; Romanowska-Dixon, B.; Szczygieł, M.; Żmijewski, M.A.; Cichorek, M.; Markiewicz, A.; Brożyna, A.A.; et al. Transplantable Melanomas in Hamsters and Gerbils as Models for Human Melanoma. Sensitization in Melanoma Radiotherapy-From Animal Models to Clinical Trials. *Int. J. Mol. Sci.* **2018**, *19*, 1048. [CrossRef]
17. Podgorska, E.; Drzal, A.; Matuszak, Z.; Swakon, J.; Slominski, A.; Elas, M.; Urbanska, K. Calcitriol and Calcidiol Can Sensitize Melanoma Cells to Low–LET Proton Beam Irradiation. *Int. J. Mol. Sci.* **2018**, *19*, 2236. [CrossRef]
18. Ciebiera, M.; Włodarczyk, M.; Ciebiera, M.; Zaręba, K.; Łukaszuk, K.; Jakiel, G. Vitamin D and Uterine Fibroids-Review of the Literature and Novel Concepts. *Int. J. Mol. Sci.* **2018**, *19*, 2051. [CrossRef]
19. Wasiewicz, T.; Piotrowska, A.; Wierzbicka, J.; Slominski, A.T.; Zmijewski, M.A. Antiproliferative Activity of Non-Calcemic Vitamin D Analogs on Human Melanoma Lines in Relation to VDR and PDIA3 Receptors. *Int. J. Mol. Sci.* **2018**, *19*, 2583. [CrossRef]
20. Slominski, A.T.; Kim, T.-K.; Janjetovic, Z.; Brożyna, A.A.; Żmijewski, M.A.; Xu, H.; Sutter, T.R.; Tuckey, R.C.; Jetten, A.M.; Crossman, D.K. Differential and Overlapping Effects of 20,23$(OH)_2$D3 and 1,25$(OH)_2$D3 on Gene Expression in Human Epidermal Keratinocytes: Identification of AhR as an Alternative Receptor for 20,23$(OH)_2D_3$. *Int. J. Mol. Sci.* **2018**, *19*, 3072. [CrossRef]

21. Ricca, C.; Aillon, A.; Bergandi, L.; Alotto, D.; Castagnoli, C.; Silvagno, F. Vitamin D Receptor Is Necessary for Mitochondrial Function and Cell Health. *Int. J. Mol. Sci.* **2018**, *19*, 1672. [CrossRef] [PubMed]
22. Piotrowska, A.; Wierzbicka, J.; Ślebioda, T.; Woźniak, M.; Tuckey, R.C.; Slominski, A.T.; Żmijewski, M.A. Vitamin D derivatives enhance cytotoxic effects of $H_2O_2$ or cisplatin on human keratinocytes. *Steroids* **2016**, *110*, 49–61. [CrossRef] [PubMed]
23. Abu el Maaty, M.A.; Almouhanna, F.; Wölfl, S. Expression of TXNIP in Cancer Cells and Regulation by 1,25(OH)$_2$D$_3$: Is It Really the Vitamin D$_3$ Upregulated Protein? *Int. J. Mol. Sci.* **2018**, *19*, 796. [CrossRef] [PubMed]
24. Wasiewicz, T.; Szyszka, P.; Cichorek, M.; Janjetovic, Z.; Tuckey, R.C.; Slominski, A.T.; Zmijewski, M.A. Antitumor effects of vitamin d analogs on hamster and mouse melanoma cell lines in relation to melanin pigmentation. *Int. J. Mol. Sci.* **2015**, *16*, 6645–6667. [CrossRef] [PubMed]

International Journal of
*Molecular Sciences*

MDPI

*Article*

# Vitamin D Status and the Relationship with Bone Fragility Fractures in HIV-Infected Patients: A Case Control Study

**Marco Atteritano [1,*], Luigi Mirarchi [1], Emmanuele Venanzi-Rullo [1], Domenico Santoro [1], Chiara Iaria [2], Antonino Catalano [1], Antonino Lasco [1], Vincenzo Arcoraci [1], Alberto Lo Gullo [1], Alessandra Bitto [1], Francesco Squadrito [1] and Antonio Cascio [3]**

1 Department of Clinical and Experimental Medicine, University of Messina, 98125 Messina, Italy; mirarchi.luigi@tiscali.it (L.M.); evenanzirullo@gmail.com (E.V.-R.); santisi@hotmail.com (D.S.); catalanoa@unime.it (A.C.); alasco@unime.it (A.L.); varcoraci@unime.it (V.A.); albertologullo@virgilio.it (A.L.G.); abitto@unime.it (A.B.); fsquadrito@unime.it (F.S.)
2 Infectious Diseases Unit-ARNAS Civico, Di Cristina, Benfratelli, 90127 Palermo, Italy; iaria.chiara@gmail.com
3 Department of Health Promotion Sciences and Mother and Child Care "G. D'Alessandro", University of Palermo, 90127 Palermo, Italy; Antonio.cascio03@unipa.it
* Correspondence: matteritano@unime.it; Tel.: +39-090-221-2388; Fax: +39-090-293-5162

Received: 5 December 2017; Accepted: 23 December 2017; Published: 2 January 2018

**Abstract:** HIV-infected patients show high risk of fracture. The aims of our study were to determine the prevalence of vertebral fractures (VFs) and their associations with vitamin D in HIV patients. 100 patients with HIV infection and 100 healthy age- and sex-matched controls were studied. Bone mineral density was measured by quantitative ultrasound at the non-dominant heel. Serum osteocalcin and C-terminal telopeptide of collagen type 1 served as bone turnover markers. Bone ultrasound measurements were significantly lower in patients compared with controls (Stiffness Index (SI): 80.58 ± 19.95% vs. 93.80 ± 7.10%, respectively, $p < 0.001$). VFs were found in 16 patients and in 2 controls. HIV patients with vertebral fractures showed lower stiffness index (SI) (70.75 ± 10.63 vs. 83.36 ± 16.19, respectively, $p = 0.045$) and lower vitamin D levels (16.20 ± 5.62 vs. 28.14 ± 11.94, respectively, $p < 0.02$). The majority of VFs (87.5%) were observed in HIV-infected patients with vitamin D insufficiency, and regression analysis showed that vitamin D insufficiency was significantly associated with vertebral fractures (OR 9.15; 95% CI 0.18–0.52, $p < 0.04$). VFs and are a frequent occurrence in HIV-infected patients and may be associated with vitamin D insufficiency.

**Keywords:** bone ultrasound; HIV; osteoporosis; vertebral fractures; vitamin D

## 1. Introduction

Low bone mineral density (BMD) is a common finding in HIV-infected patients [1,2], and a higher prevalence of osteopenia and osteoporosis up to 70% and 30%, respectively, was reported [3,4]. A systematic review concluded that the probability of osteopenia and osteoporosis could be over 6-fold and almost 4-fold higher in HIV-infected than in the non-infected population, respectively [5], and low BMD is associated with increased risk of bone fragility fractures. HIV-infected patients have been reported to show a higher rate of fragility fractures with an almost 9-fold increased risk of hip fracture and 3-fold increased risk of overall fractures [6–8]. Multiple factors appear to be involved in the pathogenesis of bone loss in HIV-infected patients, including HIV viral protein, antiretroviral therapy (ART) side effects, inflammatory cytokines, and bone turnover in addition to traditional risk factors [9–15]. As is known, vitamin D plays a prominent role in bone and mineral metabolism, and it has been suggested to be involved in the pathogenesis of decreased bone mass and fractures in the general population [16]; thus, it is plausible that vitamin D may exert a role in bone loss and fragility fractures also in HIV-infected patients. The aims of our study were to determine the prevalence of vertebral fractures and to assess the role of bone ultrasound parameters, bone turnover markers, and vitamin D in relation to vertebral fractures in HIV-infected patients.

## 2. Results

Demographic and clinical features of participants are shown in Table 1. As expected in light of the matching design of the study, chronological age ($p = 0.73$) and sex ($p = 1.00$) and family history of osteoporosis and fracture ($p = 0.79$) were comparable in HIV-infected patients and controls. Other relevant factors associated with BMD were also non-significantly different in the two groups (Table 1). The ultrasound parameter at calcaneus, SI was significantly lower in HIV-infected patients in comparison with control (SI: 80.58 ± 19.95% vs. 93.80 ± 7.10%, $p < 0.001$) (Figure 1a). Serum levels of calcium and phosphorus, urinary calcium and phosphorus (Figure 2a–d), and bone turnover markers (Figure 3c,d) did not differ significantly between the study groups. The level of PTH was higher, albeit not significantly, in patients compared with control subjects (Figure 3a). 25-Hydroxivitamin D3 (25(OH)D3) level was significant lower (20.29 ± 4.05 ng/mL vs. 35.77 ± 6.50 ng/mL, $p < 0.001$) in HIV-infected patients compared with controls (Figure 3b). Lateral spine X-ray documented single or multiple VFs in 16 patients (16%) and 2 controls, with a significantly different prevalence in the two groups ($p < 0.05$). Ten patients had a single Grade-1 fracture, 5 patients had two Grade-3 fractures, and 1 patient had multiple fractures (two Grade-1 and three Grade-3). In HIV-infected patients, the comparison between subjects without vertebral fractures (No-VFs) and subjects with VFs showed that this latter group reported a poorer measurement of SI (83.36 ± 16.19 vs. 70.75 ± 10.63; $p = 0.045$) (Table 2) and a lower 25(OH)D3 (28.14 ± 11.94 vs. 16.2 ± 5.62; $p < 0.02$) (Table 2). No statistically significant differences were found in age, BMI, smoking habits, and viro-immunological markers between fractured and non-fractured patients.

VFs were significantly associated with the insufficiency of 25(OH)D3, and these associations were significant after adjusting for age, renal function, and systemic hypertension. No other significant associations were found between 25(OH)D3 serum levels and the other clinical parameters assessed in HIV-infected patients.

**Table 1.** Anthropometric data and risk factors for fractures in two groups; results are expressed as number, means ± S.D., and percentage.

| Characteristics | HIV ($n$ = 100) | Controls ($n$ = 100) | $p$ |
|---|---|---|---|
| Age (years) | 45.36 ± 12.02 | 44.85 ± 9.30 | 0.73 |
| Body Mass Index (kg/m$^2$) | 25.01 ± 4.23 | 24.80 ± 4.65 | 0.73 |
| Sex (F/M) | 84/16 | 84/16 | 1.00 |
| Duration of disease (years) | 16.29 ± 6.48 | 0 | N/A |
| Current tobacco use, $n$ (%) | 22 (22) | 20 (20) | 0.88 |
| Alcohol intake (mL/week) | 450 ± 100 | 450 ± 80 | 1.00 |
| Calcium intake (mg/day) | 510 ± 117.88 | 520 ± 122.53 | 0.55 |
| Food energy (Kcal/day) | 1350 ± 200 | 1360 ± 150 | 0.23 |
| Low physical activity | 70 | 68 | 0.66 |
| Supplementation with calcium, $n$ | 0 | 0 | N/A |
| Supplementation with vitamin D, $n$ | 0 | 0 | N/A |
| Current steroid use, $n$ (%) | 0 | 0 | N/A |
| Antiretroviral therapy exposure | | | |
| Naive, $n$ (%) | 6 (6) | 0 | N/A |
| Experienced, $n$ (%) | 94 (94) | 0 | N/A |
| Sunlight exposure >5 h/week, $n$ (%) | 26 (26) | 26 (26) | 0.52 |
| Family history of osteoporosis and/or fractures, $n$ (%) | 12 (12) | 14 (14) | 0.79 |
| Fall history, $n$ (%) | 15 (15) | 17 (17) | 0.87 |

N/A: not applicable.

**Table 2.** Bone ultrasound parameter, laboratory data, and vertebral fractures in HIV infection and control group; results are expressed as number, means ± S.D. and percentage.

| Parameter | HIV ($n$ = 100) | Control Group ($n$ = 100) |
|---|---|---|
| Vertebral fractures (VFs), $n$ (%) | 16 (16) * | 2 (2) |
| Stiffness index (SI) (%) | 80.58 ± 19.95 § | 92.18 ± 8.06 |
| T-Score (D.S.) | −1.70 ± 0.50 § | -0.70 ± 0.30 |
| Z-Score (D.S.) | −1.20 ± 0.40 § | -0.10 ± 0.30 |
| Osteocalcin (pg/mL) | 21.31 ± 4.95 | 20.33 ± 4.35 |
| C-terminal telopeptide (ng/mL) | 0.53 ± 0.28 | 0.48 ± 0.35 |
| Parathyroid hormone level (pg/dL) | 43.72 ± 25.40 | 37.64 ± 23.60 |
| 25-hydroxivitamin D3 (ng/mL) | 20.29 ± 4.05 § | 35.77 ± 6.50 |
| Calcium (mg/dL) | 9.02 ± 0.47 | 9.12 ± 0.46 |
| Phophate (mg/dL) | 3.11 ± 0.57 | 3.14 ± 0.43 |
| Urinary calcium (mg/24 h) | 210.29 ± 84.05 | 213.93 ± 140.35 |
| Urinary phophate (mg/24 h) | 1010.29 ± 260.05 | 1063.53 ± 461.50 |
| Vitamin A (g/dL) | 62.65 ± 13.89 | 62.33 ± 13.22 |
| Creatinine (mg/dL) | 0.88 ± 0.07 | 0.89 ± 0.09 |

* $p < 0.05$; § $p < 0.001$.

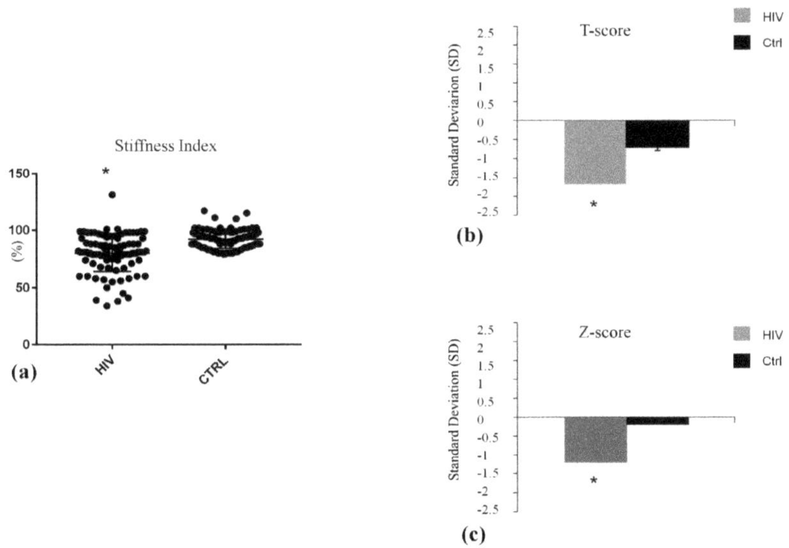


**Figure 1.** Bone Ultrasound parameters and related T and Z-score in HIV infected (HIV) and control group (Ctrl). * $p < 0.001$. Data are expressed as means and SD. (**a**) The baseline differences of stuffness index between HIV group (HIV) and control group (Ctrl) at 0, was statistically significant; (**b**) The baseline differences of T-score between HIV group and control group (Ctrl) at 0, was statistically significant; (**c**) The baseline differences of Z-score between HIV group and control group (Ctrl) at 0, was statistically significant.

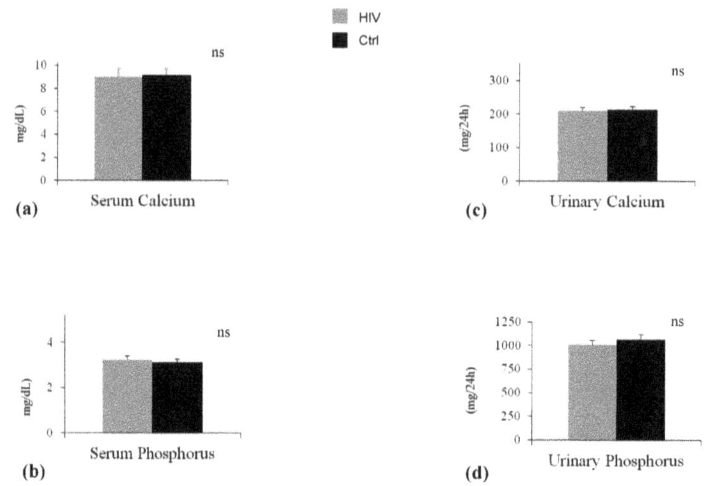


**Figure 2.** Serum and 24 hours urinary escretion of Calcium and Phosphorus in two groups. Data are expressed as means and SD. (**a**) The baseline differences of Serum Calcium between HIV group (HIV) and control group (Ctrl) was not statistically significant ($p = 0.13$); (**b**) The baseline differences of serum Phosphrous between HIV group (HIV) and control group (Ctrl) was not statistically significant ($p = 0.67$); (**c**) The baseline differences of urinary Calcium between HIV group (HIV) and control group (Ctrl) was not statistically significant ($p = 0.82$); (**d**) The baseline differences of urinary Phosphrous between HIV group (HIV) and control group (Ctrl) was not statistically significant ($p = 0.31$).

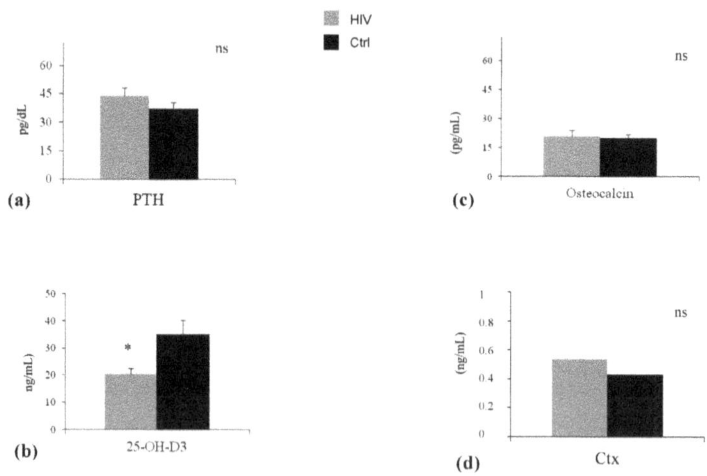


**Figure 3.** Serum parathyroid hormone (PTH), 25(OH)D3 and bone turnover markers in two groups. * $p < 0.001$. Data are expressed as means and SD. (**a**) The baseline differences of serum parathyroid hormone between HIV group (HIV) and control group (Ctrl) was not statistically significant ($p = 0.08$); (**b**) The baseline differences of Serum 25-OH-D3 between HIV group (HIV) and control group (Ctrl) was not statistically significant ($p = 0.0001$); (**c**) The baseline differences of osteocalcin between HIV group (HIV) and control group (Ctrl) was not statistically significant ($p = 0.13$); (**d**) The baseline differences of CTx between HIV group (HIV) and control group (Ctrl) was not statistically significant ($p = 0.26$).

## 3. Discussion

Our data show that HIV-infected patients exhibited lower BMD as assessed by bone quantitative ultrasound at the heel. SI measurements and the related T- and Z-score values were significantly lower in the HIV-infected patients compared with the control group, the latter being matched with patients not only for sex and age but also for other known major risk factors for osteoporosis and fractures.

The X-ray of the thoracic and lumbar spine showed single or multiple VFs, according to the Genant semiquantitative classification of VFs, in 16 HIV-infected patients with a prevalence of 16% and in only two subjects from the control group. These data are consistent with the current literature and highlight the high risk of fracture in this cohort [17–19]. Age, chronic renal insufficiency, and steroid use were previously associated with increased risk of fractures in HIV-infected patients [20,21]. However, since there are no significant differences between our HIV-infected patients and the controls, when considering these variables, it is reasonable to rule out that a low BMD and VFs are related to these factors.

Another important determinant of bone loss and fragility fractures in HIV-infected patients is represented by ART, in fact several studies, but not all, have proved the association between ART exposure and osteoporosis and/or fractures [22–24].

In our study, 94% of HIV-infected patients were treated with ART, so it was not possible to analyze the impact of the therapy, per se, on BMD and fractures. Interestingly, the HIV-infected patients with VFs reported a lower value of SI in comparison with HIV-infected patients without VFs; this result is consistent with previous data on the role of quantitative ultrasound in the detection of bone fragility. Quantitative ultrasound is able to reflect different physical properties of bone (e.g., the density, elasticity, and microarchitecture) and provide information on fracture risk [25–30].

Our data may indicate that bone ultrasound at the non-dominant heel may be used as a reliable tool to test bone health in HIV-infected patients.

Great attention has been given to the field of vitamin D in HIV-infected patients over the last few years, and epidemiological data suggest that HIV-infected patients, also from different geographical locations, including Europe, American, and Australia have a high prevalence of vitamin D insufficiency [31–33]. In our cohort, the prevalence of insufficiency of vitamin D3 in HIV-infected patients was 50%, significantly higher compared to the control group (20%). Diet and exposure to the sun are the main factors that regulate vitamin D3 levels and its activity in each subject [34]. The average self-reported sun exposure was the same in both groups, so sun exposure is not likely to be the cause of the low 25(OH)D3 level showed in HIV group than controls. Poor vitamin D status is a well-established risk factor for bone disease in the general population [35]. In addition, recent data suggest that vitamin D may exert non-skeletal functions, having a role in cardiovascular and immune regulation [36–38]. Moreover, in HIV human macrophages exposed to mycobacterium tuberculosis, vitamin D pretreatment restored critical responses, supporting a potential role for exogenous vitamin D as a therapeutic adjuvant in *M. tuberculosis* infection in HIV(+) persons [39]. The extent to which vitamin D deficiency contributes to bone loss and enhance fracture risk in the HIV-infected patients is largely unknown. Our study demonstrated that vitamin D insufficiency is associated with prevalent vertebral fractures in HIV-infected patients: 87.5% of HIV-infected patients presenting VFs had serum with a vitamin D value lower than 30 ng/mL, and vitamin D was predictive of fractures according to regression analysis. The finding that insufficient 25-hydroxyvitamin D3 levels (<30 ng/mL) are associated with VFs support the clinical relevance of vitamin D preserving bone health also in HIV-infected patients, and supports the inclusion of this modifiable risk factor as part of the screening for bone fragility in HIV-infected patients.

We acknowledge that our study has some limitations, including the small sample size and its cross-sectional and observational nature, that need to be considered when interpreting the final results. Another limitation was not to compare the population of HIV with another chronic disease to better understand the impact and specificity of HIV on bone loss and vertebral fragility fractures.

## 4. Materials and Methods

### 4.1. Subjects

The study was approved by the Local Ethics Committee for Medical Research, Messina University Hospital "G.Martino" and carried out in accordance with the Helsinki Declaration. All subjects gave their informed written consent to enter the study. One hundred consecutive patients with diagnosis of HIV infection referred from July 2012 to February 2015 to the Infection Unit of the University Hospital of Messina were evaluated. Eligibility criteria required the absence of clinical or laboratory abnormalities that suggested cardiovascular, hepatic, or renal disorders; coagulopathy, the use of oral or transdermal estrogen, progestin, androgen, or other steroids; previous or current use of bone active agents (e.g., biphosphonates), the use of cholesterol-lowering therapy, cardiovascular medications, or any other therapy that could influence bone metabolism, in particular, systemic or local corticosteroids. Of the 141 patients evaluated, 100 fulfill the inclusion criteria and were included in this cross-sectional study. One hundred healthy volunteers matched for sex and age served as the control group. All participants in both groups were white subjects from South Italy. Clinical data, including smoking status (current, former, never), physical activity, food energy, calcium intake, and sun exposure, were obtained by interview, and weight and height were determined during physical examination. The nutritional variables were determined using 24 h recall. None of the subjects in either group were on supplementation with calcium and/or vitamin D.

### 4.2. Bone Ultrasound Parameters and Vertebral Fractures Assessment

All subjects underwent a BMD evaluation by quantitative ultrasound (QUS) at the non-dominant heel by Achilles Insight, (Lunar, Madison, WI, USA). The Achilles Plus measures the speed of sound (SOS), broadband ultrasound attenuation (BUA), and a clinical index called stiffness index. Stiffness index (SI) is calculated automatically by the software according to the following formula: $SI = (0.67 \times BUA + 0.28 \times SOS) - 420$. Standardized procedures were carried out for patient positioning, data acquisition, and system calibrations. The manufacturer's phantom was used for system calibrations. The coefficient of variation for SI was 1.7%.

All participants also underwent a lateral thoracic and lumbar spine X-ray to ascertain the presence of VFs. All radiographs were assessed by two blinded physicians separately according to the quantitative methods (quantitative vertebral morphometry, QVM) using a dedicated software (MorphoXpress). Anterior, middle, and posterior heights of vertebral T4 to L4 were measured. According to the Genant classification, a vertebral fracture was defined based on reduction in anterior, middle, and/or posterior height: Grade 1: 20–25% reduction; Grade 2: 25–40% reduction; Grade 3: >40% reduction [40].

### 4.3. Biochemical Data

Blood samples were obtained by antecubital venipuncture between 8 a.m. and 11 a.m. after an overnight fast and a 10 min rest. The first 4–5 mL of blood were not used. Blood was collected in refrigerated vacutainers containing an anticoagulant mixture provided by Booehringer-Mannheim, immediately placed on ice, and centrifuged, within a few minutes, at 2000× $g$ for 20 min at 4 °C, and the plasma was frozen at −80 °C until assayed. Osteocalcin (BGP) was performed using an immunoenzymatic assay (Pantec, Turin, Italy). Serum levels of C-telopeptide of type I collagen (CTx) was assessed using the Elecys 2010 Immunoassay System (Roche, Basel, Switzerland). Serum calcium, serum phosphorus, and urinary creatinine were measured by automated routine procedures. Parathyroid hormone (PTH) and 25-hydroxyvitamin D3 were measured using high-performance liquid chromatography (Bio-Rad Laboratories S.r.l., Segrate, Milan, Italy). Vitamin D deficiency was defined as plasma levels of 25(OH)D3 <20 ng/mL, insufficiency between 20 and 30 ng/m, and normal >30 ng/mL. The intra- and inter-assay CV were <10% for both tests. The intra- and inter-assay CV were <10% for both tests.

### 4.4. Statistical Analysis

Statistical analyses were performed using Statistica 8 (Statsoft, Inc., Tulsa, OK, USA). Values are expressed as mean SD or percentage. Comparisons between groups were performed by with a Student's *t*-test. The percentage of each variable was compared between groups by Fisher's exact test. A chi-squared test ($\chi^2$) with Yates correction was conducted to assess the individual association between independent variables and the presence of insufficiency 25(OH)D3. Multivariate logistic regression analysis was used to adjust for confounders. Values of $p$ <0.05 were considered statistically significant.

## 5. Conclusions

In conclusion, our data showed a high prevalence of VFs in HIV-infected patients and highlight the role of insufficiency of vitamin D3 in increasing the risk of VFs.

**Author Contributions:** Marco Atteritano, Francesco Squadrito, and Antonio Cascio conceived and designed the experiments; Luigi Mirarchi, Emmanuele Venanzi-Rullo, Domenico Santoro, Alberto Lo Gullo, Alessandra Bitto, Chiara Iaria, Antonino Catalano, Antonino Lasco, and Vincenzo Arcoraci performer the experiments; Antonio Cascio and Marco Atteritano analyzed the data.

**Conflicts of Interest:** The authors declare no conflict of interest.

## References

1. Stone, B.; Dockrell, D.; Bowman, C.; McCloskey, E. HIV and bone disease. *Arch. Biochem. Biophys.* **2010**, *503*, 66–77. [CrossRef] [PubMed]
2. Mallon, P.W. HIV and bone mineral density. *Curr. Opin. Infect. Dis.* **2010**, *23*, 1–8. [CrossRef] [PubMed]
3. Cazanave, C.; Dupon, M.; Lavignolle-Aurillac, V.; Barthe, N.; Lawson-Ayayi, S.; Mehsen, N.; Mercié, P.; Morlat, P.; Thiébaut, R.; Dabis, F.; et al. Reduced bone mineral density in HIV-infected patients: Prevalence and associated factors. *AIDS* **2008**, *22*, 395–402. [CrossRef] [PubMed]
4. Gallant, J.E.; Staszewski, S.; Pozniak, A.L.; DeJesus, E.; Suleiman, J.M.; Miller, M.D.; Coakley, D.F.; Lu, B.; Toole, J.J.; Cheng, A.K.; et al. Efficacy and safety of tenofovir DF vs. stavudine in combination therapy in antiretroviral-naive patients: A 3-year randomized trial. *JAMA* **2004**, *292*, 191–201. [CrossRef] [PubMed]
5. Brown, T.T.; Ruppe, M.D.; Kassner, R.; Kumar, P.; Kehoe, T.; Dobs, A.S.; Timpone, J. Reduced bone mineral density in human immunodeficiency virus-infected patients and its association with increased central adiposity and postload hyperglycemia. *J. Clin. Endocrinol. Metab.* **2004**, *89*, 1200–1206. [CrossRef] [PubMed]
6. Prieto-Alhambra, D.; Güerri-Fernández, R.; de Vries, F.; Lalmohamed, A.; Bazelier, M.; Starup-Linde, J.; Diez-Perez, A.; Cooper, C.; Vestergaard, P. HIV infection and its association with an excess risk of clinical fractures: A nationwide case-control study. *J. Acquir. Immune Defic. Syndr.* **2014**, *66*, 90–95. [CrossRef] [PubMed]
7. Güerri-Fernandez, R.; Vestergaard, P.; Carbonell, C.; Knobel, H.; Avilés, F.F.; Castro, A.S.; Nogués, X.; Prieto-Alhambra, D.; Diez-Perez, A. HIV infection is strongly associated with hip fracture risk, independently of age, gender, and comorbidities: A population-based cohort study. *J. Bone Miner. Res.* **2013**, *28*, 1259–1263. [CrossRef] [PubMed]
8. Torti, C.; Mazziotti, G.; Soldini, P.A.; Focà, E.; Maroldi, R.; Gotti, D.; Carosi, G.; Giustina, A. High prevalence of radiological vertebral fractures in HIV-infected males. *Endocrine* **2012**, *41*, 512–517. [CrossRef] [PubMed]
9. Fakruddin, J.M.; Laurence, J. HIV-1 Vpr enhances production of receptor of activated NF-kappaB ligand (RANKL) via potentiation of glucocorticoid receptor activity. *Arch. Virol.* **2005**, *150*, 67–78. [CrossRef] [PubMed]
10. Malizia, A.P.; Cotter, E.; Chew, N.; Powderly, W.G.; Doran, P.P. HIV protease inhibitors selectively induce gene expression alterations associated with reduced calcium deposition in primary human osteoblasts. *AIDS Res. Hum. Retroviruses* **2007**, *23*, 243–250. [CrossRef] [PubMed]
11. Wang, M.W.; Wei, S.; Faccio, R.; Takeshita, S.; Tebas, P.; Powderly, W.G.; Teitelbaum, S.L.; Ross, F.P. The HIV protease inhibitor ritonavir blocks osteoclastogenesis and function by impairing RANKL-induced signaling. *J. Clin. Investig.* **2004**, *114*, 206–213. [CrossRef] [PubMed]

12. Bolland, M.J.; Grey, A.B.; Gamble, G.D.; Reid, I.R. CLINICAL Review: Low body weight mediates the relationship between HIV infection and low bone mineral density: A meta-analysis. *J. Clin. Endocrinol. Metab.* **2007**, *92*, 4522–4528. [CrossRef] [PubMed]
13. Hasse, B.; Ledergerber, B.; Furrer, H.; Battegay, M.; Hirschel, B.; Cavassini, M.; Bertisch, B.; Bernasconi, E.; Weber, R.; Swiss HIV Cohort Study. Morbidity and aging in HIV-infected persons: The Swiss HIV cohort study. *Clin. Infect. Dis.* **2011**, *53*, 1130–1139. [CrossRef] [PubMed]
14. Young, B.; Dao, C.N.; Buchacz, K.; Baker, R.; Brooks, J.T.; HIV Outpatient Study (HOPS) Investigators. Increased rates of bone fracture among HIV-infected persons in the HIV Outpatient Study (HOPS) compared with the US general population, 2000–2006. *Clin. Infect. Dis.* **2011**, *52*, 1061–1068. [CrossRef] [PubMed]
15. Gazzola, L.; Savoldi, A.; Bai, F.; Magenta, A.; Dziubak, M.; Pietrogrande, L.; Tagliabue, L.; del Sole, A.; Bini, T.; Marchetti, G.; et al. Assessment of radiological vertebral fractures in HIV-infected patients: Clinical implications and predictive factors. *HIV Med.* **2015**, *16*, 563–571. [CrossRef] [PubMed]
16. El Maghraoui, A.; Ouzzif, Z.; Mounach, A.; Rezqi, A.; Achemlal, L.; Bezza, A.; Tellal, S.; Dehhaoui, M.; Ghozlani, I. Hypovitaminosis D and prevalent asymptomatic vertebral fractures in Moroccan postmenopausal women. *BMC Womens Health* **2012**, *12*, 11. [CrossRef] [PubMed]
17. Arnsten, J.H.; Freeman, R.; Howard, A.A.; Floris-Moore, M.; Lo, Y.; Klein, R.S. Decreased bone mineral density and increased fracture risk in aging men with or at risk for HIV infection. *AIDS* **2007**, *21*, 617–623. [CrossRef] [PubMed]
18. Collin, F.; Duval, X.; Le Moing, V.; Piroth, L.; Al Kaied, F.; Massip, P.; Villes, V.; Chêne, G.; Raffi, F. Ten-year incidence and risk factors of bone fractures in a cohort of treated HIV1-infected adults. *AIDS* **2009**, *23*, 1021–1024. [CrossRef] [PubMed]
19. Triant, V.A.; Brown, T.T.; Lee, H.; Grinspoon, S.K. Fracture prevalence among human immunodeficiency virus (HIV)-infected versus non-HIV-infected patients in a large U.S. healthcare system. *J. Clin. Endocrinol. Metab.* **2008**, *93*, 3499–3504. [CrossRef] [PubMed]
20. Calmy, A.; Fux, C.A.; Norris, R.; Vallier, N.; Delhumeau, C.; Samaras, K.; Hesse, K.; Hirschel, B.; Cooper, D.A.; Carr, A. Low bone mineral density, renal dysfunction, and fracture risk in HIV infection: A cross-sectional study. *J. Infect. Dis.* **2009**, *200*, 1746–1754. [CrossRef] [PubMed]
21. Samaras, K.; Pett, S.; Gowers, A.; McMurchie, M.; Cooper, D.A. Iatrogenic Cushing's syndrome with osteoporosis and secondary adrenal failure in human immunodeficiency virus-infected patients receiving inhaled corticosteroids and ritonavir-boosted protease inhibitors: Six cases. *J. Clin. Endocrinol. Metab.* **2005**, *90*, 4394–4398. [CrossRef] [PubMed]
22. Hansen, A.B.; Gerstoft, J.; Kronborg, G.; Larsen, C.S.; Pedersen, C.; Pedersen, G.; Obel, N. Incidence of low and high-energy fractures in persons with and without HIV infection: A Danish population-based cohort study. *AIDS* **2012**, *26*, 285–293. [CrossRef] [PubMed]
23. Bedimo, R.; Maalouf, N.M.; Zhang, S.; Drechsler, H.; Tebas, P. Osteoporotic fracture risk associated with cumulative exposure to tenofovir and other antiretroviral agents. *AIDS* **2012**, *26*, 825–831. [CrossRef] [PubMed]
24. Bruera, D.; Luna, N.; David, D.O.; Bergoglio, L.M.; Zamudio, J. Decreased bone mineral density in HIV-infected patients is independent of antiretroviral therapy. *AIDS* **2003**, *17*, 1917–1923. [CrossRef] [PubMed]
25. Bauer, D.C.; Glüer, C.C.; Genant, H.K.; Stone, K. Quantitative ultrasound and vertebral fracture in postmenopausal women. *J. Bone Miner. Res.* **1995**, *10*, 353–358. [CrossRef] [PubMed]
26. Bauer, D.C.; Glüer, C.C.; Cauley, J.A.; Vogt, T.M.; Ensrud, K.E.; Genant, H.K.; Black, D.M. Broadband ultrasound attenuation predicts fractures strongly and independently of densitometry in older women. A prospective study. Study of Osteoporotic Fractures Research Group. *Arch. Intern. Med.* **1997**, *157*, 629–634. [CrossRef] [PubMed]
27. Bauer, D.C.; Ewing, S.K.; Cauley, J.A.; Ensrud, K.E.; Cummings, S.R.; Orwoll, E.S. Quantitative ultrasound predicts hip and non-spine fracture in men: The MrOS study. *Osteoporos. Int.* **2007**, *18*, 771–777. [CrossRef] [PubMed]
28. Glüer, C.C.; Wu, C.Y.; Genant, H.K. Broadband ultrasound attenuation signals depend on trabecular orientation: An in vitro study. *Osteoporos. Int.* **1993**, *3*, 185–191. [CrossRef] [PubMed]

29. Crisafulli, A.; Romeo, A.; Floccari, F.; Aloisi, E.; Atteritano, M.; Cincotta, M.; Aloisi, C.; Pizzoleo, M.A.; Ruello, A.; Artemisia, A.; et al. Osteoprotegerin and bone mineral density in hemodiafiltration patients. *Ren. Fail.* **2005**, *27*, 531–539. [CrossRef] [PubMed]
30. Catalano, A.; Morabito, N.; Agostino, R.M.; Basile, G.; Gaudio, A.; Atteritano, M.; Natale, G.; Xourafa, A.; Morini, E.; Adamo, V.; et al. Bone health assessment by quantitative ultrasound and dual-energy X-ray absorptiometry in postmenopausal women with breast cancer receiving aromatase inhibitors. *Menopause* **2017**, *24*, 85–91. [CrossRef] [PubMed]
31. Klassen, K.M.; Fairley, C.K.; Kimlin, M.G.; Hocking, J.; Kelsall, L.; Ebeling, P.R. Vitamin D deficiency is common in HIV-infected southern Australian adults. *Antivir. Ther.* **2016**, *21*, 117–125. [CrossRef] [PubMed]
32. Jao, J.; Freimanis, L.; Mussi-Pinhata, M.M.; Cohen, R.A.; Monteiro, J.P.; Cruz, M.L.; Sperling, R.S.; Branch, A.; Siberry, G.K.; NISDI LILAC Protocol. Low vitamin D status among pregnant Latin American and Caribbean women with HIV Infection. *Int. J. Gynaecol. Obstet.* **2015**, *130*, 54–58. [PubMed]
33. Cervero, M.; Agud, J.L.; García-Lacalle, C.; Alcázar, V.; Torres, R.; Jusdado, J.J.; Moreno Guillén, S. Prevalence of vitamin D deficiency and its related risk factor in a Spanish cohort of adult HIV-infected patients: Effects of antiretroviral therapy. *AIDS Res. Hum. Retroviruses* **2012**, *28*, 963–971. [CrossRef] [PubMed]
34. Hollis, B.W. Circulating 25-hydroxyvitamin D levels indicative of vitamin D sufficiency: Implications for establishing a new effective dietary intake recommendation for vitamin D. *J. Nutr.* **2005**, *135*, 317–322. [PubMed]
35. Parfitt, A.M. Dietary risk factors for age-related bone loss and fractures. *Lancet* **1983**, *2*, 1181–1185. [CrossRef]
36. Trochoutsou, A.I.; Kloukina, V.; Samitas, K.; Xanthou, G. Vitamin-D in the Immune System: Genomic and Non-Genomic Actions. *Mini. Rev. Med. Chem.* **2015**, *15*, 953–963. [CrossRef] [PubMed]
37. Mocanu, V.; Oboroceanu, T.; Zugun-Eloae, F. Current status in vitamin D and regulatory T cells—Immunological implications. *Rev. Med. Chir. Soc. Med. Nat. Iasi* **2013**, *117*, 965–973. [PubMed]
38. Mansueto, P.; Seidita, A.; Vitale, G.; Gangemi, S.; Iaria, C.; Cascio, A. Vitamin D Deficiency in HIV Infection: Not Only a Bone Disorder. *Biomed. Res. Int.* **2015**, *2015*, 735615. [CrossRef] [PubMed]
39. Anandaiah, A.; Sinha, S.; Bole, M.; Sharma, S.K.; Kumar, N.; Luthra, K.; Li, X.; Zhou, X.; Nelson, B.; Han, X.; et al. Vitamin D rescues impaired Mycobacterium tuberculosis-mediated tumor necrosis factor release in macrophages of HIV-seropositive individuals through an enhanced Toll-like receptor signaling pathway in vitro. *Infect. Immun.* **2013**, *81*, 2–10. [CrossRef] [PubMed]
40. Genant, H.K.; Wu, C.Y.; van Kuijk, C.; Nevitt, M.C. Vertebral fracture assessment using a semiquantitative technique. *J. Bone Miner. Res.* **1993**, *8*, 1137–1148. [CrossRef] [PubMed]

International Journal of
*Molecular Sciences*

MDPI

*Article*

# Vitamin D's Effect on the Proliferation and Inflammation of Human Intervertebral Disc Cells in Relation to the Functional Vitamin D Receptor Gene FokI Polymorphism

**Paola De Luca [1], Laura de Girolamo [1], Carlotta Perucca Orfei [1], Marco Viganò [1], Riccardo Cecchinato [2], Marco Brayda-Bruno [3] and Alessandra Colombini [1,*]**

1. Orthopaedic Biotechnology Lab, IRCCS Galeazzi Orthopaedic Institute, Via R. Galeazzi 4, 20161 Milan, Italy; deluca.paola@grupposandonato.it (P.D.L.); laura.degirolamo@grupposandonato.it (L.d.G.); carlotta.perucca@grupposandonato.it (C.P.O.); marco.vigano@grupposandonato.it (M.V.)
2. GSpine4, IRCCS Galeazzi Orthopaedic Institute, Via R. Galeazzi 4, 20161 Milan, Italy; dott.cecchinato@gmail.com
3. Scoliosis Unit, Department of Orthopedics and Traumatology-Spine Surgery III, IRCCS Galeazzi Orthopaedic Institute, Via R. Galeazzi 4, 20161 Milan, Italy; marco.brayda@spinecaregroup.it

* Correspondence: alessandra.colombini@grupposandonato.it; Tel.: +39-026-621-4067

Received: 15 June 2018; Accepted: 6 July 2018; Published: 9 July 2018

**Abstract:** Vitamin D is known to have immunomodulatory effects, is involved in osteo-cartilaginous metabolism, and may have a role in human intervertebral disc pathophysiology. Although a link between vitamin D receptor (VDR) gene variants and disc degeneration-related pathologies has been observed, its functional contribution to pathologic processes has not been assessed yet. The aim of this study was to investigate the response of disc cells to vitamin D in terms of the regulation of proliferation, metabolism, and inflammatory processes, with a particular focus on the FokI *VDR* genotype. However, although it was found that vitamin D had a pro-apoptotic effect regardless of genotype, an up-regulation of IL-1Ra and downregulation of IL-6 was found to be evident only in *Ff* cells. Regarding the metabolic effects, in *Ff* cells, vitamin D promoted an upregulation of the aggrecan in inflammatory conditions but did not have an effect on the expression of collagen-related markers. Moreover, cells bearing the *Ff* genotype were the most responsive to vitamin D in the upregulation of catabolic markers. In addition, in contrast to the *FF* genotype, vitamin D downregulated the vitamin D-dependent signaling pathway in inflamed *Ff* cells, counteracting the inflammation-mediated catabolic effects. In conclusion, *Ff* cells were found to be more responsive to the anti-inflammatory and catabolic effects of vitamin D, which is likely to be related to matrix remodeling.

**Keywords:** intervertebral disc; vitamin D; vitamin D receptor polymorphism; proliferation; inflammation

## 1. Introduction

The involvement of the vitamin D endocrine system in the pathophysiology of the human intervertebral disc is still a topic of debate which is not fully explored. Few in-vitro studies have reported that vitamin D regulates proliferation, the expression of matrix genes, production of structural proteins, cytokines, and growth factors in cells obtained from the two main anatomical components of the disc, the nucleus pulposus (NP), and the annulus fibrosus (AF), and expressing the vitamin D receptor (VDR) [1,2].

The need to perform functional studies to analyze the effects of vitamin D on the fibro-cartilaginous disc and the osteo-cartilaginous endplate (CEP) resides in a number of evidences describing the association between the four most studied, known genetic *VDR* variants (FokI, BsmI,

TaqI and ApaI), and the disc degeneration-related pathologies [3,4], although inconsistent associations have been reported [5,6].

Recently, a correlation between the aforementioned genetic variants and specific lumbar spine pathologies, such as herniation, discopathy, and osteochondrosis, has been observed [7–10]. Some *VDR* alleles and genotypes predisposed to lumbar spine pathologies have been identified in patients with a concomitant increase of type II collagen degradation products, which likely derives from the degradation of the disc's matrix [11,12]. However, there are still few findings concerning the contribution of these variants to pathologic processes. The FokI polymorphism is particularly interesting for its functional role—in fact, it is located in the start codon of the *VDR* and consists of a C to T transition, determining the transcription of a shorter, allele C (*F* allele), or longer allele T (*f* allele) polypeptide [13]. The shorter polypeptide couples more efficiently with the transcription factor II B than the longer peptides and leads to a higher transcriptional rate of vitamin D-dependent genes [14,15].

Moreover, given its involvement in osteo-cartilaginous metabolism, vitamin D might have a crucial role in the degenerative development of the disc and endplate [3].

Finally, the immunomodulatory effects of vitamin D have been also suggested [16], albeit without any clear explanation as to its possible involvement in the regulation of the inflammatory and catabolic processes present in the degenerate discs [17,18].

Based on this background, the aim of this study was to investigate the in-vitro regulation of proliferation, metabolism, and inflammatory processes of disc cells in response to vitamin D treatment, with a focus on the functional FokI *VDR* genotype. This study attempts to clarify the functional meaning of the association of this genetic variant with the predisposition to the development of disc degeneration-related pathologies.

## 2. Results

### 2.1. Anti-Proliferative Effect of Vitamin D Is Related to Induction of Apoptosis

Vitamin D treatment caused a decrease in the number of viable cells ($-2.7\%$, $p < 0.01$).

The anti-proliferative effect of vitamin D did not affect the cell cycle (Figure 1), but significantly increased the percentage of apoptotic cells ($p < 0.001$). The cells bearing the *FF* genotype showed a slight, but significant, decrease of the number of living cells ($-8\%$, $p < 0.05$) (Figure 2A). However, the rate of apoptosis was significantly increased by the vitamin D treatment in both cell types (+32% and +66%, both $p < 0.001$, for *FF* and *Ff*, respectively) (Figure 2A,B).

A

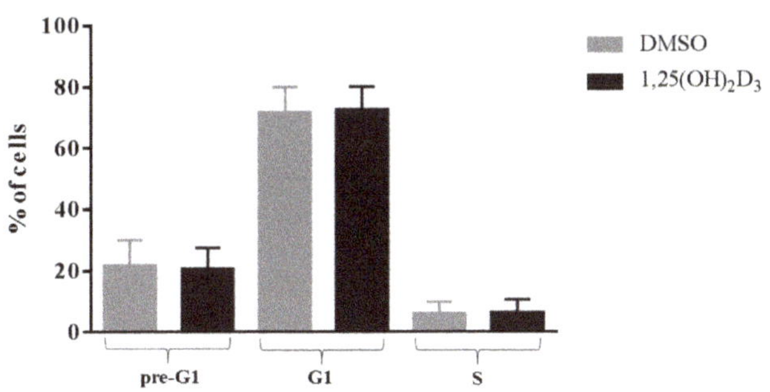


**Figure 1.** *Cont.*

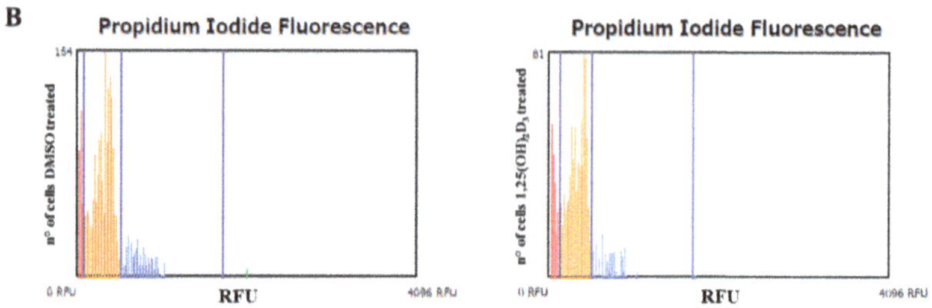


**Figure 1.** Cell cycle of disc cells in response to vitamin D treatment. (**A**) Shows the percentage of cells bearing both the *FF* and *Ff* genotype in the different cell cycle phases (pre-G1, G1 and S). Light gray and dark gray bars represent the cells treated with DMSO and vitamin D, respectively. Data are represented as mean ± SD, $n$ = 15; (**B**) Shows two representative cell cycle's graphs after DMSO and vitamin D treatment. Red, yellow, blue and green bars represent pre-G1, G1, S and G2-M phases, respectively.

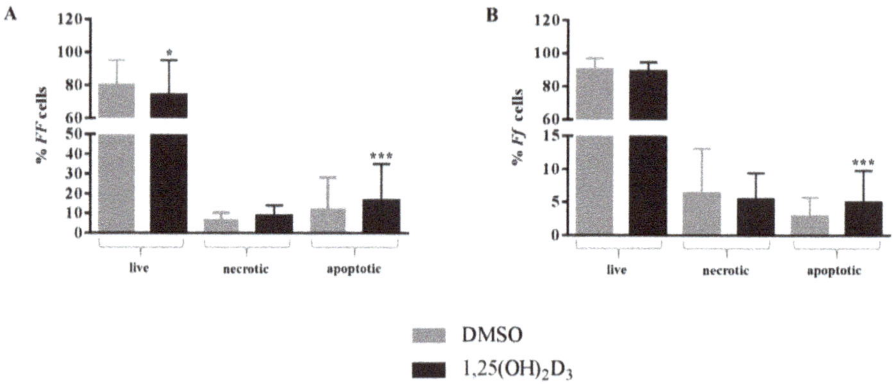


**Figure 2.** Apoptosis of disc cells in response to vitamin D treatment. Shows percentage of live, necrotic and apoptotic cells bearing *FF* (**A**) and *Ff* (**B**) genotypes. Light and dark gray bars represent the cells treated with DMSO and vitamin D, respectively. * $p < 0.05$, *** $p < 0.001$ vs. DMSO treatment. *FF* genotype $n$ = 4, *Ff* genotype $n$ = 11. Data are represented as mean ± SD.

### 2.2. *Effect of IL-1β on Disc Cells Cultured in Absence of Vitamin D*

As expected, IL-1β induced a significantly higher release of IL-1Ra ($p < 0.05$ and $p < 0.001$ for *FF* and *Ff*, respectively) (Figure 3A). Moreover, in cells bearing both the genotypes, the addition of 1 ng/mL IL-1β to the culture provoked a strong increase of the release of IL-6 up to 200-fold ($p < 0.01$ and $p < 0.001$ for *FF* and *Ff*, respectively) (Figure 3B).

In an inflamed condition, the expression of *ACAN* was strongly downregulated in the cells of both genotypes ($p < 0.01$ and $p < 0.001$ for *FF* and *Ff*, respectively) (Figure 4A). Similarly, the presence of IL-1β slightly decreased the expression of *COL1A1* with respect to the basal culture condition, although it was only in cells bearing the *Ff* genotype, which turned out to be statistically significant ($p < 0.05$) (Figure 4B). Conversely, the expression of *SOX9* was upregulated by IL-1β in cells bearing both genotypes, again only in a significant manner in cells bearing the *Ff* genotype ($p < 0.05$) (Figure 4C).

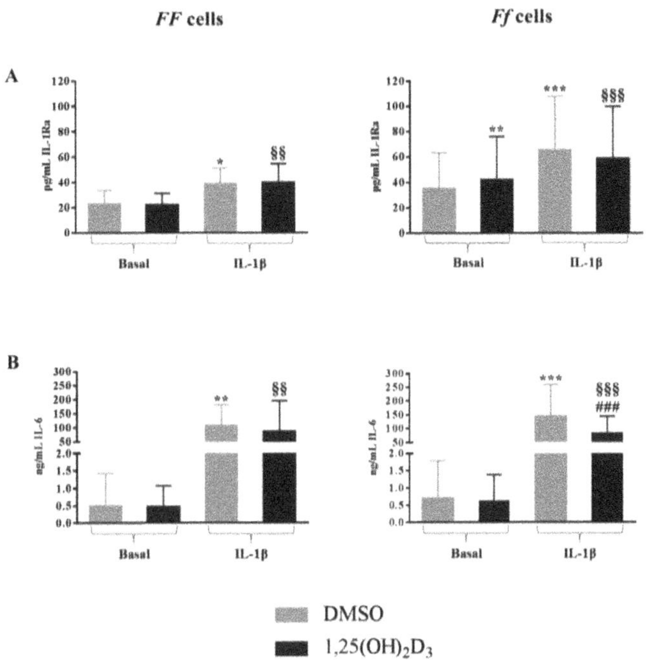


**Figure 3.** Concentrations of IL1-Ra (**A**) and IL-6 (**B**) released from *FF* and *Ff* bearing cells, in basal and inflamed (IL-1β treatment) conditions, in response to vitamin D treatment. Light and dark gray bars represent the cells treated with DMSO and vitamin D, respectively. * $p < 0.05$, ** $p < 0.01$, *** $p < 0.001$ vs. DMSO treatment in basal condition. §§ $p < 0.01$, §§§ $p < 0.001$ vs. vitamin D treatment in basal condition. ### $p < 0.001$ vs. DMSO + IL-1β treatment. *FF* genotype $n = 3$, *Ff* genotype $n = 10$. Data are represented as mean ± SD.

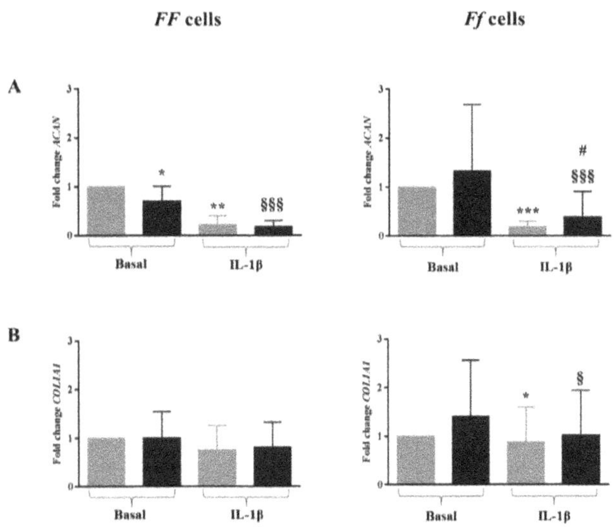


**Figure 4.** *Cont.*

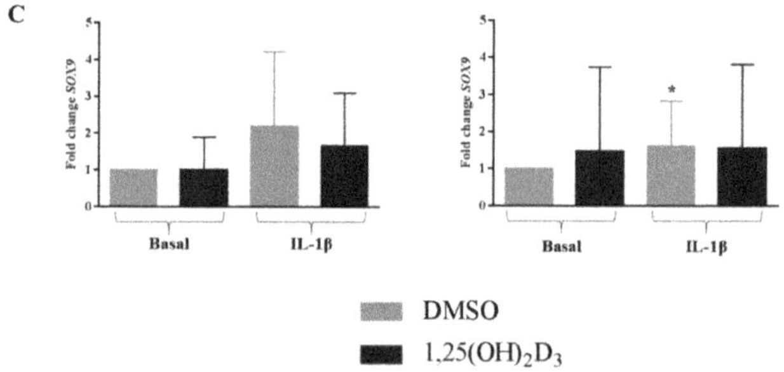


**Figure 4.** Expression of disc phenotype markers *ACAN* (**A**), *COL1A1* (**B**) and *SOX9* (**C**) in cells bearing *FF* and *Ff* genotypes, in basal and inflamed (IL-1β treatment) conditions, in response to vitamin D treatment. Light and dark gray bars represent the cells treated with DMSO and vitamin D, respectively. * $p < 0.05$, ** $p < 0.01$, *** $p < 0.001$ vs. DMSO treatment in basal condition. § $p < 0.05$, §§§ $p < 0.001$ vs. vitamin D treatment in basal condition. # $p < 0.05$ vs. DMSO + IL-1β treatment. *FF* genotype $n = 3$, *Ff* genotype $n = 7$. Data are represented as mean ± SD.

As expected, IL-1β provoked a very marked increase of *MMP1*, *MMP3* and *MMP13* expression in cells bearing both genotypes (all $p < 0.01$ for *FF* cells and all $p < 0.001$ for *Ff* cells) (Figure 5A–C).

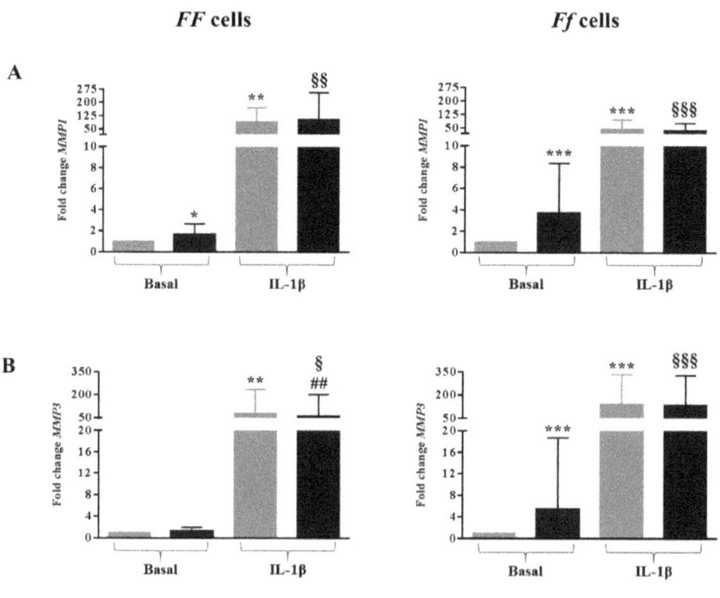


**Figure 5.** *Cont.*

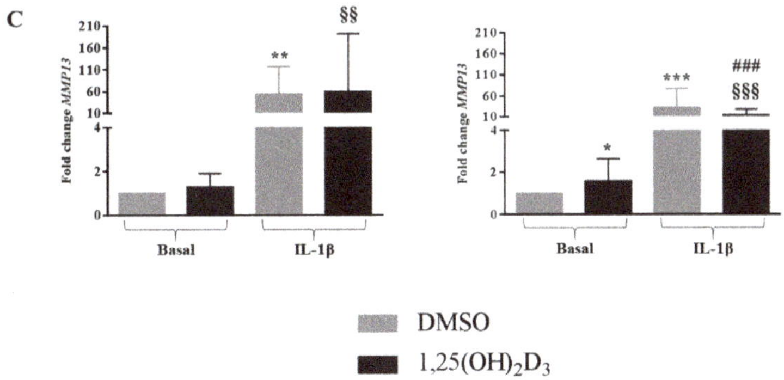


**Figure 5.** Expression of disc catabolic markers *MMP1* (**A**), *MMP3* (**B**) and *MMP13* (**C**) in cells bearing *FF* and *Ff* genotypes, in basal and inflamed (IL-1β treatment) conditions, in response to vitamin D treatment. Light and dark gray bars represent the cells treated with DMSO and vitamin D, respectively. * $p < 0.05$, ** $p < 0.01$, *** $p < 0.001$ vs. DMSO treatment in basal condition. § $p < 0.05$, §§ $p < 0.01$, §§§ $p < 0.001$ vs. vitamin D treatment in basal condition. ### $p < 0.01$, ### $p < 0.001$ vs. DMSO + IL-1β treatment. *FF* genotype $n = 3$, *Ff* genotype $n = 7$. Data are represented as mean ± SD.

In the presence of IL-1β, the expression of *CYP24* was strongly upregulated in both genotypes ($p < 0.05$ for *FF* cells and $p < 0.01$ for *Ff* cells) (Figure 6A), whereas inflammation provoked an upregulation of *VDR* expression in *Ff* cells ($p < 0.05$) (Figure 6B).

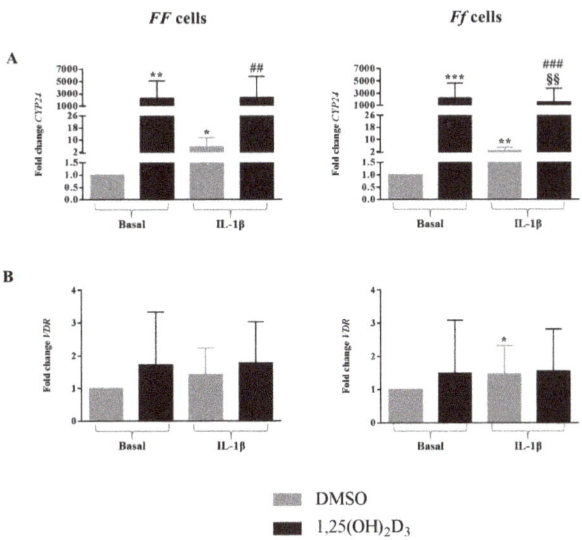


**Figure 6.** Expression of genes related to vitamin D-dependent signaling pathway *CYP24* (**A**) and *VDR* (**B**), in cells bearing *FF* and *Ff* genotypes, in basal and inflamed (IL-1β treatment) conditions, in response to vitamin D treatment. Light and dark gray bars represent the cells treated with DMSO and vitamin D, respectively. * $p < 0.05$, ** $p < 0.01$, *** $p < 0.001$ vs. DMSO treatment in basal condition. §§ $p < 0.01$, vs. vitamin D treatment in basal condition. ## $p < 0.01$, ### $p < 0.001$ vs. DMSO + IL-1β treatment. *FF* genotype $n = 3$, *Ff* genotype $n = 7$. Data are represented as mean ± SD.

### 2.3. *Vitamin D Supplementation Did Not Influence the Release of IL-1β in Disc Cells*

At basal level, no IL-1β was detected in the supernatant either in DMSO or 1,25(OH)$_2$D$_3$-treated cells. After 48 h of 1000 pg/mL IL-1β administration, an average amount of 350 ± 190 pg/mL was detected in cells, regardless of genotype or vitamin D treatment. The spontaneous degradation of IL-1β was also evaluated in culture media in the absence of cells; an average of 148.7 ± 37.4 pg/mL was found after 48 h of IL-1β administration.

### 2.4. *Vitamin D Upregulated IL-1Ra Release in Basal Condition and Downregulated IL-6 Release in Inflamed Condition in Cells Bearing the FokI VDR Ff Genotype*

The treatment with 1,25(OH)$_2$D$_3$ significantly upregulated the release of IL-1Ra only in cells bearing the *Ff* genotype in basal culture condition ($p < 0.01$) (Figure 3A). It should be noted that disc cells bearing the *Ff* genotype showed higher basal IL-1Ra concentrations—about 60 pg/mL—in comparison to the ones bearing the *FF* genotype, which had about 40 pg/mL. In inflamed conditions, 1,25(OH)$_2$D$_3$ showed to be able to downregulate the production of IL-6 only in the presence of a *VDR* FokI *Ff* genotype ($p < 0.001$) (Figure 3B).

### 2.5. *In Inflamed Condition Vitamin D Upregulated the Expression of ACAN in Ff Disc Cells, But Did Not Have an Effect on SOX9 and COL1A1 Expression*

In all conditions, with the exception of the *Ff* cells treated with 1,25(OH)$_2$D$_3$ in basal medium, a decrease in *ACAN* expression was observed. Interestingly, in the presence of IL-1β, vitamin D counteracted this effect by increasing the *ACAN* expression in *Ff* genotype cells, compared to DMSO in the same inflamed condition ($p < 0.05$) (Figure 3A).

In both basal and inflamed conditions, vitamin D did not influence the expression of *COLA1A1* and *SOX9* in cells bearing both the *FF* and *Ff* genotypes (Figure 3B,C).

### 2.6. *Vitamin D Was Able to Affect the Expression of MMPs in Cells Bearing the Ff Genotype in Both Basal and Inflamed Conditions*

Vitamin D was able to increase the *MMP1*, *MMP3* and *MMP13* expression in basal condition in cells bearing the *Ff* genotype. In contrast, 1,25(OH)$_2$D$_3$ in an inflamed condition downregulated the expression of *MMP3* in *FF* bearing cells ($p < 0.01$), and of *MMP13* in *Ff* bearing cells ($p < 0.001$) (Figure 5A–C).

### 2.7. *Vitamin D Upregulated the Vitamin D-Dependent Signaling Pathway, but This Was Downregulated in Inflamed Cells Bearing the Ff Genotype*

In general, independently from the FokI genotypes, 1,25(OH)$_2$D$_3$ upregulated *CYP24* expression, both in basal and inflamed conditions (*FF* genotype, $p < 0.01$, for both basal and inflamed conditions; *Ff* genotype, $p < 0.001$, for both basal and inflamed conditions) (Figure 6A). Interestingly, *CYP24* expression levels were downregulated by vitamin D in an inflamed environment in comparison with its levels in the presence of vitamin D alone, only in *Ff* bearing cells ($p < 0.01$).

Vitamin D did not affect *VDR* expression in either of the conditions, regardless of the cell genotype (Figure 6B).

## 3. Discussion

The main finding of this study is that the cells bearing the two different FokI *VDR* gene variants show some differences in response to the treatment with vitamin D, where the *Ff* genotype was the most responsive.

Regardless of the FokI *VDR* genotype, vitamin D had an inhibitory effect on the proliferation and metabolic activity of the disc cells while particularly favoring a pro-apoptotic induction on these kinds of cells, as already reported in previously published studies [1,2].

When the pro-inflammatory stimulus IL-1β, a cytokine well-known to induce an inflammatory and catabolic environment in the intervertebral disc [17,19–21] was added to the culture, the disc cells showed a similar IL-1β metabolism, regardless of genotype and vitamin D supplementation. Since a substantial amount of IL-1β is lost in the culture medium in absence of cells at the end of 48 h of incubation, the higher value detected in the cell culture after IL-1β administration suggests an active IL-1β cell secretion.

The disc cells bearing the FokI *VDR Ff* genotype were the most responsive to vitamin D, and showed an anti-inflammatory attitude by counteracting the upregulation induced by IL-1β of the pro-inflammatory IL-6, by upregulating the anti-inflammatory IL-1Ra release in basal condition and also by showing higher basal IL-1Ra concentrations in comparison with cells bearing the *FF* genotype, regardless of the vitamin D treatment.

Moreover, in *Ff*-bearing disc cells, vitamin D supplementation provoked an upregulation of the expression of *ACAN*, a marker of disc phenotype expressed in terminally differentiated cells, counteracting the strong downregulation of this gene observed after IL-1β treatment in absence of vitamin D. The downregulation of *ACAN* observed in basal condition after vitamin D treatment in cells bearing the *FF* genotype was in accordance with what was previously observed [2], and suggested a different behavior of the cells in response to the vitamin D treatment for what concerned the aggrecan production, depending on their genotype. In particular, the *Ff* bearing cells were, even in this case, most responsive to vitamin D in terms of preservation of the expression of this crucial marker of phenotype.

Even for what concerned the catabolic response, *Ff*-bearing cells were the most responsive, showing an upregulation of the MMP's expression of vitamin D mediated in basal condition.

On the contrary, in an inflamed environment, $1,25(OH)_2D_3$ slightly downregulated the expression of *MMP3* in *FF* bearing cells and of *MMP13* in *Ff* bearing cells, thus counteracting the increase of these MMP's expression mediated by IL-1β.

Regardless of genotype, the expression of *CYP24* was strongly upregulated by both IL-1β and $1,25(OH)_2D_3$ (even more strongly), thus showing that both stimuli influenced the vitamin D signaling pathway. Only in *Ff* bearing cells were the levels of this marker slightly downregulated by vitamin D in an inflamed environment with respect to the basal condition, making it likely to counteract the inflammation-mediated catabolic effects. A further confirmation of the influence of inflammation on the vitamin D signaling pathway was shown by the modulation of *VDR* expression. Only in *Ff* bearing cells was a slight increase of the expression of this gene observed in an inflamed condition. As already published [2], vitamin D did not affect the expression of this receptor.

The main limitation of this study is the absence of data regarding cells bearing the *ff* genotype, due to the low frequencies of this genotype in the Italian population (about 13%), as already published [7]. Moreover, the relevant data obtained at transcriptional level should also be confirmed at protein level, and further evaluation of subchondral bone cells obtained from vertebral bodies should be performed in view to confirm the anti-proliferative and catabolic effects of vitamin D observed in this study, in order to assess the potential effects of vitamin D on bone metabolism in the presence of specific functional FokI *VDR* genotypes.

These pieces of evidence suggest that calcitriol, at the pharmacological concentration used in this study, has a general anti-proliferative and catabolic effect on disc cells. For these reasons, the use of vitamin D to treat the degenerated disc with homeostatic or regenerative purposes is not suggestable, particularly for patients bearing *FF VDR* genotype which showed the highest risk of developing degenerative disc diseases [7,8,12]. Nevertheless, cells bearing the *Ff* genotype were the most responsive to the vitamin D supplementation, showing anti-inflammatory and catabolic behaviors in response to the treatment with the hormone, likely related to matrix remodeling. Starting from these pieces of evidence, the systemic or local supplementation of vitamin D could be considered as a treatment option for a particular subgroup of patients presenting disc degeneration related to osteochondrosis. In fact, this latest pathological condition involves not only the fibro-cartilaginous disc, but also the upper and lower bony-cartilaginous endplates, limiting the discs. In particular, the subchondral bone of these patients shows typical degenerative features, such as a multilevel presence of sclerosis and Schmorl's nodes [22,23]. Predisposing factors for spinal osteochondrosis are the presence of bone metabolic diseases, including osteomalacia, hyperparathyroidism, Paget's disease, infections, neoplasm and osteoporosis, which may weaken the vertebral bodies and allow Schmorl's nodes to form [24]. Although the endplates serve as the main route of nutrient supply into the disc [25], their calcification, due to the presence of chronic lesions, may lead to the loss of this role and may contribute to disc degeneration in this pathology. Also, the immune system has been postulated as having a role in the development of osteochondrosis. Discs that herniated into the vertebral endplate and eventually into the bone marrow could be recognized as foreign material, leading to an immune reaction with inflammatory cell infiltration, edema, influx of cytokines, and pain. Moreover, the herniation of the nucleus pulposus into the vertebrae can determine a cross-talk between a dysregulated immune system and bone metabolism, resulting in an imbalance of bone remodeling and consequently leading to bone loss [26]. This is a vicious circle where bone loss may predispose affected vertebrae to herniation of more disc material, exacerbating the condition.

In view of the importance of vitamin D in skeletal homeostasis [27] and of its immunomodulatory properties [16], and due to the peculiar pathological features of osteochondrosis, a vitamin D supplementation may be suggested to treat the subchondral sclerosis and the degenerated vertebral bone of the patients with spinal osteochondrosis bearing the *Ff* genotype, which showed to be the most promising subject concerning their responsiveness to the vitamin D treatment.

## 4. Materials and Methods

### 4.1. Study Population and Tissue Samples Collected

The study was approved by the ethics committee of the San Raffaele hospital (Protocol GenVDisc Version 1, 20 November 2015) and specimens were collected with patient-informed consent. Waste material from the lumbar intervertebral disc of 15 patients with a mean age of 54.6 ± 13.2 and who were affected by spine disorders was collected during discectomy. Demographic features, disc level, and patients' genotypes are listed in Table 1.

**Table 1.** Demographic features, disc level and genotypes of the recruited patients.

| Sex | Age | Spine Disorder | Disc Level | FokI *VDR* Genotype |
|---|---|---|---|---|
| F | 27 | Spondylolisthesis | L5-S1 | *FF* |
| F | 39 | Discopathy | L4-L5 | *FF* |
| F | 46 | Degenerative discopathy | L5-S1 | *Ff* |
| F | 48 | Spondylolisthesis | L5-S1 | *FF* |
| F | 48 | Degenerative discopathy | L5-S1 | *Ff* |
| F | 50 | Discopathy | L5-S1 | *Ff* |
| F | 50 | Discopathy | L5-S1 | *Ff* |
| M | 52 | Discopathy | L5-S1 | *Ff* |
| F | 55 | Discopathy | L5-S1 | *Ff* |
| M | 57 | Discopathy | L5-S1 | *Ff* |
| F | 65 | Discopathy | L5-S1 | *Ff* |
| F | 66 | Degenerative discopathy | L5-S1 | *FF* |
| F | 66 | Discopathy | L5-S1 | *Ff* |
| F | 74 | Discopathy | L5-S1 | *Ff* |
| F | 76 | Discopathy | L5-S1 | *Ff* |

### 4.2. Isolation and Expansion of Disc Cells

Disc cells were isolated by enzymatic digestion (37 °C, 22 h) using type II collagenase (Worthington Biochemical Co., Lakewood, NJ, USA) at the concentrations of 224 U/mL for NP, 560 U/mL for AF and 336 U/mL for CEP samples [28]. After digestion, the samples were filtered through a cell strainer and centrifuged (1000× g, 5 min). The cells were counted and plated at $10^5$ cells/cm$^2$, (37 °C, 5% $CO_2$) in 1 mg/mL of low-glucose Dulbecco's-modified Eagle medium (LG-DMEM, Thermo Fisher Scientific, Waltham, MA, USA) supplemented with 10% fetal bovine serum (FBS, Lonza, Basel, Switzerland), 0.29 mg/mL L-glutamine, 100 U/mL penicillin, 100 μg/mL streptomycin, 10 mM Hepes, 1 mM sodium pyruvate (all reagents from Thermo Fisher Scientific). During culture, the medium was replaced twice a week. At confluence, cells were detached using 0.05% trypsin/0.053 mM EDTA (Thermo Fisher Scientific) and plated at $5 \times 10^3$ cells/cm$^2$ for the following passages. The cells were expanded up to passage 3 and then used for the experiments.

### 4.3. Determination of FokI VDR Genotypes

Genomic DNA was extracted from disc cells according to the procedure of the Pure Link™ Genomic DNA Mini kit (Invitrogen, Carlsbad, CA, USA) and quantified spectrophotometrically (NanoDrop, Thermo Fisher Scientific). FokI *VDR* genotypes were determined by using TaqMan SNP Genotyping Assay (Thermo Fisher Scientific) for rs2228570 polymorphism and a StepOne Plus instrument (Thermo Fisher Scientific).

### 4.4. Calcitriol (1,25(OH)$_2$D$_3$) and IL-1β Treatment Protocols

Monolayer cultured cells were allowed to attach for 24 h in standard culture medium.

For cell cycle and apoptosis analysis, the cells were cultured for 24 h in low serum medium (5% FBS) to decrease the vitamin D binding protein contained in a serum, which might have interfered with the experiment. $10^{-8}$ M 1,25(OH)$_2$D$_3$, or vehicle (0.1% DMSO) (Sigma-Aldrich, St. Louis, MO, USA), were then added, and the medium changed at day 3 when the vitamin D treatment was repeated; the cells were collected after 6 days of treatment for the analysis of the cell cycle and apoptosis.

For the evaluation of the response to inflammation, $10^{-8}$ M 1,25(OH)$_2$D$_3$, or vehicle (0.1% DMSO), was added to 80% confluent cells in 5% FBS in the presence or absence of 1 ng/mL IL-1β (Sigma-Aldrich). The cells and supernatant were then evaluated after 48 h of treatment [29].

### 4.5. Cell Cycle Analysis

Cell cycle progression was evaluated by monitoring the DNA content through the Tali® Cell Cycle Kit (Thermo Fisher Scientific) consisting of an all-in-one solution containing propidium iodide, RNase A, and Triton X-100 to label cells. After 6 days of 1,25(OH)$_2$D$_3$ or DMSO treatment, the cells were detached using 0.05% trypsin/0.053 mM EDTA, centrifuged at 500× g for 5 min, washed with phosphate buffered saline solution (PBS), centrifuged again at the same previous conditions, transferred to ice and fixed into a single cell suspension with ice-cold 70% ethanol in distilled water. Later on, the cells were placed at −20 °C overnight, then centrifuged at 1000× g for 5 min at 4 °C to remove the ethanol and washed in PBS. After a centrifugation step at 500× g for 10 min at 4 °C, the PBS was removed and cells were resuspended in 200 μL of Tali® Cell Cycle Solution at RT for 30 min in the dark. The cells were briefly vortexed to gently resuspend them before the cell cycle analysis using the Tali® Image-Based Cytometer (Thermo Fisher Scientific). Using the instrument software, small cells (indicating debris) and large cells (indicating aggregates) were gated out of the analysis by setting the gate on the cell size; the threshold gates for each cell cycle phase were also set.

### 4.6. Evaluation of Apoptosis

Tali® Apoptosis Assay Kit–Annexin V Alexa Fluor® 488 and Propidium Iodide (Thermo Fisher Scientific) were used to evaluate the presence of apoptosis after the 1,25(OH)$_2$D$_3$ or DMSO treatment.

Briefly, the cells were harvested and centrifuged to discard the supernatant. Annexin binding buffer and Annexin V Alexa Fluor® 488 were added to the cells and the mixture was left at RT in the dark for 20 min. The cells were centrifuged and resuspended in Annexin binding buffer added to Tali® Propidium Iodide. Samples were incubated at RT in the dark for 5 min. The stained cells were then loaded onto the Tali® Image-Based Cytometer. Apoptotic cells showed green fluorescence, necrotic cells showed red fluorescence, dead cells (apoptosis/secondary necrosis) showed yellow fluorescence, and live cells showed little to no fluorescence.

### 4.7. Gene Expression Analysis

Total RNA was isolated from cell lysates using the PureLink® RNA Mini Kit (Thermo Fisher Scientific) and quantified spectrophotometrically (NanoDrop).

RNA were reverse-transcribed to cDNA employing the iScript cDNA Synthesis Kit (Bio-Rad Laboratories, Hercules, CA, USA). Gene expression was evaluated by real-time PCR (StepOne Plus instrument). cDNA was incubated with a PCR mixture, including the TaqMan® Gene Expression Master Mix and TaqMan® Gene Expression Assays (Thermo Fisher Scientific).

The expression of phenotype markers such as *ACAN*, Hs00153936_m1, *COL1A1*, Hs01076777_m1, *SOX9*, Hs00165814_m1, vitamin D-related genes such as *CYP24*, Hs00167999_m1, *VDR*, Hs01045840_m1, matrix metalloproteases *MMP1*, Hs00899658_m1, *MMP3*, Hs00968305_m1 and *MMP13*, Hs00233992_m1 was analyzed after IL-1β treatment.

The previously-validated *TBP*, Hs00427620_m1 was used as a housekeeping gene [14]. Data was expressed according to the dCt method.

### 4.8. Determination of Cytokines

Concentrations of soluble IL-1β, IL-1Ra and IL-6 in cell culture medium after 48 h of treatment with $1,25(OH)_2D_3$ or DMSO in the presence or absence of IL-1β were determined by commercially-available ELISA according to the manufacturers' instructions (PeproTech, Rocky Hill, NJ, USA).

The detection ranges were: 23–1500 pg/mL for IL-1β, 24–1500 pg/mL for IL-6 and 23–1500 pg/mL for IL-1Ra.

### 4.9. Statistical Analysis

Statistical analysis was performed using GraphPad Prism v5.0 software (GraphPad Software Inc., La Jolla, CA, USA). All values are expressed as the mean ± SD. Normal distribution of values were assayed by the Kolmogorov-Smirnov normality test. Paired comparisons were performed by using a two-tailed *t* test. In the case of not normally distributed values, repeated measures were compared with the Kruskal-Wallis test with the Dunns' correction. The significance level was set at a *p*-value lower than 0.05.

**Author Contributions:** Conceptualization, A.C.; Methodology, A.C.; Formal Analysis, A.C. and P.D.L.; Investigation, P.D.L., A.C., C.P.O.; Resources, R.C., M.B.-B.; Data Curation, A.C., P.D.L., M.V.; Writing—Original Draft Preparation, A.C.; Writing—Review and Editing, P.D.L., L.d.G., C.P.O., M.V., R.C., M.B.-B.; Supervision, L.d.G.; Project Administration, A.C.; Funding Acquisition, A.C.

**Funding:** This research was funded by the Italian Ministry of health (RRC-2018-2365827).

**Conflicts of Interest:** The authors declare no conflict of interest.

## Abbreviations

| | |
|---|---|
| VDR | Vitamin D receptor |
| IL-1Ra | Interleukin-1 receptor antagonist |
| IL-6 | Interleukin 6 |
| NP | Nucleus pulposus |
| AF | Annulus fibrosus |
| CEP | Osteo-cartilaginous endplate |
| IL-1β | Interleukin 1β |
| ACAN | Aggrecan |
| COL1A1 | Collagen Type I Alpha 1 Chain |
| SOX9 | SRY-Box 9 |
| MMP | Matrix metalloprotease |
| CYP24 | Cytochrome P450 Family 24 Subfamily A Member 1 |
| DMSO | Dimethyl sulfoxide |
| TBP | TATA-binding protein |

## References

1. Gruber, H.E.; Hoelscher, G.; Ingram, J.A.; Chow, Y.; Loeffler, B.; Hanley, E.N., Jr. 1,25$(OH)_2$-vitamin D3 inhibits proliferation and decreases production of monocyte chemoattractant protein-1, thrombopoietin, VEGF, and angiogenin by human annulus cells in vitro. *Spine* **2008**, *33*, 755–765. [CrossRef] [PubMed]
2. Colombini, A.; Lanteri, P.; Lombardi, G.; Grasso, D.; Recordati, C.; Lovi, A.; Banfi, G.; Bassani, R.; Brayda-Bruno, M. Metabolic effects of vitamin D active metabolites in monolayer and micromass cultures of nucleus pulposus and annulus fibrosus cells isolated from human intervertebral disc. *Int. J. Biochem. Cell Boil.* **2012**, *44*, 1019–1030. [CrossRef] [PubMed]
3. Colombini, A.; Cauci, S.; Lombardi, G.; Lanteri, P.; Croiset, S.; Brayda-Bruno, M.; Banfi, G. Relationship between vitamin D receptor gene (VDR) polymorphisms, vitamin D status, osteoarthritis and intervertebral disc degeneration. *J. Steroid Biochem. Mol. Boil.* **2013**, *138*, 24–40. [CrossRef] [PubMed]
4. Chen, L.; Zhao, S.; Niu, F.; Bi, G.B. Association between vitamin D receptor gene polymorphisms and intervertebral disc degeneration: A meta-analysis. *J. Orthop. Sci.* **2017**, *22*, 184–189. [CrossRef] [PubMed]
5. Jiang, H.; Qin, Z.L.; Zong, S.H.; He, M.L.; Zhan, X.L.; Xiao, Z.M.; Wei, Q.J. Vitamin D receptor gene polymorphisms and lumbar disc degeneration: A systematic review and meta-analysis. *Eur. Spine J.* **2017**, *26*, 267–277. [CrossRef] [PubMed]
6. Zhao, J.; Yang, M.; Shao, J.; Bai, Y.; Li, M. Association between VDR FokI Polymorphism and Intervertebral Disk Degeneration. *Genom. Proteom. Bioinform.* **2015**, *13*, 371–376. [CrossRef] [PubMed]
7. Colombini, A.; Brayda-Bruno, M.; Lombardi, G.; Croiset, S.J.; Vrech, V.; Maione, V.; Banfi, G.; Cauci, S. FokI polymorphism in the vitamin D receptor gene (VDR) and its association with lumbar spine pathologies in the Italian population: A case-control study. *PLoS ONE* **2014**, *9*, e97027. [CrossRef] [PubMed]
8. Colombini, A.; Brayda-Bruno, M.; Ferino, L.; Lombardi, G.; Maione, V.; Banfi, G.; Cauci, S. Gender differences in the VDR-FokI polymorphism and conventional non-genetic risk factors in association with lumbar spine pathologies in an Italian case-control study. *Int. J. Mol. Sci.* **2015**, *16*, 3722–3739. [CrossRef] [PubMed]
9. Colombini, A.; Brayda-Bruno, M.; Lombardi, G.; Croiset, S.J.; Ceriani, C.; Buligan, C.; Barbina, M.; Banfi, G.; Cauci, S. BsmI, ApaI and TaqI Polymorphisms in the Vitamin D Receptor Gene (VDR) and Association with Lumbar Spine Pathologies: An Italian Case-Control Study. *PLoS ONE* **2016**, *11*, e0155004. [CrossRef] [PubMed]
10. Mashayekhi, S.; Saberi, A.; Salehi, Z.; Biazar, G.; Mehrdel, R. VDR and GC gene polymorphisms modulate the risk of lumbar disc degeneration in Iran. *Clin. Neurol. Neurosurg.* **2018**, *165*, 67–71. [CrossRef] [PubMed]
11. Brayda-Bruno, M.; Vigano, M.; Cauci, S.; Vitale, J.A.; de Girolamo, L.; De Luca, P.; Lombardi, G.; Banfi, G.; Colombini, A. Plasma vitamin D and osteo-cartilaginous markers in Italian males affected by intervertebral disc degeneration: Focus on seasonal and pathological trend of type II collagen degradation. *Clin. Chim. Acta Int. J. Clin. Chem.* **2017**, *471*, 87–93. [CrossRef] [PubMed]

12. Cauci, S.; Vigano, M.; de Girolamo, L.; De Luca, P.; Perucca Orfei, C.; Banfi, G.; Lombardi, G.; Brayda-Bruno, M.; Colombini, A. High Levels of Circulating Type II Collagen Degradation Marker (CTx-II) Are Associated with Specific VDR Polymorphisms in Patients with Adult Vertebral Osteochondrosis. *Int. J. Mol. Sci.* **2017**, *18*, 2073. [CrossRef] [PubMed]
13. Uitterlinden, A.G.; Fang, Y.; Van Meurs, J.B.; Pols, H.A.; Van Leeuwen, J.P. Genetics and biology of vitamin D receptor polymorphisms. *Gene* **2004**, *338*, 143–156. [CrossRef] [PubMed]
14. Costanzo, P.; Santini, A.; Fattore, L.; Novellino, E.; Ritieni, A. Toxicity of aflatoxin B1 towards the vitamin D receptor (VDR). *Food Chem. Toxicol. Int. J. Publ. Br. Ind. Biol. Res. Assoc.* **2015**, *76*, 77–79. [CrossRef] [PubMed]
15. Jurutka, P.W.; Whitfield, G.K.; Hsieh, J.C.; Thompson, P.D.; Haussler, C.A.; Haussler, M.R. Molecular nature of the vitamin D receptor and its role in regulation of gene expression. *Rev. Endocr. Metab. Disord.* **2001**, *2*, 203–216. [CrossRef] [PubMed]
16. Guillot, X.; Semerano, L.; Saidenberg-Kermanac'h, N.; Falgarone, G.; Boissier, M.C. Vitamin D and inflammation. *Joint Bone Spine* **2010**, *77*, 552–557. [CrossRef] [PubMed]
17. Molinos, M.; Almeida, C.R.; Caldeira, J.; Cunha, C.; Goncalves, R.M.; Barbosa, M.A. Inflammation in intervertebral disc degeneration and regeneration. *J. R. Soc. Interface* **2015**, *12*, 20150429. [CrossRef] [PubMed]
18. Colombini, A.; Lombardi, G.; Corsi, M.M.; Banfi, G. Pathophysiology of the human intervertebral disc. *Int. J. Biochem. Cell Boil.* **2008**, *40*, 837–842. [CrossRef] [PubMed]
19. Le Maitre, C.L.; Freemont, A.J.; Hoyland, J.A. The role of interleukin-1 in the pathogenesis of human intervertebral disc degeneration. *Arth. Res. Ther.* **2005**, *7*, R732–R745. [CrossRef] [PubMed]
20. Le Maitre, C.L.; Pockert, A.; Buttle, D.J.; Freemont, A.J.; Hoyland, J.A. Matrix synthesis and degradation in human intervertebral disc degeneration. *Biochem. Soc. Trans.* **2007**, *35 Pt 4*, 652–655. [CrossRef] [PubMed]
21. Johnson, Z.I.; Schoepflin, Z.R.; Choi, H.; Shapiro, I.M.; Risbud, M.V. Disc in flames: Roles of TNF-alpha and IL-1beta in intervertebral disc degeneration. *Eur. Cells Mater.* **2015**, *30*, 104–116, discussion 116–117. [CrossRef]
22. McFadden, K.D.; Taylor, J.R. End-plate lesions of the lumbar spine. *Spine* **1989**, *14*, 867–869. [CrossRef] [PubMed]
23. Aufdermaur, M. Juvenile kyphosis (Scheuermann's disease): Radiography, histology, and pathogenesis. *Clin. Orthop. Relat. Res.* **1981**, 166–174.
24. Kyere, K.A.; Than, K.D.; Wang, A.C.; Rahman, S.U.; Valdivia-Valdivia, J.M.; La Marca, F.; Park, P. Schmorl's nodes. *Eur. Spine J.* **2012**, *21*, 2115–2121. [CrossRef] [PubMed]
25. Urban, J.P.; Smith, S.; Fairbank, J.C. Nutrition of the intervertebral disc. *Spine* **2004**, *29*, 2700–2709. [CrossRef] [PubMed]
26. Zhang, N.; Li, F.C.; Huang, Y.J.; Teng, C.; Chen, W.S. Possible key role of immune system in Schmorl's nodes. *Med. Hypotheses* **2010**, *74*, 552–554. [CrossRef] [PubMed]
27. Goltzman, D. Functions of vitamin D in bone. *Histochem. Cell Biol.* **2018**, *149*, 305–312. [CrossRef] [PubMed]
28. Lopa, S.; Ceriani, C.; Cecchinato, R.; Zagra, L.; Moretti, M.; Colombini, A. Stability of housekeeping genes in human intervertebral disc, endplate and articular cartilage cells in multiple conditions for reliable transcriptional analysis. *Eur. Cells Mater.* **2016**, *31*, 395–406. [CrossRef]
29. Kim, J.H.; Choi, H.; Suh, M.J.; Shin, J.H.; Hwang, M.H.; Lee, H.M. Effect of biphasic electrical current stimulation on IL-1beta-stimulated annulus fibrosus cells using in vitro microcurrent generating chamber system. *Spine* **2013**, *38*, E1368–E1376. [CrossRef] [PubMed]

International Journal of
*Molecular Sciences*

MDPI

*Article*

# Single Nucleotide Polymorphisms in the Vitamin D Receptor Gene (*VDR*) May Have an Impact on Acute Pancreatitis (AP) Development: A Prospective Study in Populations of AP Patients and Alcohol-Abuse Controls

**Anna Cieślińska [1], Elżbieta Kostyra [1], Ewa Fiedorowicz [1], Jadwiga Snarska [2], Natalia Kordulewska [1], Krzysztof Kiper [3] and Huub F. J. Savelkoul [4,*]**

1 Faculty of Biology and Biotechnology, University of Warmia and Mazury, 10-719 Olsztyn, Poland; anna.cieslinska@uwm.edu.pl (A.C.); elzbieta.kostyra@uwm.edu.pl (E.K.); ewa.kuzbida@uwm.edu.pl (E.F.); natalia.smulska@uwm.edu.pl (N.K.)
2 Department of General Surgery, Faculty of Medical Sciences, University of Warmia and Mazury, 10-719 Olsztyn, Poland; jct4444@onet.pl
3 Faculty of Medicine, Rzeszów University, 35-310 Rzeszów, Poland; krzychukiper@gmail.com
4 Cell Biology and Immunology Group, Wageningen University, 6700 AG Wageningen, The Netherlands
* Correspondence: huub.savelkoul@wur.nl

Received: 23 May 2018; Accepted: 27 June 2018; Published: 29 June 2018

**Abstract:** Vitamin D imbalance is suggested to be associated with the development of pancreatitis. Single nucleotide polymorphisms (SNPs), Apa-1, Bsm-1, Fok-1, and Taq-1, in the vitamin D receptor gene (*VDR*) are known in various diseases, but not yet in pancreatitis. The aim of this study was to explore possible associations of the four SNPs in the *VDR* receptor gene in a population of acute pancreatitis patients and alcohol-abuse controls, and to investigate the association with acute pancreatitis (AP) susceptibility. The study population ($n$ = 239) included acute pancreatitis patients ($n$ = 129) and an alcohol-abuse control group ($n$ = 110). All patients met the Diagnostic and Statistical Manual of Mental Disorders (DSM IV) criteria for alcohol dependence. DNA was extracted from peripheral leukocytes and analyzed for *VDR* polymorphisms using a polymerase chain reaction-restriction fragment length polymorphism (PCR-RFLP) method. Odd ratios (ORs) and 95% confidence intervals (CIs) were calculated using logistic regression analysis. To date, we have found allele T in Taq-1 (OR = 2.61; 95% CI: 1.68–4.03; $p < 0.0001$) to be almost three times more frequent in the AP group compared to the alcohol-abuse control patients. Polymorphism Taq-1 occurring in the vitamin D receptor may have an impact on the development of acute pancreatitis due to the lack of the protective role of vitamin D.

**Keywords:** acute pancreatitis; polymorphism; vitamin D receptor; vitamin D; SNP analysis

## 1. Introduction

Acute pancreatitis (AP) is a multifactorial disease that develops due to pancreatic ischaemia, pancreatic bile duct obstruction, or activation of pancreatic protease and production of pro-inflammatory cytokines [1,2]. It has an unpredictable course and the development of prognostic symptoms can determine patients at high risk of a severe course of this illness who require appropriate treatment and intensive care [3,4]. AP severity is related to demographic factors (age and obesity), local complications (pancreatic necrosis and fluid collection), and organ failure, as well as genetic factors [5]. Nevertheless, gallstones (cholelithiasis) and alcohol abuse are the main risk factors for acute pancreatitis.

Since vitamin D receptors have been found in several human brain structures [6–8], its role has become more important in various psychiatric issues. Low vitamin D levels have been associated with schizophrenia, depression, pancreatic cancer, and autism [8–10]. Vitamin D levels are also usually reduced in alcohol-abuse patients [11,12]. Lieber et al. [13] correlated low vitamin D serum levels in chronic alcoholics with pancreatic insufficiency. According to the research of Kim et al. [14], concentrations of the inactive form of vitamin D, 25-hydroxyvitamin D (25(OH)$D_3$), were significantly lower in dogs with AP in comparison to healthy dogs. The active vitamin D metabolite, 1,25-dihydroxyvitamin D (1,25-dihydroxycholecalciferol or 1,25$(OH)_2D_3$), selectively binds to a specific vitamin D receptor (*VDR*) [15]. This *VDR* subsequently regulates transcription of genes that are involved in calcium metabolism, cellular proliferation and differentiation, aging, and T-cell mediated immune responses [16–23].

*VDR* are also expressed on pancreatic β cells [24], and may play an essential role in maintaining normal insulin levels in accordance to glucose concentrations and to maintain glucose tolerance [25–27]. Because vitamin D acts through *VDR*, their impairment or reduced functionality, e.g., as a result of polymorphisms occurring in the *VDR* gene, may have a crucial impact on the balance in the vitamin D concentration in the circulation, and the final metabolite activity throughout the body.

According to Valdivielso and Fernandez [19], RFLP (Restriction Fragments Length Polymorphism) assays were used to identify four important polymorphisms in the *VDR* gene. Polymorphism detected in intron 8 by the restriction enzyme, Apa-1 (rs7975232), results in T (variant "A") changing into G (variant "a") [28]. Bsm-1 polymorphism (rs1544410), also located in intron 8, causes a change of A (variant "b") into G (variant "B") [29]. Taq-1 (rs731236) polymorphism (T as variant "T" changed into C as "t" variant) observed in exon 9 leads to a silent mutation in codon 352 [30]. The missense Fok-1 (rs2228570) transition, located in exon 2, results in *VDR* protein variants of 427 amino acids (f (T)) and of 424 amino acids (F (C)). In some cell types, the latter one (F variant) results in a more active form of the protein [31–34].

The aim of this study was to investigate the genetic association of the four different *VDR* polymorphisms (Apa-1, Bsm-1, Fok-1, Taq-1) with susceptibility to the development of acute pancreatitis compared to a control group of alcohol-abuse patients.

## 2. Results

The observed genotype frequencies of Apa-1 (rs7975232), Bsm-1 (rs1544410), Fok-1 (rs2228570), and Taq-1 (rs731236) polymorphisms in the *VDR* gene were studied in 110 alcohol-abuse controls with diagnosed alcohol-abuse and 129 patients with acute pancreatitis. The data obtained conformed to the Hardy-Weinberg equilibrium. In the whole study population, three genotypes at the *VDR* gene polymorphic site, Apa-1 (rs7975232), were identified: AA, Aa, and aa, with a number of genotypes of 81, 104, and 54, respectively.

At the Bsm-1 (rs1544410) *VDR* gene polymorphic site the frequency of the alleles, B and b, were determined in our study's alcohol-abuse controls and in those diagnosed with AP. Of the total 239 participants, three genotypes (BB, Bb, bb) were identified: 28 had genotype BB, 124 had Bb, and 87 had bb.

At the *VDR* gene polymorphic site, Fok-1 (rs2228570); the three FF, Ff, and ff genotypes were identified. Of the total 239 participants, 90 had genotype FF, 104 had genotype *Ff*, and 45 had genotype *ff*.

At the *VDR* gene polymorphic site, Taq-1 (rs731236); the three TT, Tt, and tt genotypes were identified, with a 0.61 T-allele frequency in the entire research population (control and AP groups). Of the total 239 participants, 83 carried the TT genotype, 127 carried the *Tt* genotype, and 29 carried the *tt* genotype.

These results suggested an association between the presence of the T-allele at position Taq-1 and the occurrence of AP. The allele T appeared almost three times more often in the AP group (OR = 2.61; 95% CI: 1.68–4.03; $p < 0.0001$) than in the alcohol-abuse control group. Also, the presence of the TT genotype was four times more frequent in the AP group (OR = 4.55; 95% CI: 1.69–12.20; $p = 0.003$) in comparison to the control group. Detailed data for the AP and control groups are shown in Tables 1 and 2.

**Table 1.** Genotype frequencies of Apa-1, Bsm-1, Fok-1, and Taq-1 in *VDR* gene polymorphisms in the studied groups and the association with AP.

| Genotype | AP *n* (%) | Control *n* (%) | OR (95% CI) AP vs. Control | *p*-Value |
|---|---|---|---|---|
| Apa-1 (rs7975232) | | | | |
| AA (TT) | 51 (39.5) | 30 (27) | 1.0 | |
| Aa (TG) | 51 (39.5) | 53 (48) | 1.76 (0.98–3.19) | 0.06 |
| aa (GG) | 27 (21) | 27 (25) | 1.70 (0.84–3.41) | 0.14 |
| AA vs. Aa + aa | | | 1.73 (1.14–2.63) | 0.009 |
| Bsm-1 (rs1544410) | | | | |
| BB (GG) | 12 (9) | 16 (15) | 1.0 | |
| Bb (GA) | 66 (51) | 58 (53) | 0.66 (0.29–1.51) | 0.32 |
| bb (AA) | 51 (40) | 36 (33) | 0.53 (0.22–1.25) | 0.15 |
| Bb + bb vs. BB | | | 1.72 (0.97–3.07) | 0.06 |
| Fok-1 (rs2228570)FF (CC) | 51 (40) | 39 (35) | 1.0 | |
| Ff (CT) | 57 (44) | 47 (43) | 1.07 (0.61–1.90) | 0.79 |
| ff (TT) | 21 (16) | 24 (22) | 1.49 (0.73–3.07) | 0.27 |
| FF vs. Ff + ff | | | 1.45(0.90–2.34) | 0.12 |
| Taq-1 (rs731236)TT (TT) | 56 (43) | 27 (25) | 1.0 | |
| Tt (TC) | 64 (50) | 63 (57) | 2.05 (1.11–3.77) | 0.02 |
| tt (CC) | 9 (7) | 20 (18) | 4.55 (1.69–12.20) | 0.003 |
| TT vs. Tt + tt | | | 2.61 (1.68–4.03) | <0.0001 |

Genotype frequencies of *VDR* SNPs were determined in the control group, alcohol group, and patients with AP, and associations with AP.

**Table 2.** *VDR* SNPs frequency of alleles in the control group and patients with AP.

| *VDR* Polymorphism | Allele | AP $n = 129$ | Control $n = 110$ | Study Population (AP + Control) $n = 239$ |
|---|---|---|---|---|
| Apa-1 (rs7975232) | A (T) | 0.59 | 0.51 | 0.56 |
| | a (G) | 0.41 | 0.49 | 0.44 |
| Bsm-1 (rs1544410) | B (G) | 0.35 | 0.41 | 0.38 |
| | b (A) | 0.65 | 0.59 | 0.62 |
| Fok-1 (rs2228570) | F (C) | 0.62 | 0.57 | 0.59 |
| | f (T) | 0.38 | 0.43 | 0.41 |
| Taq-1 (rs731236) | T (T) | 0.68 | 0.53 | 0.61 |
| | t (C) | 0.32 | 0.47 | 0.39 |

## 3. Discussion

Combined genetic, metabolic, and environmental factors all contribute to the development and re-occurrence of acute and chronic pancreatitis [35]. To the best of our knowledge, this is the first examination and comparison of *VDR* gene polymorphism in patients diagnosed with acute pancreatitis. Our study showed higher levels than the accepted reference points for bilirubin, alanine transaminase (ALT), and aspartate transaminase (AST) in the AP-patients (1.9 mg/dL, 155.7 IU/L, and 155.2 IU/L, respectively). This may be associated with liver damage resulting from AP-group alcohol abuse, as high alcohol consumption/addiction is considered a major cause of AP [4]. AP patients also had increased amylase activity and significantly higher lipase activity that indicate pancreatic dysfunction. Their additional significantly increased C-reactive protein (CRP) levels demonstrate that this process can be enhanced by ongoing inflammation.

The average level of vitamin D in the AP and control groups (44.2 and 47.7 nmol/L, respectively) was similar, as noted in alcohol-abuse patients [12]. Although vitamin D deficiency in people abusing alcohol was noted before, the review of Tardelli et al. [36] suggests that data concerning vitamin D levels in alcohol-use patients are controversial. Ogunsakin et al. [37] found that vitamin D ($25(OH)D_3$), and its active form, ($1,25(OH)_2D_3$), were significantly reduced in alcohol-abuse patients. Also, ethanol-fed mice showed reduced levels of ($1,25(OH)_2D_3$). These conflicting data suggest that, in selected cases, the analyzed SNPs may be responsible for the observed deficiency of vitamin D responsiveness [38,39].

Our study complements the analysis of AP being a metabolic disorder with genetic factors. We have focused on *VDR* polymorphism rather than vitamin D concentration because of its potential final effect on glucose metabolism. Forouhi et al. [40] suggested that the direct effect of vitamin D on the secretory function of pancreatic cells is through their *VDR*s, and this is suggested to be the explanation for the association between a lower serum vitamin D status and a high risk of hypoglycemia and insulin resistance. Vitamin D action through its receptors may also be responsible for the regulation of insulin secretion by the β-cells in a glucose-dependent manner [41]. Abnormalities in these processes, therefore, lead to glucose metabolism disorders [42,43]. Unexpectedly, we found an increased glucose concentration to a mean of 127.4 mg/dL in AP-patients compared to the reference values (Table 3). This may be a result of glucose homeostasis dysfunction, leading to subsequent inflammation, which is consistent with the clinical parameters established in our AP group where we detected high CRP levels in the 4.4 mg/dL range (Table 3).

**Table 3.** Distribution of selected characteristics in acute pancreatitis patients and alcohol-abuse control group.

| Characteristic | Alcohol | AP Group | *p*-Value |
|---|---|---|---|
| | $n$ = 110 | $n$ = 129 | |
| Age (years) | 44.2 (±8.0) | 52.4 (±10.3) | 0.09 |
| Body mass (Kg) | 71.3 (±8.6) | 75.6 (±8.7) | 0.73 |
| Amylase activity in serum (IU/L) | 114.4 (±61.1) | 1647.5 (±636.5) | <0.001 |
| Lipase (IU/L) | 131.1 (±45.8) | 1446.7 (±814.6) | <0.001 |
| Bilirubin (mg/dL) | 0.9 (±0.7) | 1.9 (±0.87) | <0.001 |
| Glucose (mg/dL) | 85.5 (±12.1) | 127.4 (±33.9) | <0.001 |
| AST (IU/L) | 70.2 (±51.7) | 155.2 (±71.6) | <0.001 |
| ALT (IU/L) | 55.2 (±9.7) | 155.7 (±30.2) | <0.001 |
| CRP (mg/dL) | 0.58 (±1.0) | 4.4 (±1.4) | <0.001 |
| 25-hydroxyvitamin D (nmol/L) | 44.2 (±17.1) | 46.7 (±18.4) | 0.35 |
| Female (%) | 17.0 | 29.0 | |
| APACHE II scale | n.a. | 4.4 (±1.4) | |

Values are as expressed as mean ± SD; n.a.: not applicable.

Vitamin D was also suggested as an immune modulator because of the existence of *VDR*s in activated T lymphocytes, macrophages, and thymus tissue [44,45]. Palomer et al. stated that inflammatory factors have often been associated with insulin resistance and β-cell failure [46]. Moreover, vitamin D reduces oxidative stress through the induction of an antioxidant activity by itself [47,48], and, importantly, oxidative damage has been implicated in acute pancreatitis initiation [49,50]. Because *VDR* is strongly expressed in pancreatic beta cells [51], its signaling may be reduced or even activated dependent on the occurrence of specific SNPs in AP similar to what has been described in the progression of some tumors [52–54]. Reduced vitamin D receptor signaling might be a potential mechanism underlying increased foam cell formation, resulting in an accelerated development of cardiovascular disease in diabetic subjects [55].

Here, we used SNP analysis to identify differences in the frequency of Apa-1, Taq-1, Bsm-1, and Fok-1 genotypes/alleles in the *VDR* gene between AP and alcohol-abuse patients. These SNPs are located near the untranslated region and are possibly linked to a poly-A-microsatellite repeat that could affect *VDR* mRNA stability. The frequency of alleles and the distribution of genotypes in AP and alcohol-abuse controls for Bsm-1 (0.38 for B), Apa-1 (0.56 for A), and Fok-1 (0.59 for F) (Tables 1 and 2) was comparable with the data in a Caucasian population presented by Uitterlinden et al. [33]. These authors found polymorphisms of 0.66 of F in Fok-1, 0.42 of B in Bsm-1, and 0.44 of A in Apa-1. In our study, only the polymorphism in Taq-1 was increased (0.61 in T) in AP (Table 2) compared to the value of 0.43 for T as described previously [33]. Polymorphisms affecting the vitamin D and *VDR* axis are associated with an ongoing degree of inflammation associated with the release of pro-inflammatory cytokines, possibly resulting from modulation of the inflammasome, alterations of gut permeability, and microbial translocation, as suggested by Al-Daghri et al. [56].

Our results suggest that allele *T* in the Taq-1 polymorphic site of the *VDR* gene is almost three times more frequent (OR = 2.61 (95% CI: 1.68–4.03, $p < 0.0001$) in acute pancreatitis patients than in alcohol-abuse controls. *VDR* polymorphism may thus play an important role in vitamin D metabolism independently of the actual vitamin D plasma concentration. Polymorphisms in Bsm-1, Taq-1, Apa-1, and Fok-1 were associated with renal diseases, cancer, neurolithiasis diabetes, asthma, atopic dermatitis, and autism [33,34,57–63]. To our best knowledge, there was no research focused on *VDR* polymorphism in acute pancreatitis. Only Fok-1 polymorphism was noted in correlation with pancreas allograft [64].

From our results, it can be inferred that, despite a polymorphism in the *VDR* receptor gene (Taq-1), both AP and the control groups showed no difference in plasma concentrations of vitamin D (Table 3). This implies that the interaction between the active metabolite of vitamin D (1,25-dihydroxyvitamin D) and the receptor is responsible for the final effect. Therefore, polymorphisms in the vitamin D receptor gene may have an impact on the development of acute pancreatitis due to the lack of a protective role of vitamin D.

There are several limitations in our study that we will address in subsequent studies. Our results were obtained in a group of 129 AP patients, and we need to repeat and expand these studies in larger replication analyses. Moreover, the *VDR* is part of the nuclear receptor family of transcription factors. Upon activation by vitamin D, a heterodimer is formed between the *VDR* and the retinoid-X receptor, and this complex interacts with vitamin D responsive elements in the DNA, thereby driving expression or transrepression of selected genes. It would, therefore, be of value to analyze also the retinoid-X receptor (*RXR*) gene for the occurrence of SNPs. However, so far, a polymorphism in the *RXR* gene was only found in hyperlipidemia and type 2 diabetes [65]. In addition, we could not extensively compare vitamin D plasma concentrations in controls and AP groups because of their different lifestyles, including diet, sun exposure, and vitamin D supplementation. However, our earlier research [34] did show a lack of a significant correlation between serum 25-hydrohyvitamin D (25(OH)$D_3$) levels and *VDR* polymorphism, probably reflecting that plasma vitamin D levels reflect an inactive form of the functional vitamin D3. Therefore, subsequent work will include polymorphisms in other genes affecting the metabolism of vitamin D3.

The prevalence of acute pancreatitis is still increasing and patients are insufficiently diagnosed, necessitating the development of new analytical methods. Here, we added new insights into the putative role of SNPs in *VDR* in pancreatitis development, and, therefore, this study contributes to individualized research on the interaction and impact of environmental factors, including nutrition, in pancreatic secretory disorders.

## 4. Materials and Methods

### 4.1. Ethics and General Study Information

Our study comprised of 239 individuals (57 females and 182 males) assigned to control (C) or test groups (AP—acute pancreatitis group). Patients were recruited by specialists at the Department of General Surgery and Oncology of the Warmia and Mazury University Hospital in Olsztyn in 2011–2017. Controls were recruited by specialists at the Department for the treatment of alcohol dependent patients, including patients with other dysfunctions, and the Department for the treatment of withdrawal syndromes of the Jozef Babinski Specialist Psychiatric Hospital in Krakow in 2015. Test patients and controls were of Polish ethnic origin. All patients were treated according to the Patient Right Protection Act of our institution and international guidelines, and the Local Bioethics Committee approved our study (No 13/2016, 27 April 2016).

### 4.2. Controls and AP Group Characteristic

The test group included 129 patients (38 females and 91 males) with diagnosed AP (mean age ranging from 28 to 76 years; average 52.4). These patients presented for treatment 8–36 h after the onset of associated pain and vomiting or emetic reflex. Alcohol abuse was the etiological factor in all AP patients. Exclusions from the study included patients with chronic circulatory system, liver, kidney, and lung disease. Blood samples were taken from the forearm vein for the panel of biochemical tests evaluating pancreatic enzymes, and additional tests assessed general health condition and performance of individual systems and organs.

Laboratory tests were performed upon arrival at the hospital and at 48 h after admission. A computed tomography (CT) scan was performed on all patients within 48 h after arrival at the hospital for detection of the development of fluid collections, the extent of inflammation, and necrotic changes. Acute Physiology and Chronic Health Evaluation (APACHE) II scores were calculated using data from the first 24 h after admission. Serum CRP levels were measured at admission.

The general health of included patients and the course of their AP disease was assessed by evaluating 4 to 16 points on the APACHE II scale. Abdominal CT with contrast (intravenous and per os) evaluated morphological changes in the pancreas and surrounding environment. Patient evaluation via imaging scales predicting acute pancreatitis severity and the development of pancreatic complications was performed 3–4 days after the onset of symptoms, then again after 10–12 days treatment. Microbiological analyses of stool samples of both the control and AP group did not reveal any significant difference. Possible small remaining fluid reservoirs prior to surgical intervention and puncture and self-absorption mechanisms were monitored by ultrasound. The control group comprised of 110 individuals (19 females and 91 males), with a mean age ranging from 23 to 58 years (average 44.2). All patients met the DSM IV criteria for alcohol dependence. All participants (AP and Control group) in this study were Caucasians. The characteristics of AP patients and controls and laboratory parameters established during hospitalization are shown in Table 3.

### 4.3. Polymorphism of VDR Genes in Control and AP Groups

DNA was isolated from peripheral blood using a GeneJET™ Whole Blood Genomic DNA Purification Mini Kit (Thermo Scientific, Waltham, MA, USA) according to the manufacturer's instructions. Bsm-1, Taq-1, Apa-1, and Fok-1 *VDR* polymorphisms were assessed by polymerase chain reaction-restriction fragment length polymorphism (PCR-RFLP). Primers examining the polymorphism in Fok-1 were as previously described [34,66] with slight modifications [67], while primers for Bsm-1, Taq-1, and Apa-1 were designed with the Primer3 application (http://bioinfo.ut.ee/primer3-0.4.0/). The primer specificity was verified with the BLAST algorithm, and primer sequences used for amplification of Fok-1, Bsm-1, Taq-1, and Apa-1 restriction enzyme polymorphisms are listed in Table 4.

**Table 4.** Primers for *VDR* SNPs and polymerase chain reaction-restriction fragment length polymorphism (PCR-RFLP) conditions.

| SNP | Primer Sequence | Restriction Enzyme | PCR/RFLP Products (bp) |
|---|---|---|---|
| Apa-1 | TaqF: 5′-ggatcctaaatgcacggaga-3′<br>TaqR: 5′-aggaaaggggttaggttgga-3′ | FastDigest Apa-1 | aa: 484, 146<br>AA: 630<br>Aa: 630, 484, 146<br>PCR product: 630 |
| Bsm-1 | BsmF: 5′-cggggagtatgaaggacaaa-3′<br>BsmR: 5′-ccatctctcaggctccaaag-3′ | FastDigest Bsm-1 | bb: 243, 105<br>BB: 348<br>Bb: 348, 243, 105<br>PCR product: 348 |
| Fok-1 | FokR: 5′-atggaaacaccttgcttcttctccctc-3′<br>FokF: 5′-agctggccctggcactgactctggctct-3′ | FastDigest Fok-1 | ff: 198, 69<br>FF: 267<br>Ff: 267, 198, 69<br>PCR product: 267 |
| Taq-1 | ATaq1F: 5′-ggatcctaaatgcacggaga-3′<br>ATaq1R: 5′-aggaaaggggttaggttgga-3′ | FastDigest Taq-1 | tt: 225, 200, 205<br>TT: 425, 205<br>Tt: 425, 225, 200, 205<br>PCR product: 630 |

PCR amplification was conducted in a thermal cycler according to the following program: Initial denaturation: 94 °C for 3 min, proper denaturation: 94 °C for 30 s, attaching the starters at 61 °C for all genes for 30 s, synthesis: 72 °C for 30 s, final synthesis: 72 °C for 5 min, number of cycles: 40, cooling: 4 °C. The mixture in the volume of 25 μL consisted of DreamTaq™ Green Master Mix (Thermo Scientific), specific primers, the DNA matrix, and ultrapure water (Sigma-Aldrich, Saint Louis, MO, USA). The yield and specificity of PCR products were evaluated by electrophoresis in 1.5% agarose gel (Promega, Fitchburg, MA, USA) and staining with GelGreen Nucleic Acid Gel Stain (Biotium, Hayward, CA, USA).

Amplified fragments were digested with the appropriate restriction enzyme (Thermo Scientific) according to the manufacturer's instructions, and visualized on a 2.5% agarose gel. DNA sequencing of random chosen samples after amplification was used to confirm proper genotyping.

### 4.4. Statistical Analysis

The mean values in the control and AP groups were compared using a student's *t*-test. The frequency distribution of common risk factors for AP are presented as the mean ± SD. The genotype distribution among subjects was analyzed for Hardy-Weinberg equilibrium (HWE) using the chi-square test, and genotype and SNP allele frequencies were compared in AP patients and control groups by a Fisher's test. Odds ratios (ORs) and 95% confidence intervals (CIs) were calculated using logistic regression analysis, and used to compare both allele frequencies in alcohol-abuse controls and AP patients, and allele frequencies between females and males. The risk of AP development was estimated via the wild-type genotype vs. the wild/mutant and mutant-type genotypes. Only all 4-genotyped SNPs probes were used in the calculations. Statistical analysis was conducted on GraphPad Prism software (v 6.01; San Diego, CA, USA), with $\leq$0.01 $p$-value considered statistically significant.

**Author Contributions:** H.F.J.S. and A.C.: Conceptualization, Methodology, Validation, Formal Analysis, Investigation, Data Curation, Writing, Original Draft Preparation; E.K.: Supervision, Writing, Review & Editing; E.F.: Writing, Review & Editing; J.S.: Writing, Review & Editing; N.K.: Writing, Review & Editing; K.K.: Writing, Review & Editing.

**Acknowledgments:** The authors sincerely thank all the patients who participated in this study.

**Conflicts of Interest:** The authors declare that there are no conflicts of interest.

## References


1. Weiss, F.U.; Simon, P.; Mayerle, J.; Kraft, M.; Lerch, M.M. Germline mutations and gene polymorphism associated with human pancreatitis. *Endocrinol. Metab. Clin. N. Am.* **2006**, *35*, 289–302. [CrossRef] [PubMed]
2. Nieminen, A.; Maksimow, M.; Mentula, P.; Kyhälä, L.; Kylänpää, L.; Puolakkainen, P.; Salmi, M. Circulating cytokines in predicting development of severe acute pancreatitis. *Crit. Care* **2014**, *18*, R104. [CrossRef] [PubMed]
3. Wojtuń, S.; Gil, J.; Błaszak, A. Dietary treatment in acute pancreatitis. *Pol. Merkur. Lekarski.* **2007**, *22*, 469–473. [PubMed]
4. Lankisch, P.G.; Apte, M.; Banks, P.A. Acute pancreatitis. *Lancet* **2015**, *4*, 85–96. [CrossRef]
5. Yadav, D.; Vege, S.S.; Chari, S.T. Epidemiology of Pancreatitis. In *GI Epidemiology: Diseases and Clinical Methodology*, 2nd ed.; Wiley Blackwell: Oxford, UK, 2014; pp. 306–312.
6. Eyles, D.W.; Smith, S.; Kinobe, R.; Hewison, M.; McGrath, J.J. Distribution of the vitamin D receptor and 1α-hydroxylase in human brain. *J. Chem. Neuroanat.* **2005**, *29*, 21–30. [CrossRef] [PubMed]
7. Eyles, D.W.; Burne, T.H.; McGrath, J.J. Vitamin D, effects on brain development, adult brain function and the links between low levels of vitamin D and neuropsychiatric disease. *Front. Neuroendocrinol.* **2013**, *34*, 47–64. [CrossRef] [PubMed]
8. Anglin, R.E.; Samaan, Z.; Walter, S.D.; McDonald, S.D. Vitamin D deficiency and depression in adults: Systematic review and meta-analysis. *Br. J. Psychiatry* **2013**, *202*, 100–107. [CrossRef] [PubMed]
9. Raimondi, S.; Maisonneuve, P.; Lowenfels, A.B. Epidemiology of pancreatic cancer: An overview. *Nat. Rev. Gastroenterol. Hepatol.* **2009**, *6*, 699. [CrossRef] [PubMed]
10. Bhimani, M.M. Vitamin D: Does it play a role in psychiatry. *J. Pak. Med. Assoc.* **2012**, *62*, 181. [PubMed]
11. Wagnerberger, S.; Kanuri, G.; Bergheim, I. Alcohol drinking patterns and nutrition in alcoholic liver disease. In *Trends in Alcoholic Liver Disease Research-Clinical and Scientific Aspects*; InTech: Rijeka, Croatia, 2012; pp. 1–22.
12. Neupane, S.P.; Lien, L.; Hilberg, T.; Bramness, J.G. Vitamin D deficiency in alcohol-use disorders and its relationship to comorbid major depression: A cross-sectional study of inpatients in Nepal. *Drug Alcohol. Depend.* **2013**, *133*, 480–485. [CrossRef] [PubMed]
13. Lieber, C.S. Alcohol: Its metabolism and interaction with nutrients. *Annu. Rev. Nutr.* **2000**, *20*, 395–430. [CrossRef] [PubMed]
14. Kim, D.I.; Kim, H.; Son, P.; Kang, J.H.; Kang, B.T.; Yang, M.P. Serum 25-hydroxyvitamin D concentrations in dogs with suspected acute pancreatitis. *J. Vet. Med. Sci.* **2017**, *79*, 1366–1373. [CrossRef] [PubMed]
15. Holick, M.F. Vitamin D deficiency. *N. Engl. J. Med.* **2007**, *357*, 266–281. [CrossRef] [PubMed]
16. Haussler, M.R.; Whitfield, G.K.; Haussler, C.A.; Hsieh, J.C.; Thompson, P.D.; Selznick, S.H.; Jurutka, P.W. The nuclear vitamin D receptor: Biological and molecular regulatory properties revealed. *J. Bone Miner. Res.* **1998**, *13*, 325–349. [CrossRef] [PubMed]
17. Mehta, R.G.; Mehta, R.R. Vitamin D and cancer. *J. Nutr. Biochem.* **2002**, *13*, 252–264. [CrossRef]
18. Mathieu, C.; van Etten, E.; Decallonne, B.; Guilietti, A.; Gysemans, C.; Bouillon, R.; Overbergh, L. Vitamin D and 1, 25-dihydroxyvitamin D3 as modulators in the immune system. *J. Steroid Biochem. Mol. Biol.* **2004**, *89*, 449–452. [CrossRef] [PubMed]
19. Valdivielso, J.M.; Fernandez, E. Vitamin D receptor polymorphisms and diseases. *Clin. Chim. Acta* **2006**, *371*, 1–12. [CrossRef] [PubMed]
20. Sayeed, I.; Stein, D.G. Progesterone as a neuroprotective factor in traumatic and ischemic brain injury. *Prog. Brain Res.* **2009**, *175*, 219–237. [CrossRef] [PubMed]
21. Ramagopalan, S.V.; Heger, A.; Berlanga, A.J.; Maugeri, N.J.; Lincoln, M.R.; Burrell, A.; Watson, C.T. A ChIP-seq defined genome-wide map of vitamin D receptor binding: Associations with disease and evolution. *Genome Res.* **2010**, *20*, 1352–1360. [CrossRef] [PubMed]
22. Sigmundsdottir, H. From the bench to the clinic: New aspects on immunoregulation by vitamin D analogs. *Dermatoendocrinology* **2011**, *3*, 187–192. [CrossRef]
23. Harms, L.R.; Burne, T.H.; Eyles, D.W.; McGrath, J.J. Vitamin D and the brain. *Best Pract. Res. Clin. Endocrinol. Metab.* **2011**, *25*, 657–669. [CrossRef] [PubMed]
24. Johnson, J.D.; Kuang, S.; Misler, S.; Polonsky, K.S. Ryanodine receptors in human pancreatic β cells: Localization and effects on insulin secretion. *FASEB J.* **2004**, *18*, 878–880. [CrossRef] [PubMed]

25. Zaki, M.; Kamal, S.; Basha, W.A.; Youness, E.; Ezzat, W.; El-Bassyouni, H.; Amr, K. Association of vitamin D receptor gene polymorphism (VDR) with vitamin D deficiency, metabolic and inflammatory markers in Egyptian obese women. *Genes Dis.* **2017**, *4*, 176–182. [CrossRef]
26. Ishida, H.; Norman, A.W. Demonstration of a high affinity receptor for 1, 25-dihydroxyvitamin D3 in rat pancreas. *Mol. Cell. Endocrinol.* **1988**, *60*, 109–117. [CrossRef]
27. Johnson, J.A.; Grande, J.P.; Roche, P.C.; Kumar, R. Immunohistochemical localization of the $1,25(OH)_2D_3$ receptor and calbindin D28k in human and rat pancreas. *Am. J. Physiol.* **1994**, *267*, E356–E360. [CrossRef] [PubMed]
28. Faraco, J.H.; Morrison, N.A.; Baker, A.; Shine, J.; Frossard, P.M. ApaI dimorphism at the human vitamin D receptor gene locus. *Nucleic Acids Res.* **1989**, *17*, 2150. [CrossRef] [PubMed]
29. Morrison, N.A.; Yeoman, R.; Kelly, P.J.; Eisman, J.A. Contribution of trans-acting factor alleles to normal physiological variability: Vitamin D receptor gene polymorphism and circulating osteocalcin. *Proc. Natl. Acad. Sci. USA* **1992**, *89*, 6665–6669. [CrossRef] [PubMed]
30. Morrison, N.A.; Qi, J.C.; Tokita, A.; Kelly, P.J.; Crofts, L.; Nguyen, T.V.; Eisman, J.A. Prediction of bone density from vitamin D receptor alleles. *Nature* **1994**, *367*, 284–287. [CrossRef] [PubMed]
31. Saijo, T.; Ito, M.; Takeda, E.; Huq, A.H.; Naito, E.; Yokota, I.; Kuroda, Y. A unique mutation in the vitamin D receptor gene in three Japanese patients with vitamin D-dependent rickets type II: Utility of single-strand conformation polymorphism analysis for heterozygous carrier detection. *Am. J. Hum. Genet.* **1991**, *49*, 668. [PubMed]
32. Gross, C.; Eccleshall, T.R.; Malloy, P.J.; Villa, M.L.; Marcus, R.; Feldman, D. The presence of a polymorphism at the translation initiation site of the vitamin D receptor gene is associated with low bone mineral density in postmenopausal Mexican-American women. *J. Bone Miner. Res.* **1996**, *11*, 1850–1855. [CrossRef] [PubMed]
33. Uitterlinden, A.G.; Fang, Y.; van Meurs, J.B.; van Leeuwen, H.; Pols, H.A. Vitamin D receptor gene polymorphisms in relation to Vitamin D related disease states. *J. Steroid Biochem. Mol. Biol.* **2004**, *89*, 187–193. [CrossRef] [PubMed]
34. Cieślińska, A.; Kostyra, E.; Chwała, B.; Moszyńska-Dumara, M.; Fiedorowicz, E.; Teodorowicz, M.; Savelkoul, H.F. Vitamin D Receptor Gene Polymorphisms Associated with Childhood Autism. *Brain Sci.* **2017**, *7*, 115. [CrossRef] [PubMed]
35. Whitcomb, D.C. Genetic risk factors for pancreatic disorders. *Gastroenterology* **2013**, *144*, 1292–1302. [CrossRef] [PubMed]
36. Tardelli, V.S.; do Lago, M.P.P.; da Silveira, D.X.; Fidalgo, T.M. Vitamin D and alcohol: A review of the current literature. *Psychiatry Res.* **2017**, *248*, 83–86. [CrossRef] [PubMed]
37. Ogunsakin, O.; Hottor, T.; Mehta, A.; Lichtveld, M.; McCaskill, M. Chronic ethanol exposure effects on vitamin D levels among subjects with alcohol use disorder. *Environ. Health Insights* **2016**, *10*, EHI-S40335. [CrossRef] [PubMed]
38. Becker, K.G. Autism, immune dysfunction and Vitamin D. *Acta Psychiatr. Scand.* **2011**, *124*, 74. [CrossRef] [PubMed]
39. Eyles, D.; Burne, T.; McGrath, J. Vitamin D in fetal brain development. *Semin. Cell Dev. Biol.* **2011**, *22*, 629–636. [CrossRef] [PubMed]
40. Forouhi, N.G.; Luan, J.A.; Cooper, A.; Boucher, B.J.; Wareham, N.J. Baseline serum 25-hydroxy vitamin D is predictive of future glycemic status and insulin resistance: The Medical Research Council Ely Prospective Study 1990–2000. *Diabetes* **2008**, *57*, 2619–2625. [CrossRef] [PubMed]
41. Knekt, P.; Laaksonen, M.; Mattila, C.; Härkänen, T.; Marniemi, J.; Heliövaara, M.; Reunanen, A. Serum vitamin D and subsequent occurrence of type 2 diabetes. *Epidemiology* **2008**, *19*, 666–671. [CrossRef] [PubMed]
42. Dobrian, A.D.; Ma, Q.; Lindsay, J.W.; Leone, K.A.; Ma, K.; Coben, J.; Nadler, J.L. Dipeptidyl peptidase IV inhibitor sitagliptin reduces local inflammation in adipose tissue and in pancreatic islets of obese mice. *Am. J. Physiol. Endocrinol. Metab.* **2010**, *300*, E410–E421. [CrossRef] [PubMed]
43. Zhao, H.; Wei, R.; Wang, L.; Tian, Q.; Tao, M.; Ke, J.; Hong, T. Activation of glucagon-like peptide-1 receptor inhibits growth and promotes apoptosis of human pancreatic cancer cells in a cAMP-dependent manner. *Am. J. Physiol. Endocrinol. Metab.* **2014**, *306*, E1431–E1441. [CrossRef] [PubMed]
44. Chiu, K.C.; Chu, A.; Go, V.L.W.; Saad, M.F. Hypovitaminosis D is associated with insulin resistance and $\beta$ cell dysfunction. *Am. J. Clin. Nutr.* **2004**, *79*, 820–825. [CrossRef] [PubMed]

45. Hewison, M.; Gacad, M.A.; Lemire, J.; Adams, J.S. Vitamin D as a cytokine and hematopoetic factor. *Rev. Endocr. Metab. Disord.* **2001**, *2*, 217–227. [CrossRef] [PubMed]
46. Palomer, X.; González-Clemente, J.M.; Blanco-Vaca, F.; Mauricio, D. Role of vitamin D in the pathogenesis of type 2 diabetes mellitus. *Diabetes Obes. Metab.* **2008**, *10*, 185–197. [CrossRef] [PubMed]
47. Garcion, E.; Wion-Barbot, N.; Montero-Menei, C.N.; Berger, F.; Wion, D. New clues about vitamin D functions in the nervous system. *Trends Endocrinol. Metab.* **2002**, *13*, 100–105. [CrossRef]
48. Berk, M.; Jacka, F.N.; Williams, L.J.; Ng, F.; Dodd, S.; Pasco, J.A. Is this D vitamin to worry about? Vitamin D insufficiency in an inpatient sample. *Aust. N. Z. J. Psychiatry* **2008**, *42*, 874–878. [CrossRef] [PubMed]
49. Sweiry, J.H.; Mann, G.E. Role of oxidative stress in the pathogenesis of acute pancreatitis. *Scand. J. Gastroenterol. Suppl.* **1996**, *31*, 10–15. [CrossRef]
50. Tsai, K.; Wang, S.S.; Chen, T.S.; Kong, C.W.; Chang, F.Y.; Lee, S.D.; Lu, F.J. Oxidative stress: An important phenomenon with pathogenetic significance in the progression of acute pancreatitis. *Gut* **1998**, *42*, 850–855. [CrossRef] [PubMed]
51. Wang, Y.; Zhu, J.; DeLuca, H.F. Where is the vitamin D receptor? *Arch. Biochem. Biophys.* **2012**, *523*, 123–133. [CrossRef] [PubMed]
52. Brożyna, A.A.; Jóźwicki, W.; Slominski, A.T. Decreased VDR expression in cutaneous melanomas as marker of tumor progression: New data and analyses. *Anticancer Res.* **2014**, *34*, 2735–2743.
53. Jóźwicki, W.; Brożyna, A.A.; Siekiera, J.; Slominski, A.T. Expression of vitamin D receptor (VDR) positively correlates with survival of urothelial bladder cancer patients. *Int. J. Mol. Sci.* **2015**, *16*, 24369–24386. [CrossRef] [PubMed]
54. Agliardi, C.; Guerini, F.R.; Saresella, M.; Caputo, D.; Leone, M.A.; Zanzottera, M.; Al-Daghri, N. Vitamin D receptor (VDR) gene SNPs influence VDR expression and modulate protection from multiple sclerosis in HLA-DRB1* 15-positive individuals. *Brain Behav. Immun.* **2011**, *25*, 1460–1467. [CrossRef] [PubMed]
55. Oh, J.; Weng, S.; Felton, S.K.; Bhandare, S.; Riek, A.; Butler, B.; Bernal-Mizrachi, L. 1,25$(OH)_2$ vitamin D inhibits foam cell formation and suppresses macrophage cholesterol uptake in patients with type 2 diabetes mellitus. *Circulation* **2009**, *120*, 687–698. [CrossRef] [PubMed]
56. Al-Daghri, N.M.; Guerini, F.R.; Al-Attas, O.S.; Alokail, M.S.; Alkharfy, K.M.; Draz, H.M.; Biasin, M. Vitamin D receptor gene polymorphisms are associated with obesity and inflammosome activity. *PLoS ONE* **2014**, *9*, e102141. [CrossRef] [PubMed]
57. Poon, A.H.; Laprise, C.; Lemire, M.; Montpetit, A.; Sinnett, D.; Schurr, E.; Hudson, T.J. Association of vitamin D receptor genetic variants with susceptibility to asthma and atopy. *Am. J. Respir. Crit. Care Med.* **2004**, *170*, 967–973. [CrossRef] [PubMed]
58. Raby, B.A.; Lazarus, R.; Silverman, E.K.; Lake, S.; Lange, C.; Wjst, M.; Weiss, S.T. Association of vitamin D receptor gene polymorphisms with childhood and adult asthma. *Am. J. Respir. Crit. Care Med.* **2004**, *170*, 1057–1065. [CrossRef] [PubMed]
59. Kim, J.S.; Kim, Y.I.; Song, C.; Yoon, I.; Park, J.W.; Choi, Y.B.; Lee, K.S. Association of vitamin D receptor gene polymorphism and Parkinson's disease in Koreans. *J. Korean Med. Sci.* **2005**, *20*, 495–498. [CrossRef] [PubMed]
60. Saadi, A.; Gao, G.; Li, H.; Wei, C.; Gong, Y.; Liu, Q. Association study between vitamin D receptor gene polymorphisms and asthma in the Chinese Han population: A case-control study. *BMC Med. Genet.* **2009**, *10*, 71. [CrossRef] [PubMed]
61. Knekt, P.; Kilkkinen, A.; Rissanen, H.; Marniemi, J.; Sääksjärvi, K.; Heliövaara, M. Serum vitamin D and the risk of Parkinson disease. *Arch. Neurol.* **2010**, *67*, 808–811. [CrossRef] [PubMed]
62. Heine, G.; Hoefer, N.; Franke, A.; Nöthling, U.; Schumann, R.R.; Hamann, L.; Worm, M. Association of vitamin D receptor gene polymorphisms with severe atopic dermatitis in adults. *Br. J. Dermatol.* **2013**, *168*, 855–858. [CrossRef] [PubMed]
63. Maalmi, H.; Sassi, F.H.; Berraies, A.; Ammar, J.; Hamzaoui, K.; Hamzaoui, A. Association of vitamin D receptor gene polymorphisms with susceptibility to asthma in Tunisian children: A case control study. *Hum. Immunol.* **2013**, *74*, 234–240. [CrossRef] [PubMed]
64. Rahsaz, M.; Azarpira, N.; Nikeghbalian, S.; Geramizadeh, B.; Aghdaie, M.H.; Daraie, M.; Karimi, M.H. Vitamin D receptor genotype in pancreas allograft: A pilot study. *Exp. Clin. Transplant.* **2012**, *10*, 487–491. [CrossRef] [PubMed]

65. Vimaleswaran, K.S.; Cavadino, A.; Berry, D.J.; Mangino, M.; Andrews, P.; Moore, J.H.; Hyppönen, E. Interaction between allelic variations in vitamin D receptor and retinoid X receptor genes on metabolic traits. *BMC Genet.* **2014**, *15*, 37. [CrossRef] [PubMed]
66. Pani, M.A.; Knapp, M.; Donner, H.; Braun, J.; Baur, M.P.; Usadel, K.H.; Badenhoop, K. Vitamin D receptor allele combinations influence genetic susceptibility to type 1 diabetes in Germans. *Diabetes* **2000**, *49*, 504–507. [CrossRef]
67. Maalmi, H.; Berraïes, A.; Tangour, E.; Ammar, J.; Abid, H.; Hamzaoui, K.; Hamzaoui, A. The impact of vitamin D deficiency on immune T cells in asthmatic children: A case-control study. *J. Asthma Allergy* **2012**, *5*, 11. [CrossRef] [PubMed]

International Journal of
*Molecular Sciences*

MDPI

*Review*

# The Impact of Vitamin D in the Treatment of Essential Hypertension

**Christian Legarth [1], Daniela Grimm [1,*], Markus Wehland [2], Johann Bauer [3] and Marcus Krüger [2]**

1 Institute of Biomedicine, Pharmacology, Aarhus University, Wilhelm Meyers Allé 4, DK-8000 Aarhus C, Denmark; chr_brod@hotmail.com
2 Clinic for Plastic, Aesthetic and Hand Surgery, Otto von Guericke University, Leipziger Str. 44, 39120 Magdeburg, Germany; markus.wehland@med.ovgu.de (M.W.); marcus.krueger@med.ovgu.de (M.K.)
3 Max Planck Institute for Biochemistry, Am Klopferspitz 18, 82152 Martinsried, Germany; jbauer@biochem.mpg.de
* Correspondence: dgg@biomed.au.dk; Tel.: +45-871-67693

Received: 12 January 2018; Accepted: 31 January 2018; Published: 3 February 2018

**Abstract:** The aim of this review is to investigate, whether there is a possible link between vitamin D supplementation and the reduction of blood pressure in hypertensive patients. The renin-angiotensin -aldosterone system is known for being deeply involved in cardiovascular tonus and blood pressure regulation. Hence, many of the pharmaceutical antihypertensive drugs inhibit this system. Interestingly, experimental studies in mice have indicated that vitamin D supplementation significantly lowers renin synthesis and blood pressure. It is conceivable that similar mechanisms may be found in the human organism. Regarding this, large-scale cross-sectional studies suggest the serum 25(OH)D-level to be inversely correlated to the prevalence of hypertension. However, randomized controlled trials (RCTs) have not found a clear association between vitamin D supplementation and improvements in hypertension. Nevertheless, the missing association of vitamin D and hypertension in clinical trials can be due to suboptimal study designs. There are hints that restoration of serum 25(OH)D levels during vitamin D therapy is essential to achieve possible beneficial cardiovascular effects. It is important to perform long-term trials with a short dose interval and a high bioavailability of supplementation. Taken together, more RCTs are required to further investigate if vitamin D can be beneficial for the reduction of blood pressure.

**Keywords:** vitamin D; hypertension; essential hypertension; renin-angiotensin-aldosterone system (RAAS); cholecalciferol

---

## 1. Introduction

According to the World Health Organization (WHO), one in three adults worldwide has raised blood pressure—a condition that causes around half of all deaths from stroke and heart disease [1]. Adiposity, lack of physical activity and excessive salt intake are some of the best-known environmental factors associated with hypertension. In recent years, yet another cause has been postulated: vitamin D deficiency [2–6]. Vitamin D is a key player in calcium homeostasis, in maintaining optimal bone metabolism and reducing fracture risk [7]. Several studies indicate that vitamin D also seems to play a protective role against the development of hypertension [5,8]. In this review, we summarize the existing literature that is concerned with vitamin D and hypertension and investigate if vitamin D (supplements) could be a beneficial treatment agent for hypertensive individuals.

For the literature search, the online databases PubMed (http://www.ncbi.nlm.nih.gov/pubmed/), Scopus (http://www.scopus.com/) and clinicaltrials.gov (http://clinicaltrials.gov/) were used up to January 2018. Search terms such as; "(Cholecalciferol OR vitamin D) and hypertension"; "Essential hypertension and (cholecalciferol OR ergocalciferol)"; "Vitamin D deficiency and hypertension", have been applied in the online databases. There were no restrictions in language set up. In PubMed, a search for "Essential hypertension" gave 30986 items, "Vitamin D" 74983 items and "hypertension and vitamin D" gave 1950 items accessed on 29 January 2018.

## 2. Arterial Hypertension

### 2.1. Definition, Causes and Risks

Arterial hypertension belongs to the most prevalent diseases and accounts for about 7.5 million deaths per year (about 13% of all deaths) worldwide. According to the WHO, hypertension is a major risk factor for the development of a variety of diseases, including cardiovascular diseases (CVD), kidney failure, cognitive impairment etc. [1]. It is estimated that approximately 22% (about 1 billion individuals) of the adult population above 25 years of age worldwide suffer from arterial hypertension [1]. In Europe, hypertension affects about 30–45% of people as reported in 2013 [9]; a large proportion is still unaware of their condition and left untreated [10].

Hypertension is a condition, in which either the systolic blood pressure (SBP) and/or diastolic blood pressure (DBP) stays elevated persistently. An exact definition has been difficult to establish. However, in 2017 the American College of Cardiology has newly defined hypertension in the Guidelines for the Prevention, Detection, Evaluation, and Management of High Blood Pressure in Adults [11]. The new classification in given in Table 1.

**Table 1.** Classification of hypertension.

| Classification | SBP (mmHg) | Header | DBP (mmHg) |
|---|---|---|---|
| Normal | <120 | and | <80 |
| Elevated | 120–129 | and | <80 |
| Hypertension, Stage 1 | 130–139 | or | 80–89 |
| Hypertension, Stage 2 | ≥140 | or | ≥90 |

Approximately 95% of all cases of arterial hypertension can be classified as essential hypertension (EH). This condition is characterized by an imbalance between vascular tonus and cardiac output without any identifiable cause.

The mean arterial BP (MAP) can be derived analogously from Ohm's law [12], which states:

$$\text{MAP} = \text{Cardiac Output (CO)} \times \text{Total Peripheral Resistance (TPR)} \quad (1)$$

An increase in MAP can lead to small artery remodelling, which increases the media-to-lumen ratio, TPR, causes wall damage and reduces distensibility in large arteries. These alterations finally lead to vessel wall hypertrophy and further elevate SBP—a vicious circle is created [13].

Age, smoking, high Body Mass Index (BMI), a salty diet and genetic dispositions are some of the known environmental and hereditary precursors linked to EH. Hence, it is important to eliminate the triggers, increase level of physical exercise and normalize BMI. The importance of preventing hypertension is underpinned by the findings of Lewington et al. [14]. They found that a 2 mmHg lowering of normal SBP could give an approximately 10% reduced risk of stroke mortality [14].

### 2.2. Management of Hypertension

In addition to lifestyle changes, pharmaceutical treatment may be relevant. Thiazide diuretics where the first group of effective antihypertensive drugs to be introduced in 1958 [15]. Since then, many medications have been developed. The treatment goals are to reduce the number of CVD-events and thereby to improve mortality. Different drug classes are currently available for the treatment of hypertension: 1. Diuretics (thiazides) [16]. Some of the most extensively used drugs in this class include bendroflumethiazide, hydrochlorothiazide and indapamide [17]; 2. Calcium channel blockers:

Dihydropyridines such as amlodipine and nifedipine have specific impact on the vessels and induce a compensatory activation of baroreceptors [18]; 3. β-adrenoceptor antagonists: Metoprolol and atenolol with selectivity to $\beta_1$-adrenoceptors and propranolol with no selectivity are some of the most common drugs of this class [19]; 4. Blockers of the renin-angiotensin-aldosterone system (RAAS): The group of angiotensin-converting-enzyme inhibitors (ACEi) includes drugs such as enalapril and captopril [20], angiotensin II receptor blockers (ARBs) include losartan, azilsartan and valsartan [21], and 5. Direct renin inhibitors comprise drugs like aliskiren [22].

## 3. Vitamin D

Vitamin D contains a steroid scaffold and possesses lipophilic properties. About 80–90% is endogenously synthetized and the remaining 10–20% come from nutritional intake. The metabolic pathways of vitamin D are shown in Figure 1. The inactive vitamin D is found in two distinct forms in the organism, cholecalciferol ($D_3$) and ergocalciferol ($D_2$) respectively. These inactive forms must undergo two hydroxylations to become the active form $1,25(OH)_2D_3$. The endogenous synthesis of cholecalciferol is catalysed by sun-exposure of the skin with UVB radiation. In this process, 7-dehydrocholesterol is transformed to pre-vitamin $D_3$, which isomerises to cholecalciferol. In the blood stream these two prohormones ($D_3$ and $D_2$) are bound to vitamin D binding protein (DBP). A hydroxylase enzyme in the liver, CYP2R1 (known as cytochrome P450 2R1), is responsible for converting $D_2$ and $D_3$ into 25(OH)D. The final activating step mainly takes place in the in proximal tubule cells, were 1α-hydroxylase (CYP27B1) converts 25(OH)D into $1,25(OH)_2D_3$ named calcitriol. The last conversion step is highly regulated by feedback mechanisms [23].

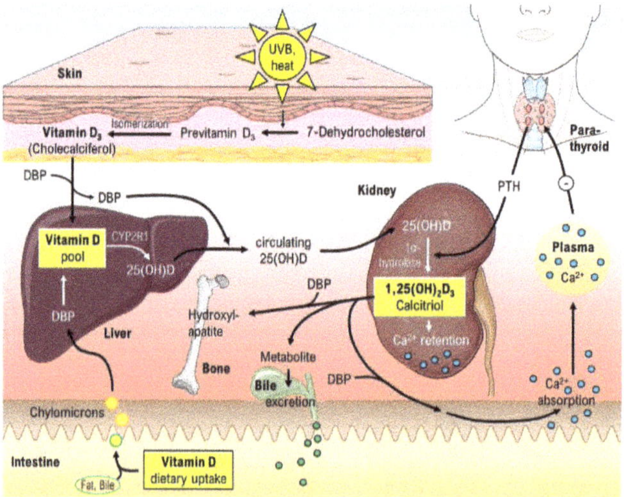


**Figure 1.** Overview of the vitamin D metabolism. 25(OH)D, Calcidiol; $1,25(OH)_2D_3$, Calcitriol; CYP2R1, cytochrome P450 2R1; DBP, Vitamin D binding protein. Parts of the figure were drawn by using pictures from Servier Medical Art.

### 3.1. The Vitamin D Receptor (VDR)

Brumbaugh et al. were the first to find evidence for the presence of a vitamin D receptor (VDR) [24]. This VDR is an intracellular receptor, able to bind $1,25(OH)_2D_3$ and subsequently stimulate VDR to heterodimerize with retinoid X receptor (RXR) [23]. The VDR-RXR complex may translocate to the nucleus and bind to the regulatory site in the promoter region of DNA sequence elements. Thus, the gene expression of specific target genes will be regulated and facilitate synthesis of vitamin D-regulated proteins. The well-known biological effects of $1,25(OH)_2D_3$ include absorption of $Ca^{2+}$ from the small intestine, bone metabolism and calcium- and phosphorus homeostasis (Figure 1). The global function of the vitamin D system becomes clear, when we take a look at the VDR distribution in various tissues and cells of the human body (Figure 2) [25].

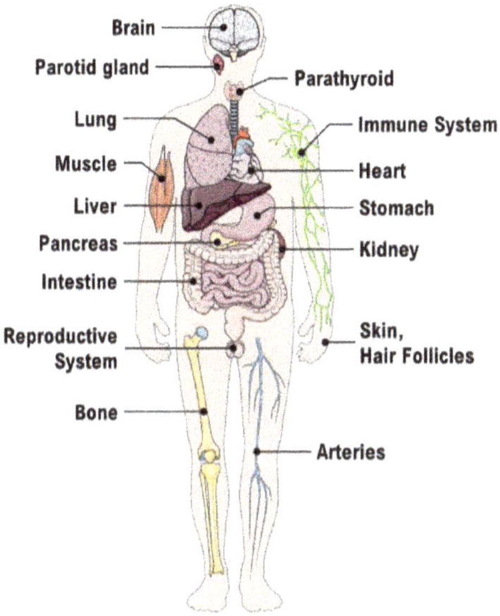


**Figure 2.** Tissue distribution of the Vitamin D receptor. Parts of the figure were drawn by using pictures from Servier Medical Art.

### 3.2. Vitamin D Status

The lipophilic nature of the vitamin $D_3$ and the high binding affinity of $1,25(OH)_2D_3$ to DBP in serum make these two substances difficult to use as markers. Thus, 25(OH)D concentration in plasma is considered the best measurement for vitamin D status [26].

Table 2 shows recommended doses of vitamin D intake and ideal serum 25(OH)D levels. However, there is an ongoing debate about the optimal levels of vitamin D. Intake reference values for vitamin D were developed by the Food and Nutrition Board (FNB) at the Institute of Medicine of The National Academies, USA. The Institute recommends 600 international units (IU) (15 μg) of vitamin D a day for adults ages 19 to 70. For adults age >70, the recommendation increases to 800 IU (20 μg) a day.

**Table 2.** Vitamin D recommendations.

| Recommendation | Children and Adolescents | Adults |
|---|---|---|
| The optimal concentration of 25(OH)D in plasma | 20–60 ng/mL<br>≈ 50–150 nM | 30–80 ng/mL<br>≈ 75–200 nM |
| Supplementations: recommended dose when severe deficiency | up to 5000 IU/day = 125 μg/d | up to 7000 IU/day ≈ 175 μg/d |

Adverse effects of vitamin D supplementation are rare. The initial signs of vitamin D intoxication are hypercalcemia, hypercalciuria and hyperphosphatemia [27]. In case of an overdose, hypercalcemia and/or hyperphosphatemia-associated symptoms can occur. Examples are weakness, fatigue, headache, appetite loss, dry mouth, nausea, vomiting, cardiovascular symptoms and others.

### *3.3. Suppression of Renin Production*

In 2002, Li et al. designed an in vivo study of renin expression in VDR-null mice [28]. One group of wildtype mice ($n$ = 9) were compared with a group of VDR-null mice ($n$ = 8). The mice were given optimal growing conditions. After two months of age, they were put on a special diet for five weeks to normalize the calcium levels in plasma. Afterwards, the BP was measured under anaesthesia, renin-expression as well as the angiotensin (ANG) II activity were analysed. Interestingly, these analyses revealed a significantly higher (>20 mmHg) diastolic BP and SBP in VDR-null mice vs. wildtype mice. To obtain quantitative values of mRNA renin-expression, the Northern blot method was used. A significant 3.5-fold higher renin-expression and 2.5-fold higher serum level of ANG II in VDR-null mice vs. wildtype mice was seen ($p < 0.001$) [28].

To investigate the direct effects of active vitamin D on renin synthesis, another group of wildtype mice had five injections of 30 pmol 1,25(OH)$_2$D$_3$ in three following days. It turned out that the 1,25(OH)$_2$D$_3$ treatment gave a 50% reduction in renin-mRNA, when compared to the control group ($n \geq 3$ in each group). Taken together, these findings imply the importance of vitamin D as an effective suppressor of the renin synthesis.

To elucidate the molecular pathways behind the downregulating effect of vitamin D on renin-transcription, Yuan et al. prepared an in vitro study [29]. In this study, specific juxtaglomerular As4.1 tumour cells from mice kidneys were analysed. In As4.1 cells the cyclic adenosine monophosphate (cAMP)/protein kinase A (PKA) pathway (shown in Figure 3) is deeply involved in the transcription of prorenin-mRNA. A G$\alpha_S$-coupled protein activates adenylate cyclase (AC), which converts adenosine triphosphate (ATP) into cAMP. The elevated level of cAMP activates protein kinase A (PKA). The catalytic subunit of PKA then translocates to the nucleus, where it phosphorylates the cAMP response element-binding protein (CREB). Subsequently, CREB binds to its response element (CRE) in the promoter region of the *Ren1C* gene. The other co-activators CBP and p300 are recruited to form a CREB-CBP-CRE complex that promotes the gene transcription of pro-renin-mRNA.

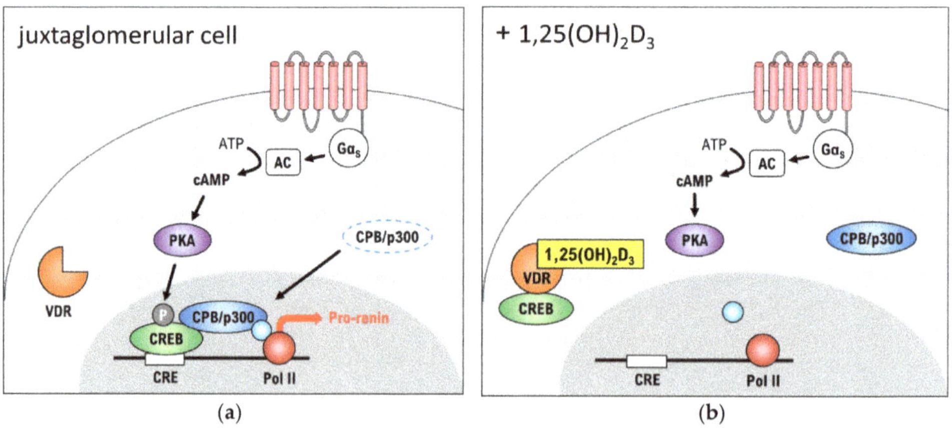


**Figure 3.** cAMP-PKA pathway. (**a**) Signalling in a juxtaglomerular cell in absence of 1,25(OH)$_2$-vitamin D$_3$; (**b**) Signalling in presence of 1,25(OH)$_2$-vitamin D$_3$. cAMP: cyclic adenosine monophosphate, CBP: CREB-binding protein, CRE: cAMP-dependent response element, CREB: cAMP response element-binding protein, G$\alpha_S$: G$_S$-protein alpha subunit, P: phosphate, PKA: protein kinase A, Pol II: RNA polymerase II, VDR: vitamin D receptor. The "+" stands for "JG in presence of".

Remarkably, 1,25(OH)$_2$D$_3$ liganded-VDR can interact directly with CREB to blunt its binding to CRE. It seems that these actions are carried out without the heterodimerization of liganded-VDR to RXR. Hence, it is believed that higher plasma levels of vitamin D can suppress the renin formation in juxtaglomerular cells [29]. However, another study indicates that decreased levels of renin in VDR-knockout mice will not cause a fall in BP [30]. Hence, it is important to state that more mechanisms must be implicated in a potential BP reducing effect of vitamin D.

## 4. Effects of Vitamin D on the Local Renin-Angiotensin System (RAS)

A local RAS is situated in several tissues, including heart, vessels, kidneys, lung, adrenal gland and nervous system [31]. The RAS acts in the control of cardiovascular, renal, and adrenal functions that regulate BP, body fluid and electrolyte status. It is known that the cardiac RAS is activated in cardiac pressure-overload and hypertensive rat models [32–35], and that the RAS plays an autocrine-paracrine role in the development of cardiac hypertrophy. VDR-deficient mice develop hypertension accompanied by an increase in heart weight. This finding reflects, at least in part, effects from activation of the systemic RAS [31].

Vitamin D is involved in cardiovascular protection, but only few studies examined the impact of the VDR in atherosclerosis. Macrophages express all components of the RAS, and are therefore involved in the process of atherosclerosis. It has been shown that macrophages in atherosclerotic lesions contain ANG II [36]. Another study showed that macrophage VDR signalling, in part by suppressing the local RAS, inhibits atherosclerosis in mice [37].

The antiatherosclerotic role of the VDR signalling in leukocytes/macrophages, and at least part of the antiatherosclerotic mechanism is to block the activation of the local RAS in macrophages and within the atherosclerotic lesion. Therefore, further studies to investigate the benefits of vitamin D and its analogues in atherosclerosis in both preclinical and clinical studies are desirable.

The intrarenal RAS is a key player for renal damage. Vitamin D deficiency activates the local RAS in the kidney, and thus it causes renal injury [38]. It is known that vitamin D is a negative RAS regulator by suppressing renin expression [28]. Moreover, vitamin D deficiency promotes the RAS [28], and a chronic RAS activation impacts lung function and induces a lung fibrosis [39]. Shi et al. showed in a recent paper that Vitamin D deficiency can induce profibrotic factors and activate the fibrotic cascade. These RAS-mediated effects are independent of an increased blood pressure [40]. In addition, there exists a local RAS in bone, which is involved in bone metabolism [41]. A recent study demonstrated that $1,25(OH)_2D_3$ influences bone metabolism by downregulating the local bone RAS in a mice model with glucocorticoid-induced osteoporosis [42]. Vitamin D influences also the local pancreatic islet RAS and improves the islet beta cell secretory function [43]. A mice animal model with vitamin D receptor ablation revealed an activation of the islet RAS [43]. The application of calcitriol showed beneficial effects on the RAS activation under high-glucose conditions. In addition, it revealed a positive calcitriol effect on elevating the islet beta cell secretory response to glucose [43].

These animal studies show that vitamin D deficiency is a key player in different diseases. It influences the local RAS in various tissues. Vitamin D deficiency is an important health problem. Therefore, it is necessary to perform future studies to establish clinical guidelines for vitamin D supplementation required to achieve adequate vitamin D levels in people who are at risk for hypertension, atherosclerosis, cardiovascular disease, diabetes, pulmonary fibrosis, osteoporosis and others.

In particular, vitamin D deficiency might be linked to cardiovascular disease. There is a higher risk of high blood pressure (hypertension). However, more studies in this field are necessary. It is still too early to confirm and there is an ongoing discussion, whether a low vitamin D level causes hypertension or whether vitamin D supplementation will play a role in the treatment of hypertension. The role of vitamin D supplementation in the management or treatment of these diseases mentioned above must be studied in the future.

## 5. Vitamin D and Essential Hypertension

National Health and Nutrition Examination Survey (NHANES) is a survey providing information about health statistics in a representative sample of the U.S. population [44]. Based on the NHANES III data (survey period 1988–1994), Martins et al. examine the association between serum 25(OH)D-level and the prevalence of hypertension [45]. In this cross-sectional study, all individuals with available data and above 20 years of age ($n$ = 15,088) are included. The study population is then divided into quartiles according to their serum 25(OH)-level. The 1st Quartile (25(OH)D < 21 ng/mL) shows a 20.46% prevalence of hypertension, while the 4th Quartile (25(OH)D $\geq$ 37 ng/mL) has a prevalence of 15.10%. Comparing the 1st to the 4th Quartile gives an odds ratio (OR) = 1.30 (95%CI: 1.13–1.48), adjusted for race, sex and age.

Scragg et al. use the same NHANES III data [46]. Here, individuals receiving anti-hypertensive medication are excluded and data is adjusted for physical activity, BMI, age, sex and race ($n$ = 12,644). Both studies find a significant inverse correlation between BP and serum 25(OH)D level ($p < 0.01$). Ke et al. prepared a systematic review of the available observational studies in the period 2007–2014. It only includes published studies with healthy adults ($n$ = 90,535) [2]. Comparing odds of hypertension from the top category to the bottom category of serum 25(OH)D level gives OR = 0.79 (95%CI: 0.73–0.87). This supports the hypothesis of an inverse relationship between hypertension and vitamin D status.

The strength of these studies is in particular due to the great sample size and it can be considered representative for the population. The disadvantages include the lack of temporal separation between data collection of exposure (vitamin D status) and outcome (hypertension). Hence, a causal relation can be difficult to establish even though a significant association is found. Furthermore, it can be hard to exclude the possibility of an inverse causality between exposure and outcome.

To meet the challenges of possible inverse causality between serum 25(OH)D level and hypertension, Kunutskor et al. reviewed all prospective studies in this field up until 2012 [47]. The included studies had to be representative for the population, with at least one year of follow-up and without prevalent hypertension at baseline ($n$ = 283,537). Participants had serum 25(OH)D measured at baseline ($n$ = 48,633) or vitamin D status assessed from dietary intake ($n$ = 238,199). Hypertension during follow-up time was considered the endpoint.

Pooled relative risk (RR) in this review showed RR = 0.88 (95%CI: 0.81–0.97) i.e., a 12% reduction in risk of incident hypertension for every increase of 10 ng/mL in serum 25(OH)D level [47]. The group of participants with their vitamin D status based only upon the dietary intake had RR = 1.0 (95%CI: 0.95–1.05). Thus, no association between dietary vitamin D intake and risk of incident hypertension appeared. It is worth noting that dietary intake is a source of only 5–20% of vitamin D and its metabolites in plasma [48]. Therefore, it can be difficult to use the dietary intake as marker of vitamin D status, which can explain the lack of inverse correlation in this group.

To examine whether vitamin D supplementation is beneficial in hypertensive patients, the relevant studies are listed in Table 3. The included studies must meet the following criteria, to be discussed in review: participants should be adults (aged $\geq$18 years) with diagnosed vitamin D deficiency and/or arterial hypertension at baseline. Identified studies must be published or last updated in the time period 1 January 2012 until 27 November 2017. The intervention group must be administered cholecalciferol and participants may not suffer from preeclampsia. To minimize confounding, only RCT will be assessed.

**Table 3.** Overview of recent clinical trials with vitamin D intervention in hypertensive patients.

| Title | Design | Objective | Conclusions |
| --- | --- | --- | --- |
| The Styrian Vitamin D Hypertension Trial: effects of vitamin D on blood pressure and cardiovascular risk factors. [49] NCT02136771 | Randomized, Double-blind, Placebo-controlled, $n = 200$ | To assess the effects on 24-h systolic BP and diastolic BP of vitamin $D_3$ supplementation of 2800 IU/day for 8 weeks in vitamin D deficient individuals. | There was no significant effect on systolic and diastolic BP after treatment with vitamin D supplementation. |
| Vitamin D therapy in individuals with prehypertension or hypertension: the DAYLIGHT trial. [50] NCT01240512 | Randomized, Double-blind, Parallel assignment, Multi-center, $n = 534$ | To compare the BP-lowering effects of high-dose (4000 IU/day) vs. low-dose (400 IU/day) of cholecalciferol for 6 months in vitamin D deficient individuals. The participants were prehypertensive or hypertensive at baseline and had not been taking antihypertensive drugs. | No significant changes in BP-measures was observed in the two groups. Nevertheless, a non-significant ($p = 0.71$) decrease in 24-h SBP was observed −0.8 mmHg and −1.6 mmHg in the two groups, respectively. |
| Effect of Vitamin D replacement During Winter Months in Patients With Hypertension. [51] NCT01166165 | Randomized, Double-blind, Placebo-controlled, $n = 130$ | To investigate therapeutic effects of 3000 IU/day cholecalciferol for 20 weeks in hypertensive patients. | In the overall group, no significant reductions in 24-h-BP, when compared to placebo.<br>A subgroup analysis containing only deficient plasma−25(OH)D (<32 ng/mL) individuals at baseline, showed a significant reduction in 24-h systolic/diastolic BP −4/−3 mmHg. |
| The effect of vitamin D supplementation on arterial stiffness in an elderly community-based population. [52] EUDRA number: 2010–024417–31 | Randomized, Double-blind, Parallel assignment, $n = 119$ | To compare the effects of 50,000 IU vs. 100,000 IU single-dose intramuscular injection of cholecalciferol in a group of elderly people. | 8 weeks after treatment the group receiving high-dose of cholecalciferol had a significant improvement in arterial stiffness compared to the low-dose group. At the same time, systolic BP seemed to elevate in high-dose group. |
| Vitamin D therapy to reduce blood pressure and left ventricular hypertrophy in resistant hypertension. [53] EUDRA number: 2008-002681-63 | Randomized, Double-blind, Placebo-controlled, $n = 68$ | To assess effects of high-dose cholecalciferol supplementation (100,000 IU every 2[nd] month for 6 months) in patients treated with ≥3 antihypertensive drugs. | The study showed no improvements in systolic or diastolic BP after 6 months of treatment. |
| Cholecalciferol treatment to reduce blood pressure in older patients with isolated systolic hypertension: the VitDISH randomized controlled trial. [54] ISRCTN92186858 | Randomized, Double-blind, Placebo-controlled, $n = 159$ | To explore the effects on BP of high-dose cholecalciferol treatment (100,000 IU every 3rd month for 1 year) in elderly patients with isolated systolic hypertension | Treatment did not reduce BP or improve other secondary cardiovascular outcomes. |
| Vitamin D and nifedipine in the treatment of Chinese patients with grades I-II essential hypertension: a randomized placebo-controlled trial. [55] ChiCTR-ONC-13003840 | Randomized, Double-blind, Placebo-controlled, $n = 126$ | To assess the effects of cholecalciferol (2000 IU/day for 6 months) as 'add on' to nifedipine in essential hypertensive patients. | Cholecalciferol as 'add on' gave a significant systolic/diastolic BP reduction (−6.2/−4.2 mmHg).<br>In subgroup analysis of vitamin D insufficient (at baseline) patients, showed −7.1/−5.7 mmHg BP reduction ($p = 0.001$). |
| The effect of vitamin D supplementation on blood pressure in patients with elevated blood pressure and vitamin D deficiency: a randomized, double-blind, placebo-controlled trial. [56] | Randomized, Double-blind, Placebo-controlled, $n = 42$ | To assess BP lowering effect of cholecalciferol 50,000 IU/week for 8 weeks in hypertensive, vitamin D deficient patients. | In the vitamin D deficient group (VDG) 92.7% of individuals recovered from insufficiency.<br>Middle arterial pressure (MAP) decreased in average −3.7 mmHg in VDG and increased 0.9 mmHg in place-controls ($p < 0.001$). |

In summary, animal studies had demonstrated that vitamin D deficiency is associated with high blood pressure by mechanisms revealing a direct effect on the renin-angiotensin-aldosterone axis. An altered vitamin D signalling in different animal models might be involved in cardiovascular diseases such as hypertension, cardiac hypertrophy, and atherosclerosis. Lower serum vitamin D was reported to be associated with higher BP levels in cross-sectional studies [46] and was associated inversely with blood pressure in a large sample representative of the US population [46]. This was supported by basic research. However, there are controversial results of basic research to randomized clinical trials.

A systematic review investigated whether vitamin D supplementation or its analogues may reduce blood pressure [57]. The trial search comprised a period from 1 January 1966, through 31 March 2014. 46 trials with 4541 participants were included in the trial-level meta-analysis [57]. The authors concluded that vitamin D supplementation is not effective as a drug for decreasing blood pressure and is not recommended as an antihypertensive drug [57].

It also has to be taken into account that genetic animal models may not completely reflect vitamin D deficiency of the human organism. Patients with vitamin D deficiency might suffer from chronic illnesses or have unknown cardiovascular risk factors. There might be dietary differences (diet with low vitamin D), differences in outdoor activities or sun-exposure. There is evidence suggesting that vitamin D application has little or no influence on BP, but to answer this point completely new trials are necessary. One example is the ongoing Vitamin D and Omega-3 Hypertension Trial (VITAL Hypertension) (NCT01653678) at Brigham and Women's Hospital, Harvard Medical School, Boston, MA, USA. VITAL is an ongoing research study in 25,874 men and women across the U.S. The VITAL Hypertension study is being conducted among participants in VITAL and will examine whether vitamin D or omega-3 fatty acids are related to changes in BP and hypertension.

Another trial which is recruiting is the Vitamin D and Immune Mechanisms of Hypertension in Type 2 Diabetics (VDIM) trial (NCT03348280). This observational study will evaluate whether a particular type of circulating white blood cell, monocytes, from type 2 diabetics with high blood pressure and vitamin D deficiency vs. sufficiency will induce hormones that increase BP.

New international trials can inform about the best vitamin D supplementation dose, and to definitively evaluate the clinical utility of vitamin D therapy. Future studies should consider a design with a large sample size, control of medication, dietary salt and glucose intake, to avoid an activation of the RAAS. Large randomized and long-term trials focusing on patients with severe vitamin D deficiency and hypertension are needed before vitamin D can be recommended for prevention or treatment of hypertension.

The question of whether vitamin D deficiency can cause high BP unfortunately cannot be answered yet.

## 6. Discussion

The possible beneficial effects of Vitamin D supplementation on BP are the main focus of this review. As mentioned in the previous sections, epidemiological studies find some proof of an inverse correlation between serum vitamin D-status and prevalence of EH. However, the overview of interventional studies in Table 3 does not give a clear picture of vitamin D supplementation in relation to cardiovascular health. It is of interest whether vitamin D therapy in the intervention group (vitamin D deficiency at baseline) is able to raise serum 25(OH)D to the optimal levels (30–80 ng/mL) [58]. Common to the three RCTs is that they all show no effects from vitamin D supplementation in 24-h SBP [52–54]. These three studies share that the intervention groups at time endpoint are still insufficient in serum 25(OH)D levels. Common to Witham et al. trials is the high-dosage of oral administered cholecalciferol (100,000 IU) given at least two months apart [53,54]. This does not seem to restore the 25(OH)D levels, which can possibly explain the lack of effect on BP. In order to obtain positive effects of vitamin D therapy, there is something that suggests the dose interval to be short (e.g., daily or weekly administration).

McGreevy et al. trial participants were given intramuscular injections of 100,000 IU cholecalciferol, which did not restore serum 25(OH)D levels [52]. One possible explanation could be a diminished bioavailability. However, the primary outcome 'arterial stiffness' was improved after intervention. This could probably be carried out by an unknown VDR-RXR-protective mechanism in the endothelial wall, independently of renin secretion and the RAAS. Cipriani et al. conclude high-dosage of p.o. administered cholecalciferol to be more efficiently in raising serum 25(OH)D levels, when compared to equivalent intramuscular injections [59]. Hence, future study designs should rather use oral administration than intramuscular injections in the intervention groups.

The DAYLIGHT trial [50] and Styrian trial [49] have the biggest sample size ($n$ = 534 and $n$ = 200) among the RCTs for this review. These trials find no significant changes in measured BP outcomes after a daily administration of vitamin D in the study period.

Larsen et al. found in the overall group no significant BP beneficial effects of vitamin D. Nevertheless, subgroup analysis revealed significant 24-h BP reductions in a subgroup with vitamin D-insufficiency at baseline [51]. One might suggest that there could be a certain threshold in serum 25(OH)D to achieve cardiovascular benefits. Interestingly, Mozaffari-Khosravi et al. observed a remarkably increment in serum 25(OH)D levels and significant MAP reductions at same time [56]. On average, serum levels of 25(OH)D were raised to 51.7 ng/mL in intervention group, which could give support to the idea of a dose-response relation. Hence, study designs with lack of satisfying restoration of serum 25(OH)D levels will fail to improve BP.

Chen et al. investigate the administration of vitamin D as 'add-on' to a nifedipine treatment [55]. This combined therapy seems to markedly reduce systolic and diastolic BP—but further studies are needed to elucidate the effects of cholecalciferol and antihypertensive drugs combined.

## 7. Conclusions and Outlook

Essential hypertension is a condition, in which an imbalance between vasoconstriction and vasodilation occurs. More factors, both epigenetic and environmental, are related to the development of EH. Many epidemiological studies find that vitamin D deficiency can be associated to prevalent hypertension. Furthermore, experimental studies in mice can explain a possible mechanism behind a CVD-protective effect of vitamin D supplementation. Hence, vitamin D injections were shown to inhibit the renin synthesis. This was mediated by the liganded-VDR, which could interact with specific transcription factors to reduce renin transcription. Cholecalciferol can thus be a potentially cost-effective antihypertensive drug.

However, clinical trials in this area show conflicting results—which in part can be attributed to suboptimal study designs. Hence, further studies that efficiently restore serum vitamin D status, are desirable.

Future aspects in this research field include investigating if vitamin D analogues with higher selectivity to VDR could have BP lowering effects. The potential synergistic effect of cholecalciferol and antihypertensive drugs also needs to be further investigated.

**Acknowledgments:** This review was the bachelor thesis of Christian Legarth.

**Author Contributions:** Christian Legarth, Marcus Krüger and Daniela Grimm wrote the review; Christian Legarth, Johann Bauer and Markus Wehland searched the trials and checked the references, Marcus Krüger and Christian Legarth made the figures.

**Conflicts of Interest:** The authors declare no conflict of interest.

## Abbreviations

| | |
|---|---|
| $1,25(OH)_2D_3$ | Calcitriol |
| 25(OH)D | Calcidiol |
| 95%-CI | 95% confidence interval |
| AC | Adenylate cyclase |

| | |
|---|---|
| ACE | Angiotensin-converting enzyme |
| ACEi | Angiotensin-converting enzyme inhibitor |
| ANG | Angiotensin |
| ARB | Angiotensin II receptor blocker |
| $AT_1/AT_2$ | Angiotensin receptor type 1 or 2 |
| BMI | Body mass index |
| BP | Blood pressure |
| cAMP | cyclic adenosine monophosphate |
| CBP | CREB-binding protein |
| CO | Cardiac output |
| CRE | cAMP response element |
| CREB | cAMP-dependent response element binding protein |
| CVD | Cardiovascular disease |
| CYP | Cytochrome P450 |
| DBP | Vitamin D binding protein |
| EH | Essential hypertension |
| HR | Heart rate |
| IU | International units |
| MAP | Mean arterial pressure |
| mRNA | messenger ribonucleic acid |
| NHANES | National Health and Nutrition Examination Survey |
| OR | Odds-ratio |
| PKA | Protein kinase A |
| RAAS | Renin-angiotensin-aldosterone system |
| RCT | Randomized controlled trial |
| RR | Relative risk |
| RXR | Retinoid X receptor |
| SBP | Systolic blood pressure |
| TPR | Total periphery resistance |
| VDR | Vitamin D receptor |
| WHO | World Health Organization |

## References


1. A Global Brief on Hypertension. Available online: http://apps.who.int/iris/bitstream/10665/79059/1/WHO_DCO_WHD_2013.2_eng.pdf (accessed on 30 October 2017).
2. Ke, L.; Mason, R.S.; Kariuki, M.; Mpofu, E.; Brock, K.E. Vitamin D status and hypertension: A review. *Integr. Blood Press. Control.* **2015**, *8*, 13–35. [PubMed]
3. Mehta, V.; Agarwal, S. Does Vitamin D Deficiency Lead to Hypertension? *Cureus* **2017**, *9*, e1038. [CrossRef] [PubMed]
4. Qi, D.; Nie, X.L.; Wu, S.; Cai, J. Vitamin D and hypertension: Prospective study and meta-analysis. *PLoS ONE* **2017**, *12*, e0174298. [CrossRef] [PubMed]
5. Geleijnse, J.M. Vitamin D and the Prevention of Hypertension and Cardiovascular Diseases: A Review of the Current Evidence. *Am. J. Hypertens.* **2011**, *24*, 253–262. [CrossRef] [PubMed]
6. Vaidya, A.; Forman, J.P. Vitamin D and Hypertension: Current Evidence and Future Directions. *Hypertension* **2010**, *56*, 774–779. [CrossRef] [PubMed]
7. Delmi, M.; Rapin, C.H.; Bengoa, J.M.; Delmas, P.D.; Vasey, H.; Bonjour, J.P. Dietary supplementation in elderly patients with fractured neck of the femur. *Lancet* **1990**, *335*, 1013–1016. [CrossRef]
8. Shamardl, H.A.; El-Ashmony, S.M.; Kamel, H.F.; Fatani, S.H. Potenial Cardiovascular and Renal Protective Effects of Vitamin D and Coenzyme Q10 in l-NAME-Induced Hypertensive Rats. *Am. J. Med. Sci.* **2017**, *354*, 190–198. [CrossRef] [PubMed]

9. Mancia, G.; Fagard, R.; Narkiewicz, K.; Redon, J.; Zanchetti, A.; Bohm, M.; Christiaens, T.; Cifkova, R.; De Backer, G.; Dominiczak, A.; et al. 2013 ESH/ESC guidelines for the management of arterial hypertension: The Task Force for the Management of Arterial Hypertension of the European Society of Hypertension (ESH) and of the European Society of Cardiology (ESC). *Eur. Heart J.* **2013**, *34*, 2159–2219. [PubMed]
10. Kronborg, C.N.; Hallas, J.; Jacobsen, I.A. Prevalence, awareness, and control of arterial hypertension in Denmark. *J. Am. Soc. Hypertens.* **2009**, *3*, 19.e2–24.e2. [CrossRef] [PubMed]
11. Whelton, P.K.; Carey, R.M.; Aronow, W.S.; Casey, D.E., Jr.; Collins, K.J.; Dennison Himmelfarb, C.; DePalma, S.M.; Gidding, S.; Jamerson, K.A.; Jones, D.W.; et al. 2017 ACC/AHA/AAPA/ABC/ACPM/AGS/APhA/ASH/ASPC/NMA/PCNA Guideline for the Prevention, Detection, Evaluation, and Management of High Blood Pressure in Adults: Executive Summary: A Report of the American College of Cardiology/American Heart Association Task Force on Clinical Practice Guidelines. *Hypertension* **2017**. [CrossRef]
12. Hill, L.K.; Sollers Iii, J.J.; Thayer, J.F. Resistance reconstructed estimation of total peripheral resistance from computationally derived cardiac output—Biomed 2013. *Biomed. Sci. Instrum.* **2013**, *49*, 216–223. [PubMed]
13. Laurent, S.; Boutouyrie, P. The structural factor of hypertension: Large and small artery alterations. *Circ. Res.* **2015**, *116*, 1007–1021. [CrossRef] [PubMed]
14. Lewington, S.; Clarke, R.; Qizilbash, N.; Peto, R.; Collins, R. Age-specific relevance of usual blood pressure to vascular mortality: A meta-analysis of individual data for one million adults in 61 prospective studies. *Lancet* **2002**, *360*, 1903–1913. [CrossRef]
15. Saklayen, M.G.; Deshpande, N.V. Timeline of History of Hypertension Treatment. *Front. Cardiovasc. Med.* **2016**, *3*, 3. [CrossRef] [PubMed]
16. Hughes, A.D. How do thiazide and thiazide-like diuretics lower blood pressure? *J. Renin Angiotensin Aldoster. Syst.* **2004**, *5*, 155–160. [CrossRef] [PubMed]
17. Brøsen, K.; Simonsen, U.; Kampmann, J.P.; Thirstrup, S. *Basal Og Klinisk Farmakologi*, 5th ed.; FADL's Forlag: Copenhagen, Denmark, 2014; ISBN 978-87-7749-683-7.
18. Striessnig, J.; Ortner, N.J.; Pinggera, A. Pharmacology of L-type Calcium Channels: Novel Drugs for Old Targets? *Curr. Mol. Pharmacol.* **2015**, *8*, 110–122. [CrossRef] [PubMed]
19. Mason, R.P.; Giles, T.D.; Sowers, J.R. Evolving mechanisms of action of beta blockers: Focus on nebivolol. *J. Cardiovasc. Pharmacol.* **2009**, *54*, 123–128. [CrossRef] [PubMed]
20. Van Vark, L.C.; Bertrand, M.; Akkerhuis, K.M.; Brugts, J.J.; Fox, K.; Mourad, J.J.; Boersma, E. Angiotensin-converting enzyme inhibitors reduce mortality in hypertension: A meta-analysis of randomized clinical trials of renin-angiotensin-aldosterone system inhibitors involving 158,998 patients. *Eur. Heart J.* **2012**, *33*, 2088–2097. [CrossRef] [PubMed]
21. Abramov, D.; Carson, P.E. The role of angiotensin receptor blockers in reducing the risk of cardiovascular disease. *J. Renin Angiotensin Aldoster. Syst.* **2012**, *13*, 317–327. [CrossRef] [PubMed]
22. Pool, J.L. Direct renin inhibition: Focus on aliskiren. *J. Manag. Care Pharm.* **2007**, *13*, 21–33. [CrossRef] [PubMed]
23. Baeke, F.; Takiishi, T.; Korf, H.; Gysemans, C.; Mathieu, C. Vitamin D: Modulator of the immune system. *Curr. Opin. Pharmacol.* **2010**, *10*, 482–496. [CrossRef] [PubMed]
24. Brumbaugh, P.F.; Haussler, M.R. Nuclear and cytoplasmic binding components for vitamin D metabolites. *Life Sci.* **1975**, *16*, 353–362. [CrossRef]
25. Wang, Y.; Zhu, J.; DeLuca, H.F. Where is the vitamin D receptor? *Arch. Biochem. Biophys.* **2012**, *523*, 123–133. [CrossRef] [PubMed]
26. Farrell, C.J.; Herrmann, M. Determination of vitamin D and its metabolites. *Best Pract. Res. Clin. Endocrinol. Metab.* **2013**, *27*, 675–688. [CrossRef] [PubMed]
27. Razzaque, M.S. Can adverse effects of excessive vitamin D supplementation occur without developing hypervitaminosis D? *J. Steroid Biochem. Mol. Biol.* **2017**. [CrossRef] [PubMed]
28. Li, Y.C.; Kong, J.; Wei, M.; Chen, Z.F.; Liu, S.Q.; Cao, L.P. 1,25-Dihydroxyvitamin D(3) is a negative endocrine regulator of the renin-angiotensin system. *J. Clin. Investig.* **2002**, *110*, 229–238. [CrossRef] [PubMed]
29. Yuan, W.; Pan, W.; Kong, J.; Zheng, W.; Szeto, F.L.; Wong, K.E.; Cohen, R.; Klopot, A.; Zhang, Z.; Li, Y.C. 1,25-dihydroxyvitamin D3 suppresses renin gene transcription by blocking the activity of the cyclic AMP response element in the renin gene promoter. *J. Biol. Chem.* **2007**, *282*, 29821–29830. [CrossRef] [PubMed]

30. Kong, J.; Qiao, G.; Zhang, Z.; Liu, S.Q.; Li, Y.C. Targeted vitamin D receptor expression in juxtaglomerular cells suppresses renin expression independent of parathyroid hormone and calcium. *Kidney Int.* **2008**, *74*, 1577–1581. [CrossRef] [PubMed]
31. Xiang, W.; Kong, J.; Chen, S.; Cao, L.P.; Qiao, G.; Zheng, W.; Liu, W.; Li, X.; Gardner, D.G.; Li, Y.C. Cardiac hypertrophy in vitamin D receptor knockout mice: Role of the systemic and cardiac renin-angiotensin systems. *Am. J. Physiol. Endocrinol. Metab.* **2005**, *288*, E125–E132. [CrossRef] [PubMed]
32. Bruckschlegel, G.; Holmer, S.R.; Jandeleit, K.; Grimm, D.; Muders, F.; Kromer, E.P.; Riegger, G.A.; Schunkert, H. Blockade of the renin-angiotensin system in cardiac pressure-overload hypertrophy in rats. *Hypertension* **1995**, *25*, 250–259. [CrossRef] [PubMed]
33. Grimm, D.; Jabusch, H.C.; Kossmehl, P.; Huber, M.; Fredersdorf, S.; Griese, D.P.; Kramer, B.K.; Kromer, E.P. Experimental diabetes and left ventricular hypertrophy: Effects of beta-receptor blockade. *Cardiovasc. Pathol.* **2002**, *11*, 229–237. [CrossRef]
34. Jurkovicova, D.; Dobesova, Z.; Kunes, J.; Krizanova, O. Different expression of renin-angiotensin system components in hearts of normotensive and hypertensive rats. *Physiol. Res.* **2001**, *50*, 35–42. [PubMed]
35. Rothermund, L.; Kreutz, R.; Kossmehl, P.; Fredersdorf, S.; Shakibaei, M.; Schulze-Tanzil, G.; Paul, M.; Grimm, D. Early onset of chondroitin sulfate and osteopontin expression in angiotensin II-dependent left ventricular hypertrophy. *Am. J. Hypertens.* **2002**, *15*, 644–652. [CrossRef]
36. Potter, D.D.; Sobey, C.G.; Tompkins, P.K.; Rossen, J.D.; Heistad, D.D. Evidence that macrophages in atherosclerotic lesions contain angiotensin II. *Circulation* **1998**, *98*, 800–807. [CrossRef] [PubMed]
37. Szeto, F.L.; Reardon, C.A.; Yoon, D.; Wang, Y.; Wong, K.E.; Chen, Y.; Kong, J.; Liu, S.Q.; Thadhani, R.; Getz, G.S.; et al. Vitamin D receptor signaling inhibits atherosclerosis in mice. *Mol. Endocrinol.* **2012**, *26*, 1091–1101. [CrossRef] [PubMed]
38. Carey, R.M.; Siragy, H.M. The intrarenal renin-angiotensin system and diabetic nephropathy. *Trends Endocrinol. Metab.* **2003**, *14*, 274–281. [CrossRef]
39. Wang, J.; Chen, L.; Chen, B.; Meliton, A.; Liu, S.Q.; Shi, Y.; Liu, T.; Deb, D.K.; Solway, J.; Li, Y.C. Chronic Activation of the Renin-Angiotensin System Induces Lung Fibrosis. *Sci. Rep.* **2015**, *5*, 15561. [CrossRef] [PubMed]
40. Shi, Y.; Liu, T.; Yao, L.; Xing, Y.; Zhao, X.; Fu, J.; Xue, X. Chronic vitamin D deficiency induces lung fibrosis through activation of the renin-angiotensin system. *Sci. Rep.* **2017**, *7*, 3312. [CrossRef] [PubMed]
41. Kalupahana, N.S.; Moustaid-Moussa, N. The renin-angiotensin system: A link between obesity, inflammation and insulin resistance. *Obes. Rev.* **2012**, *13*, 136–149. [CrossRef] [PubMed]
42. Shen, L.; Ma, C.; Shuai, B.; Yang, Y. Effects of 1,25-dihydroxyvitamin D3 on the local bone renin-angiotensin system in a murine model of glucocorticoid-induced osteoporosis. *Exp. Ther. Med.* **2017**, *13*, 3297–3304. [CrossRef] [PubMed]
43. Cheng, Q.; Li, Y.C.; Boucher, B.J.; Leung, P.S. A novel role for vitamin D: Modulation of expression and function of the local renin-angiotensin system in mouse pancreatic islets. *Diabetologia* **2011**, *54*, 2077–2081. [CrossRef] [PubMed]
44. Chyba, M.M.; Washington, L.R. Plan and operation of the Third National Health and Nutrition Examination Survey, 1988–1994. Series 1: Programs and collection procedures. *Vital. Health Stat 1* **1994**, *32*, 1–407.
45. Martins, D.; Wolf, M.; Pan, D.; Zadshir, A.; Tareen, N.; Thadhani, R.; Felsenfeld, A.; Levine, B.; Mehrotra, R.; Norris, K. Prevalence of cardiovascular risk factors and the serum levels of 25-hydroxyvitamin D in the United States: Data from the Third National Health and Nutrition Examination Survey. *Arch. Intern. Med.* **2007**, *167*, 1159–1165. [CrossRef] [PubMed]
46. Scragg, R.; Sowers, M.; Bell, C. Serum 25-hydroxyvitamin D, ethnicity, and blood pressure in the Third National Health and Nutrition Examination Survey. *Am. J. Hypertens.* **2007**, *20*, 713–719. [CrossRef] [PubMed]
47. Kunutsor, S.K.; Apekey, T.A.; Steur, M. Vitamin D and risk of future hypertension: Meta-analysis of 283,537 participants. *Eur. J. Epidemiol.* **2013**, *28*, 205–221. [CrossRef] [PubMed]
48. Mohr, S.B.; Garland, C.F.; Gorham, E.D.; Garland, F.C. The association between ultraviolet B irradiance, vitamin D status and incidence rates of type 1 diabetes in 51 regions worldwide. *Diabetologia* **2008**, *51*, 1391–1398. [CrossRef] [PubMed]

49. Pilz, S.; Gaksch, M.; Kienreich, K.; Grubler, M.; Verheyen, N.; Fahrleitner-Pammer, A.; Treiber, G.; Drechsler, C.; Obermayer-Pietsch, B.; Schwetz, V.; et al. Effects of vitamin D on blood pressure and cardiovascular risk factors: A randomized controlled trial. *Hypertension* **2015**, *65*, 1195–1201. [CrossRef] [PubMed]
50. Arora, P.; Song, Y.; Dusek, J.; Plotnikoff, G.; Sabatine, M.S.; Cheng, S.; Valcour, A.; Swales, H.; Taylor, B.; Carney, E.; et al. Vitamin D therapy in individuals with prehypertension or hypertension: The DAYLIGHT trial. *Circulation* **2015**, *131*, 254–262. [CrossRef] [PubMed]
51. Larsen, T.; Mose, F.H.; Bech, J.N.; Hansen, A.B.; Pedersen, E.B. Effect of cholecalciferol supplementation during winter months in patients with hypertension: A randomized, placebo-controlled trial. *Am. J. Hypertens.* **2012**, *25*, 1215–1222. [CrossRef] [PubMed]
52. McGreevy, C.; Barry, M.; Davenport, C.; Byrne, B.; Donaghy, C.; Collier, G.; Tormey, W.; Smith, D.; Bennett, K.; Williams, D. The effect of vitamin D supplementation on arterial stiffness in an elderly community-based population. *J. Am. Soc. Hypertens.* **2015**, *9*, 176–183. [CrossRef] [PubMed]
53. Witham, M.D.; Ireland, S.; Houston, J.G.; Gandy, S.J.; Waugh, S.; Macdonald, T.M.; Mackenzie, I.S.; Struthers, A.D. Vitamin D therapy to reduce blood pressure and left ventricular hypertrophy in resistant hypertension: Randomized, controlled trial. *Hypertension* **2014**, *63*, 706–712. [CrossRef] [PubMed]
54. Witham, M.D.; Price, R.J.; Struthers, A.D.; Donnan, P.T.; Messow, C.M.; Ford, I.; McMurdo, M.E. Cholecalciferol treatment to reduce blood pressure in older patients with isolated systolic hypertension: The VitDISH randomized controlled trial. *JAMA Intern. Med.* **2013**, *173*, 1672–1679. [CrossRef] [PubMed]
55. Chen, W.R.; Liu, Z.Y.; Shi, Y.; Yin, D.W.; Wang, H.; Sha, Y.; Chen, Y.D. Vitamin D and nifedipine in the treatment of Chinese patients with grades I-II essential hypertension: A randomized placebo-controlled trial. *Atherosclerosis* **2014**, *235*, 102–109. [CrossRef] [PubMed]
56. Mozaffari-Khosravi, H.; Loloei, S.; Mirjalili, M.R.; Barzegar, K. The effect of vitamin D supplementation on blood pressure in patients with elevated blood pressure and vitamin D deficiency: A randomized, double-blind, placebo-controlled trial. *Blood Press. Monit.* **2015**, *20*, 83–91. [CrossRef] [PubMed]
57. Beveridge, L.A.; Struthers, A.D.; Khan, F.; Jorde, R.; Scragg, R.; Macdonald, H.M.; Alvarez, J.A.; Boxer, R.S.; Dalbeni, A.; Gepner, A.D.; et al. Effect of Vitamin D Supplementation on Blood Pressure: A Systematic Review and Meta-analysis Incorporating Individual Patient Data. *JAMA Intern. Med.* **2015**, *175*, 745–754. [CrossRef] [PubMed]
58. Franczyk, A.; Stolarz-Skrzypek, K.; Wesolowska, A.; Czarnecka, D. Vitamin D and vitamin D receptor activators in treatment of hypertension and cardiovascular disease. *Cardiovasc. Hematol. Disord. Drug Targets* **2014**, *14*, 34–44. [CrossRef] [PubMed]
59. Cipriani, C.; Romagnoli, E.; Pepe, J.; Russo, S.; Carlucci, L.; Piemonte, S.; Nieddu, L.; McMahon, D.J.; Singh, R.; Minisola, S. Long-term bioavailability after a single oral or intramuscular administration of 600,000 IU of ergocalciferol or cholecalciferol: Implications for treatment and prophylaxis. *J. Clin. Endocrinol. Metab.* **2013**, *98*, 2709–2715. [CrossRef] [PubMed]

International Journal of
*Molecular Sciences*

MDPI

*Review*

# Vitamin D in Neurological Diseases: A Rationale for a Pathogenic Impact

**Rita Moretti *, Maria Elisa Morelli and Paola Caruso**

Neurology Clinic, Department of Medical, Surgical and Health Sciences, University of Trieste, Strada di Fiume, 447, 34149 Trieste, Italy; mariaelisa.morelli@gmail.com (M.E.M.); caruso.paola1983@libero.it (P.C.)

* Correspondence: moretti@units.it

Received: 27 June 2018; Accepted: 26 July 2018; Published: 31 July 2018

**Abstract:** It is widely known that vitamin D receptors have been found in neurons and glial cells, and their highest expression is in the hippocampus, hypothalamus, thalamus and subcortical grey nuclei, and substantia nigra. Vitamin D helps the regulation of neurotrophin, neural differentiation, and maturation, through the control operation of growing factors synthesis (i.e., neural growth factor [NGF] and glial cell line-derived growth factor (GDNF), the trafficking of the septohippocampal pathway, and the control of the synthesis process of different neuromodulators (such as acetylcholine [Ach], dopamine [DA], and gamma-aminobutyric [GABA]). Based on these assumptions, we have written this review to summarize the potential role of vitamin D in neurological pathologies. This work could be titanic and the results might have been very fuzzy and even incoherent had we not conjectured to taper our first intentions and devoted our interests towards three mainstreams, demyelinating pathologies, vascular syndromes, and neurodegeneration. As a result of the lack of useful therapeutic options, apart from the disease-modifying strategies, the role of different risk factors should be investigated in neurology, as their correction may lead to the improvement of the cerebral conditions. We have explored the relationships between the gene-environmental influence and long-term vitamin D deficiency, as a risk factor for the development of different types of neurological disorders, along with the role and the rationale of therapeutic trials with vitamin D implementation.

**Keywords:** neuro-degeneration; MS; demyelination; vascular disease; stroke; AD; vitamin D-OH 25; VDR; VDH; calcium

## 1. Introduction

Low levels of vitamin D, considering serum 25-hydroxyvitamin D (25(OH)D), have been recognized as a widespread health problem, affecting approximately one billion people worldwide [1]. Latitude, season, cultural norms, religious practices, limited awareness, lack of knowledge, indoor lifestyles, urban living, skin pigmentation, malnutrition, diet, co-morbidities (like tuberculosis, malnutrition, chronic inflammatory conditions, etc.), and drugs, may contribute to vitamin D deficiency, especially in the developing world, even if some genetic predispositions can interfere with it (i.e., blacks tend to have lower levels of 25(OH)D compared to whites) [2,3].

Vitamin D is a lipid-soluble vitamin, which can be synthesized and act as a hormone [4]. The active form of vitamin D, 1,25(OH)$_2$D, known as calcitriol, has chemical similarities to typical hormones, such as testosterone, estrogen, and cortisol.

Vitamin D yields both genomic and non-genomic actions; the Vitamin D Receptors (VDR) mediates the former, one of the representative of the steroid hormone superfamily, which are evident in more than 30 human tissues, therefore regulating 3% of the human genome (approximately 700 genes) [5]. Nuclear VDRs are found in most cells, and support the role for the extra-skeletal benefits of vitamin D. The VDR are found in almost all human tissues, participating in the classic actions of vitamin D in the bone, gut, and kidney, but are also involved in immune functions, hormone secretion, and cellular proliferation and differentiation [6]. To be biologically active, vitamin D undergoes hydroxylations in the liver, mediated by the 25-hydroxylase, and in the kidney, mediated by 1$\alpha$-hydroxylase. The 1,25(OH)$_2$D is recognized by VDR, in various cells, mainly in the intestine. As vitamin D plays a significant role in modulating the immune system in the intestine, it is possible that its deficiency could deteriorate the gut barrier function, favoring the translocation of endotoxins. Vitamin D deficiency has been associated with intestinal dysbiosis and increased susceptibility to intestinal diseases; moreover, vitamin D could be contributing to disturbances of the glucose metabolism by modulating the composition of the gut microbiota. The changed intestinal microbiome has become epidemic, in parallel with the epidemic of vitamin D deficiency, suggesting that they might be linked. Proper supplemental doses of vitamin D plus all of the B vitamins appears to return the intestinal microbiome back to normal after few months [7,8]. Moreover, there are more and more evidence that links the microbiome to neurologic disorders [9], which can share a common auto-immune involvement; the potential link between the three actors (VDR, neurological disorders, and vitamin D) is quite fascinating and many studies should be done to gain knowledge on them.

The non-genomic actions of vitamin D help to cooperate with the classical genomic pathway, to transactivate VDR, and exert the effects of calcitriol; the signal does not depend on the transcription phases, but it can operate via cross-talk with different signaling pathways, such as rapid membrane response binding proteins (see later). The genomic and the non-genomic pathways support the emerging not calcemic effects of the vitamin D metabolites, trying to define a role in the autoimmune pathologies, infectious diseases, diabetes mellitus, obesity, and cardio-metabolic disorders [10].

It has also been demonstrated that Vitamin D is involved in different neurological disorders, such as multiple sclerosis, Alzheimer's disease, Parkinson's disease, and stroke [11–14].

The presence of VDR in the hippocampus, hypothalamus, thalamus, cortex, and substantia nigra [15] prompted many studies on the possible determinant role of vitamin D in different neurological conditions [16,17]. It has been evidenced that calcitriol is a fundamental actor in the neuronal differentiation and the neural maturation [18]. Vitamin D normalizes the trafficking of the septo-hippocampal pathways, mainly via the neural growth factor (NGF); moreover, it is an active controller of the genetic regulation of the synthesis of acetylcholine (Ach), dopamine (DA), serotonin (5HT3), and gamma-aminobutyric (GABA) [19–21].

The knock-out model of VDR$^{-/-}$ has an accelerated aging process in all of the organs, as well as in the brain [22,23], with a significant in-brain decrease of NGF [24] and other neurotransmitters, such as Ach [25,26]. The congenital deficiency of vitamin D significantly reduces the activity of glutamic acid decarboxylase (GAD) 65/67 (critical enzymes in GABAergic inter-neurons) and the levels of glutamate and glutamine in the brain tissue [27].

For these reasons, we reckon that a vitamin D deficiency could contribute to the complex relationship between genetics and the environment, the two common poles that many neurological pathologies debate between. It could intervene in the exacerbation of precipitating factors or in demodulating the repair process; therefore, we consider that this topic should be approached from a neurological perspective. Therefore, we have started this review. We searched MEDLINE using the following search terms: "vitamin D central nervous system", both "vitamin D" and "central nervous system"; "vitamin D immune system/response", both "vitamin D" and "immune system" or "immune response"; "vitamin D multiple sclerosis risk", both "vitamin D" and "multiple sclerosis risk"; "vitamin D multiple sclerosis relapse", both "vitamin D" and "multiple sclerosis relapse"; "vitamin D multiple sclerosis magnetic resonance imaging", both "vitamin D" and "multiple sclerosis magnetic resonance imaging"; "vitamin D multiple sclerosis disability", both "vitamin D" and "multiple sclerosis disability"; "vitamin D supplementation/therapy/treatment multiple sclerosis", both "vitamin D supplementation" or "vitamin D therapy" or "vitamin D treatment" and "multiple sclerosis"; "vascular dementia", both "vascular" and "dementia"; "subcortical vascular dementia", both "subcortical" and "dementia"; "Alzheimer's disease"; "pathogenesis neural-degeneration"; "amyloid"; "cholinergic afferents"; "arteriolosclerosis"; "cerebral flow regulation"; and "stroke". The publications that were selected were mostly from the past 20 years, but they did not exclude the frequently referenced and highly regarded older publications. The research has been extended, with the same strings, EMBASE, COCHRANE LIBRARY, and LILACS. All of the searches were done from 1 January 1993 up to 31 May 2018. We considered papers published in English, French, German, and Italian. Congress abstracts and isolated case reports (even if the total cases were under 10) were not considered. Review articles and book chapters are cited to provide additional details. The authors carefully read all of the selected articles.

## 2. Vitamin D Deficiency and Multiple Sclerosis: Role in the Susceptibility, Activity, and Treatment of the Disease

Multiple Sclerosis (MS) is a multifactorial demyelinating pathology affecting the brain and the spinal cord, firmly relying on an altered immune response. Activated autoreactive T cells invade the blood–brain barrier and initiate an inflammatory response that leads to myelin destruction and axonal loss. The etiology of MS, the mechanisms associated with its onset, the unpredictable clinical course, and the different rates of progression leading to disability over time, remain unresolved questions [28].

Immunological research makes two significant observations to explain the link between vitamin D and the immune system.

Firstly, most immune cells, of both the innate and the adaptive immune system, express the VDR [29,30]. Secondly, the immune cells exhibit an active vitamin D metabolism, with the expression of the rate-limiting enzyme for vitamin D synthesis, 1$\alpha$-hydroxylase (CYP27B1) [31]. Immune cells are, therefore, able to synthesize and secrete vitamin D in an autocrine and paracrine condition [32]. The immune cell types, targeted by vitamin D, include monocytes and macrophages, dendritic cells (DCs), and T and B cells [33]. 1,25(OH)$_2$ vitamin D induces monocytes proliferation and differentiation into macrophages [34], the potentiate the expression of interleukin-1 (IL-1), and the antimicrobial peptides (cathelicidin, $\beta$-defensin-2, and hepcidin) [35]. Vitamin D inhibits DC differentiation and maturation, their major histocompatibility complex (class II) expression (MHC), CD40, CD80, CD86, and IL-12 (while inducing the production of IL-10), leading to reduced T cell stimulatory capacity [36,37]. Vitamin D decreases the production of nitric oxide (NO), mediating the downregulation of the inducible nitric oxide synthase (iNOS) expression [38]. Moreover, it stimulates the development of natural killer T (NKT) cells, and it increases IL-4 and interferon (IFN)-$\gamma$ production [39]; vitamin D attenuates the proliferation of CD8+ T cells and reduces their cytotoxic activity by decreasing the production of IL-2, IL-17, and IFN-$\gamma$.

Vitamin D exerts its immunomodulatory effects on T lymphocytes, by inhibiting the production of pro-inflammatory Th1 cytokines (IL-1, IL-2, IL-6, IL-12, IFN-$\gamma$, TNF-$\alpha$, and TNF-$\beta$), and stimulating the production of anti-inflammatory regulatory Th2 cytokines (IL-4, IL-5, and IL-10) [40,41].

Thus, vitamin D potentiates the innate immune system and regulates the adaptive immune system, mainly by inducing the split to Th2 and regulatory T cells (Tregs), over the Th1 and Th17 lymphocytes differentiation [42], via direct and indirect actions on naive CD4+ cells. The general result is an evident effect of switching capacity, from a pro-inflammatory autoimmune to an anti-inflammatory tolerogenic immunological response.

In multiple sclerosis (MS) patients, the blood level of 25(OH)D or 1,25(OH)$_2$D has been linked with the suppressive activity of Tregs [43], and the number of Tregs has been correlated with the serum levels of 25(OH)D or 1,25(OH)$_2$D. The Tregs are increased in the MS patients that have supplemented with vitamin D [44,45].

According to many studies, vitamin D may have an impact on the balance between the inflammatory and anti-inflammatory mechanisms, which might help/regulate the remyelination process. Vitamin D increases the microglial activation, promoting the clearance of myelin debris, and consequently activating the remyelination process [46]. In the oligodendrocyte precursor cells (OPC) cultures, vitamin D up-regulates the transcription of VDR and NGF mRNA, but not of myelin basic protein (MBP) [47–49]. The remyelination of demyelinated lesions has been observed in the early stages of MS [50,51], and it has been supported by neuroimaging findings [52,53]. However, remyelination might be incomplete [54], and eventually, it might cease [55], because of the OPC inability to migrate and reach the site of demyelination [56], or for the OPC differentiation inability [57]. In fact, an inflammatory microenvironment prevents OPC maturation and the differentiation into oligodendrocytes, and subsequently prevents axon remyelination. This is relevant for MS treatment, as current treatments are only useful for controlling immune mechanisms (i.e., in the early stages of the disease), but do not affect remyelination. In vitro studies show that when blocking VDR, there is a reduction of OPC differentiation, with a consequent blockage of myelination and remyelination; on the other hand, by activating VDR, via vitamin D, there is an increase in the differentiation of OPC and the consequent remyelination [58]. Likewise, neural stem cells (NSC) express VDR and 1,25(OH)$_2$D, which seems to promote the NCS proliferation and differentiation into neurons and oligodendrocytes, and to reduce astrogliosis [59,60].

As vitamin D deficiency has been proposed as a significant risk factor in MS development, most epidemiologic observational studies have suggested that adequate vitamin D levels may reduce the risk of MS onset and modify the course of the disease. MS is a disease that is virtually unknown at the equator, and the prevalence of the disease increases in populations that live farther away from the equator [61]. The prevalence of MS is higher at higher latitudes, and tends to peak in the areas with the lowest exposure to ultraviolet (UV) light [62,63]; however, in these areas, diets rich in vitamin D-containing oily fish may offset this risk to some degree [62–65]. Moreover, the risk of MS has been found to decrease among people who migrate from higher to lower latitudes [66]. This latitudinal finding has been declining in recent decades, instead of an associated increasing trend towards avoiding sun exposure by staying indoors for more extended periods of the day, even in warmer climates [67,68]. In fact, higher levels of sun exposure (past, recent, and cumulative) were independently associated with higher levels of vitamin D and with a significantly reduced risk of developing demyelinating events [69]. Sunlight seems to have an immunosuppressive effect and, therefore, the effects of sunlight on MS risk could be related to the sunlight itself, or to an increase of vitamin D [70]. The association between the calculated vitamin D intake from diet or supplements, and the risk of developing MS has been prospectively evaluated in two large cohorts involving more than 187,000 women [71]. Women who had a higher intake of dietary vitamin D (approximately 700 IU/day) had a 33% lower incidence of MS compared with those with a lower intake. Moreover, the women who used vitamin D supplements (more than 400 UI/day) had a 41% reduced risk of developing MS compared with non-users. Higher levels of 25(OH)D (independently, from dietary vitamin D intake) also seem to predict a lower risk of MS onset. A more recent prospective study confirmed these findings and reported that the levels of vitamin D over 30 ng/mL were associated with a decreased MS risk [72]. Adiposity has been associated with lower vitamin D levels [73,74], and a higher body mass index (BMI) has been associated with higher incidence of MS in adolescent women, but not in adult women [75].

A crucial question related to a primary prevention trial or from vitamin D supplementation in MS is the relevant age of exposure, which can range from in utero to adolescence and adulthood [76]. Mirzaei et al. studied a large cohort and analyzed the association between maternal dietary vitamin D intake and the predicted maternal serum 25(OH)D during pregnancy, and their daughters' risk of developing MS [77]. The study showed that the relative risk of MS was significantly lower in the women whose mothers had high vitamin D intake during pregnancy compared with the women born to low-intake mothers. A diminished in utero exposure to vitamin D, coupled with the solar cycle and latitudinal differences, may be an environmental risk factor for the development of MS. Similarly, albeit not statistically significant, a reduced MS risk was reported among the women reporting an increased vitamin D intake from supplements in adolescence [78]. These results suggest that MS risk is related not only to new vitamin D levels, but it might also be related to its levels during childhood or even in utero. Several studies, including a meta-analysis, demonstrated that spring-born have a significantly higher lifetime MS risk than autumn-born, which has been attributed, at least in part, to insufficient in utero vitamin D levels, because of low maternal serum vitamin D levels during winter [79,80].

In a large population-based case-control study [81], children born with 25(OH)D levels <10 ng/mL seemed to be at a high risk of developing MS. Likewise, the level of sun exposure in childhood and adolescence (e.g., by outdoor leisure activities), which may serve as a proxy for vitamin D supply in early life, has been inversely linked to the risk of MS in adulthood [82,83]. In a recent longitudinal Canadian study of 302 children with the acute demyelinating syndrome, low vitamin D levels were significantly associated with MS risk in the subsequent three years [84]. One report showed that children with higher serum 25(OH)D concentrations at presentation, with an acquired demyelinating syndrome, had a lower risk of early MS diagnosis [85]. Gender- and sex-related immunological differences may influence the association between vitamin D and MS. The disproportional increase in the incidence of MS in women is likely to be caused by sex-specific exposure or susceptibility to environmental factors [86]. The data supporting an interaction between female sex, possibly mediated by estrogen, and vitamin D for MS risk is accumulating. A protective effect of sun exposure was only observed in female monozygotic twins [87], and the association of sun-sensitive skin types with a disability was only found in untreated female MS patients [88]. In vitro studies of MBP-specific T cell proliferation have shown sex differences in the metabolism of vitamin D, which were confirmed by treating male MBP-specific T cells with 17β-estradiol in the assay [89]. In one animal study, vitamin D resulted in fewer clinical, histopathologic, and immunologic signs of Experimental Autoimmune Encephalomyelitis (EAE) in female mice compared with ovariectomized females and intact or castrated males [90]. A synergy between $E_2$ and vitamin D occurred through the VDR-mediated enhancement of $E_2$ synthesis, as well as through the $E_2$-mediated enhancement of VDR expression due to the inflammation of central nervous system (CNS). In males, $E_2$ did not enable vitamin D to inhibit EAE [91], possibly suggesting that vitamin D-mediated protection in EAE is female-specific, and that MS tends to have a more aggressive course in men than in women.

Several genome-wide association studies (GWAS) and gene-candidate studies have investigated the influence of the specific genetic polymorphisms of vitamin D metabolites on the 25(OH)D levels, and their susceptibility to MS; one study found [92] an association between the short variant of the VDR protein (F allele) and a genetic predisposition to lower 25(OH)D levels, but not to a higher risk of MS; in a GWAS of 4501 European patients [93], single-nucleotide polymorphisms (SNPs) of the gene encoding components of the vitamin D binding protein were associated with 25(OH)D concentrations, or with the genes involved in vitamin D synthesis or activation; moreover, a genetically lowered 25(OH)D level was strongly associated with increased MS risk and progression in two studies [94,95].

The role of vitamin D in MS disease progression has also been assessed: it has been hypothesized that 25(OH)D levels can predict a later development of MS in acute optic neuritis (ON) [96], but the result is inconclusive. A discrete quantity of studies demonstrated that vitamin D levels affect clinical relapses and MS disease activity. In a retrospective study of 110 patients with pediatric-onset MS, the authors found that each increase of 10 ng/mL in the 25(OH)D level was associated with a 34% decrease in the relapse risk [97]. Similar findings were seen in a prospective cohort study, whose authors concluded that raising 25(OH)D by 20 ng/mL could decrease the hazard of relapse by up to 50% [98]. In a prospective longitudinal study, the relapse risk was significantly reduced in those patients with medium (20–40 ng/mL) and high (>40 ng/mL) serum vitamin D levels, compared with those with low levels (<20 ng/mL) [99]. Moreover, the same authors found that for each doubling of the serum vitamin D concentration from a baseline of 10, 20, and 30 ng/mL, MS relapse risk decreased by 27%. In another study, lower vitamin D levels predicted a conversion from Clinically Isolated Syndrome (CIS) to clinically definite MS [100,101]. In the study by Embry et al., low serum 25(OH)D levels predicted an increased likelihood of gadolinium-enhancing lesions (Gd+) in the Magnetic Resonance Imaging (MRI) scans performed in the subsequent two months period [102]. In the EPIC study [103], a five-year MS cohort study in which subjects had clinical and MRI evaluations and gave a blood sample annually. The authors concluded that individuals with CIS/relapsing-remitting multiple sclerosis (RRMS) with higher vitamin D levels have a lower risk of the subsequent development of new T2 lesions and Gd+ lesions on a brain MRI, even after accounting for potential confounding factors. Moreover, an increment of 10 ng/mL of 25(OH)D was associated with a 15% lower risk of new T2 lesions and a 32% lower risk of Gd+ lesions [103]. In a post hoc analysis, including up to two years of follow-up of participants treated with interferon beta (IFNB)-1b in the BENEFIT trial [104], Gd+ lesions development was inversely associated with 25(OH)D levels, those patients whose 25(OH)D levels were >20 ng/mL had a 39% lower risk of new Gd+ lesions. Unfortunately, across all of the analyses, associations with lower vitamin D were generally stronger for MRI than for the clinical outcomes. Moreover, the participants of the Betaferon Efficacy Yielding Outcomes of a New Dose (BEYOND) study (BEYOND study) [105], treated with IFNB-1b, with higher serum 25(OH)D levels, had lower numbers of new T2 and Gd+ lesions during the first 12 months of follow-up. Moreover, a 20 ng/mL higher serum 25(OH)D level was associated with a 31% lower rate of new lesions, and the patients with 25(OH)D $\geq$ 40 ng/mL showed 47% lower rate of new T2 lesions and new Gd+ lesions, when compared with the patients who had serum levels of 20–32 ng/mL. Vitamin D and disease-modifying therapies (DMTs) may positively influence each other and produce an additive, or even synergistic, effect on MS disease activity. In an observational cohort study, which included 178 patients with MS [106], the patients who were treated by IFN had significantly higher 25(OH)D levels than those who were not. Interestingly, the IFN treatment was protective only against relapses among the patients with higher vitamin D levels. The authors hypothesized that treatment with IFNB might increase the serum vitamin D levels through an enhanced responsiveness to sun exposure [106]. The same authors did not find similar associations for glatiramer acetate (GA) therapy and vitamin D. More recently, Laursen et al. [107] found that higher vitamin D levels in CIS may slow neurodegeneration evaluated by brain volume measures. In fact, they found that each 10 ng/mL increase in 25(OH)D was significantly associated with a 7.8 mL higher gray matter volume [108]. Variations in the relapse rate and the number of MRI brain lesions have shown a seasonal pattern that can be related to a variation in Ultra Violet Radiation (UVR) exposure and vitamin D status [109,110], with some exceptions [111,112]. Most cross-sectional studies have concluded a negative correlation between the 25(OH) D level and disability [103,113,114], and, interestingly, even a direct correlation between the 25(OH) D level and poorer memory performance [115]. However, the causality is considered uncertain, at the moment.

Given the previous reviewed findings, the assessment of vitamin D supplementation for a possible disease-modifying course of MS is obviously of crucial interest. Unfortunately, the current evidence does not offer a definite consensus for supplementation. Kimball et al. [115] performed a six-month safety study with escalating doses of vitamin D, and they found a significant reduction in the mean number of Gd+ lesions at the end of the study. In an open-label randomized trial, patients randomized to a vitamin D supplementation had an annualized relapse rate (ARR) significantly lower in the treatment, with a prolonged relapse-free time and with a persistent reduction in the T cell proliferation [44]. In a one year double-blind, randomized placebo-controlled trial with vitamin D3 as an add-on treatment to IFNB-1b, the MRI T2 lesion burden and the new/enlarging T2 lesions, tended to increase more in the placebo group than in the vitamin D group, however, without statistical significance [116]. A preliminary Iranian study assessed the safety and efficacy of high-dose vitamin D supplementation during pregnancy in women with MS [117]. The women in the vitamin D group had significantly fewer relapses during pregnancy, a tendency for fewer relapses up six months after delivery, and a more stable Expanded Disability Status Scale (EDSS) than those without supplementation [107]. In a longitudinal study [118], in which 170 natalizumab-treated patients were followed for one year between two winter seasons, patients with insufficient serum 25(OH)D levels at baseline (<20 ng/mL) were advised to take vitamin D supplements, and a significant inverse relationship with the ARR was found, as, for each nmol/L increase in 25(OH)D, a 0.014 decrease in ARR was observed. The double-blind, multicentre, 48-week study, named SOLAR, of supplementation of high-dose oral vitamin D has been the most extensive study to date [119]. An insignificant trend toward lower ARR in the treatment group was found in vitamin D groupersus placebo, and no statistically significant differences in the disease activity were found between the two groups. Vitamin D3 was associated with a statistically significant reduction in the combined unique lesions (secondary endpoint). The effect of the addition of vitamin D3 to IFNB-1a over 96 weeks was investigated with appropriate cognitive tests [120]. It did not show a significant trend toward a lower ARR (primary endpoint) among those patients receiving vitamin D treatment, which became statistically significant when the analysis was restricted to those who completed the study. Among the completers, there were also significantly fewer new T2 lesions in the vitamin D group. It is unclear whether the findings of these trials are related to insufficient power or other issues leading to an inability to detect a treatment effect for all outcomes [120].

Larger randomized controlled trails (RCTs) are currently underway to reveal the role of vitamin D supplementation as an add-on to IFNB therapy in the treatment of RRMS, and even CIS, patients, namely, the VITADEM study in Spain, EVIDIMS study in Germany, PrevANZ study in Australia, D-Lay-MS study in France, and the VIDAMS trial in the United States. One study investigated the cognitive effects of vitamin D supplementation on patients with MS who were treated with IFNB [121], and after the follow-up period they scored better on the Brief Visuospatial Memory test (BVMT) for delayed recall.

In conclusion, numerous studies suggest that vitamin D supplementation may benefit MS patients, although larger RCTs are needed to establish this supplementation as a standard of care for MS. Moreover, there is no consensus on the definition of 'sufficient' vitamin D levels in MS, and many physicians question whether people with MS should empirically be supplemented while waiting for more conclusive results of vitamin D clinical trials. In the view of the Institute of Medicine (IOM), 25(OH) D levels greater than 20 ng/mL (50 nmol/L) are sufficient. The Endocrine Society, considering skeletal and non-skeletal health, argues for levels of almost 30 ng/mL (75 nmol/L). In the MS field, numerous studies suggest that serum 25(OH)D levels of approximately 40 ng/mL (100 nmol/L) are the lower limit for controlling MRI and clinical activity. Some experts favor maintaining 25(OH)D levels between 30 and 50 ng/mL in MS patients, as immune-modulatory effects have been observed in these ranges in experimental studies. Longitudinal RCTs are needed to establish the recommended levels of vitamin D supplementation necessary to reduce the risk of MS onset and MS pathological activity. According to other authors [122], for individuals with CIS or MS and vitamin D deficiency (<20 ng/mL), an 'attack' supplementation with 50,000 IU/week of vitamin D2 (ergocalciferol) for eight weeks, and a subsequent evaluation of the serum level, is recommended. The continuation of this therapy until the serum 25(OH)D is higher than 30 ng/dL may be necessary. A maintenance supplementation with vitamin D3 (cholecalciferol) at 1000 to 2000 IU/day may be started when the deficiency has been corrected, and at once in case of vitamin D insufficiency (20–29 ng/mL). As current evidence also suggests both an obstetric and pediatric benefit from vitamin D against the risk of developing MS, vitamin D supplementation in children and pregnant women at risk of developing or being affected by MS should be considered [123]. According to other authors, as a goal, the general aim is 25(OH)D levels between 40 and 60 ng/mL, and most patients should take a dose that arises from 2000 to 5000 IU/day vitamin D3 [123]. Many studies have been conducted to show the potential relationships between genome-wide association studies (GWAS) and gene-candidate studies, in order to find out the specific genetic polymorphisms of the vitamin D metabolites on the 25(OH)D levels, and their susceptibility to MS and to vitamin D supplementation efficacy in MS [92–95]. Moreover, the Human Leukocyte Antigen (HLA) haplotype might influence epistasis, trans and cis effects, and parent-of-origin effects, as well as the Epstein Barr Virus (EBV) and vitamin D effects on the course of MS [124]. Le Mokry et al. (2015) identified SNPs associated with 25(OH)D level from SUNLIGHT, the largest ($n$ = 33,996) genome-wide association study to date for vitamin D. Four SNPs were genome-wide significant for 25(OH)D levels, and all four SNPs lay in, or near, genes strongly implicated in separate mechanisms influencing 25(OH)D. The authors found out that the count of the 25(OH)D-decreasing alleles across these four SNPs was strongly associated with a lower 25(OH)D level. Other recent perspectives [95] confirmed the research by Ahn et al. [93]; it seems that there are strong links between the 25(OH)D levels and the genotyping of CYP2R1- and NAD-synthetase (NADSYN1-SNPs) in MS patients. The analysis revealed lower 25(OH)D concentrations in MS patients and an association of rs10766197 CYP2R1 SNP with MS risk. After stratifying the MS patients according to gender, the authors [95] found that the minor allele A of rs10766197 had a higher frequency in men in comparison with the women affected by MS. Additionally, the presence of allele A in men was associated with disease progression, assessed by EDSS and Multiple Sclerosis Severity Score (MSSS) scores [96]. The results are not univocal at all; a recent study focussed on VDR polymorphisms (Fok-I, Bsm-I, Taq-I, and Apa-I) that were genotyped by a polymerase chain reaction (PCR), followed by the restriction fragment length polymorphism (RFLP) analyses in both groups, and the serum 25(OH) D levels were determined in the MS patients by high-performance liquid chromatography (HPLC). The distribution of genotype and allele frequencies of the four VDR polymorphisms did not differ significantly between the MS patients and healthy controls, and were unrelated to the forms and the course of MS. Low serum levels of 25(OH) D were observed in MS patients, but no association was observed between the VDR and 25(OH) D levels, except for Fok-I. Moreover, the MS patients with the *FF* and *Ff* genotype had a significantly lower serum level of 25(OH) D [92]. Moreover, further studies are necessary to develop potential relationships between the genome-wide association studies (GWAS) and gene-candidate studies for determining the influence of specific genetic polymorphisms of the vitamin D metabolites on the 25(OH) D levels, and the susceptibility to MS and to vitamin D supplementation efficacy in MS.

## 3. Vitamin D Deficiencies and Ischemic Stroke

Stroke is the leading cause of significant long-term disability all over the world, and is an enormous source of global disease burden. Ischemic stroke recognizes a heterogeneous etiology, caused by not modifiable risk factors (genetic, age, and sex), and modifiable risk factors, like hypertension, diabetes mellitus (DM), dyslipidemia, sedentary lifestyle, and smoking [125].

Vitamin D deficiency is found to increase the risk of vascular disease and ischemic stroke in healthy subjects; the risk is higher for ischemic stroke than for hemorrhagic ones. Moreover, vitamin D deficiency is associated with other contributing factors for ischemic stroke (i.e., hypertension, hyperlipidemia, diabetes mellitus, and ischemic heart disease). In a stroke, vitamin D deficiency might relate with higher disease severity and adverse outcomes, including death; moreover, hypovitaminosis D is independently associated with larger ischemic infarct volume [126]; finally, vitamin D deficiency is related to slower recovery after stroke. Moreover, recently, it has been found [125,126] that lower concentrations of 25(OH) D were associated with a higher risk of incident stroke in models adjusted for age, race, age and race interaction, and sex. The magnitude and strength of the associations were unchanged after the adjustment for the season of blood draw; systolic blood pressure; BMI; diabetes; current cigarette smoking; atrial fibrillation; and use of anti-hypertensive medications, aspirin, and statins. In their work, no statistically significant differences in the association of lower 25(OH) D with a higher risk of incident stroke were observed in black subjects when compared with white subjects. Vitamin D can inhibit the development of thrombosis, which may provide a rational explanation for the relationship between vitamin D and ischemic stroke. Nevertheless, vitamin D might induce hemorrhagic stroke through other mechanisms, such as inflammation and endothelium shear stress. Even less is known regarding the relationship between vitamin D deficiency and other cerebrovascular diseases, like vessel dissection. Vianello et al. [127] reported that in acute aortic dissection (AAD), hypovitaminosis D is not associated with changes in bone-related metabolic pathways, but is inversely related to osteocalcin, which could be an exciting molecule that is able to mediate the effect of an inadequate 25(OH) D level at vascular level.

Moderate to strong associations between lower serum 25(OH)D concentrations and stroke were identified in different analytic approaches, even after controlling for traditional demographic and lifestyle covariates. Such associations were most evident among young females, younger than 50 years. The mechanisms behind the associations between vitamin D and cerebrovascular and cardiologic profiles have been widely examined in both animal and human studies. Accumulating evidence has shown adverse regulatory effects of vitamin D on the renin-angiotensin system. Renin expression was significantly increased among the vitamin D receptor knockout mice and suppressed among the wild-type mice after being injected with $1,25(OH)_2D$. The treatment of vitamin D in rats may lead to increased endothelium-dependent vascular relaxation and the inhibition of vascular smooth muscle cell growth and proliferation. Vitamin D supplementation in human subjects may contribute to improved insulin sensitivity and beta-cell function, as well as lower levels of inflammatory markers [128–130].

Moreover, vitamin D deficiency has been shown in several pathologies related to higher incidence of stroke. Shah Sanket et al. reported that a vitamin D deficiency was observed in a significant number of patients with chronic obstructive pulmonary disease (COPD), and in more than half of the study subjects there is an increasing frequency across the combined COPD class of cardiovascular and cerebrovascular disease [130]. Moreover, the stroke patients with enough vitamin D had more favorable outcomes, including improved muscle strength and bone density. Manouchehri et al. [131] showed that a vitamin D deficiency had an increased risk of ischemic stroke by nearly seven-fold compared to the controls. They reported a 13-fold increase in the risk of large vessel stroke, and a 4.37-folded increase for the small vessel stroke was observed. The risk of stroke increases with a concomitant deficit of vitamin D, vitamin B12, and homocysteine. Serum homocysteine, vitamin B12, and vitamin D levels are associated with baseline first-ever stroke severity, but also contribute, to some extent, to its prognosis in the early period after stroke. The early detection and management of these laboratory parameters may contribute both to primary and to secondary stroke prevention. Hyperhomocysteinemia increases the likelihood of stroke and is mostly dependent on folic acid (vitamin B9), vitamin B12, and vitamin B6 serum levels. Vitamin B12 deficiency can be detected in 10–40% of the general population and may contribute to stroke and cognitive decline. The status of homocysteine, vitamin B12, and vitamin D can go further to assess the importance of these correctable factors during the functional recovery after stroke.

The patients with a low level of serum vitamin D at the onset of the stroke showed more severe disability [126]. Qiu et al. [132] also reported a strict association between lower serum levels of 25(OH) D, and the stroke recurrence and stroke mortality at 24 months. Also, the inverse association between serum 25(OH) D levels and functional outcome in patients with acute ischemic stroke had been reported. Mortality in stroke patients is higher for those subjects <75 years old with a low serum 25(OH) D level at stroke onset [133–135]. Few studies have also reported vitamin D deficiency after a stroke. The mean level of 25(OH) D was lower in the subacute and chronic patients than in the healthy controls; furthermore, the level of 25(OH)D was lower in the patients with a longer duration of illness after stroke, for one month. These data suggest that it might be explained by an exhaustion of vitamin D, stored before stroke onset, and insufficient synthesis after stroke. Finally, the patients with an independent gait have higher levels of vitamin D than the immobilized patients. Restricted outdoor activity may have affected the synthesis of vitamin D after the stroke and resulted in a vitamin D deficiency.

The pathophysiology of stroke is quite complex and still not completely understood. Cerebral ischemic damage is due to the activation of several inflammatory events, including the infiltration of circulating immune cells and the activation of microglia, astrocytes, and endothelial cells [136]. In stroke individuals, the mechanism involved in vessel disease, mediated by a vitamin D deficit, might consist of a release of atherogenic pro-inflammatory cytokines, which foster atherosclerotic vascular changes and might induce plaque instability [137,138]. Vitamin D plays an additional role for the regulation of the inflammation process, through prostaglandin inhibition, reduction of mitogen-activated protein kinase (MAPK), and by reducing the expression of the nuclear factor kappa B (NF-κB) pathways [41,139]. Moreover, the down-regulation of pro-inflammatory cytokines, such as tumor necrosis factor (TNF)-$\alpha$, interleukin (IL)-6, IL-12, and interferon (IFN)-$\gamma$, and the up-regulation of anti-inflammatory T regulatory (Treg) and Th2 cells and their cytokines have been reported [140,141]. Thus, the deficit of vitamin D is responsible for endothelial dysfunction [142] and is considered an independent risk factor for the occurrence of acute ischemic stroke [143,144]. As previously reported, VDD is strictly associated with higher levels of high sensitivity C-reactive protein (hsCRP), remarking the anti-inflammatory activities of vitamin D, including the inhibition of IL-6 synthesis by monocytes. CRP acts on endothelial cells inducing tissue factor expression, and promotes smooth muscle and endothelial cell proliferation [145]. Moreover, CRP increases the plasminogen activator inhibitor–1 expression [146] and induces several inflammatory genes via NF-κB activation in human endothelial cells [147]. Usually, vitamin D deficiency in stroke patients preceded stroke, and the prevalence of vitamin D deficiency is more evident in stroke patients than general medical patients. On the other hand, vitamin D supplementation in post-stroke patients may play a role in the prevention of recurrent stroke and improves the functional outcome after stroke [147–150].

Zhao-Nan Wei reported that a vitamin D deficiency is associated with a 3.2-fold increased risk of poor functional outcome events. The adjustment for established cardiovascular risk factors, including glucose level, age, and National Institute of Health Stroke Scale (NIHSS) score, did not attenuate this association. For the vitamin D deficiency, the adjusted risk of mortality increased by 290% [151]. A high dose oral vitamin D supplementation produced a short-term improvement in the endothelial function in stroke patients, with a better management of blood hypertension. Moreover, a possible cardiovascular protective role of vitamin D has been proposed, in reducing the mortality risk in patients with renal failure.

A meta-analysis of 50 RCTs of supplemental vitamin D, administered for a median of two years, involving predominantly senior women who were mainly in institutions and dependent care, showed that vitamin D decreased mortality [152]. Moreover, vitamin D3 (cholecalciferol) has been shown to decrease mortality significantly. Gupta et al. [152], in a randomized controlled open-label trial, found that patients with acute ischaemic stroke that received vitamin D and calcium supplementation, along with the usual care, compared with those receiving usual care alone, presented a higher probability of survival, achieving a better outcome at six months compared to the controls [153]. Concerning stroke, the VITamin D and Omega-3 triaL (VITAL) studied the effect of supplementation on the total cancer and major cardiovascular events (a composite of myocardial stroke, and death due to cardiovascular events), thus reporting significant effects only on a reduction of bone fractures [154,155]. Similarly, the benefit of vitamin D supplementation for stroke-related depressive symptoms is still being debated [156]. If vitamin D contributes to post-stroke recovery by pathophysiological mechanisms that should still be elucidated (some could be related to a reduction in the volume of the cerebral infarct and to neuroprotective properties), vitamin D supplementation could bring the hope of a benefit that is not known yet.

In summary, the precise relationship between vitamin D and stroke is still unclear, a revision of the data suggests that vitamin D status is associated with ischemic stroke and with the injury volume. The vitamin D levels should be measured in all stroke patients and should be considered as an independent stroke risk factor. The supplementation of vitamin D could be considered as a fundamental part of stroke therapy, but new studies should be done.

## 4. Vitamin D Deficiency and Neurodegenerative Diseases

Different important studies have implicated amyloid beta accumulation, hyperphosphorylation of tau, oxidative stress, mitochondrial dysfunction, and inflammation as the major responsible factors for a neurodegenerative process, which underlies Alzheimer's disease (AD) [157,158]. Nevertheless, the calcium excitotoxic hypothesis and glutamate currents theories can support and amplify the Alzheimer's cascade of events [159]. Moreover, many doubts still merge, especially for sporadic AD cases pathogenesis. It has been established that the etiology of the sporadic AD might involve multiple gene-environment relationships, and probably many epigenetic mechanisms [158,160]. The most debated aspects are those concerning the reasons for Abeta accumulation and the tauopathy consequences in the aging brain being seriously possible, and the relationship between their accumulation and various environmental risk factors, which the brain gathers throughout its life. Epigenetic modifications can act as first hits, with a consequence that remains latent for many years, until a second hit (probably determined by metabolic factors, i.e., altered nutrition, pro-inflammatory cytokines, and aging by itself) can promote the degenerative progression [161]. As a result of these premises, VDR, the major effector of vitamin D, polymorphisms have been studied in AD [162]. It has been demonstrated that there is a link between the altered gene expression of VDR and 1,25MARRS (membrane-associated rapid response steroid-binding); this association endorses the less active employment of vitamin D inside neurons, and lets them be more prone to degeneration [163]. Brewer et al. [163] and confirmed by Thibault et al. [164] recognized that cultured hippocampal cells treated with adequate concentrations of VDH (1–100 nM) were protected against excitotoxic insults, probably because of a modulation of L-type voltage-sensitive calcium channels, whose increase has been documented in hippocampus aged cells [165], and the aged long-term cultured hippocampus cells [166]. VDH down-regulates mRNA expression for different subunits of L-type voltage-sensitive calcium channels. Consequently, VDH has a fundamental role in the homeostasis of calcium-mediated activities [164], such as neuronal death and apoptosis. The $VDR^{-/-}$ knock-out model shows a higher sensitivity to neuro-degeneration, with a rapid increase of calcium currents and neural death [166,167]. Many different VDR polymorphisms have been described as increasing the susceptibility to AD [168], mainly due to the altered expression of neurotrophins. Genome analyses, transcriptomics, and proteomics have pointed out the role of VDR polymorphism in late-onset AD susceptibility [162]. In fact, even in animal models, vitamin D seems to interfere with cognitive functions, even in other ways, probably through its different polymorphic gene expression of VDR [26]. It seems, for instance, that Bsm I and Taq I altered carriers are more prone to manifest memory and cognitive dysfunctions, as well as a vitamin D defect; on the contrary, the APA-I haplotype is associated with an increased risk of fractures, but not memory alterations [169]. Therefore, in line with Buell and Dawson-Hughes [169], calcium concentrations are not likely to vary in the different haplotypes, indicating a protective effect on the brain, beyond the calcium homeostasis. Thus, in the Alzheimer disease culture models, VDH stimulates the amyloid plaques [170,171], supporting the phagocytosis induced by macrophages of a soluble amyloid beta protein [171,172], and reduces the inflammation response induced by amyloid deposition [21]. Lipopolysaccharide-induced levels of mRNA encoding for macrophage colony stimulating factors and tumor necrosis factors in cultured astrocytes are partially reduced after vitamin D treatment [21]. Additionally, vitamin D has neuroprotective properties against glutamate toxicity [173]. It inhibits the synthesis of inducible nitric oxide synthase and regulates the gamma-glutamyl transpeptidase, fundamental in the metabolism of glutathione [174]. Furthermore, vitamin D enhances the protein phosphatase 2A activity, modulating the redox state, and thus reducing the age-related tau hyperphosphorylation, limiting the cascade of neuronal dying back and the promotions of collateral inflammatory potentiation [175]. To summarize, it seems that vitamin D acts in the brain, through the regulation of NGF and neurotransmitters, regulating calcium homeostasis, promoting anti-inflammatory responses, interfering with amyloid beta metabolism, and implementing a brain oxidative response. On the contrary, clinical practice does not give univocal results. A vitamin D deficit has been widely detected in the frail old population [176,177], and calcium homeostasis is heavily lowered in neurodegenerative pathologies, such as Alzheimer disease [177,178]. In a mild cognitive impairment population, not including Alzheimer disease, a low level of vitamin D was reported [179]. Nevertheless, the vitamin D implementation in clinical practice does not give univocal results. A seven-year follow-up study confirmed that a larger intake of vitamin D is associated with a lower risk of developing AD in normal aging women [180,181]. Some recent works have evidenced that the combined effect of vitamin D and docosahexaenoic acid can enhance the neural protection towards the different effects of beta-amyloid deposition [182,183], and a six-month trial determined that the patient obtains better results, with even limited effects, when the memantine was prescribed in association with vitamin D, rather than alone, in the AD patients [182]. On the other hand, two studies showed the opposite results [183,184]. Even though many different studies are needed to demonstrate a definite role in the real clinical practice of the supplementation of vitamin D, many questions remain without a proper answer, such as:

1. When should the implementation begin?
2. Is it a possible preventive therapy for reducing the cascade of events of the AD?
3. Should it be considered as one of the multifactor agents (along with folate, vitamin B12, and antioxidant substances) that can procrastinate the AD second hit process?

### 4.1. Vascular Dementia

Emergent evidence from experimental and clinical studies suggests that vitamin D may be a causal effect for neurodegeneration, concerning vascular dementia, supporting the underlying/concomitant atherosclerotic pathology [185]. Some studies link the vitamin defect to an increased risk of hypertension, diabetes, congestive heart failure, myocardial infarction, and stroke [186]. Moreover, some very recent studies link it to small vessel disease and vascular dementia as well [187,188]. The vitamin D receptors are hugely expressed by the endothelial cells, and their activation induces the promotion towards the maturation of immature cells, via Vasoactive Endothelium derived Growth Factor (VEGF) [189]. The vitamin D receptor gene is up-regulated during inflammation in endothelial cells [190–192], and the vitamin D analogs protect against advanced glycation products derived insults [193]. Enriched vitamin D diet models promote an anti-lymphoproliferative effect on the endothelium and a diminished response to inflammatory cytokines [194,195]. Moreover, as previously described, vitamin D regulates the expression of 74 genes and 36 proteins, connected with the correct development of the cytoskeleton and exerting a regulation on post-transcriptional controls for L-type voltage-sensitive calcium channels [196,197]. All of the effects of vitamin D on the vascular system link it to vascular dementia in many different ways, such as: epidemiological [198], based on vascular pressure control [191], based on biological properties of vitamin D, above described [199], or simply considering vitamin D deficiency as a risk-modifiable factor [170,200,201]. It seems quite interesting that the three works that link the vitamin D defect to small vessel disease-related dementia [188,189] stand on two axiomatic biological properties, the anti-oxidative capacity of vitamin D (therefore a loss of protection against Reactive Oxygene Species (ROS), whenever it lacks), and the control of the smooth vessel, which is fundamental for auto-regulation in brain circulation. Many studies evidenced an excess of superoxide by Nicotinamide Adenine Dinucleotide Phosphate Hydrogen (NADPH) oxidase in small vessel disease [202]. This causes an increase of ROS, indirectly evidenced by a hyper-expression of the Nitrogen Oxide (NOX) 2 and NOX 4 oxidase isoforms [202,203]; ROS induced damage is one of the triggers for apoptosis. In vitro, vitamin D down-regulates the activity of NF-κB activity [204] and stimulates the anti-inflammatory cytokines [205]. Additionally, vitamin D-binding proteins are more evident in proximity to endothelium injury [206], inhibiting the expression of Matrix Metalloproteinase (MMP-2, MMP-9), and of the endothelium growth factor [207]; they probably diminish the activity of the platelet-derived growth factor, conversely up-regulating thrombomodulin [208,209]. The second study, entirely dedicated to small vessel disease and vitamin D [198], lights some shadows in a fascinating problem; it presents a public association between bone small vessel disease and osteoporosis. Authors hypothesize a common method of alteration, in the peripheral autonomous nervous system, which deteriorates the small-bone vessels as well as brain vessels [210]. The first and the third study on the topic [188,189], starting from two different perspectives have some common points. Chung et al. [188] demonstrated that vitamin D is inversely associated with lacunes, white matter hyperintensities, and deep micro-bleeding in the deep white matter, not elsewhere, suggesting a high impact on small vessel disease. Moretti et al. [198] hypothesized that vitamin D might act on the altered control of cerebral blood flow (CBF) [211,212] and in the distorted neurovascular coupling system [213,214], both heavily impaired in small vessel dementia. Intracerebral calcium, intimately regulated by vitamin D, interferes with the vessel activation. Vessel relaxation is heavily influenced by ATP-sensitive potassium channels (delayed rectifier and inward rectifier potassium channel), but also by calcium-activated potassium channels [215,216]. Mediated by cAMP, calcium-activated-potassium channels seem to be involved in the negative "feedback system to regulate vascular tone" [215]. It seems that in atherosclerosis models, there is a major impairment of calcium-activated potassium channels in mainstream vessels [215]. On the other hand, the neurovascular coupling system (smooth vascular cells, neurons, and astrocytes) seems to be intimately regulated by vitamin D, which is determinant for the glutamate release. Glutamate passes from the synapse, activating Nicotinamide Adenine Dinucleotide Phosphate Hydrogen- Receptors (NMDAR) in the neurons and metabotropic glutamate receptors in the astrocytes [214]. Moreover, by interfering with the calcium influx, vitamin D indirectly mediates the neural nitric oxide synthase. NO activates phospholipase A2 in the astrocytes, enabling the prostaglandin cascade, which widens the arteries. The presence of vitamin D increases, moreover, acetylcholine, and VIP, which potentiates the mechanism. Low levels of vitamin D might, therefore, interfere with smooth cells control, with NO and neuropeptides synthesis, and with defective neurovascular coupling, leading to some aspects of small vessel disease. There are very few the studies concerning vitamin D supplementation in vascular dementia.

### 4.2. Parkinson Disease (PD)

A significant concentration of vitamin D receptors were found in the hippocampus, in the prefrontal cortex–brain, and the substantia nigra [217]. There is evidence of low levels of vitamin D and increased bone turnover markers, such as bone alkaline phosphatase, compared to the controls [218]. All of the non-genomic roles of vitamin D are evident in the PD models [219,220]; there is an active regulation of calcium inside the neurons and astrocytes; there is a substantial regulation of L-type voltage-sensitive calcium channels; it has been described an induction and promotion of NGF and other growth factors, with a down-regulation of MMP and other inflammatory activity, such as prostaglandins and COX-2 activity; and there is also a reduction of oxidative stress, by interfering with NO production by lipopolysaccharide-stimulated macrophages [221]. There are many data that even relate the vitamin-related genomic factors to PD progression. Increased levels of MHC class II expressions were detected in the monocytes of the cerebrospinal fluid of the PD patients [222], together with Human Leukocyte Antigen Complex- antigen D related (HLA-DR) positive reactive microglia, found in the *substantia nigra* and in nigro-striatal tract of PD patients [223]. Recent evidence demonstrated that vitamin D suppresses the MHC class II antigens [224] and IFN-gamma induced HLA-DR antigen expression in human cells [225]. Additionally, Cytochrome P450 (CYP), and in particular, CYP2D6, which has a polymorphic expression, and which is expressed in neurons and the gut, has a different expression in PD patients; here, it has been found preponderantly as the CYP2D6*4 allele [226]. CYP2D6 acts as a 25-hydroxylase, which can convert vitamin D3 into 25OHD, being the key enzyme to determine a deficiency of vitamin D. In the mouse MPTP treated model (inducing PD phenotype), Singh et al. [225] reported the stable expression of the animal ortholog of CYP2D6, CYP2D22. Very interestingly, the CYP2D loci are located on chromosome 22, where many genes related to PD are segregated [227,228]. It seems surprising that the deletion of chromosome 22q11 was reported to be associated with PD [228], but also with the reduced serum calcium and a reduced level of vitamin D [228].

Moreover, there are three more links between vitamin D and PD. The first one is the so-called Sp1 transcription factor. The protein encoded by this gene is a zinc finger transcription factor that binds to the guanine-cytosine (GC)-rich motifs of many promoters. The encoded protein is involved in many cellular processes, including cell differentiation, cell growth, apoptosis, immune responses, response to DNA damage, and chromatin remodeling. Post-translational modifications, such as phosphorylation, acetylation, glycosylation, and proteolytic processing, significantly affect the activity of this protein, which can be an activator or a repressor. The Sp1 transcriptions factor is a DNA binding protein that mediates the accurate response to oxidative stress in neurons [229,230], controls the expression of the dopamine transporter gene, and of the dopamine receptor gene [231,232]. The link between vitamin D and Sp1 is represented by the fact that many hormones have different binding sites for the transcription factors for Sp1. In fact, two vitamin D-responsive elements (which act by inducing the expression of CYP24 genes, previously described, and 25-OHD-24 hydroxylase) are helped by the synergic activation of Sp1 [229,233]. As Sp1 controls the expression of the dopamine transporter gene within the dopaminergic neurons, VDR conditions and promotes Sp1 trigger gene expression.

The second genomic link factor between vitamin D and PD is represented by Heme oxygenase, a stress protein with anti-oxidant properties. In the healthy brain, its expression is limited to neuroglia [234], whereas it is overexpressed in PD brains, but not in AD patients [235]. In PD brains, Heme oxygenase 1 is overexpressed in astrocytes within the *substantia nigra* and the deteriorated dopaminergic neurons [236]. Entirely unpredicted, calcitriol seems able to delay the effect of Heme oxygenase, reducing the glial fibrillary immunoreactivity [237].

Finally, vitamin D seems to interfere with the activity of Poly (ADP-Ribose) Polymerase-1, also called PARP1. It is a stress protein, which can, however, promote neuronal death. In fact, there is an overexpression of PARP1 in the *substantia nigra* of PD patients [238]; increased levels of vitamin D down regulate the PARP-1 expression, probably mediated by a diminishment of microglial activation [239]. All of these aspects considered in the clinical trials on vitamin D supplementation in Parkinson's disease have been carried out. Although some data demonstrated that low levels of vitamin D might influence the speed of the disease, the supplementation of vitamin D did not modify the disease outcomes. That conclusion probably lies in a clinical time of supplementation, far from the initial ones of labs models.

## 5. Conclusions

Accumulating evidence shows that vitamin D is important for accurate brain functioning. Moreover, it is important to note that many important neurological diseases are related to low levels of vitamin D. These expressions are supported by in vivo, in vitro, and in human trials. Unfortunately, at the moment, the results for its supplementation in neurological disorders, in particular concerning data on vitamin supplementation and their progression or eventual real clinical benefits are inconclusive. As such, there is a strong need for randomized clinical trials on neurodegenerative patients and vitamin D supplementation, to optimize time, efficacy, and appropriate monitoring [240,241].

It should be said that vitamin D supplementation might be safe and inexpensive. However, physicians may be uncomfortable recommending larger doses of vitamin D, because of its potential toxicity. The rarity of reports of vitamin D toxicity can be explained, in part, by the kidney's ability to limit the production of active calcitriol. Increased calcitriol levels inhibit parathormone (PTH), causing the calcitriol production in the kidney to decrease. The renal 24-hydroxylase activity further limits the availability of calcitriol by creating inert metabolites of both calcitriol (1,24,25-trihydroxyvitamin D) and calcidiol (24,25-dihydroxyvitamin D). The 24-hydroxylase gene is under the transcriptional control of calcitriol, thereby providing a tight negative feedback.

Concerning vitamin D replacement, several studies have shown that both vitamin D2 and vitamin D3 are effective in maintaining the serum 25(OH)D levels. Both D2 (ergocalciferol) and D3 (cholecalciferol) are available as dietary supplements, and both appear to be effective. A single dose of 50,000 IU of D2 or D3 produces a similar increase in the total 25(OH)D concentration, but the apparent longer half-life of D3 suggests that less frequent dosing may be needed. To replenish the serum 25(OH)D levels in persons with a vitamin D deficiency, one cost-effective regimen is oral ergocalciferol at 50,000 IU per week for eight weeks. The optimal time for rechecking the serum levels after being replenished has not been clearly defined, but the goal is to achieve a minimum level of 30 ng per mL.

Special mention is needed for patients who require tube feeding or parenteral nutrition. In those cases, ergocalciferol capsules containing D2 in oil, which can clog the feeding tube, should not be used. Cholecalciferol capsules and tablets contain D3 in powder form and can be used.

New approaches are emerging in vitamin D deficiency. The combined treatment using bisphosphonates and vitamin D has been suspected to be more effective compared with bisphosphonate or vitamin D alone in the prevention of atherosclerosis. In a recent study, the combination of bisphosphonates and vitamin D has been seen to be safe and effective in systemic lupus erythematosus. Bisphosphonates are expected to inhibit the arterial plaque development and calcification through several mechanisms (mostly inducing decreased levels of inflammatory cytokines and matrix metalloproteinases). Bisphosphonates also decrease a variety of mature vascular cells, which migrate into the vessel walls and injure vascular endothelial cells. Moreover, treatment with statins in combination with bisphosphonates has been shown to be more effective in terms of reducing atherosclerotic plaque, compared with either monotherapy in patients with hypercholesterolemia [242,243].

Reading what the literature has reported, it seems that the supplementation of vitamin D might reduce the risks and improve the pathological features in different neurological conditions, but, because of different and unknown reasons, it seems that its supplementation might not be sufficient to change the outcomes and the disease phenotype. All of this considered, there is a strong need for further research in this field, even to account for the best quality of vitamin implementation.

**Author Contributions:** R.M. designed the study, supervised the entire project and directly studied neurodegeneration and vitamin D; P.C. revised the manuscript and wrote the vascular paragraphs; M.E.M. wrote the multiple sclerosis paragraph.

**Funding:** This Research received no external funding.

**Acknowledgments:** The authors thank Hellen Davies, for her precious help and assistance.

**Conflicts of Interest:** The authors declare no conflict of interest.

## References

1. Holick, M.F. Vitamin D Deficiency. *N. Engl. J. Med.* **2007**, *357*, 266–281. [CrossRef] [PubMed]
2. Arabi, A.; El Rassi, R.; El-Hajj Fuleihan, G. Hypovitaminosis D in developing countries-prevalence, risk factors and outcomes. *Nat. Rev. Endocrinol.* **2010**, *6*, 550–561. [CrossRef] [PubMed]
3. Prentice, A.; Schoenmakers, I.; Jones, K.; Jarjou, L.A.; Goldberg, G. Vitamin D Deficiency and Its Health Consequences in Africa. *Clin. Rev. Bone Miner. Metab.* **2009**, *7*, 94–106. [CrossRef] [PubMed]
4. Ross, A.C.; Manson, J.E.; Abrams, S.A.; Aloia, J.F.; Brannon, P.M.; Clinton, S.K.; Durazo-Arvizu, R.A.; Gallagher, J.C.; Gallo, R.L.; Jones, G.; et al. The 2011 report on dietary reference intakes for calcium and vitamin D from the Institute of Medicine: What clinicians need to know. *J. Clin. Endocrinol. Metab.* **2011**, *96*, 53–58. [CrossRef] [PubMed]
5. Di Somma, C.; Scarano, E.; Barrea, C.; Zhukouskay, V.V.; Savastano, S.; Mele, C.; Scacchi, M.; Aimaretti, G.; Colao, A.; Marzullo, P. Vitamin D and neurological disease: An endocrine view. *Int. J. Mol. Sci.* **2017**, *18*, 2482. [CrossRef] [PubMed]
6. Yang, L.; Ma, J.; Zhang, X.; Fan, Y.; Wang, L. Protective role of the vitamin D receptor. *Cell. Immunol.* **2012**, *279*, 160–166. [CrossRef] [PubMed]
7. Gominak, S.C. Vitamin D deficiency changes the intestinal microbiome reducing B vitamin production in the gut. The resulting lack of pantothenic acid adversely affects the immune system, producing a "pro-inflammatory" state associated with atherosclerosis and autoimmunity. *Med. Hypotheses* **2016**, *94*, 103–107. [PubMed]
8. Luthold, R.V.; Fernandes, G.R.; Franco-de-Moraes, A.C.; Folchetti, L.G.; Ferreira, S.R. Sandra Roberta G Ferreira. Gut microbiota interactions with the immunomodulatory role of vitamin D in normal individuals. *Metabolism* **2017**, *69*, 76–86. [CrossRef] [PubMed]
9. Tremlett, H.; Bauer, K.C.; Appel-Cresswell, S.; Finlay, B.B.; Waubant, E. The gut microbiome in human neurological disease: A review. *Ann. Neurol.* **2017**, *81*, 369–382. [CrossRef] [PubMed]
10. Holick, M.F.; Chen, T.C. Vitamin D deficiency: A worldwide problem, with health consequences. *Am. J. Clin. Nutr.* **2008**, *87*, 1080S–1086S. [CrossRef] [PubMed]
11. DeLuca, G.C.; Kimball, S.M.; Kolasinski, J.; Ramagopalan, S.V.; Ebers, G.C. Review: The role of vitamin D in nervous system health and disease. *Neuropathol. Appl. Neurobiol.* **2013**, *39*, 458–484. [CrossRef] [PubMed]
12. Wang, L.; Song, Y.; Manson, J.E.; Pilz, S.; März, W.; Michaëlsson, K.; Lundqvist, A.; Jassal, S.K.; Barrett-Connor, E.; Zhang, C.; et al. Circulating 25-hydroxy-vitamin D and risk of cardiovascular disease: A meta-analysis of prospective studies. *Circ. Cardiovasc. Qual. Outcomes* **2012**, *5*, 819–829. [CrossRef] [PubMed]
13. Brondum-Jacobsen, P.; Nordestgaard, B.G.; Schnohr, P.; Benn, M. 25-Hydroxyvitamin D and symptomatic ischemic stroke: An original study and meta-analysis. *Ann. Neurol.* **2013**, *73*, 38–47. [CrossRef] [PubMed]

14. Chowdhury, R.; Stevens, S.; Ward, H.; Chowdhury, S.; Sajjad, A.; Franco, O.H. Circulating vitamin D, calcium and risk of cerebrovascular disease: A systematic review and meta-analysis. *Eur. J. Epidemiol.* **2012**, *27*, 581–591. [CrossRef] [PubMed]
15. Fleet, J.C. Rapid, membrane-initiated actions of 1,25 dihydroxyvitamin D: What are they and what do they mean? *J. Nutr.* **2004**, *134*, 3215–3218. [CrossRef] [PubMed]
16. Veenstra, T.D.; Prufer, K.; Koenigsberger, C.; Brimijoin, S.W.; Grande, J.P.; Kumar, R. 1,25-Dihydroxyvitamin D3 receptors in the central nervous system of the rat embryo. *Brain Res.* **2005**, *804*, 193–205. [CrossRef]
17. Annweiler, C.; Montero-Odasso, M.; Hachinski, V.; Seshadri, S.; Bartha, R.; Beauchet, O. Vitamin D concentration and lateral cerebral ventricle volume in older adults. *Mol. Nutr. Food Res.* **2013**, *57*, 267–276. [CrossRef] [PubMed]
18. Brown, J.I.; Bianco, J.; McGrath, J.; Eyles, D.W. 1,25-Dihydroxyvitamin D3 induces nerve growth factor, promotes neurite outgrowth and inhibits mitosis in embryonic rat hippocampal neurons. *Neurosci. Lett.* **2003**, *343*, 139–143. [CrossRef]
19. Baksi, S.N.; Hughes, M.J. Chronic vitamin D deficiency in the weanling rat alters catecholamine metabolism in the cortex. *Brain Res.* **1982**, *242*, 387–390. [CrossRef]
20. Puchacz, E.; Stumpf, W.E.; Stachowiak, E.K.; Stachowiak, M.K. Vitamin D increases expression of the tyrosine hydroxylase gene in adrenal medullary cells. *Mol. Brain Res.* **1996**, *36*, 193–196. [CrossRef]
21. Garcion, E.; Wion-Barbot, N.; Montero-Menei, C.N.; Berger, F.; Wion, D. New clues about vitamin D functions in the nervous system. *Trends Endocrinol. Metab.* **2002**, *13*, 100–105. [CrossRef]
22. Minasyan, A.; Keisala, T.; Lou, Y.R.; Kalueff, A.V.; Tuohimaa, P. Neophobia, sensory and cognitive functions, and hedonic responses in vitamin D receptor mutant mice. *J. Steroid Biochem. Mol. Biol.* **2007**, *104*, 274–280. [CrossRef] [PubMed]
23. Burne, T.H.J.; Johnston, A.N.B.; McGrath, J.J.; MacKay-Sim, A. Swimming behavior and post-swimming activity in Vitamin D receptor knockout mice. *Brain Res. Bull.* **2006**, *69*, 74–78. [CrossRef] [PubMed]
24. Gezen-Ak, D.; Dursun, E.; Yilmazer, S. The effects of vitamin D receptor silencing on the expression of LVSCC-A1C and LVSCC-A1D and the release of NGF in cortical neurons. *PLoS ONE* **2011**, *6*, e17553. [CrossRef] [PubMed]
25. Eyles, D.W.; Liu, P.Y.; Josh, P.; Cui, X. Intracellular distribution of the of the vitamin D receptor in the brain. A comparison with classical target tissues and redistribution with development. *Neuroscience* **2014**, *268*, 1–9. [CrossRef] [PubMed]
26. Becker, A.; Eyles, D.W.; McGarth, J.J.; Grecksc, G. Transient prenatal vitamin D deficiency is associated with subtle alterations in learning and memory functions in adult rats. *Behav. Brain Res.* **2005**, *161*, 306–312. [CrossRef] [PubMed]
27. Byrne, J.H.; Voogt, M.; Turner, K.M.; Eyles, D.W.; McGrath, J.J.; Burne, T.H. The impact of adult vitamin D deficiency on behavior and brain function in male Sprague-Dawley rats. *PLoS ONE* **2013**, *8*, e71593. [CrossRef] [PubMed]
28. Achiron, A.; Gurevich, M.; Magalashvili, D.; Kishner, I.; Dolev, M.; Mandel, M. Understanding autoimmune mechanisms in multiple sclerosis using gene expression microarrays: Treatment effect and cytokine-related pathways. *Clin. Dev. Immunol.* **2004**, *11*, 299–305. [CrossRef] [PubMed]
29. Bhalla, A.K.; Amento, E.P.; Clemens, T.L.; Holick, M.F.; Krane, S.M. Specific high-affinity receptors for 1,25-dihydroxyvitamin D3 in human peripheral blood mononuclear cells: Presence in monocytes and induction in T lymphocytes following activation. *J. Clin. Endocrinol. Metab.* **1983**, *57*, 1308–1310. [CrossRef] [PubMed]
30. Veldman, C.M.; Cantorna, M.T.; DeLuca, H.F. Expression of 1,25-dihydroxyvitamin D3 receptor in the immune system. *Arch. Biochem. Biophys.* **2000**, *374*, 334–338. [CrossRef] [PubMed]
31. Chun, R.F.; Liu, P.T.; Modlin, R.L.; Adams, J.S.; Hewison, M. Impact of vitamin D on immune function: Lessons learned from genome-wide analysis. *Front. Physiol.* **2014**, *5*, 151. [CrossRef] [PubMed]
32. Gottfried, E.; Rehli, M.; Hahn, J.; Holler, E.; Andreesen, R.; Kreutz, M. Monocyte-derived cells express CYP27A1 and convert vitamin D3 into its active metab-olite. *Biochem. Biophys. Res. Commun.* **2006**, *349*, 209–213. [CrossRef] [PubMed]
33. Mora, J.; Iwata, M.; von Andrian, U. Vitamin effects on the immune system: Vitamins A and D take centre stage. *Nat. Rev. Immunol.* **2008**, *8*, 685–698. [CrossRef] [PubMed]
34. Lips, P. Vitamin D physiology. *Prog. Biophys. Mol. Biol.* **2006**, *92*, 4–8. [CrossRef] [PubMed]

35. Martineau, A.R.; Wilkinson, K.A.; Newton, S.M.; Floto, R.A.; Norman, A.W.; Skolimowska, K.; Davidson, R.N.; Sørensen, O.E.; Kampmann, B.; Griffiths, C.J.; et al. IFN-gamma- and TNF-independent vitamin D-inducible human suppression of mycobacteria: The role of cathelicidin LL-37. *J. Immunol.* **2007**, *178*, 7190–7198. [CrossRef] [PubMed]
36. Penna, G.; Adorini, L. 1 Alpha, 25-dihydroxyvitamin D3 inhibits differentiation, maturation, activation, and survival of dendritic cells leading to impaired alloreactive T cell activation. *J. Immunol.* **2000**, *164*, 2405–2411. [CrossRef] [PubMed]
37. Garcion, E.; Nataf, S.; Berod, A.; Darcy, F.; Brachet, P. 1,25-Dihydroxyvitamin D3 inhibits the expression of inducible nitric oxide synthase in rat central nervous system during experimental allergic encephalomyelitis. *Brain Res. Mol. Brain Res.* **1997**, *45*, 255–267. [CrossRef]
38. Yu, S.; Cantorna, M.T. The vitamin D receptor is required for iNKT cell development. *Proc. Natl. Acad. Sci. USA* **2008**, *105*, 5207–5212. [CrossRef] [PubMed]
39. Van Etten, E.; Mathieu, C. Immunoregulation by 1,25-dihydroxyvitamin D3: Basic concepts. *J. Steroid Biochem. Mol. Biol.* **2005**, *97*, 93–101. [CrossRef] [PubMed]
40. Chen, S.; Sims, G.P.; Chen, X.X.; Gu, Y.Y.; Chen, S.; Lipsky, P.E. Modulatory effects of 1,25-dihydroxyvitamin D3 on human B cell differentiation. *J. Immunol.* **2007**, *179*, 1634–1647. [CrossRef] [PubMed]
41. Boonstra, A.; Barrat, F.J.; Crain, C.; Heath, V.L.; Savelkoul, H.F.; O'Garra, A. 1α,25-Dihydroxyvitamin D3 has a direct effect on naive CD4(+) T cells to enhance the development of Th2 cells. *J. Immunol.* **2001**, *167*, 4974–4980. [CrossRef] [PubMed]
42. Smolders, J.; Thewissen, M.; Peelen, E.; Menheere, P.; Cohen Tervaert, T.W.; Hupperts, R.; Hupperts, R. Vitamin D status is positively correlated with regulatory T cell function in patients with multiple sclerosis. *PLoS ONE* **2009**, *4*, e6635. [CrossRef] [PubMed]
43. Smolders, J.; Menheere, P.; Thewissen, M.; Peelen, E.; Cohen Tervaert, J.W.; Hupperts, R.; Damoiseaux, J. Regulatory T cell function correlates with serum 25-hydroxyvitamin D, but not with 1, 25-dihydroxyvita-min D, parathyroid hormone and calcium levels in patients with relapsing remitting multiple sclerosis. *J. Steroid Biochem. Mol. Biol.* **2010**, *121*, 243–246. [CrossRef] [PubMed]
44. Burton, J.M.; Kimball, S.; Vieth, R.; Bar-Or, A.; Dosch, H.M.; Cheung, R.; Gagne, D.; D'Souza, C.; Ursell, M.; O'Connor, P. A phase I/II dose-escalation trial of vitamin D3 and calcium in multiple sclerosis. *Neurology* **2010**, *74*, 1852–1859. [CrossRef] [PubMed]
45. Mahon, B.D.; Gordon, S.A.; Cruz, J.; Cosman, F.; Cantorna, M.T. Cytokine profile in patients with multiple sclerosis following vitamin D supplementation. *J. Neuroimmunol.* **2003**, *134*, 128–132. [CrossRef]
46. Kotter, M.R.; Li, W.W.; Zhao, C.; Franklin, R.J.M. Myelin impairs CNS remyelination by inhibiting oligodendrocyte precursor cell differentiation. *J. Neurosci.* **2006**, *26*, 328–332. [CrossRef] [PubMed]
47. Newmark, H.L.; Newmark, J. Vitamin D and Parkinson's disease—A hypothesis. *Mov. Disord.* **2007**, *22*, 461–468. [CrossRef] [PubMed]
48. Baas, D.; Prüfer, K.; Ittel, M.E.; Kuchler-Bopp, S.; Labourdette, G.; Sarliève, L.L.; Brachet, P. Rat oligodendrocytes express the vitamin D(3) receptor and respond to 1,25-dihydroxyvitamin D(3). *Glia* **2000**, *31*, 59–68. [CrossRef]
49. Eyles, D.W.; Smith, S.; Kinobe, R.; Hewison, M.; McGrath, J.J. Distribution of the vitamin D receptor and 1-hydroxylase in human brain. *J. Chem. Neuroanat.* **2005**, *29*, 21–30. [CrossRef] [PubMed]
50. Lassmann, H.; Bruck, W.; Lucchinetti, C.; Rodriguez, M. Remyelination in multiple sclerosis. *Mult. Scler.* **1997**, *3*, 33–136. [CrossRef] [PubMed]
51. Patrikios, P.; Stadelmann, C.; Kutzelnigg, A.; Rauschka, H.; Schmidbauer, M.; Laursen, H.; Sorensen, P.S.; Brück, W.; Lucchinetti, C.; Lassmann, H. Remyelination is extensive in a subset of multiple sclerosis patients. *Brain* **2006**, *129*, 3165–3172. [CrossRef] [PubMed]
52. Bruck, W.; Bitsch, A.; Kolenda, H.; Bruck, Y.; Stiefel, M.; Lassmann, H. Inflammatory central nervous system demyelination: Correlation of magnetic resonance imaging findings with lesion pathology. *Neurology* **1997**, *42*, 783–793. [CrossRef] [PubMed]
53. Franklin, R.J.; Ffrench-Constant, C.; Edgar, J.M.; Smith, K.J. Neuroprotection and repair in multiple sclerosis. *Nat. Rev. Neurol.* **2012**, *8*, 624–634. [CrossRef] [PubMed]
54. Ludwin, S.K.; Maitland, M. Long-term remyelination fails to reconstitute normal thickness of central myelin sheaths. *J. Neurol. Sci.* **1984**, *64*, 193–198. [CrossRef]
55. Hanafy, K.A.; Sloane, J.A. Regulation of remyelination in multiple sclerosis. *FEBS Lett.* **2011**, *585*, 3821–3828. [CrossRef] [PubMed]

56. Pluchino, S.; Zanotti, L.; Brini, E.; Ferrari, S.; Martino, G. Regeneration and repair in multiple sclerosis: The role of cell transplantation. *Neurosci. Lett.* **2009**, *456*, 101–106. [CrossRef] [PubMed]
57. Fancy, S.P.; Kotter, M.R.; Harrington, E.P.; Huang, J.K.; Zhao, C.; Rowitch, D.H.; Franklin, R.J. Overcoming remyelination failure in multiple sclerosis and other myelin disorders. *Exp. Neurol.* **2010**, *225*, 18–23. [CrossRef] [PubMed]
58. De la Fuente, A.G.; Errea, O.; van Wijngaarden, P.; Gonzalez, G.A.; Kerninon, C.; Jarjour, A.A.; Lewis, H.J.; Jones, C.A.; Nait-Oumesmar, B.; Zhao, C.; et al. Vitamin D receptor-retinoid X receptor heterodimer signaling regulates oligodendrocyte progenitor cell differentiation. *J. Cell Biol.* **2015**, *211*, 975–985. [CrossRef] [PubMed]
59. Shirazi, H.A.; Rasouli, J.; Ciric, B.; Rostami, A.; Zhang, G.X. 1,25-Dihydroxyvitamin D3 enhances neural stem cell proliferation and oligodendrocyte differentiation. *Exp. Mol. Pathol.* **2015**, *98*, 240–245. [CrossRef] [PubMed]
60. Ferguson, B.; Matyszak, M.K.; Esiri, M.M.; Perry, V.H. Axonal damage in acute multiple sclerosis lesions. *Brain* **1997**, *120*, 393–399. [CrossRef] [PubMed]
61. Hayes, C.E. Vitamin D: A natural inhibitor of multiple sclerosis. *Proc. Nutr. Soc.* **2000**, *59*, 531–535. [CrossRef] [PubMed]
62. Swank, R.; Lerstad, O.; Strøm, A.; Backer, J. Multiple Sclerosis in Rural Norway. *N. Eng. J. Med.* **1952**, *246*, 721–728. [CrossRef]
63. Simpson, S.; Blizzard, L.; Otahal, P.; van der Mei, I.; Taylor, B. Latitude is significantly associated with the prevalence of multiple sclerosis: A meta-analysis. *J. Neurol. Neurosurg. Psychiatry* **2011**, *82*, 1132–1141. [CrossRef] [PubMed]
64. Westlund, K. Distribution and mortality time trend of multiple sclerosis and some other diseases in Norway. *Acta Neurol. Scand.* **1970**, *46*, 455–483. [CrossRef] [PubMed]
65. Baarnhielm, M.; Olsson, T.; Alfredsson, L. Fatty fish intake is associated with decreased occurrence of multiple sclerosis. *Mult. Scler.* **2014**, *20*, 726–732. [CrossRef] [PubMed]
66. Kurtzke, J.; Beebe, G.; Norman, J. Epidemiology of multiple sclerosis in US veterans: III. Migration and the risk of MS. *Neurology* **1985**, *35*, 672. [CrossRef] [PubMed]
67. Ascherio, A.; Munger, K. Environmental risk factors for multiple sclerosis. Part II: Noninfectious factors. *Ann. Neurol.* **2007**, *61*, 504–513. [CrossRef] [PubMed]
68. Ascherio, A.; Munger, K.; Simon, K. Vitamin D and multiple sclerosis. *Lancet Neurol.* **2010**, *9*, 599–612. [CrossRef]
69. Lucas, R.; Ponsonby, A.; Dear, K.; Valery, P.C.; Pender, M.P.; Taylor, B.V.; Kilpatrick, T.J.; Dwyer, T.; Coulthard, A.; Chapman, C.; et al. Sun exposure and vitamin D are independent risk factors for CNS demyelination. *Neurology* **2011**, *76*, 540–548. [CrossRef] [PubMed]
70. Norval, M.; Halliday, G.M. The consequences of UV-induced immunosuppression for human health. *Photochem. Photobiol.* **2011**, *87*, 965–977. [CrossRef] [PubMed]
71. Munger, K.; Zhang, S.; O'Reilly, E.; Hernán, M.A.; Olek, M.J.; Willett, W.C.; Ascherio, A. Vitamin D intake and incidence of multiple sclerosis. *Neurology* **2004**, *62*, 60–65. [CrossRef] [PubMed]
72. Salzer, J.; Hallmans, G.; Nystrom, M.; Stenlund, H.; Wadell, G.; Sundstrom, P. Vitamin D as a protective factor in multiple sclerosis. *Neurology* **2012**, *79*, 2140–2145. [CrossRef] [PubMed]
73. Dong, Y.; Pollock, N.; Stallmann-Jorgensen, I.S.; Gutin, B.; Lan, L.; Chen, T.C.; Keeton, D.; Petty, K.; Holick, M.F.; Zhu, H. Low 25-hydroxyvitamin D levels in adolescents: Race, season, adiposity, physical activity, and fitness. *Pediatrics* **2010**, *125*, 1104–1111. [CrossRef] [PubMed]
74. Bischof, M.G.; Heinze, G.; Vierhapper, H. Vitamin D status and its relation to age and body mass index. *Hormone Res.* **2006**, *66*, 211–215. [CrossRef] [PubMed]
75. Munger, K.L.; Chitnis, T.; Ascherio, A. Body size and risk of MS in two cohorts of US women. *Neurology* **2009**, *73*, 1543–1550. [CrossRef] [PubMed]
76. Handel, A.E.; Giovannoni, G.; Ebers, G.C.; Ramagopalan, S.V. Environmental factors and their timing in adult-onset multiple sclerosis. *Nat. Rev. Neurol.* **2010**, *6*, 156–166. [CrossRef] [PubMed]
77. Mirzaei, F.; Michels, K.B.; Munger, K.; O'Reilly, E.; Chitnis, T.; Forman, M.R.; Giovannucci, E.; Rosner, B.; Ascherio, A. Gestational vitamin D and the risk of multiple sclerosis in offspring. *Ann. Neurol.* **2011**, *70*, 30–40. [CrossRef] [PubMed]

78. Munger, K.L.; Chitnis, T.; Frazier, A.L.; Giovannucci, E.; Spiegelman, D.; Ascherio, A. Dietary intake of vitamin D during adolescence and risk of multiple sclerosis. *J. Neurol.* **2011**, *258*, 479–485. [CrossRef] [PubMed]
79. Dobson, R.; Giovannoni, G.; Ramagopalan, S. The month of birth effect in multiple sclerosis: Systematic review, meta-analysis and effect of latitude. *J. Neurol. Neurosurg. Psychiatry* **2013**, *84*, 427–432. [CrossRef] [PubMed]
80. Munger, K.; Aivo, J.; Hongell, K.; Soilu-Hänninen, M.; Surcel, H.; Ascherio, A. Vitamin D status during pregnancy and risk of multiple sclerosis in offspring of women in the Finnish maternity cohort. *JAMA Neurol.* **2016**, *73*, 515. [CrossRef] [PubMed]
81. Nielsen, N.; Munger, K.; Stenager, E.; Ascherio, A. Author response: Neonatal vitamin D status and risk of multiple sclerosis: A population-based case-control study. *Neurology* **2017**, *89*, 411. [CrossRef] [PubMed]
82. Van der Mei, I.A.F.; Ponsonby, A.-L.; Dwyer, T.; Blizzard, L.; Simmons, R.; Taylor, B.V.; Butzkueven, H.; Kilpatrick, T. Past exposure to sun, skin phenotype, and risk of multiple sclerosis: Case-control study. *BMJ* **2003**, *327*, 316. [CrossRef] [PubMed]
83. Alter, M.; Kahana, E.; Loewenson, R. Migration and risk of multiple sclerosis. *Neurology* **1978**, *28*, 1089–1093. [CrossRef] [PubMed]
84. Banwell, B.; Bar-Or, A.; Arnold, D.L.; Sadovnick, D.; Narayanan, S.; McGowan, M.; O'Mahony, J.; Magalhaes, S.; Hanwell, H.; Vieth, R.; et al. Clinical, environmental, and genetic determinants of multiple sclerosis in children with acute demyelination: A prospective national cohort study. *Lancet Neurol.* **2011**, *10*, 436–445. [CrossRef]
85. Hanwell, H.; Vieth, R.; Magalhaes, S.; McGowan, M.; Marrie, R.A.; Bar-Or, A.; Sadovnick, A.D.; Arnold, D.; Banwell, B. Vitamin D status as a predictor of MS outcome following an initial paediatric demyelinating event. *Mult. Scler.* **2009**, *15*, S40–S41.
86. Orton, S.M.; Herrera, B.M.; Yee, I.M.; Valdar, W.; Ramagopalan, S.V.; Sadovnick, A.D.; Ebers, G.C. Sex ratio of multiple sclerosis in Canada: A longitudinal study. *Lancet Neurol.* **2006**, *5*, 932–936. [CrossRef]
87. Islam, T.; Gauderman, W.J.; Cozen, W.; Hamilton, A.S.; Burnett, M.E.; Mack, T.M. Differential twin concordance for multiple sclerosis by latitude of birthplace. *Ann. Neurol.* **2006**, *60*, 56–64. [CrossRef] [PubMed]
88. Woolmore, J.A.; Stone, M.; Pye, E.M.; Partridge, J.M.; Boggild, M.; Young, C.; Jones, P.W.; Fryer, A.A.; Hawkins, C.P.; Strange, R.C. Studies of associations between disability in multiple sclerosis, skin type, gender and ultraviolet radiation. *Mult. Scler.* **2007**, *13*, 369–375. [CrossRef] [PubMed]
89. Correale, J.; Ysrraelit, M.C.; Gaitan, M.I. Oestrogens and 1,25 dihydroxyvitamin D3 show synergistic effect on multiple sclerosis immunomodulation. *Mult. Scler.* **2009**, *15*, S196.
90. Eikelenboom, M.J.; Killestein, J.; Kragt, J.J.; Uitdehaag, B.M.; Polman, C.H. Gender differences in multiple sclerosis: Cytokines and vitamin D. *J. Neurol. Sci.* **2009**, *286*, 40–42. [CrossRef] [PubMed]
91. Nashold, F.E.; Spach, K.M.; Spanier, J.A.; Hayes, C.E. Estrogen controls vitamin D3-mediated resistance to experimental autoimmune encephalomyelitis by controlling vitamin D3 metabolism and receptor expression. *J. Immunol.* **2009**, *183*, 3672–3681. [CrossRef] [PubMed]
92. Agnello, L.; Scazzone, C.; Ragonese, P.; Salemi, G.; Lo Sasso, B.; Schillaci, R.; Musso, G.; Bellia, C.; Ciaccio, M. Vitamin D receptor polymorphisms and 25-hydroxyvitamin D in a group of Sicilian multiple sclerosis patients. *Neurol. Sci.* **2016**, *37*, 261–267. [CrossRef] [PubMed]
93. Ahn, J.; Yu, K.; Stolzenberg-Solomon, R.; Simon, K.C.; McCullough, M.L.; Gallicchio, L.; Jacobs, E.J.; Ascherio, A.; Helzlsouer, K.; Jacobs, K.B.; et al. Genome-wide association study of circulating vitamin D levels. *Hum. Mol. Genet.* **2010**, *2739–2745*, 2739–2745. [CrossRef] [PubMed]
94. Mokry, L.E.; Ross, S.; Ahmad, O.S.; Forgetta, V.; Smith, G.D.; Goltzman, D.; Leong, A.; Greenwood, C.M.; Thanassoulis, G.; Richards, J.B. Vitamin D and Risk of Multiple Sclerosis: A Mendelian Randomization Study. *PLoS Med.* **2015**, *12*, e1001866. [CrossRef] [PubMed]
95. Scazzone, C.; Agnello, L.; Ragonese, P.; Lo Sasso, B.; Bellia, C.; Bivona, G.; Schillaci, R.; Salemi, G.; Ciaccio, M. Association of CYP2R1 rs10766197 with MS risk and disease progression. *J. Neurosci. Res.* **2018**, *96*, 297–304. [CrossRef] [PubMed]
96. Pihl-Jensen, G.; Frederiksen, J. 25-Hydroxyvitamin D levels in acute monosymptomatic optic neuritis: Relation to clinical severity, paraclinical findings and risk of multiple sclerosis. *J. Neurol.* **2015**, *262*, 1646–1654. [CrossRef] [PubMed]

97. Mowry, E.; Krupp, L.; Milazzo, M.; Chabas, D.; Strober, J.B.; Belman, A.L.; McDonald, J.C.; Oksenberg, J.R.; Bacchetti, P.; Waubant, E. Vitamin D status is associated with relapse rate in pediatric-onset, M.S. *Ann. Neurol.* **2010**, *67*, 618–624. [PubMed]
98. Simpson, S.; Taylor, B.; Blizzard, L.; Ponsonby, A.L.; Pittas, F.; Tremlett, H.; Dwyer, T.; Gies, P.; van der Mei, I. Higher 25-hydroxyvitamin D is associated with lower relapse risk in, M.S. *Ann. Neurol.* **2010**, *68*, 193–203. [PubMed]
99. Runia, T.; Hop, W.; de Rijke, Y.; Buljevac, D.; Hintzen, R. Lower serum vitamin D levels are associated with a higher relapse risk in multiple sclerosis. *Neurology* **2012**, *79*, 261–266. [CrossRef] [PubMed]
100. Martinelli, V.; Dalla Costa, G.; Colombo, B.; Dalla Libera, D.; Rubinacci, A.; Filippi, M.; Furlan, R.; Comi, G. Vitamin D levels and risk of multiple sclerosis in patients with clinically isolated syndromes. *Mult. Scler. J.* **2013**, *20*, 147–155. [CrossRef] [PubMed]
101. Kuhle, J.; Disanto, G.; Dobson, R.; Adiutori, R.; Bianchi, L.; Topping, J.; Bestwick, J.P.; Meier, U.C.; Marta, M.; Dalla Costa, G.; et al. Conversion from clinically isolated syndrome to multiple sclerosis: A large multicentre study. *Mult. Scler.* **2015**, *21*, 1013–1024. [CrossRef] [PubMed]
102. Embry, A.F.; Snowdon, L.R.; Vieth, R. Vitamin D and seasonal fluctuations of gadolinium-enhancing magnetic resonance imaging lesions in multiple sclerosis. *Ann. Neurol.* **2000**, *48*, 271–272. [CrossRef]
103. Mowry, E.; Waubant, E.; McCulloch, C.; Okuda, D.T.; Evangelista, A.A.; Lincoln, R.R.; Gourraud, P.A.; Brenneman, D.; Owen, M.C.; Qualley, P.; et al. Vitamin D status predicts new brain magnetic resonance imaging activity in multiple sclerosis. *Ann Neurol.* **2012**, *72*, 234–240. [CrossRef] [PubMed]
104. Ascherio, A.; Munger, K.; White, R.; Köchert, K.; Simon, K.C.; Polman, C.H.; Freedman, M.S.; Hartung, H.P.; Miller, D.H.; Montalbán, X.; et al. Vitamin D as an early predictor of multiple sclerosis activity and progression. *JAMA Neurol.* **2014**, *71*, 306–314. [CrossRef] [PubMed]
105. Fitzgerald, K.; Munger, K.; Ko¨chert, K.; Arnason, B.G.; Comi, G.; Cook, S.; Goodin, D.S.; Filippi, M.; Hartung, H.P.; Jeffery, D.R.; et al. Association of Vitamin D Levels With multiple sclerosis activity and progression in patients receiving interferon Beta-1b. *JAMA Neurol.* **2015**, *72*, 1458. [CrossRef] [PubMed]
106. Stewart, N.; Simpson, S.; van der Mei, I.; Ponsonby, A.L.; Blizzard, L.; Dwyer, T.; Pittas, F.; Eyles, D.; Ko, P.; Taylor, B.V. Interferon-and serum 25-hydroxyvitamin D interact to modulate relapse risk in, M.S. *Neurology* **2012**, *79*, 254–260. [CrossRef] [PubMed]
107. Laursen, J.; Søndergaard, H.; Sørensen, P.; Sellebjerg, F.; Oturai, A. Vitamin D supplementation reduces relapse rate in relapsing-remitting multiple sclerosis patients treated with natalizumab. *Mult. Scler. Relat. Dis.* **2016**, *10*, 169–173. [CrossRef] [PubMed]
108. Mowry, E.; Pelletier, D.; Gao, Z.; Howell, M.; Zamvil, S.; Waubant, E. Vitamin D in clinically isolated syndrome: Evidence for possible neuroprotection. *Eur. J. Neurol.* **2015**, *23*, 327–332. [CrossRef] [PubMed]
109. Auer, D.P.; Schumann, E.M.; Kumpfel, T.; Gossl, C.; Trenkwalder, C. Seasonal fluctuations of gadolinium-enhancing magnetic resonance resonance imaging lesions in multiple sclerosis. *Ann. Neurol.* **2000**, *47*, 276–277. [CrossRef]
110. Rovaris, M.; Comi, G.; Sormani, M.P.; Wolinsky, J.S.; Ladkani, D.; Filippi, M. Effects of seasons on magnetic resonance imagingmeasured disease activity in patients with multiple sclerosis. *Ann. Neurol.* **2001**, *49*, 415–416. [CrossRef] [PubMed]
111. Killestein, J.; Rep, M.H.; Meilof, J.F.; Adèr, H.J.; Uitdehaag, B.M.; Barkhof, F.; van Lier, R.A.; Polman, C.H. Seasonal variation in immune measurements and MRI markers of disease activity in, M.S. *Neurology* **2002**, *58*, 1077–1080. [CrossRef] [PubMed]
112. Smolders, J.; Menheere, P.; Kessels, A.; Damoiseaux, J.; Hupperts, R. Association of vitamin D metabolite levels with relapse rate and disability in multiple sclerosis. *Mult. Scler. J.* **2008**, *14*, 1220–1224. [CrossRef] [PubMed]
113. Weinstock-Guttman, B.; Zivadinov, R.; Qu, J.; Cookfair, D.; Bang, E.; Bergsland, N.; Hussein, S.; Cherneva, M.; Willis, L.; et al. Vitamin D metabolites are associated with clinical and MRI outcomes in multiple sclerosis patients. *J. Neurol. Neurosurg. Psychiatry* **2011**, *82*, 189–195. [CrossRef] [PubMed]
114. Koven, N.; Cadden, M.; Murali, S.; Ross, M. Vitamin D and long-term memory in multiple sclerosis. *Cognit. Behav. Neurol.* **2013**, *26*, 155–160. [CrossRef] [PubMed]
115. Kimball, S.M.; Ursell, M.R.; O'Connor, P.; Vieth, R. Safety of vitamin D3 in adults with multiple sclerosis. *Am. J. Clin. Nutr.* **2007**, *86*, 645–651. [CrossRef] [PubMed]

116. Soilu-Hanninen, M.; Aivo, J.; Lindstro¨m, B.; Elovaara, I.; Sumelahti, M.L.; Färkkilä, M.; Tienari, P.; Atula, S.; Sarasoja, T.; Herrala, L.; et al. A randomised, double blind, placebo controlled trial with vitamin D3as an add on treatment to interferon b-1b in patients with multiple sclerosis. *J. Neurol. Neurosurg. Psychiatry* **2012**, *83*, 565–571. [CrossRef] [PubMed]
117. Etemadifar, M.; Janghorbani, M. Efficacy of high-dose vitamin D3 supplementation in vitamin D deficient pregnant women with multiple sclerosis: Preliminary findings of a randomized-controlled trial. *Iran. J. Neurol.* **2015**, *14*, 67–73. [PubMed]
118. Smolders, J.; Hupperts, R.M.M.; Vieth, R.; Grimaldi, L.M.; Holmoy, T.; Killestein, J.; Rieckmann, P.; Schluep, M.; Vieth, R.; Hostalek, U.; et al. High dose cholecalciferol (vitamin D3) oil as add-on therapy in subjects with relapsing-remitting multiple sclerosis (RRMS) receiving subcutaneous interferon s-1a (scIFNs-1a). *ECTRIMS Online Libr.* **2016**, *16*, 147013.
119. Camu, W. Cholecalciferol supplementation in relapsing multiple sclerosis patients treated with subcutaneous interferon beta-1a: A randomized controlled trial. *Neurology* **2017**, *88* (Suppl. 16), S44.004.
120. Darwish, H.; Haddad, R.; Osman, S.; Ghassan, S.; Yamout, B.; Tamim, H.; Khoury, S. Effect of vitamin D replacement on cognition in multiple sclerosis patients. *Sci. Rep.* **2017**, *7*, 45926. [CrossRef] [PubMed]
121. Solomon, A.J.; Whitham, R.H. Multiple Sclerosis and Vitamin D: A Review and Recommendations. *Curr. Neurol. Neurosci. Rep.* **2010**, *10*, 389–396. [CrossRef] [PubMed]
122. Shoemaker, T.J.; Mowry, E.M. A review of vitamin D supplementation as disease-modifying therapy. *Mult. Scler.* **2018**, *24*, 6–11. [CrossRef] [PubMed]
123. Mozaffarian, D.; Benjamin, E.J.; Go, A.S.; Arnett, D.K.; Blaha, M.J.; Cushman, M.; Das, S.R.; de Ferranti, S.; Després, J.P.; Fullerton, H.J.; et al. Heart Disease and Stroke Statistics-2016 Update: A Report from the American Heart Association. *Circulation* **2016**, *133*, e38–e360. [CrossRef] [PubMed]
124. Sadovnick, A.D. Genetic background of multiple sclerosis. *Autoimmun. Rev.* **2012**, *11*, 163–166. [CrossRef] [PubMed]
125. Daubail, B.; Jacquin, A.; Guilland, J.-C.; Hervieu, M.; Osseby, G.-V.; Rouaud, O.; Giroud, M.; Béjot, Y. Serum 25-hydroxyvitamin D predicts severity and prognosis in stroke patients. *Eur. J. Neurol.* **2013**, *20*, 57–61. [CrossRef] [PubMed]
126. Judd, S.E.; Morgan, C.J.; Panwar, B.; Howard, V.J.; Wadley, V.G.; Jenny, N.S.; Kissela, B.M.; Gutiérrez, O.M. Vitamin D deficiency and incident stroke risk in community-living black and white adults. *Int. J. Stroke* **2016**, *11*, 93–102. [CrossRef] [PubMed]
127. Vianello, E.; Dozio, E.; Barassi, A.; Tacchini, L.; Lamont, J.; Trimarchi, S.; Marrocco-Trischitta, M.M.; Romanelli, M.C. Vitamin D Deficiency Is Associated with Increased Osteocalcin Levels in Acute Aortic Dissection: A Pilot Study on Elderly Patients. *Hindawi Mediat. Inflamm.* **2017**, 6412531. [CrossRef] [PubMed]
128. Kato, S. The function of vitamin D receptor in vitamin D action. *J. Biochem.* **2000**, *127*, 717–722. [CrossRef] [PubMed]
129. Borges, A.C.; Feres, T.; Vianna, L.M.; Therezinha, B.P. Effect of cholecalciferol treatment on the relaxant responses of spontaneously hypertensive rat arteries to acetylcholine. *Hypertension* **1999**, *34*, 897–901. [CrossRef] [PubMed]
130. Sanket, S.; Madireddi, J.; Stanley, W.; Sura, P.; Prabhu, M. Relation between Vitamin D Deficiency and Severity of Chronic Obstructive Pulmonary Disease-A Case Control Study. *J. Clin. Diagn. Res.* **2016**, *10*, OC16–OC19. [CrossRef] [PubMed]
131. Manouchehri, N.; Vakil-Asadollahi, M.; Zandifar, A.; Rasmani, F.; Saadatnia, M. Vitamin D Status in Small Vessel and Large Vessel Ischemic Stroke Patients: A Case–control Study. *Adv. Biomed. Res.* **2017**, *6*, 146. [PubMed]
132. Qiu, H.; Wang, M.; Mi, D.; Zhao, J.; Tu, W.; Liu, Q. Vitamin D Status and the Risk of Recurrent Stroke and Mortality in Ischemic Stroke Patients: Data from a 24-Month Follow-Up Study in China. *J. Nutr. Health Aging* **2017**, *21*, 766–771. [CrossRef] [PubMed]
133. Daubail, B.; Jacquin, A.; Guilland, J.-C.; Khoumri, C.; Aboa-Eboulé, C.; Giroud, M.; Béjot, Y. Association between serum concentration of vitamin D and 1-year mortality in stroke patients. *Cerebrovasc. Dis.* **2014**, *37*, 364–367. [CrossRef] [PubMed]
134. Alfieri, D.F.; Lehmann, M.F.; Oliveira, S.R.; Flauzino, T.; Delongui, F.; de Araújo, M.C.; Dichi, I.; Delfino, V.D.; Mezzaroba, L.; Simão, A.N.; et al. Vitamin D deficiency is associated with acute ischemic stroke, C-reactive protein, and short-term outcome. *Metab. Brain Dis.* **2017**, *32*, 493–502. [CrossRef] [PubMed]

135. Kim, K.; Cho, K.H.; Im, S.H.; Choi, J.; Yu, J.; Kim, M. Decrement of Serum Vitamin D Level After Stroke. *Ann. Rehabil. Med.* **2017**, *41*, 944–950. [CrossRef] [PubMed]
136. Artaza, J.N.; Mehrotra, R.; Norris, K.C. Vitamin D and the cardiovascular system. *Clin. J. Am. Soc. Nephrol.* **2009**, *4*, 1515–1522. [CrossRef] [PubMed]
137. Cannell, J.J.; Grant, W.B.; Holick, M.F. Vitamin D and inflammation. *Dermatoendocrinology* **2014**, *6*, e983401. [CrossRef] [PubMed]
138. Cohen-Lahav, M.; Shany, S.; Tobvin, D.; Chaimovitz, C.; Douvdevani, A. Vitamin D decreases NFkappaB activity by increasing IkappaBalpha levels. *Nephrol. Dial. Transplant.* **2006**, *21*, 889–897. [CrossRef] [PubMed]
139. Liu, X.; Nelson, A.; Wang, X.; Farid, M.; Gunji, Y.; Ikari, J.; Iwasawa, S.; Basma, H.; Feghali-Bostwick, C.; Rennard, S.I. Vitamin D modulates prostaglandin E2 synthesis and degradation in human lung fibroblasts. *Am. J. Respir. Cell Mol. Biol.* **2014**, *50*, 40–50. [PubMed]
140. Penna, G.; Roncari, A.; Amuchastegui, S.; Daniel, K.C.; Berti, E.; Colonna, M.; Adorini, L. Expression of the inhibitory receptor ILT3 on dendritic cells is dispensable for induction of CD4 + Foxp3 + regulatory T cells by 1,25-dihydroxyvitamin D3. *Blood* **2005**, *106*, 3490–3497. [CrossRef] [PubMed]
141. Tarcin, O.; Yavuz, D.G.; Ozben, B.; Telli, A.; Ogunc, A.V.; Yuksel, M.; Toprak, A.; Yazici, D.; Sancak, S.; Deyneli, O.; et al. Effect of Vitamin D Deficiency and Replacement on Endothelial Function in Asymptomatic Subjects. *J. Clin. Endocrinol. Metab.* **2009**, *94*, 4023–4030. [CrossRef] [PubMed]
142. Marniemi, J.; Alanen, E.; Impivaara, O.; Seppänen, R.; Hakala, P.; Rajala, T.; Rönnemaa, T. Dietary and serum vitamins and minerals as predictors of myocardial infarction and stroke in elderly subjects. *Nutr. Metab. Cardiovasc. Dis.* **2005**, *15*, 188–197. [CrossRef] [PubMed]
143. Sun, Q.; Pan, A.; Hu, F.B.; Manson, J.E.; Rexrode, K.M. 25-Hydroxyvitamin D Levels and the Risk of Stroke: A Prospective Study and Meta-analysis. *Stroke* **2012**, *43*, 1470–1477. [CrossRef] [PubMed]
144. Cirillo, P.; Golino, P.; Calabro, P.; Calì, G.; Ragni, M.; De Rosa, S.; Cimmino, G.; Pacileo, M.; De Palma, R.; Forte, L.; et al. C-reactive protein induces tissue factor expression and promotes smooth muscle and endothelial cell proliferation. *Cardiovasc. Res.* **2005**, *68*, 47–55. [CrossRef] [PubMed]
145. Chen, C.; Nan, B.; Lin, P.; Yao, Q. C-reactive protein increases plasminogen activator inhibitor-1 expression in human endothelial cells. *Thromb. Res.* **2008**, *122*, 125–133. [CrossRef] [PubMed]
146. Hattori, Y.; Matsumura, K.K.; Kasai, K. Vascular smooth muscle cell activation by C-reactive protein. *Cardiovasc. Res.* **2003**, *58*, 186–195. [CrossRef]
147. Pilz, S.; Tomarchitz, A.; Drechsler, C.; Zittermann, A.; Dekker, J.M.; März, W. Vitamin [8]D supplementation: A promising approach for the prevention and treatment of strokes. *Curr. Drug Targets* **2011**, *12*, 88–96. [CrossRef] [PubMed]
148. Park, K.Y.; Chung, P.W.; Kim, Y.B.; Moon, H.S.; Suh, B.C.; Won, Y.S.; Kim, J.M.; Youn, Y.C.; Kwon, O.S. Serum vitamin [9]D status as a predictor of prognosis in patients with acute ischemic stroke. *Cerebrovasc. Dis.* **2015**, *40*, 73–80. [CrossRef] [PubMed]
149. Ladeby, R.; Wirenfeldt, M.; Garcia-Ovejero, D.; Fenger, C.; Dissing-Olesen, L.; Dalmau, I.; Finsen, B. Microglial cell population dynamics in the injured adult central nervous system. *Brain Res. Brain Res. Rev.* **2005**, *48*, 196–206. [CrossRef] [PubMed]
150. Wei, Z.; Kuang, J. Research Article: Vitamin D deficiency in relation to the poor functional outcomes in nondiabetic patients with ischemic stroke. *Biosci. Rep.* **2018**, *38*, BSR20171509. [CrossRef] [PubMed]
151. Bjelakovic, G.; Gluud, L.L.; Nikolova, D.; Whitfield, K.; Wetterslev, J.; Simonetti, R.G.; Bjelakovic, M.; Gluud, C. Vitamin D supplementation for prevention of mortality in adults. *Cochrane Database Syst. Rev.* **2011**, *7*, CD007470.
152. Gupta, A.; Prabhakar, S.; Modi, M.; Bhadada, S.K.; Kalaivani, M.; Lal, V.; Khurana, D. Effect of Vitamin D and calcium supplementation on ischaemic stroke outcome: A randomised controlled open-label trial. *Int. J. Clin. Pract.* **2016**, *70*, 764–770. [CrossRef] [PubMed]
153. Manson, J.E. Vitamin D and the heart: Why we need large-scale clinical trials. *Cleve Clin. J. Med.* **2010**, *77*, 903–910. [CrossRef] [PubMed]
154. LeBoff, M.S.; Yue, A.Y.; Copeland, T.; Cook, N.R.; Buring, J.E.; Manson, J.E. VITAL-Bone Health: Rationale and design of two ancillary studies evaluating the effects of vitamin D and/or omega-3 fatty acid supplements on incident fractures and bone health outcomes in the VITamin D and OmegA-3 TriaL (VITAL). *Contemp. Clin. Trials* **2015**, *41*, 259–268. [CrossRef] [PubMed]

155. Shaffer, J.A.; Edmondson, D.; Wasson, L.T.; Falzon, L.; Homma, K.; Ezeokoli, N.; Li, P.; Davidson, K.W. Vitamin D supplementation for depressive symptoms: A systematic review and meta-analysis of randomized controlled trials. *Psychosom. Med.* **2014**, *76*, 190–196. [CrossRef] [PubMed]
156. Karran, E.; Mercken, M.; Strooper, B.D. The amyloid cascade hypothesis for Alzheimer's disease: An appraisal for the development of therapeutics. *Nat. Rev. Drug Discov.* **2011**, *10*, 698–712. [CrossRef] [PubMed]
157. Roher, A.E.; Lowenson, J.D.; Clarke, S.; Woods, A.S.; Cotter, R.J.; Gowing, E.; Ball, M.J. Beta amyloid (1–42) is a major component of cerebrovascular amyloid deposits: Implications for the pathology of Alzheimer's disease. *Proc. Natl. Acad. Sci. USA* **1993**, *90*, 10836–10840. [CrossRef] [PubMed]
158. Choi, S.H.; Kim, Y.H.; Hebisch, M.; Sliwinski, C.; Lee, S.; D'Avanzo, C.; Chen, H.; Hooli, B.; Asselin, C.; Muffat, J.; et al. A three dimensional human neural cell culture model of Alzheimer's disease. *Nature* **2014**, *515*, 274–278. [CrossRef] [PubMed]
159. Lahiri, D.K.; Maloney, B. The LEARn (Latent early life associated regulation) model integrates environmental risk factors and the developmental basis of Alzheimer's disease, and proposes remedial steps. *Exp. Gerontol.* **2010**, *45*, 291–296. [CrossRef] [PubMed]
160. Banerjee, A.; Khemka, V.K.; Gaungly, A.; Roy, D.; Gauguly, U.; Chakrabarti, S. Vitamin D and Alzheimer's Disease: Neurocognition to therapeutics. *Int. J. Alzheimers Dis.* **2015**, *2015*. [CrossRef] [PubMed]
161. Gezen, A.K.D.; Dursun, B.; Bilgic, B.; Hanağasi, H.; Ertan, T.; Gürvit, H.; Emre, M.; Eker, E.; Ulutin, T.; Uysal, O.; et al. Vitamin D receptor gene haplotype is associated with late-onset Alzheimer's Disease. *Tohoku J. Exp. Med.* **2012**, *228*, 189–196. [CrossRef]
162. Dursun, E.; Gezen Ak, D.; Ylmazer, S. A new mechanism for amyloid Beta induction of iNOS: Vitamin D-VDR pathwy disruption. *J. Alzheimers Dis.* **2013**, *36*, 459–474. [CrossRef] [PubMed]
163. Brewer, L.D.; Thibault, V.; Chen, K.C.; Langub, M.C.; Landfield, P.W.; Porter, N.M. Vitamin D hormone confers neuroprotection in parallel with downregulation of L-type calcium channel expression in hippocampal neurons. *J. Neurosci.* **2001**, *21*, 98–108. [CrossRef] [PubMed]
164. Thibault, O.; Landfield, P.W. Increase in single L-type calcium channels in hippocampal neurons during aging. *Science* **1996**, *272*, 1017–1020. [CrossRef] [PubMed]
165. Porter, N.M.; Thibault, O.; Thibault, V.; Chen, K.C.; Landfield, P.W. Calcium-channel density and hippocampal cell death with age in long-term culture. *J. Neurosci.* **1997**, *17*, 5629–5639. [CrossRef] [PubMed]
166. Gezen-Ak, D.; Dursun, E.; Yilmazer, S. Vitamin D inquiry in hippocampal neurons: Consequences of vitamin D-VDR pathway disruption on calcium channel and the vitamin D requirement. *Neurol. Sci.* **2013**, *34*, 1453–1458. [CrossRef] [PubMed]
167. Wang, L.; Har, K.; Van Baaren, J.M.; Price, J.C.; Beecham, G.W.; Gallins, P.J.; Whitehead, P.L.; Wang, G.; Lu, C.; Slifer, M.A.; et al. Vitamin D receptor and AD: A genetic and functional study. *Neurobiol. Aging* **2012**, *33*, 1844.e1–1844.e9. [CrossRef] [PubMed]
168. Kuningans, M.; Mooijart, S.P.; Jolles, J.; Slagboom, P.E.; Westendorp, R.G.; van Heemst, D. VDR genes variants associate with cognitive functions and depressive symptom in old age. *Neurobiol. Aging* **2009**, *30*, 466–473. [CrossRef] [PubMed]
169. Buell, J.S.; Dawson-Hughes, B.; Scott, T.M.; Weiner, D.E.; Dallal, G.E.; Qui, W.Q.; Bergethon, P.; Rosenberg, I.H.; Folstein, M.F.; Patz, S.; et al. 25-Hydroxyvitamin D, dementia and cerebrovascular pathology in elders receiving home services. *Neurology* **2010**, *74*, 18–26. [CrossRef] [PubMed]
170. Grimm, M.O.W.; Lehmann, J.; Mett, J.; Zimmer, V.C.; Grösgen, S.; Stahlmann, C.P.; Hundsdörfer, B.; Haupenthal, V.J.; Rothhaar, T.L.; Herr, C.; et al. Impact of vitamin D on amyloid precursor protein processing and amyloid beta peptide degradation in Alzheimer's Disease. *Neurodegen. Dis.* **2014**, *13*, 75–81. [CrossRef] [PubMed]
171. Miziwicki, M.T.; Menegaz, D.; Zhang, J.; Barrientos-Durán, A.; Tse, S.; Cashman, J.R.; Griffin, P.R.; Fiala, M. Genomic and nongenomic signaling induced by 1 alpha 25 OH2 vitamin D3 promotes the recovery of amyloid phagocytosis by Alzheimer's disease macrophages. *J. Alzheimers Dis.* **2012**, *29*, 51–62. [CrossRef] [PubMed]
172. Taniura, H.; Ito, M.; Sanada, N.; Kuramoto, N.; Ohno, Y.; Nakamichi, N.; Yoneda, Y. Chronic vitamin d3 treatment protects against neurotoxicity by glutamate in association with upregulation of vitamin D receptor mRNA expression in cultured rat cortical neurons. *J. Neurosci. Res.* **2006**, *83*, 1179–1189. [CrossRef] [PubMed]

173. Garcion, E.; Sindij, L.; Leblondel, G.; Brachet, P.; Darcy, F. 1–25 dihydroxyvitamin D3 regulates the synthesis of gamma glutamyl transpeptidase and gluthatione levels in rat primary astrocytes. *J. Neurochem.* **1999**, *73*, 859–866. [CrossRef] [PubMed]
174. Briones, T.L.; Darwish, H. Decrease in age related tau hyperphosphorylation and cognitive improvement following vitamin D supplementation are associated with modulation of brain energy metabolism and redox state. *Neuroscience* **2014**, *262*, 143–155. [CrossRef] [PubMed]
175. Thomas, M.K.; Lloyd-Jones, D.M.; Thadani, R.I.; Shaw, A.C.; Deraska, D.J.; Kitch, B.T.; Vamvakas, E.C.; Dick, I.M.; Prince, R.L.; Finkelstein, J.S. Hypovitaminosis D in medical inpatients. *N. Engl. J. Med.* **1998**, *338*, 777–783. [CrossRef] [PubMed]
176. Gloth, F.M.; Gundberg, C.M.; Hollis, B.W.; Haddad, J.G.; Tobin, J.D. Vitamin D deficiency in homebound elderly persons. *JAMA* **1995**, *274*, 1683–1686. [CrossRef] [PubMed]
177. Moon, J.H.; Lim, S.; Han, J.W.; Kim, K.M.; Choi, S.H.; Kim, K.W.; Jang, H.C. Serum 25-hydroxyvitamin D level and the risk of mild cognitive impairment and dementia: The Korean Longitudinal Study on Health and Aging (KloSHA). *Clin. Endocrinol.* **2015**, *83*, 36–42. [CrossRef] [PubMed]
178. Khachaturian, Z.S. The role of calcium regulation in brain aging: Reexamination of a hypothesis. *Aging* **1989**, *1*, 17–34. [CrossRef] [PubMed]
179. Annweiler, C.; Montero-Odasso, M.; Llewellyn, D.J.; Richard Devantoy, S.; Duque, G.; Beauchet, O. Meta-analysis of memory and executive dysfunctions in relation to vitamin D. *J. Alzheimers Dis.* **2013**, *37*, 147–171. [CrossRef] [PubMed]
180. Annweiler, C.; Rolland, Y.; Schott, A.M.; Blain, H.; Vellas, B.; Herrmann, F.R.; Beauchet, O. Higher vitamin D dietary intake is associated with lower risk of AD: A 7-year follow-up. *J. Gerontol. Ser. A Biol. Sci. Med. Sci.* **2012**, *67*, 1205–1211. [CrossRef] [PubMed]
181. Fiala, M.; Miziwiki, M.T. Neuroprotective and immune effects of active forms of vitamin D3 and docosahexaenoic acid in AD patients. *Funct. Foods Health Dis.* **2011**, *1*, 545–554.
182. Annweiler, C.; Herrmann, F.R.; Fantino, B.; Brugg, B.; Beauchet, O. Effectiveness of the combination of memantine plus vitamin D on cognition in patients with AD: A pre–post pilot study. *Cognit. Behav. Neurol.* **2012**, *25*, 121–127. [CrossRef] [PubMed]
183. Stein, M.S.; Schere, S.C.; Ladd, K.S.; Harrison, L.C. A randomized controlled trial of high dose vitamin D2 followed by intranasal insulin in, A.D. *J. Alzheimers Dis.* **2011**, *26*, 477–484. [CrossRef] [PubMed]
184. Przybelski, R.; Agrawal, S.; Krueger, D.; Engelke, J.A.; Walbrun, F.; Binkley, N. Rapid correction of low vitamin D status in nursing home residents. *Osteoporos. Int.* **2008**, *19*, 1621–1628. [CrossRef] [PubMed]
185. LaPrincess, C.B.; Michos, E.D.; Reis, J.P. Vitamin D in atherosclerosis, vascular disease, and endothelial function. *Curr. Drug Targets* **2011**, *12*, 54–60.
186. Prabhakar, P.; Chandra, S.R.; Supriya, M.; Issac, T.G.; Prasad, C.; Christopher, R. Vitamin D status and vascular dementia due to cerebral small vessel disease in the elderly Asian Indian population. *J. Neurol. Sci.* **2015**, *359*, 108–111. [CrossRef] [PubMed]
187. Alagiakrishnan, K.; Hsueh, J.; Zhang, E.; Khan, K.; Senthilselvan, A. Small vessel disease/white matter disease of the brain and its association with osteoporosis. *J. Clin. Med. Res.* **2015**, *7*, 297–302. [CrossRef] [PubMed]
188. Chung, P.W.; Park, K.Y.; Lim, J.M.; Shin, D.W.; Park, M.S.; Chung, Y.J.; Ha, S.Y.; Ahn, S.W.; Shin, H.W.; Kim, Y.B.; et al. 25-Hydroxyvitmain D status is associated with chronic cerebral small vessel disease. *Stroke* **2015**, *46*, 248–251. [CrossRef] [PubMed]
189. Mancuso, P.; Rahman, A.; Hershey, S.D.; Dandu, L.; Nibbelink, K.A.; Simpson, R.U. 1,25-Dihydroxyvitamin-D3 treatment reduces cardiac hypertrophy and left ventricular diameter in spontaneously hypertensive heart failure-prone (cp/+) rats independent of changes in serum leptin. *J. Cardiovasc. Pharmacol.* **2008**, *51*, 559–564. [CrossRef] [PubMed]
190. Bodyak, N.; Ayus, J.C.; Achinger, S.; Shivalingappa, V.; Ke, Q.; Chen, Y.S.; Debra, L.R.; Isaac, S.; Hector, T.; Paul, E.K.; et al. Activated vitamin D attenuates left ventricular abnormalities induced by dietary sodium in DAHL salt-sensitive animals. *Proc. Natl. Acad. Sci. USA* **2007**, *104*, 16810–16815. [CrossRef] [PubMed]
191. Wang, T.J.; Pencina, M.J.; Booth, S.L.; Jacques, P.F.; Ingelsson, E.; Lanier, K.; Benjamin, E.J.; D'Agostino, R.B.; Wolf, M.; Vasan, R.S. Vitamin D deficiency and risk of cardiovascular disease. *Circulation* **2008**, *117*, 503–511. [CrossRef] [PubMed]

192. Zittermann, A.; Schleithoff, S.S.; Tenderich, G.; Berthold, H.K.; Körfer, R.; Stehle, P. Low vitamin D status a contributing factor in the pathogensis of congestive heart failure? *J. Am. Coll. Cardiol.* **2003**, *41*, 105–112. [CrossRef]
193. Somjen, D.; Weisman, Y.; Kohen, F.; Gayer, B.; Limor, R.; Sharon, O.; Jaccard, N.; Knoll, E.; Stern, N. 25-hydroxyvitamin D3–1alpha-hydroxylase is expressed in human vascular smooth muscle cells and is upregulated by parathyroid hormone and estrogenic compounds. *Circulation* **2005**, *111*, 1666–1671. [CrossRef] [PubMed]
194. Danielsson, C.; Nayeri, S.; Wiesinger, H.; Thieroff-Ekerdt, R.; Carlberg, C. Potent gene regulatory and antiproliferative activities of 20-methyl analogues of 1,25 dihydroxyvitamin D3. *J. Cell. Biochem.* **1996**, *63*, 199–206. [CrossRef]
195. Eyles, D.W.; Burne, T.H.J.; McGrath, J.J. Vitamin D effects on brain development, adult brain function and the links between low levels of vitamin D and neuropsychiatric disease. *Front. Neuroendocrinol.* **2013**, *34*, 47–64. [CrossRef] [PubMed]
196. Almeras, L.; Eyles, D.; Benech, P.; Laffite, D.; Villard, C.; Patatian, A.; Boucraut, J.; Mackay-Sim, A.; McGrath, J.; Féron, F. Developmental vitamin D deficiency alters brain protein expression in the adult rat: Implications for neuropsychiatric disorders. *Proteomics* **2007**, *7*, 769–780. [CrossRef] [PubMed]
197. Sutherland, M.K.; Somerville, M.J.; Yoong, L.K.K.; Bergeron, C.; Haussler, M.R.; Crapper McLachlan, D.R. Reduction of vitamin D hormone receptor mRNA levels in Alzheimer as compared to Huntington hippocampus: Correlation with calbindin-28k mRNA levels. *Mol. Brain Res.* **1992**, *3*, 239–250. [CrossRef]
198. Moretti, R.; Furlanis, G.; Caruso, P.; Signori, R.; Gazzin, S.; Tiribelli, C. Vitamin D in the Small Vessel disease. *Curr. Neurobiol.* **2017**, *8*, 112–124.
199. Li, Y.C.; Kong, J.; Wei, M.; Chen, Z.F.; Liu, S.Q.; Cao, L.P. 1,25 Dihydroxyvitamin D(3) is a negative endocrine regulator of the renin-angiotensin system. *J. Clin. Investig.* **2002**, *110*, 229–238. [CrossRef] [PubMed]
200. Buell, J.S.; Dawson-Hughes, B. Vitamin D and neurocognitive dysfunction: Preventing "D"ecline. *Mol. Asp. Med.* **2008**, *29*, 415–422. [CrossRef] [PubMed]
201. De Silva, T.M.; Miller, A.A. Cerebral small vessel disease: Targeting oxydative stress as a novel therapeutic strategy. *Front. Pharmacol.* **2016**. [CrossRef] [PubMed]
202. Miller, A.A.; Drummond, G.R.; De Silva, T.M.; Mast, A.E.; Hickey, H.; Williams, J.P.; Broughton, B.R.; Sobey, C.G. NADPH oxidase activity is higher in cerebral versus systemic arteries of four animal species: Role of Nox2. *Am. J. Physiol. Heart Circ. Physiol.* **2009**, *296*, 220–225. [CrossRef] [PubMed]
203. Kong, X.F.; Zhi, X.H.; Pei, Y.L.; Jackson, D.M.; Holick, M.F. Molecular cloning, characterization, and promoter analysis of the human 25-hydroxyvitamin D3-1aplha.hydroxyalse gene. *Proc. Natl. Acad. Sci. USA* **1999**, *96*, 6988–6993. [CrossRef] [PubMed]
204. Timms, P.M.; Mannan, N.; Hitman, G.A.; Noonan, K.; Mills, P.G.; Syndercombe-Court, D.; Aganna, E.; Price, C.P.; Boucher, B.J. Circualtion MMMP9, vitamin D, and variation in the TIMP-1 response with VDR genotype: Mechanism for inflammatory damage in chronic disorders. *QJM* **2002**, *95*, 787–796. [CrossRef] [PubMed]
205. Raymond, M.A.; Desrmeaux, A.; Labelle, A.; Soulez, M.; Soluez, G.; Langelier, Y.; Pshezhetsky, A.V.; Hébert, M.J. Endothelial stress induces the release of vitamin D binding protein, a novel growth factor. *Biochem. Biophys. Res. Commun.* **2005**, *338*, 1374–1382. [CrossRef] [PubMed]
206. Nakagawa, K.; Sasaki, Y.; Kato, S.; Kubodera, N.; Okano, T. 22-oxa-1-alpha,25 dyhydroxyvitamin D3 inhibits metastastasis and angiogenesis in lung cancer. *Carcinogensesis* **2005**, *26*, 1044–1054. [CrossRef] [PubMed]
207. Zittermnann, A.; Koerfer, R. Protective and toxic effects of vitamin D on vascular calcification: Clinical implications. *Mol. Asp. Med.* **2008**, *29*, 423–432. [CrossRef] [PubMed]
208. Andress, D.I. Non classical aspects of differential vitamin D receptor activation. *Drugs* **2007**, *67*, 1999–2012. [CrossRef] [PubMed]
209. Williamson, J.B.; Lewis, G.F.; Nyenhuis, D.L.; Stebbins, G.T.; Murphy, C.; Handelman, M.; Harden, E.; Harden, E.; Heilman, K.M.; Gorelick, P.B.; Porges, S.W. The effects of cerebral white matter changes on cardiovascular responses to cognitive and physical activity in a stroke population. *Psychophysiology* **2012**, *49*, 1618–1628. [CrossRef] [PubMed]
210. Laurent, S.; Katsahian, S.; Fassot, C.; Tropeano, A.I.; Gautier, I.; Laloux, B.; Boutouyrie, P. Aortic stiffness is an independent predictor of fatal stroke in essential hypertension. *Stroke* **2003**, *34*, 1203–1206. [CrossRef] [PubMed]

211. Baumbach, G.L.; Heistad, D.D. Adaptive changes in cerebral blood vessels during chronic hypertension. *J. Hypertens.* **1991**, *9*, 987–991. [CrossRef] [PubMed]
212. Huneau, C.; Benali, H.; Chabriat, H. Investigating human neurovascular coupling using functional neuroimaging: A critical review of dynamic models. *Front. Neurosci.* **2015**, *9*, 467. [CrossRef] [PubMed]
213. Phillips, A.A.; Chan, F.H.N.; Zheng, M.M.Z.; Krassioukov, A.V.; Ainslie, P.N. Neurovascular coupling in human: Physiology, methodological advances and clinical implications. *JCBFM* **2016**, *36*, 647–664. [CrossRef] [PubMed]
214. Kitazono, T.; Faraci, F.M.; Tagchi, H.; Heistad, D.D. Role of potassium channels in cerebral blood vessels. *Stroke* **1995**, *26*, 1713–1723. [CrossRef] [PubMed]
215. Nelson, M.T.; Patlak, J.B.; Worley, J.F.; Standen, N.B. Calcium channels, potassium channels, and voltage dependence of arterial smooth muscle tone. *Am. J. Physiol.* **1990**, *259*, 3–18. [CrossRef] [PubMed]
216. Bolotina, V.; Gericke, M.; Bregoestovski, P. Kinetic differences between calcium activate potassium channels in smooth muscle isolated from normal and atherosclerotic human aorta. *Proc. R. Soc. Lond. B Biol. Sci.* **1991**, *244*, 51–55. [CrossRef] [PubMed]
217. Cui, X.; Pelekanos, M.; Liu, P.Y.; Burne, T.H.J.; McGrath, J.J.; Eyles, D.W. The vitamin D receptor in dopamine neurons; its presence in human substantia nigra and its ontogenesis in rat midbrain. *Neuroscience* **2013**, *236*, 77–87. [CrossRef] [PubMed]
218. Abou Raya, S.; Helmii, M.; Abou Raya, A. Bone and mineral metabolism in older adults with Parkinson's disease. *Age Ageing* **2009**, *38*, 675–680. [CrossRef] [PubMed]
219. Lu'o'ng, K.; Nguyen, L. Role of vitamin D in Parkinson's Disease. *Int. Sch. Res. Netw.* **2012**, *2012*, 124289.
220. Fiszer, U.; Mix, E.; Fredrikson, S.; Kostulas, V.; Link, H. Parkinson's disease and immunological abnormalities increase of HLA-DR expression on monocytes in cerebrospinal fluid and of CD45ro+ T cells in peripheral blood. *Acta Neurol. Scand.* **1994**, *90*, 160–166. [CrossRef] [PubMed]
221. McGeer, P.L.; Itagaki, S.; Boyes, B.E.; McGeer, E.G. Reactive microglia are positive for HLA-DR in the substantia nigra of PD and AD' disease brain. *Neurology* **1988**, *38*, 1285–12919. [CrossRef] [PubMed]
222. Tokuda, N.; Levy, R.B. 1,25 Dihydroxyvitamin D3 stimulates phagocytosis but suppresses HLA-DR and CD13 antigen expression in human mononuclear phagocytes. *Proc. Soc. Exp. Biol. Med.* **1996**, *211*, 244–250. [CrossRef] [PubMed]
223. Tone, T.; Eto, K.; Katsuoka, K.; Nishioka, K.; Nishiyama, S. Suppression of gamma interferon induced HLA-DR antigen expression on normal and transformed keratocytes by 1,25 $(OH)_2$ vitamin D3. *Jpn. J. Dermatol.* **1991**, *101*, 519–525.
224. Stefanovic, M.; Topic, E.; Khanna, V.K.; Shukla, R.; PArmar, D. Association of polymorphism in cytochrome P450 2D6 and N-Acetyltransferase-2 with Parkinson's disease. *Dis. Mark.* **2010**, *28*, 87–93.
225. Singh, S.; Singh, K.; Patel, D.K.; Singh, C.; Nath, C.; Singh, V.K.; Singh, R.K.; Singh, M.P. The expression of cyp2d33, an ortholog of human cyp2D6 in mouse striatum and its modulation in 1-MPTP induced PD phenotype and nicotine-mediated neuroprotection. *Rejuv. Res.* **2009**, *12*, 185–197. [CrossRef] [PubMed]
226. Shojaee, S.; Sina, F.; Bnihossein, S.; Kazemi, M.H.; Kalhor, R.; Shahidi, G.A.; Fakhrai-Rad, H.; Ronaghi, M.; Elahi, E. Genome-wide linkage analysis of a parkinsonian pyramidal syndrome pedigree by 500 K SNP arrays. *Am. J. Hum. Genet.* **2008**, *82*, 1375–1384. [CrossRef] [PubMed]
227. Wilhelmsen, K.; Mirel, D.; Marder, K.; Bernstein, M.; Naini, A.; Leal, S.M.; Cote, L.J.; Tang, M.X.; Freyer, G.; Graziano, J.; et al. Is there a genetic susceptibility locus for PD on chromosome 22q13? *Ann. Neurol.* **1997**, *41*, 813–817. [CrossRef] [PubMed]
228. Stagi, S.; Lapi, E.; Gambiner, E.; Manoni, C.; Genuardi, M.; Colarusso, G.; Conti, C.; Chiarelli, F.; de Martino, M.; Azzari, C. Bone density and metabolism in subjects with microdeletion of chromosome 22q11. *Eur. J. Endocr.* **2010**, *163*, 329–337. [CrossRef] [PubMed]
229. Ryu, H.; Lee, J.; Zamam, K.; Kubilis, J.; Ferrante, R.J.; Ross, B.D.; Neve, R.; Ratan, R.R. Sp1 and Sp3 are oxidative stress-inducible, anti-death transcription factors in cortical neurons. *J. Neurosci.* **2003**, *23*, 3597–3606. [CrossRef] [PubMed]
230. Wang, J.; BAnnon, M.J. Sp1 and Sp3 activate transcription of the human dopamine transporter gene. *J. Neurochem.* **2005**, *93*, 474–482. [CrossRef] [PubMed]
231. Yajima, S.; Lee, S.H.; Minowa, T.; Mouradian, M.M. Sp family transcription factors regulate expression of rat D2 dopamine receptor gene. *DNA Cell Biol.* **1998**, *17*, 471–479. [CrossRef] [PubMed]

232. Choi, J.M.; Hong, J.H.; Chae, M.J.; Ngyuen, P.H.; Kang, H.S.; Ma, H.I.; Kim, Y.J. Analysis of mutations and the association between polymorphisms in the cerebral dopamine neurotrophic factor (CDNF) gene and, P.D. *Neurosci. Lett.* **2011**, *493*, 97–101. [CrossRef] [PubMed]
233. Baranano, D.; Snyder, S.H. Neural roles for heme oxygenase: Contrasts to nitric oxide synthase. *Proc. Natl. Acad. Sci. USA* **2001**, *98*, 10996–11002. [CrossRef] [PubMed]
234. Mateo, I.; Infante, J.; Sanchez-Juan, P.; García-Gorostiaga, I.; Rodríguez-Rodríguez, E.; Vázquez-Higuera, J.L.; Berciano, J.; Combarros, O. Serum heme oxygenase-1 levels are increased in PD but not in, A.D. *Acta Neurol. Scand.* **2010**, *121*, 136–138. [CrossRef] [PubMed]
235. Castellani, R.; Smith, M.A.; Richey, P.L.; Perry, G. Glycoxidation and oxidative stress in PD and diffuse Lewy body disease. *Brain Res.* **1996**, *737*, 195–200. [CrossRef]
236. Oermann, E.; Bidmon, H.J.; Witte, O.W.; Zilles, K. Effects of 1 alpha, 25 dihydroxyvitamin D3 on the expression of HO1 and GFAP in glial cells of the photothrombotically lesioned cerebral cortex. *J. Chem. Neuroanat.* **2004**, *28*, 225–238. [CrossRef] [PubMed]
237. Soos, J.; Engelhardt, I.; Siklos, L.; Havas, L.; Majtenyi, K. The expression of PARP, NF-κB and parvalbumin is increased in, O.D. *Neuroreport* **2004**, *15*, 1715–1718. [CrossRef] [PubMed]
238. Mabley, J.G.; Wallace, R.; Pacher, P.; Murphy, K.; Szabo, C. Inhibition of PARP polymerase by the active form of vitamin D. *Int. J. Mol. Med.* **2007**, *19*, 947–952. [PubMed]
239. Suzuki, M.; Yoshioka, M.; Hashimoto, M.; Murakami, M.; Noya, M.; Takahashi, D.; Urashima, M. Randomized, double-blind, placebo-controlled trial of vitamin D supplement in Parkinson's disease. *Am. J. Clin. Nutr.* **2013**, *97*, 1004–1013. [CrossRef] [PubMed]
240. Ohmura, K.; Kato, M.; Watanabe, T.; Oku, K.; Bohgaki, T.; Horita, T.; Yasuda, S.; Ito, Y.M.; Sato, N.; Atsumi, T. Effect of combined treatment with bisphosphonate and vitamin D on atherosclerosis in patients with systemic lupus erythematosus: A propensity scorebased analysis. *Arthrit. Res. Ther.* **2018**, *20*, 72. [CrossRef] [PubMed]
241. Aydoğan, N.H.; Özel, İ.; İltar, S.; Kara, T.; Özmeriç, A.; Alemdaroğlu, K.B. The effect of vitamin D and bisphosphonate on fracture healing: An experimental study. *J. Clin. Orthop. Trauma* **2016**, *7*, 90–94. [CrossRef] [PubMed]
242. Kang, Y.M.; Hong, S.H.; Yang, J.H.; Oh, J.C.; Park, J.O.; Lee, B.H.; Lee, S.Y.; Kim, H.S.; Lee, H.M.; Moon, S.H. Pamidronate down-regulates tumor necrosis factor-alpha induced matrix metalloproteinases expression in human intervertebral disc cells. *J. Bone Metab.* **2016**, *23*, 165–173. [CrossRef] [PubMed]
243. Kennel, K.A.; Drake, M.T.; Hurley, D.L. Vitamin D Deficiency in Adults: When to Test and How to Treat. *Mayo Clin. Proc.* **2010**, *85*, 752–758. [CrossRef] [PubMed]

International Journal of
*Molecular Sciences*

MDPI

*Review*

# Vitamin D: Effect on Haematopoiesis and Immune System and Clinical Applications

**Mayte Medrano, Estrella Carrillo-Cruz, Isabel Montero and Jose A Perez-Simon ***

Department of Hematology, University Hospital Virgen del Rocio, Instituto de Biomedicina de Sevilla (IBIS/CSIC/CIBERONC), Universidad de Sevilla, 41013 Sevilla, Spain; mayte_md26@hotmail.com (M.M.); estrellacarrillocruz@gmail.com (E.C.-C.); icas3760@hotmail.com (I.M.)

* Correspondence: josea.perez.simon.sspa@juntadeandalucia.es; Tel.: +34-955013260

Received: 6 August 2018; Accepted: 4 September 2018; Published: 8 September 2018

**Abstract:** Vitamin D is a steroid-like hormone which acts by binding to vitamin D receptor (VDR). It plays a main role in the calcium homeostasis and metabolism. In addition, vitamin D display other important effects called "non-classical actions." Among them, vitamin D regulates immune cells function and hematopoietic cells differentiation and proliferation. Based on these effects, it is currently being evaluated for the treatment of hematologic malignancies. In addition, vitamin D levels have been correlated with patients' outcome after allogeneic stem cell transplantation, where it might regulate immune response and, accordingly, might influence the risk of graft-versus-host disease. Here, we present recent advances regarding its clinical applications both in the treatment of hematologic malignancies and in the transplant setting.

**Keywords:** vitamin D; haematopoiesis; leukaemia; myelodysplastic syndrome; allogeneic stem cell transplantation

## 1. Vitamin D: Metabolism and Mechanisms of Action

### 1.1. Production and Metabolism of Vitamin D

Vitamin D (vit D) is a fat-soluble steroid synthesized in several steps which ends with the formation of vit D3 or calcitriol, the most active form of vit D. It is known for its role in the regulation of calcium and phosphorus levels as well as bone mineralization. The term vitamin D is imprecise and groups together different components that are part of this family of steroid hormones. Firstly, pre-D3 is produced in the skin from 7-dehydrocholesterol after exposure to ultraviolet irradiation. Besides, this irradiation also produces lumisterol and tachysterol from pre-D3. The synthesis of lumisterol is a reversible process and can be converted back to pre-D, which binds to Vitamin D-binding protein (DBP) and is subsequently removed from the skin. Pre-D3 has to be hydroxylated twice to became fully activated. The first hydroxylation occurs in the liver and, to a lesser extent, in other tissues by 25-hydroxylase (CYP2R1) to produce 25(OH)D. This is the principal circulating form of vit D and provides a clinically useful marker for vit D status. In the kidney, the enzyme CYP27B1 hydroxylates 25(OH)D, which is then converted into 1,25(OH)D. This process is stimulated by parathyroid hormone and inhibited by calcium, phosphate and fibroblast growth factor 23 (FGF-23). CYP27B1 can be also found in other tissues including epithelial cells, immune cells and parathyroid glands. The extrarenal production of 1,25(OH)D is under a different control, mainly by cytokines such as tumour necrosis factor $\alpha$ (TNF$\alpha$) and interferon $\gamma$ (IFN$\gamma$). In the kidney, the catalytic enzyme CYP24A1 is the responsible for the inactivation of 1,25(OH)D producing 24,25(OH)D [1]. Several oxidative reactions after 24-hydroxilation and conjugation with glucuronic acid generate some compounds that are excreted through the bile.

The balance between 1α-hydroxylase and 24-hydroxylase activities is regulated by calcitriol, calcium and phosphate serum levels. Parathyroid hormone (PTH) stimulates the synthesis of the 1α-hydroxylase under low serum calcium conditions or low levels of vitamin D, resulting in the increase of 1,25(OH)D activation. PTH also inhibits 24-hydroxylase and can induce osteoclast and osteocytes synthesis of the FGF-23, which acts by reducing the expression of renal sodium-phosphate transporters and regulates vitamin D homeostasis by suppressing renal expression of 1α-hydroxylase and inducing 24-hydroxylase, thus reducing serum calcitriol levels and subsequently serum calcium under hyperphosphatemia conditions [2].

After ultra violet (UV) exposure, maximum levels of vit D are achieved. From that moment, UV irradiation further converts pre-D to lumisterol and tachysterol preventing higher levels to be reached (Figure 1) [3,4]. Also, the amount of melanin in the epidermis can modify the effectiveness of the sunlight in producing pre-D3 [5].

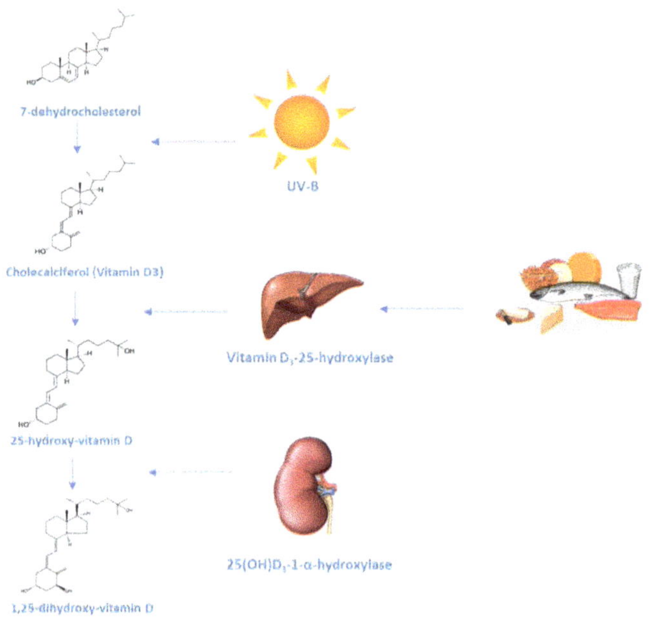


**Figure 1.** Synthesis of vitamin D and the vitamin D metabolic pathway. The vitamin D metabolites are transported in blood bound primarily to vitamin D binding protein (DBP) (85–88%) and albumin (12–15%).

### 1.2. Vitamin D Receptor (VDR)

Vit D operates via binding vitamin D receptor (VDR). Vit D-VDR heterodimerize with retinoid-X-receptor (RXR) within the cell nucleus where it binds vitamin D responsive elements (VDRE) to serve as a transcription factor for numerous target genes. VDR/RXR complexes can attract co-activators or co-repressors to induce or repress gene transcription, depending on the target gene. Steroid receptor coactivator (SRC) and vitamin D receptor interacting protein (DRIP) complex have been identified as coactivators [6]. SRC coactivators recruit histone acetyl transferases (HAT) to the gene promoting transcription. In addition to acetylation, histone methylation also occurs. Recent studies have shown that methyltransferases may also play a role in VDR-mediated transcription [7]. The DRIP complex acts as a bridge between VDRE and the initiation complex (TATA box, RNA polymerase II and other proteins) to facilitate transcription. Co-repressors of VDR function act in the absence of ligand or in the presence of antagonists. The most studied corepressors for VDR are the nuclear receptor corepressor (NCoR) and the silencing mediator for retinoid or thyroid-hormone receptors (SMRT) [8,9].

1,25(OH)D can also inhibit gene transcription via VDR or inhibiting directly other transcription factors. VDR can also functions un-liganded to 1,25(OH)D. As an example, it modulates mammalian hair cycling in keratinocytes by regulating genes such as *CASP14*, *S100A8*, *SOSTDC1* [10].

The VDR has a very short N-terminal domain when compared to other nuclear hormone receptors. The human VDR has two potential starting sites. A common polymorphism (FokI) alters the first ATG start site to ACG when contains the C genotype (formerly F) instead of T (formerly f). Individuals with C genotype begin translation three codons downstream resulting in a VDR protein three amino acids shorter (424 instead of 427 aa). It has been demonstrated in transfection experiments that the short isoform produces a more potent immune response as it resulted in a higher nuclear factor κB (NF-κB)- and nuclear factor of activated T-cells (NFAT)-driven transcription and a higher IL-12 expression in dendritic cells and monocytes [11]. This polymorphism has also been correlated with a reduced bone [12]. The DNA binding domain is comprised of two zinc fingers. The proximal (N-terminal) zinc finger is a specific site to bind to the VDREs whereas the second zinc finger serves for heterodimerization to the retinoid X receptor (RXR). The second half of the molecule is the ligand binding domain, where it binds 1,25(OH)D, which also contains regions required for heterodimerization to RXR. The major activation domain, AF-2, is located at the C-terminal end, which is critical for the binding to coactivators [6].

## 2. Effect of Vitamin D on the Immune System

In addition to its effect on calcium metabolism, vit D plays other important physiological roles. These effects are called "non-classical actions" and were identified 30 years ago, when receptors for 1,25-dihydroxyvitamin D3 (1,25(OH)$_2$D3) were detected in several cell lines [13–15]. In fact, VDR is present in most cells types, which explains its wide range of effects [16]. As previously mentioned, vit D is primarily involved in calcium and phosphate homeostasis. Besides, vit D has other functions, such as regulation of hormone secretion. More specifically, 1,25(OH)D stimulates insulin and thyroid-stimulating hormone (TSH) secretion. Pancreatic β cells have CYP27B1, VDR and calbindin-D. Studies with calbindin-D null mice suggest that it modulates depolarization-stimulated insulin release [17,18].

Vitamin D is also implicated in the regulation of proliferation and differentiation of cells. As an example, vit D is involved in normal breast development and in hepatic cell growth. 1,25(OH)D collaborates in the maturation of type II epithelial pneumocytes by increasing phospholipid production and surfactant release and stimulates the innate immune response in bronchial epithelial cells [19]. Vit D stimulates calcium uptake by cardiac muscle cell [20], which is necessary to the contractility. It has been reported that vit D deficiency is associated with increased risk of myocardial infarction in men [21]. Vit D is also essential for skeletal muscle function. Therefore, vit D deficiency produce proximal muscle weakness [22].

Within the spectrum of non-classical effects of vitamin D, actions on cells of the immune system are included [23,24]. In patients with granulomatous diseases such as sarcoidosis, high levels of 1,25(OH)$_2$D3 and hypercalcemia are observed. An increase in the enzyme 25-hydroxyvitamin D-1-α-hydroxylase (1α-hydroxylase) is also observed. Unlike normal subjects, in whom the activity of this enzyme is located in the kidney, in patients with sarcoidosis, activity is also observed in macrophages [25–28]. The deregulation of 1,25(OH)$_2$D3 is not restricted to sarcoidosis but is a common feature in many granulomatous diseases [29]. The precise nature of the interaction between vitamin D and the immune system took many years to identify and there are still many questions about this interaction.

### 2.1. Vitamin D and Innate Immunity

#### 2.1.1. Macrophages, Vitamin D and Cathelicidin

Macrophages and monocytes play a key role in the initiation of non-specific responses to pathogenic organisms or tissue damage. This role consists in phagocytizing pathogens or cellular debris and then eliminating or assimilating the material.

For many years it was thought that the most important action of vitamin D on macrophages was due to its ability to stimulate the differentiation of monocytic precursors to mature macrophages [23,30–32]. This concept was supported by observations that showed different expression of the vitamin D receptor (VDR) and $\alpha$-1 hydroxylase in the different stages of differentiation of macrophages. Some studies show that human macrophages are able to synthesize 1,25(OH)$_2$D3 upon exposure to IFN$\gamma$ [33,34]. The confirmation of this effect on monocytes was obtained by Modlin et al. who described genes involved in innate immunity regulation which are specifically modulated in monocytes by *Mycobacterium tuberculosis*. VDR and the gene coding for 1-$\alpha$-hydroxylase (CYP27B1) are induced through the "toll like" 2/1 receptor pathway (TLR 2/1). The interaction of TLR 2/1 with the precursor of vitamin D (25(OH)D3) stimulates the expression of the antibacterial protein cathelicidin [35]. Regulation of the transcription of cathelicidin by binding VDR to 25(OH)D3 is possible because its promoter gene contains a functional VDRE. This element is only present in the most developed primates, which is suggestive that the regulation of vitamin D in this facet of innate immunity is a recent event of evolutionary development [36,37]. The precise mechanism by which TLR activation induces the expression of VDR and 1-$\alpha$-hydroxylase is not clear. The analysis of the events involved in the transcriptional regulation of CYP27 B1 suggests that the interaction with TLR involves the JAK/STAT pathway, the MAP kinases and NF-$\kappa$B and that all of this occurs in synergy with the induction of CYP27B1 by IFN$\gamma$ [38]. Other studies propose that the induction of CYP27B1 by TLR2/1 occurs indirectly through IL-15 and IL17A. The enzyme 24-hydroxylase, whose function is to inactivate 1,25(OH)$_2$D3, is encoded by a gene (CYP24) that is induced by 25(OH)D3 after activation by TLR2/1 in monocytes [38,39].

In summary, vitamin D is a potent stimulator of the mechanisms associated with the elimination of pathogens and the appearance of this system in primates (including early Homo sapiens), exposed to abundant amounts of sunlight, was an evolutionary advantage. In fact, it is a system that has key control mechanisms; not only has a catabolic enzyme, 24-hydroxylase, which attenuates the responses to 1,25(OH)$_2$D3 but also feedback mechanisms. In this regard, 1,25(OH)$_2$D3 is a potent down-regulator of TLR2 and TLR4 in monocytes, decreasing inflammatory responses. Hence, using CYP24 and TLR regulatory mechanisms, vitamin D helps to promote innate immune responses by preventing tissue damage associated with excessive inflammation [40,41].

#### 2.1.2. Dendritic Cells and Antigenic Presentation

Dendritic cells (DC) are responsible for the presentation of antigens, resulting from the elimination of pathogens and tissue debris, to cells belonging to adaptive immunity.

Myeloid dendritic cells (M-DCs) produce high levels of IL-12, whereas high levels of IFN are secreted by plasmacytoid dendritic cells (P-DCs) with distinct effects on activation and differentiation of T-cell [42].

Dendritic cells participate in immune responses producing both immunogenic and tolerogenic phenotype. Immature DCs, by the expression of inhibitory receptors, induce anergy among CD4+ cells and elicit generation of IL-10 producing T regs cells [43]. The tolerogenic activity of immature DC may also be related to the expression of endocytic manose receptors (MR), which can deliver negative signals to T-cells. MR levels are up-regulated by anti-inflammatory molecules such as corticosteroids, vit D, Th2 cytokines (IL-4 and IL-3) and are down-regulated by proinflammatory stimuli (IFN-$\alpha$). At the same time, Tregs are capable of maintaining the tolerogenic state by inhibiting myeloid DC maturation, reducing their antigen-presenting function and decreasing IL-12 secretion [44].

The expression of VDR in purified dendritic cells was reported for the first time in 1987 [45]. Studies carried out subsequently purifying skin dendritic cells (Langerhans cells) showed that 1,25(OH)$_2$D3 could attenuate the effect of antigen presentation [46]. However, the role of vitamin D and its metabolites was not elucidated until the advent of dendritic cell models derived from monocytes.

In the year 2000, Gauzzi et al. showed that 1,25(OH)$_2$D3 and its synthetic analogues inhibited the maturation of dendritic cells derived from monocytes, suppressing their ability to present antigens to T cells, suggesting vitamin D could promote tolerance [47–49]. This concept was evaluated in pancreatic islet transplant models in which a lower rejection rate was observed in mice treated with vitamin D [50]. This response seems to be due to a decreased maturation of DC and a concomitant involvement of suppressor T lymphocytes or regulatory T cells [51].

Overall, vit D induces DCs tolerogenicity due to the capacity to inhibit differentiation, maturation, costimulatory molecule expression and IL-12 production, leading to decreased allostimulatory capacity while enhancing IL-10 secretion which favours the induction of regulatory T cells. Vit D not only inhibits DC differentiation and maturation but also transform differentiated IFN-DC into a more immature stage [47].

1,25(OH)$_2$D3 preferentially affects myeloid dendritic cells. Although some studies suggest an apparent insensitivity of plasmacytoid dendritic cells to 1,25(OH)$_2$D3, it is possible that the local synthesis of vitamin D by these cells could participate in the mechanism of tolerogenicity through the VDR receptor expressed in T cells [52,53].

### 2.2. *Vitamin D and Adaptive Immunity*

#### 2.2.1. Vitamin D and T Lymphocytes

Non-activated T cells express undetectable levels of VDR, while receptor expression increases upon T cell activation. Overall, vit D could limit the potential tissue damage associated with Th1 immune responses. However, the validity of this generalization was questioned in mouse models in which vit D was able to inhibit the cytokines associated with both Th1 (INF$\gamma$) and Th2 (IL-4) [54–58]. In addition, it is interesting to note that in animal models of inflammatory bowel disease, the treatment with vit D reduces the expression of IL-17 [59,60]. Hence, it is possible that vit D exerts its effects on inflammation and autoimmune disease also through the regulation of Th17 cells. Finally, vitamin D also affect regulatory T cells. In this regard, in 2002, it was shown that 1,25(OH)$_2$D3 in conjunction with steroids stimulated the production of IL-10 by the CD4+ CD25+ regulatory T cells [51,61]. Other studies indicate that vitamin D produces a preferential T-regulatory differentiation [62,63]. Accordingly, vitamin D stimulates the secretion of IL-10 and TLR9 by regulatory T cells, which connects immune and adaptive responses [64].

In contrast to the wide effect on CD4+ T cells, CD8+ T cells show a poor response to 1,25(OH)$_2$D3 [65–68]. Despite their significant expression of VDR, 1,25(OH)$_2$D3 does not have a significant impact on CD8+ lymphocytes, as shown in animal models of autoimmune encephalomyelitis [69,70].

In addition to its effect on certain T cell populations, recent studies show that vitamin D may also have an effect on the phenomenon of "homing" or tropism of T cells by specific tissues [71]. Some studies suggest that vitamin D inhibits the migration of T cells to lymph nodes. It might also have an effect on T cell homing into the skin by upregulating the cytokine receptor (CCR10), whose ligand CCL27 is expressed by epidermal keratinocytes. This effects on the phenomenon of "homing" is produced by both 25(OH)D3 and by 1,25(OH)$_2$D3 and several authors suggest that both dendritic cells and T lymphocytes are the source of 1$\alpha$ hydroxylase activity [72]. By contrast, vitamin D exert a negative effect on cytokines and their receptors in the gastrointestinal tract [73].

#### 2.2.2. Vitamin D and B Lymphocytes

In addition to the classical function of B lymphocytes as precursors of plasma cells that produce antibodies and, therefore, as a cellular subtype in which humoral immunity rests, populations of B lymphocytes whose main function is the production of IL-10 have been described that could correspond to regulatory B subpopulations. Some animal models of autoimmune diseases have revealed that the regulatory B cells produce IL-10 and can suppress inflammatory processes. IL-10 derived from B cells seems to be important for the generation and maintenance of regulatory T cells. In models in which the maturation of B cells is blocked, there is a reduced capacity for the induction of tolerance and this inability is associated with inadequate potential for the generation of FoxP3+ regulatory T cells. These findings are corroborated by murine models of arthritis in which the deficiency of IL-10 derived from B cells exacerbates the disease and correlates with low levels of regulatory T cells, which are restored with the transfer of IL-10 producing B cells [74–81]. Deficits in the function and frequency of regulatory B cells have been reported in multiple sclerosis and systemic lupus erythematosus [82–85].

Like T cells, active B cells express VDR. As shown in experimental studies, B cells at rest have low but detectable levels of mRNA for VDR [86,87]. After incubation with anti CD40/IL-21, an increase in VDR expression is observed. The addition of 1,25(OH)$_2$D3 increases the expression of VDR mRNA. VDR is functional in activated B cells [85] while the mRNA for CYP24A1 is not detectable in B cells at rest [88]. Activated B cells are therefore able to metabolize vitamin D and respond to its active component.

Initial studies indicated that 1,25(OH)$_2$D3 could regulate the production of B cells and the secretion of immunoglobulins [89]. Studies in which purified B cells are cultivated in the presence of vitamin D show a decrease in the percentage and absolute number of plasma cells as well as IgA, IgG and IgM levels, the results being contradictory with IgE [90–95]. The formation of memory B lymphocytes is also inhibited by 1,25(OH)$_2$D3. Therefore, the effects of vitamin D on B cells could be summarized as a decrease in the "pool" of memory B cells and inhibition of the generation of plasma cells, with the consequent decrease in the secretion of antibodies. On naïve B cells, constitutively expressing VDR and CYP27B1, the response to vitamin D consists in an increase of the expression of VDR, of 25(OH)D3 and of the degrading enzyme of 1,25(OH)$_2$D3 CYP24A1. After activation, the expression of VDR and CYP27B1 does increase, which also results in an increase of the level of 1,25(OH)$_2$D3 and stimulates the negative feedback mechanism through the increase of CYP24A1 [85]. The participation of regulatory B cells in the immune homeostasis exerted by vitamin D seems to be confirmed in murine models of autoimmune diseases in which a functioning pathway mediated by IL-10 is required to guarantee the effect of 1,25(OH)$_2$D3 on the disease [96,97].

### 2.3. *Genetic Fingerprint of Vitamin D in the Immune System*

Deep genomic analysis has allowed to define a new perspective on vitamin D and its function in humans. In macrophages, the description of increased levels of VDR and 1$\alpha$ hydroxylase (CYP27B1) after a pathogenic challenge underscored the importance of the intracrine system of vitamin D as a mediator of the immune response [34,98]. It is now known that macrophages and dendritic cells are capable of responding to 25(OH)D3, the largest circulating metabolite of vitamin D, which provides a link between these cells and vitamin D in humans. The identification of hundreds of target genes for 1,25(OH)$_2$D3 in immune cells has also provided a new perspective on the role of vitamin D in the adaptive immune system [33].

The first study based on genome assays focused on vitamin D was published by John White and colleagues at McGill University in Montreal and used a combination of DNA arrays on genes regulated by 1,25(OH)$_2$D3 and strategies in silico [26,27,33,99–103]. The deep analysis of the target sequences that are capable of binding to VDR reveals response elements (VDRE) that are located next to genes that promote antibacterial proteins such as cathelicidin (CAMP) and $\beta$ defensin 2 (DEFB4). Only CAMP seems to be induced transcriptionally in monocytes [104]. The underlying mechanism for the differential regulation of CAMP and DEFB4 by 1,25(OH)$_2$D3 was determined in subsequent studies. The first one described the increased expression in monocytes of DEFB4 after treatment with 1,25(OH)$_2$D3 and IL-1. This required the cooperative occupation of VDRE by NF-κB and, on the other hand, the binding of VDRE with the DEFB4 promoter gene [104].

The importance of NF-κB and VDR as co-inductors of the transcription of β defensin 2 (DEFB4) was subsequently reinforced in studies focused on the nucleotide-binding oligomerization domain-containing protein 2 (NOD2) protein [105]. Cells treated with 1,25(OH)$_2$D3 and the NOD ligand multidrug resistance (MDR), derived from microorganisms, show a potent induction of DEFF4 dependent on NF-κB [36,37]. However, the induction of CAMP is primarily dependent on the binding of VDR to the VDRE promoter. The VDRE element initially identified as the CAMP promoter appears to be specific to human and subhuman primates [106]. The acquisition of a VDRE element for the CAMP gene seems to have occurred due to the introduction of a nuclear element (SINE) that put CAMP under the control of the VDR receptor. This specific adaptation of primates has been conserved in humans and in primates of the Old and New World, suggesting that CAMP's regulatory transcriptional mechanism for vitamin D confers biological advantages. It is assumed that this mechanism could be potently activated by the relatively high levels of 25(OH)D3 and 1,25(OH)$_2$D3 that are characteristic of non-human primates.

Recognition and response to pathogens involves the identification of molecular patterns associated with pathogen surveillance (PAMPs) through pattern recognition receptors (PRRs), including the extensive family of "Toll like" receptors (TLR), noncatalytic transmembrane proteins that interact with specific PAMPs [35]. In genomic studies on models of tuberculous infection, the TLR 2/1 stimulus induces the expression of CY27B1 and VDR, suggesting that the endocrine system of vitamin D is involved in the macrophagic response to *Mycobacterium tuberculosis*. The macrophages treated with ligand TLR1/2 are reactive to the 1,25(OH)$_2$D3 and 25(OH)D3 forms of vitamin D, confirming the functional efficacy of the intracrine model.

The stimulation of TLR1/2 by *Mycobacterium tuberculosis* also produces the induction of the catabolic enzyme of vitamin D (CYP24A1) and the antibacterial protein CAMP. The expression of the other antibacterial protein DEFB4 is a result of the cooperative action between the TLR1/2, Il-1, NOD2 MDP pathway. Antibacterial proteins such as CAMP and DEFB4 play a crucial role in bacterial intracellular death mediated by vitamin D. Monocytes treated with increasing concentrations of the CAMP peptide show a dose-dependent reduction of internalization of *Mycobacterium tuberculosis* and a similar inhibition in macrophages occurs in presence of 25(OH)D3, this effect being interrupted by the VDR antagonists [37]. Vitamin D and its analogues are capable of promoting autophagy, this induction being very important to provoke antibacterial responses through vitamin D in the tuberculous infection [107].

The intracrine synthesis of 1,25(OH)$_2$D3 seems to regulate the expression of another antibacterial protein, hepcidin (HAMP) [107,108]. The major function attributed to HAMP seems to be the suppression of the membrane protein ferroportin, the only intracellular iron exporter. This link in cells such as enterocytes, hepatocytes and monocytes play a key role in the so-called chronic process anaemia [107]. Iron restriction from the circulation provides an important host response to systemic infection, although for pathogens such as *Mycobacterium tuberculosis*, that evades immune surveillance at the intracellular level, iron accumulation might favour the growth of intracellular pathogens. Vitamin D in its forms 25(OH)D3 and 1,25(OH)$_2$D3 suppress the transcription of HAMP in monocytes and hepatocytes, which leads to the release of the blockade to ferroportin dependent on HAMP, favouring the transport of iron and decreasing its intracellular concentration [109].

Neutrophils express VDR but, unlike monocytes and macrophages, they do not seem to express a functional 1α-hydroxylase enzyme and are not subject to an intracrine activation of the vitamin D system [110].

Dendritic cells, which belong to the same hematopoietic lineage of monocytes and macrophages, express VDR and CYP27B1 and exhibit an active intracrine system of vitamin D [45,110,111]. As discussed above, dendritic cells use a paracrine system of vitamin D, in which differentiation into antigen-presenting cells implies an increase in CYP27 B1 and, paradoxically, a decrease in VDR expression [112]. Therefore, immature cells express VDR and respond to 1,25(OH)$_2$D3 produced by mature cells, with low expression of VDR. Such mechanism pursues the maturation of the dendritic cell and the promotion of T activation and prevents the over-elaboration of the immune response.

In mice, CD8 cells express the activating enzyme of vitamin D, 1-$\alpha$-hydroxylase [67]. However, in experimental models of autoimmune diseases mediated by CD8 lymphocytes, vitamin D does not seem to play any role. CD8$\alpha\alpha$, a variant of CD8 T cells and vitamin D are tied and play a role in the suppression of gastrointestinal inflammation [113].

### 2.4. Vitamin D Levels and Immune Function

Most of the deep genomic analyses that explore the immunomodulatory effects of vitamin D in vitro have focused on the use of 1,25(OH)$_2$D3 or its synthetic analogues. However, the induction by pathogens of an intracrine system such as that of monocytes/macrophages suggests that in vivo regulation is independent of 1,25(OH)$_2$D3. Probably, this system is initially driven by the local activation of 25(OH)D3, the predominant circulating form of vitamin D.

Epidemiological studies have shown that insufficient levels of vitamin D (serum levels below 30 ng/mL) are associated with an increased risk of tuberculosis [114–117]. Clinical trials with vitamin D supplements added to conventional antibiotics have shown variable success. When 10,000 IU of vitamin D were used daily, serum vitamin D levels increased in tuberculous patients but it did not improve the efficacy of the treatment as compared to patients included in the placebo arm. However, in a specific subgroup of patients with a single Taq1 nucleotide polymorphism in the VDR gene, this reduction was demonstrated, suggesting that genetic factors may influence the immune response to vitamin D supplementation [118].

The link between vitamin D and infection is not restricted to patients with tuberculosis. Among patients with sepsis, circulating levels of 25(OH)$_2$D3 are correlate with serum concentrations of CAMP and also correlates with poor prognosis [119,120]. Low levels of 25(OH)D3 are linked to respiratory infections such as influenza and, in patients with chronic renal failure, are correlated with an increased rate of infection and mortality [121]. The application of deep genomic analysis to assess the impact of serum vitamin D status on immune function is limited [122]. In animal models, deep analysis of immune responses has been attempted [123]. Mice deficient in vitamin D showed a decreased expression of angiogenin 4, an antibacterial protein that acts by minimizing the invasion of tissues by enteric bacteria. This leads to higher levels of bacteria in the colon epithelium [124]. This deregulation is related with tissue inflammation in inflammatory bowel diseases and, accordingly, vitamin D could protect from this inflammation by inducing the antibacterial protein angiogenin 4 [125].

A recent study of almost 34,000 individuals shows that genetic variations in DBP influence on the serum concentrations of the DBP protein which, in turn, are linked to the total serum levels of 25(OH)D3 and 1,25(OH)$_2$D3. The genetic variations of the DBP protein could be related to different affinities of 25(OH)D3 with the DBP protein [126–128]. The antibacterial responses to 25(OH)D3 are more pronounced with forms of low affinity BPD involving high free levels of 25(OH)D3 [129,130].

## 3. Vitamin D in Haematopoiesis and Hematopoietic Stem Cells

The physiologically active form of vitamin D, 1,25-dihydroxycholecalciferol or 1,25(OH)$_2$D3 promote monocytic differentiation of HL60, a human promyelocytic leukaemia cell line [131,132]. It is also well recognized that 1,25(OH)$_2$D3 induces normal mononuclear blood cells to differentiate towards the monocyte-macrophage route of maturation [133]. Studies in vitro shown that vitamin D suppresses colony formation of normal human granulocyte macrophage progenitors (CFU-GM) and, by contrast, induces differentiation of colonies into monocyte-macrophages [134]. Experiments with hematopoietic stem cells and leukemic cell lines treated with the active form of vitamin D demonstrate an increase in monocyte/macrophage differentiation and an increase in the number of mature cells that is not evident in mice lacking VDR [135–137]. After its binding to the vitamin D analogue, VDR forms a homodimer or binds to the retinoid X receptor (RXR), which proceeds to interact with VDRE originating a transcription signal on several effector RNAs [56,136]. In addition to RXR, VDR can also bind to the retinoic acid receptor (RAR) which promotes the differentiation of mature granulocytes [138]. Experiments with cell cultures suggest that activated RAR and VDR compete for their binding to RXR and the relative balance between the RAR/RXR and VDR/RXR dimers influences the relative activity of granulopoiesis or monopoiesis [139]. In animal models, the lack of VDR and RAR allows the generation of appropriate monocyte and granulocyte colony forming units, with which it is suspected that the primary effect of vitamin D occurs on cytokine signalling and on the final stages of differentiation of these two cell types.

The complexity of the interaction between VDR and RXR has been reflected in recent observations in which it is shown that retinoic acid and vitamin D can potentiate their action mutually, so that VDR stimulation seems to increase in the presence of RAR-binding. RXR [136].

Studies with VDR knockout (KO) mice showed that the lack of VDR does not affect the normal haematopoiesis and mice presented normal relative numbers of red and white blood cells. By contrast, the addition of vitamin D derivatives can influence at later stages of haematopoiesis [56,140]. 1,25(OH)$_2$D3 activates certain intracellular signalling pathways, which have been suggested to intersect at a common nodal point, Raf-1, such as: lipid signalling pathways (protein kinase C pathway), the phosphatidylinositol-3-kinase (PI3K)-AKT pathway and mitogen activated kinase (MAPK) pathways (Figure 2) [131,141].

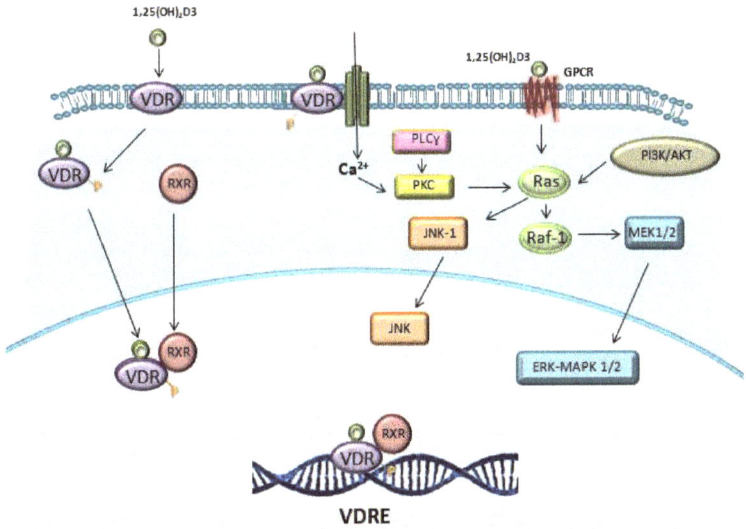


**Figure 2.** 1,25(OH)$_2$D3 acts mediating lipid signalling pathways (protein kinase C pathway), phosphatidylinositol-3-kinase (PI3K)-AKT pathway and mitogen activated kinase (MAPK) pathways.

Regarding the lipid signalling pathway, vitamin D can increase the activity of sphingomyelinase and protein kinase C (PKC) [131,141]. The latter can influence on signal transduction through MAPK pathways. In addition, PKC is an important mediator of hematopoietic cell differentiation [142].

Lipid signalling pathways involve sphingomyelinase, whose activity increases in HL-60 cells treated with 1,25(OH)$_2$D3, leading to increased ceramide levels and enhancement of vitamin D-induced differentiation [143,144].

1,25(OH)$_2$D3 can activate PI3K-AKT pathway, involved in the formation of a VDR/PI3K complex in a signalling pathway that parallels the MAPK pathways, which mediate cell differentiation.

LY294002, a PI3K inhibitor, inhibits 1,25$(OH)_2$D3-induced monocytic cell marker CD14 and CD11b expression in THP-1 cells, demonstrating the implication of PI3K pathway in promoting differentiation [145,146]. Vitamin D has also been proposed to induce differentiation by disassembling AKT-Raf1 complex, upregulating Raf1 and activating the Raf/MEK/ERK MAPK pathway.

Three different MAPK signalling cascades are implicated in 1,25$(OH)_2$D3-induced cell growth arrest and differentiation: Raf-1/MEK/ERK MAKP, JNK/MAPK and p38/MAPK pathways [131].

With regards to lymphopoiesis, Yu et al. demonstrated, using a VDR KO mice lacking vitamin D, that the expression of the vitamin D receptor (VDR) is required for normal thymic development and function of invariant Natural Killer T (iNKT) cells, which are intrinsically defective and lack T-bet expression. In vitro studies showed an inhibitory effect of vitamin D on NK cell development, while promoting myeloid differentiation. However, analysis of CD4 and CD8 T cells and regulatory T cells numbers in the thymus identified no differences between wild type (WT) and VDR KO mice [73,147,148].

VDR KO mice present an extramedullary haematopoiesis because of abnormal bone mineralization [149]. Hematopoietic defects such as anaemia, extramedullary haematopoiesis, thrombocytopenia, myelofibrosis and myelodysplasia was exhibited by children with vitamin D deficiency-associated rickets [150]. 1,25(OH)D3 affects embryonic hematopoietic stem and progenitor cell (HSPC) numbers in vivo and in vitro via VDR-mediated regulation of pro-proliferative responses independent of $Ca^{2+}$ flux [150].

1,25(OH)D3 negatively influences hemogenic endothelial formation independent of VDR activation by antagonizing Hedgehog signalling [151]. Ex vivo treatment with 1,25(OH)D3 increased the proliferation, survival and multi-lineage colony forming activity of CD34+ human umbilical cord blood hematopoietic stem cells (HSCs) [150].

## 4. Clinical Applications

### 4.1. Use of Vitamin D in the Treatment of Hematologic Malignancies

Due to the aforementioned effects on maturation, vitamin D and its analogues have been used as treatment in myeloid neoplasms, particularly myelodysplastic syndromes (MDS) and acute myeloblastic leukaemias (AML). Specific preclinical experiences with HL-60 and other leukemic lines, such as U-937 and THP-1, have shown differentiation and apoptosis of blasts with vitamin D, suggesting that these components, such as all-*trans* retinoic acid (ATRA) in acute promyelocytic leukaemia (APL), reverse the blockade of the differentiation of myeloblasts [131,136,141].

The anti-leukemic activity of vitamin D was described almost three decades ago as Tanaka showed that treatment with a vitamin D analogue, improved survival in leukemic mice [32]. Muto demonstrated that calcitriol can inhibit cell cycle and induces differentiation of leukaemia cells through VDR [152]. Other numerous studies on AML blasts or leukemic cell lines showed that vitamin D induce cell differentiation and growth inhibition [30,31,141,145,153–155]. The exact mechanism by which the activation of the vitamin D receptor induces this effect is not completely clear and investigations have revealed complex cross-signals involving P13 kinase, MAPK pathway and probably the upregulation of factors such as p53 [131].

Unfortunately, early preclinical trials using supraphysiological doses that induced differentiation also induced hypercalcemia [136,156]. However, other studies showed that fractionated doses could achieve the same effect on differentiation, maintaining the level of vitamin D in the physiological range [137].

There are several studies in which vitamin D therapy has been used as the single agent in myelodysplastic syndromes (MDS). The first study was carried out in 1985 by Koeffler and colleagues, reporting 18 patients treated with 1,25$(OH)_2$D3 with a dose greater than 2 mg/day. Although 8 patients presented minor haematological responses, the response did not persist for more than 12 weeks and hypercalcemia was a common toxicity [156].

In a retrospective study conducted by Hermine et al. [157], it was shown that VDR expression in AML is correlated to prognosis. Accordingly, patients presenting higher VDR expression have an increased survival. Moreover, patients' prognosis is correlated to the expression of VDR-targeted genes. Patients with higher CAMP expression presented an increased event free survival (EFS) compared to patients with lower levels of CAMP expression.

Other studies with vitamin D analogues failed to demonstrate haematological responses in MDS although tolerance was adequate [158]. Recently, Motomura et al. randomized a series of 30 patients to receive 25(OH)D3 versus supportive treatment. Only one of the 15 patients in the vitamin D group progressed to AML versus seven in the control group [159].

There are several attempts to combine vitamin D with other cytotoxic agents. Siitonen et al. reported a series of 19 patients with MDSs treated with a combination of 13-*cis* retinoic acid, $1,25(OH)_2D3$ (1 mg/day) and valproic acid, used as a histone deacetylase inhibitor [160]. Three patients had haematological response but intolerance was recorded in eight patients, due to 13-*cis* retinoic acid and valproic acid.

In 2008, a study was conducted with 63 patients with MDS with a combination of erythropoietin, 13-*cis* retinoic acid, $1,25(OH)_2D3$ and thioguanine in the presence of blasts. An overall erythroid response of 60% was obtained, reaching 93% in low risk patients [161]. Subsequently, vitamin D analogues were combined with cytotoxic chemotherapy in a series of 53 patients with MDS in whom 13-*cis* retinoic acid (20–40 mg daily) was added to $1,25(OH)_2D3$ (1–1, 5 mg daily) with or without thioguanine. The overall response rate was 60% and 50% of patients achieved transfusional independence [162].

The only study that used vitamin D monotherapy in AML was conducted in the eighties on 5 patients, four of whom presented a transient reduction in the number of blasts and only one described a brief normalization of the spinal study [163]. In 1992, Slapak et al. reported the use of continuous infusion of cytarabine for 21 days (20 $mg/m^2$ daily), hydroxyurea (500 mg twice daily) and $1,25(OH)_2D3$ (0.5 mg twice a day) [162]. An overall response rate of 79% was obtained with 45% complete response rate. The toxicity was primarily haematological consisting of neutropenia and thrombocytopenia, only two patients developed asymptomatic hypercalcemia that did not require treatment. These results were considered at least not inferior to those achieved with cytarabine as the sole agent. Ferrero and colleagues carried out another study in 2008 with a similar population of 30 patients (24 AML, 6 MDS) in which they used subcutaneous cytarabine (8 $mg/m^2$ twice a day), $1,25(OH)_2D3$ (1 mg daily), 13-*cis* retinoic acid (20–40 mg daily) and thioguanine (40 mg daily). The median survival was 7.5 months, being 16 months for the responders. The toxicity was mainly related to cytopenias [164].

The only study published in the literature comparing combination treatment with vitamin D analogues ($1,25(OH)_2D3$ 1 µg/day) and chemotherapy (cytarabine 15 $mg/m^2$ daily subcutaneously until the blasts were lower 50%) compared to a control group was performed in Sweden on a total of 78 patients, 68 with MDS and 15 with AML. Half of the patients also received 13-*cis* retinoic acid (1 mg/kg daily). The mean survival was 10.5 months, with no significant difference between both groups [165].

Interestingly, vitamin D serum levels have also been correlated to response rate to 5-azacytidine (AZA) among patients with MDS or AML. In this regard, Radujkovic et al. analysed serum levels before starting AZA in 58 patients. Estimated probability of 2-year overall survival in the low versus high vitamin D levels group was 14% versus 40% ($p < 0.05$). In multivariable analysis, adverse cytogenetics and vitamin D levels were independent predictors of survival [166]. Similarly, Lee et al. reported a series of 97 patients diagnosed with AML who received intensive chemotherapy; in this study, a significantly worse outcome was observed among those patients with low vitamin D levels [167]. These data are in contrast to those reported by Pardanani et al. who did not find any relationship between vitamin D levels and prognosis in a series of 409 patients diagnosed with different myeloid neoplasms and MDS [168].

There are also studies evaluating the role of vitamin D in lymphoid malignancies. In this regard, several preclinical studies have demonstrated activity of the vitamin D analogue EB1089 in the multiple myeloma cell line H929. This agent promotes apoptosis and induce cell cycle arrest by downregulation of cyclin-dependent kinases [169–171]. There are preclinical studies that show that vitamin D has an inhibitory effect on neoplastic lymphoid cells but to date there have been no studies in humans [172].

A retrospective study performed by Kelly J et al. analysed 183 patients who were enrolled in three SWOG trials and had 25(OH)D serum levels available. There was no association between vitamin D deficiency and clinical response. After a median follow-up of 5.4 years, vitamin D-deficient patients had a significantly worse progression free survival (PFS) (hazard ratio (HR), 2.00; $p = 0.011$) and overall survival (OS) (HR, 3.57; $p = 0.003$) as compared with those with higher levels. Multivariable analysis suggested that lower levels of vitamin D were associated with a higher risk of either progression or death but neither result was significant. In addition, 240 patients enrolled onto the parent PRIMA clinical trial were analysed. After a median follow-up of 6.6 years, vitamin D-deficient patients had significantly lower PFS (HR, 1.66; $p = 0.013$) but not OS (HR, 1.84; $p = 0.14$) as compared with those with higher levels Multivariable analysis confirmed that lower levels of vitamin D were associated with a higher risk of either progression or death [173].

Similar results have been reported by Tracy SI et al. in a series of 642 patients with follicular lymphoma. The authors evaluated whether vitamin D insufficiency was associated with adverse outcomes; with a median follow-up of 59 months, 297 patients (46%) had an event (progression, treatment failure), 78 had died and 42 (6.5%) had a lymphoma-related death. Vitamin D deficiency was associated with inferior EFS at 12 months (OR = 2.05; 95% confidence interval (CI) 1.18–3.54), OS (hazards ratio (HR) = 2.35; 95% CI 1.37–4.02) and lymphoma-specific survival (HR = 2.97; 95% CI 1.52–5.80) for the full cohort [174].

In diffuse large cell lymphoma Hohaus et al. analysed 128 patients. 25(OH)D levels below 20 ng/mL at diagnosis and IPI were independently associated with a worse EFS. Moreover, patients with normalized 25(OH)D levels following supplementation showed a better EFS as compared to those patients with persistent insufficient 25(OH)D levels [175].

### *4.2. Vitamin D as a Modulator of the Immune Response in Allogeneic Transplantation*

There is considerable interest in vitamin D analogues for their immunomodulatory effects, which could be considered an effective approach among patients undergoing allogeneic hematopoietic stem cell transplantation (HSCT) to prevent graft-versus-host disease (GVHD). The VDR genes are polymorphic in the human population and this genetic variation in VDR has been investigated in patients undergoing HSCT. Cho and colleagues conducted an analysis of 147 patients. They analysed the polymorphisms for VDR and evaluated the association with the prognosis of the patients. They showed a correlation between the polymorphisms of the anchor site of the restriction enzyme Taq1 and survival, so that heterozygotes (who have at least one copy of the C allele) had a better overall and disease-free survival than homozygous TT. The functional significance of this allelic variation is unknown and no direct association with a higher or lower VDR activity was reported. This study also found that recipients with two copies for the "A" allele, related to polymorphisms for the Apal anchor site, had a lower risk of acute GVHD and infections [176]. These polymorphisms have been related to VDR activity so that homozygosity for the "a" allele translates into greater activity [177].

Middleton and colleagues studied a cohort of 88 patients with myeloid malignancies undergoing HSCT, correlating VDR polymorphisms of both recipients and donors with prognosis [178]. Like Cho et al., they detected a marked trend towards a decreased risk of acute GVHD in recipients with AA genotype (low VDR activity). Receptors with aa genotype and high VDR activity showed a trend toward a higher risk of acute GVHD, although differences were not statistically significant. However, recipients of donors with low VDR activity (AA) had a higher risk of death.

Bogunia-Kubik et al. published an analysis on 123 patients [177]. They found an association between the FokI FF genotype, which is associated with increased VDR activity and patient's prognosis.

If the donor and the recipient had the FF genotype, the recipients had a higher risk of GVHD. The Apal genotype, as in other studies, also had an impact on the risk of GVHD. Contrary to the data described in the study by Middleton et al., the AA donor genotype (low VDR activity) was associated with a higher risk of GVHD as compared to the genotype that had at least one an allele. At the same time, aa receptors (high VDR activity) had a higher risk of death and GVHD compared to the genotype with low VDR activity, which is consistent with other studies.

Thus, the vitamin D receptor and its mediation on immune signalling appear to have an impact on immune reconstitution after HSCT and the risk of infection and graft versus host disease.

Remarkably, a significant proportion of patients display low levels of vit D before HSCT [179,180]. In this regard, several studies have been reported describing the impact of the vitamin D levels before HSCT and post-transplant outcomes. More specifically, von Bahr et al. described an association between low levels of vit D and an increased risk of GVHD and CMV reactivation [180]. Similar results have been described by Hansson et al., who described an increased risk of death, relapse and cGVHD among patients with low vit D levels although, strikingly, grades 2 to 4 aGVHD occurred more frequently among patients with normal levels of Vitamin D [181]. Moreover, a higher risk of relapse has also been described among patients with low vitamin D levels. In this regard, Radujkovic et al. analysed a series of 492 patients undergoing HSCT; results were validated in an independent cohort of 398 patients. 396 (80%) and 348 (87%) patients had vit D deficiency before transplant in the training and validation cohort, respectively. Vit D deficiency was significantly associated with inferior overall survival, which was mainly attributed to a higher risk of relapse (HR, 1.96; $p = 0.006$) in patients diagnosed with myeloid (HR, 2.55; $p = 0.014$) but not with lymphoid malignancies (HR, 1.60; $p = 0.147$) [166].

With these data in mind we designed a phase I/II prospective trial in which 150 patients were included in three consecutive cohorts of 50 patients each group: control group (who did not receive vitamin D); low dose group (1000 UI vitamin D daily) and high dose group (5000 UI vitamin D daily). No significant differences were observed in terms of acute GVHD, relapse, non-relapse mortality and overall survival. By contrast, a significantly lower cumulative incidence of both overall and moderate plus severe chronic GVHD at 1 year was observed in patients receiving low (37.5% and 19.5%, respectively) or high doses of vitamin D (42.4% and 27%, respectively) as compared to the control group (67.5% and 44.7%, respectively) ($p < 0.05$). In multivariable analysis, treatment with vitamin D significantly decreased the risk of both overall (for low dose (HR = 0.31, $p = 0.002$) and for high dose of vitamin D (HR = 0.36, $p = 0.006$)) and moderate plus severe cGVHD (for low dose (HR = 0.22, $p = 0.001$) and for high dose vitamin D (HR = 0.33, $p = 0.01$)). There were no adverse events attributed to the vitamin D, more specifically, no case of hypercalcemia was observed. With this low toxicity profile, a prospective randomized trial would be required to confirm the potential efficacy of vitamin D as immune-modulatory agent after HSCT [182].

## 5. Conclusions

Our knowledge of vitamin D effects has grown in the past 20 years. The mechanisms of action and the role of vitamin D receptor in addition to its classic effects on calcium and bone homeostasis is well stablished. Vitamin D receptor is expressed on immune cells, which are all capable of synthesizing the active vitamin D metabolite. Moreover, vitamin D has the capability to act in an autocrine manner and can modulate the innate and adaptive immune responses.

As far as the clinical applications of vitamin D is concerned, several studies have been reported both in myeloid as well as in lymphoid malignancies suggesting that vitamin D may promote tumour cells differentiation and might play a role, in combination with other agents, for the treatment of these disorders, although no prospective randomized study is available to confirm these findings. In addition, in the transplant setting, the effect of vitamin D on the immune system might also influence patient's outcome and, in this regard, different studies have evaluated the relationship between vitamin D levels pre-HSCT and risk of infections, graft-versus-host disease and relapse after transplantation.

A phase I/II prospective trial suggests the potential benefit of the use of vitamin D to prevent GvHD. Prospective randomized trials would be required to confirm these findings.

**Author Contributions:** Writing-Original Draft Preparation, I.M., J.A.P.-S., M.M., E.C.-C.; Writing-Review & Editing, J.A.P.-S., M.M.; Supervision, J.A.P.-S.; Project Administration, M.M.; Funding Acquisition, J.A.P.-S.

**Funding:** This work was supported by research grants from ISCIII-FEDER (PI17/02177) M.M. has a predoctoral grant PFIS-ISCIII (FI16/00412).

**Conflicts of Interest:** The authors declare no conflict of interests.

## Abbreviations

| | |
|---|---|
| vit D | Vitamin D |
| 25(OH)D | Calcidiol, 25 hydroxycholecalciferol or 25-hidroxivitamin D |
| $1,25(OH)_2D3$ | 1,25-Dihydroxyvitamin D3 |
| DBP | Vitamin D-binding protein |
| CYP2R1 | Vitamin D 25-hydroxylase or cytochrome P450 2R1 |
| CYP27B1 | 1-$\alpha$-Hydroxylase (1$\alpha$-hydroxylase) |
| CYP24A1 | Cytochrome P450 family 24 subfamily A member 1 |
| FGF-23 | Fibroblast growth factor 23 |
| PTH | Parathyroid hormone |
| TNF$\alpha$ | Tumour necrosis factor $\alpha$ |
| IFN$\gamma$ | Interferon gamma |
| UV | Ultra violet |
| VDR | Vitamin D receptor |
| RXR | Retinoid-X-receptor |
| VDRE | Vitamin D responsive elements |
| HAT | Histone acetyl transferases |
| NCoR | Nuclear receptor corepressor |
| SMRT | Retinoid or thyroid-hormone receptors |
| NFAT | Nuclear factor of activated T-cells |
| TSH | Thyroid-stimulating hormone |
| TRH | Thyrotropin-releasing hormone |
| TLR | Toll like receptor |
| NF-$\kappa$B | Nuclear factor $\alpha$ B |
| DC | Dendritic cells |
| M-DCs | Myeloid dendritic cells |
| P-DCs | Plasmacytoid dendritic cells |
| MR | Manose receptors |
| CAMP | Cathelicidin |
| DEFB4 | $\beta$ defensin 2 |
| CCR10 | C-C chemokine receptor type 10 |
| NOD2 | Nucleotide-binding oligomerization domain-containing protein 2 |
| SINE | Introduction of a nuclear element |
| PAMPs | Molecular patterns associated with pathogen surveillance |
| PRRs | Pattern recognition receptors |
| MDR | Multidrug resistance |
| HAMP | Hepcidin antimicrobial peptide |
| CFU-GM | Granulocyte macrophage colony forming cell |
| RAR | Retinoic acid receptor |
| RXR | Retinoid X receptor |
| SRC | Steroid receptor coactivator |
| DRIP | Vitamin D receptor interacting protein |
| PKC | Protein kinase C |
| MAPK | Mitogen activated kinase |
| iNKT | Invariant natural killer T cells |

| | |
|---|---|
| HSPC | Hematopoietic stem progenitor cell |
| HSC | Hematopoiectic stem cell |
| MDS | Myelodysplastic syndromes |
| AML | Acute myeloblastic leukaemias |
| ATRA | All-trans retinoic acid |
| APL | Acute promyelocytic leukaemia |
| EFS | Event free survival |
| AZA | 5-Azacytidine |
| HR | Hazard ratio |
| CI | Confidence interval |
| OS | Overall survival |
| HSCT | Hematopoietic stem cell transplantation |
| CMV | Cytomegalovirus |
| cGVHD | Chronic graft-versus-host disease |
| aGVHD | Acute graft-versus-host disease |

## References


1. Aranow, C. Vitamin D and the immune system. *J. Investig. Med.* **2011**, *59*, 881–886. [CrossRef] [PubMed]
2. Gil, A.; Plaza-Diaz, J.; Mesa, M.D. Vitamin D: Classic and novel actions. *Ann. Nutr. Metab.* **2018**, *72*, 87–95. [CrossRef] [PubMed]
3. Holick, M.F.; Richtand, N.M.; McNeill, S.C.; Holick, S.A.; Frommer, J.E.; Henley, J.W.; Potts, J.T.J. Isolation and identification of previtamin D3 from the skin of rats exposed to ultraviolet irradiation. *Biochemistry* **1979**, *18*, 1003–1008. [CrossRef] [PubMed]
4. Holick, M.F.; McLaughlin, J.A.; Clark, M.B.; Holick, S.A.; Potts, J.T.J.; Anderson, R.R.; Blank, I.H.; Parrish, J.A.; Elias, P. Photosynthesis of previtamin D3 in human and the physiologic consequences. *Science* **1980**, *210*, 203–205. [CrossRef] [PubMed]
5. Bell, N.H.; Greene, A.; Epstein, S.; Oexmann, M.J.; Shaw, S.; Shary, J. Evidence for alteration of the vitamin D endocrine system in blacks. *J. Clin. Investig.* **1985**, *76*, 470–473. [CrossRef] [PubMed]
6. Rachez, C.; Gamble, M.; Chang, C.P.; Atkins, G.B.; Lazar, M.A.; Freedman, L.P. The DRIP complex and SRC-1/p160 coactivators share similar nuclear receptor binding determinants but constitute functionally distinct complexes. *Mol. Cell Biol.* **2000**, *20*, 2718–2726. [CrossRef] [PubMed]
7. Seth-Vollenweider, T.; Joshi, S.; Dhawan, P.; Sif, S.; Christakos, S. Novel mechanism of negative regulation of 1,25-dihydroxyvitamin D3-induced 25-hydroxyvitamin D3 24-hydroxylase (Cyp24a1) transcription: Epigenetic modification involving cross-talk between protein-arginine methyltransferase 5 and the SWI/SNF complex. *J. Biol. Chem.* **2014**, *289*, 33958–33970. [CrossRef] [PubMed]
8. Perissi, V.; Jepsen, K.; Glass, C.K.; Rosenfeld, M.G. Deconstructing repression: Evolving models of co-repressor action. *Nat. Rev. Genet.* **2010**, *11*, 109–123. [CrossRef] [PubMed]
9. Horlein, A.J.; Naar, A.M.; Heinzel, T.; Torchia, J.; Gloss, B.; Kurokawa, R.; Ryan, A.; Kamei, Y.; Söderström, M.; Glass, C.K.; et al. Ligand-independent repression by the thyroid hormone receptor mediated by a nuclear receptor co-repressor. *Nature* **1995**, *377*, 397–404. [CrossRef] [PubMed]
10. Haussler, M.R.; Whitfield, G.K.; Kaneko, I.; Haussler, C.A.; Hsieh, D.; Hsieh, J.C.; Jurutka, P.W. Molecular mechanisms of vitamin D action. *Calcif. Tissue Int.* **2013**, *92*, 77–98. [CrossRef] [PubMed]
11. Van Etten, E.; Verlinden, L.; Giulietti, A.; Ramos-Lopez, E.; Branisteanu, D.D.; Ferreira, G.B.; Overbergh, L.; Verstuyf, A.; Bouillon, R.; Roep, B.O.; et al. The vitamin D receptor gene FokI polymorphism: Functional impact on the immune system. *Eur. J. Immunol.* **2007**, *37*, 395–405. [CrossRef] [PubMed]
12. Harris, S.S.; Eccleshall, T.R.; Gross, C.; Dawson-Hughes, B.; Feldman, D. The vitamin D receptor start codon polymorphism (FokI) and bone mineral density in premenopausal American black and white women. *J. Bone Miner. Res.* **1997**, *12*, 1043–1048. [CrossRef] [PubMed]
13. Hewison, M. Vitamin D and the immune system: New perspectives on an old theme. *Endocrinol. Metab. Clin. N. Am.* **2010**, *39*, 365–379. [CrossRef] [PubMed]
14. Eisman, J.A.; Martin, T.J.; MacIntyre, I.; Moseley, J.M. 1,25-dihydroxyvitamin-D-receptor in breast cancer cells. *Lancet* **1979**, *314*, 1335–1336. [CrossRef]

15. Makishima, M.; Lu, T.T.; Xie, W.; Whitfield, G.K.; Domoto, H.; Evans, R.M.; Haussler, M.R.; Mangelsdorf, D.J. Vitamin D receptor as an intestinal bile acid sensor. *Science* **2002**, *296*, 1313–1316. [CrossRef] [PubMed]
16. Bouillon, R.; Carmeliet, G.; Verlinden, L.; Van Etten, E.; Verstuyf, A.; Luderer, H.F.; Lieben, L.; Mathieu, C.; Demay, M. Vitamin D and human health: Lessons from vitamin D receptor null mice. *Endocr. Rev.* **2008**, *29*, 726–776. [CrossRef] [PubMed]
17. Sooy, K.; Schermerhorn, T.; Noda, M.; Surana, M.; Rhoten, W.B.; Meyer, M.; Fleischer, N.; Sharp, G.W.; Christakos, S. Calbindin-D(28k) controls [Ca($^{2+}$)](i) and insulin release. Evidence obtained from calbindin-d(28k) knockout mice and β cell lines. *J. Biol. Chem.* **1999**, *274*, 34343–34349. [CrossRef] [PubMed]
18. Tornquist, K. Pretreatment with 1,25-dihydroxycholecalciferol enhances thyrotropin-releasing hormoneand inositol 1,4,5-trisphosphate-induced release of sequestered $Ca^{2+}$ in permeabilized GH4C1 pituitary cells. *Endocrinology* **1992**, *131*, 1677–1681. [CrossRef] [PubMed]
19. Herscovitch, K.; Dauletbaev, N.; Lands, L.C. Vitamin D as an anti-microbial and anti-inflammatory therapy for Cystic Fibrosis. *Paediatr. Respir. Rev.* **2013**, *15*, 154–162. [CrossRef] [PubMed]
20. Selles, J.; Boland, R. Rapid stimulation of calcium uptake and protein phosphorylation in isolated cardiac muscle by 1,25-dihydroxyvitamin D3. *Mol. Cell Endocrinol.* **1991**, *77*, 67–73. [CrossRef]
21. Girgis, C.M.; Clifton-Bligh, R.J.; Hamrick, M.W.; Holick, M.F.; Gunton, J.E. The roles of vitamin D in skeletal muscle: Form, function, and metabolism. *Endocr. Rev.* **2013**, *34*, 33–83. [CrossRef] [PubMed]
22. Boland, R. Role of vitamin D in skeletal muscle function. *Endocr. Rev.* **1986**, *7*, 434–448. [CrossRef] [PubMed]
23. Abe, E.; Miyaura, C.; Tanaka, H.; Shiina, Y.; Kuribayashi, T.; Suda, S.; Nishii, Y.; DeLuca, H.F.; Suda, T. 1α,25-dihydroxyvitamin D3 promotes fusion of mouse alveolar macrophages both by a direct mechanism and by a spleen cell-mediated indirect mechanism. *Proc. Natl. Acad. Sci. USA* **1983**, *80*, 5583–5587. [CrossRef] [PubMed]
24. Bhalla, A.K.; Amento, E.P.; Serog, B.; Glimcher, L.H. 1,25-Dihydroxyvitamin D3 inhibits antigen-induced T cell activation. *J. Immunol.* **1984**, *133*, 1748–1754. [PubMed]
25. Adams, J.S.; Gacad, M.A. Characterization of 1α-hydroxylation of vitamin D3 sterols by cultured alveolar macrophages from patients with sarcoidosis. *J. Exp. Med.* **1985**, *161*, 755–765. [CrossRef] [PubMed]
26. Adams, J.S.; Sharma, O.P.; Gacad, M.A.; Singer, F.R. Metabolism of 25-hydroxyvitamin D3 by cultured pulmonary alveolar macrophages in sarcoidosis. *J. Clin. Investig.* **1983**, *72*, 1856–1860. [CrossRef] [PubMed]
27. Barbour, G.L.; Coburn, J.W.; Slatopolsky, E.; Norman, A.W.; Horst, R.L. Hypercalcemia in an anephric patient with sarcoidosis: Evidence for extrarenal generation of 1,25-dihydroxyvitamin D. *N. Engl. J. Med.* **1981**, *305*, 440–443. [CrossRef] [PubMed]
28. Hewison, M. Vitamin D and the immune system. *J. Endocrinol.* **1992**, *132*, 173–175. [CrossRef] [PubMed]
29. Hewison, M.; Burke, F.; Evans, K.N.; Lammas, D.A.; Sansom, D.M.; Liu, P.; Modlin, R.L.; Adams, J.S. Extra-renal 25-hydroxyvitamin D3-1α-hydroxylase in human health and disease. *J. Steroid Biochem. Mol. Biol.* **2007**, *103*, 316–321. [CrossRef] [PubMed]
30. Abe, E.; Miyaura, C.; Sakagami, H.; Takeda, M.; Konno, K.; Yamazaki, T.; Yoshiki, S.; Suda, T. Differentiation of mouse myeloid leukemia cells induced by 1α,25-dihydroxyvitamin D3. *Proc. Natl. Acad. Sci. USA* **1981**, *78*, 4990–4994. [CrossRef] [PubMed]
31. Koeffler, H.P.; Amatruda, T.; Ikekawa, N.; Kobayashi, Y.; DeLuca, H.F. Induction of macrophage differentiation of human normal and leukemic myeloid stem cells by 1,25-dihydroxyvitamin D3 and its fluorinated analogues. *Cancer Res.* **1984**, *44*, 5624–5628. [PubMed]
32. Tanaka, H.; Abe, E.; Miyaura, C.; Shiina, Y.; Suda, T. 1α,25-dihydroxyvitamin D3 induces differentiation of human promyelocytic leukemia cells (HL-60) into monocyte-macrophages, but not into granulocytes. *Biochem. Biophys. Res. Commun.* **1983**, *117*, 86–92. [CrossRef]
33. Koeffler, H.P.; Reichel, H.; Bishop, J.E.; Norman, A.W. Gamma-Interferon stimulates production of 1,25-dihydroxyvitamin D3 by normal human macrophages. *Biochem. Biophys. Res. Commun.* **1985**, *127*, 596–603. [CrossRef]
34. Kreutz, M.; Andreesen, R.; Krause, S.W.; Szabo, A.; Ritz, E.; Reichel, H. 1,25-dihydroxyvitamin D3 production and vitamin D3 receptor expression are developmentally regulated during differentiation of human monocytes into macrophages. *Blood* **1993**, *82*, 1300–1307. [PubMed]
35. Liu, P.T.; Stenger, S.; Li, H.; Wenzel, L.; Tan, B.H.; Krutzik, S.R.; Ochoa, M.T.; Schauber, J.; Wu, K.; Meinken, C. Toll-like receptor triggering of a vitamin D-mediated human antimicrobial response. *Science* **2006**, *311*, 1770–1773. [CrossRef] [PubMed]

36. Gombart, A.F.; Borregaard, N.; Koeffler, H.P. Human cathelicidin antimicrobial peptide (CAMP) gene is a direct target of the vitamin D receptor and is strongly up-regulated in myeloid cells by 1,25-dihydroxyvitamin D3. *FASEB J.* **2005**, *19*, 1067–1077. [CrossRef] [PubMed]
37. Wang, T.T.; Nestel, F.P.; Bourdeau, V.; Nagai, Y.; Wang, Q.; Liao, J.; Tavera-Mendoza, L.; Lin, R.; Hanrahan, J.W.; Mader, S. Cutting edge: 1,25-dihydroxyvitamin D3 is a direct inducer of antimicrobial peptide gene expression. *J. Immunol.* **2004**, *173*, 2909–2912. [CrossRef] [PubMed]
38. Krutzik, S.R.; Hewison, M.; Liu, P.T.; Robles, J.A.; Stenger, S.; Adams, J.S.; Modlin, R.L. IL-15 links TLR2/1-induced macrophage differentiation to the vitamin D-dependent antimicrobial pathway. *J. Immunol.* **2008**, *181*, 7115–7120. [CrossRef] [PubMed]
39. Liu, N.; Kaplan, A.T.; Low, J.; Low, J.; Nguyen, L.; Liu, G.Y.; Equils, O.; Hewison, M. Vitamin D induces innate antibacterial responses in human trophoblasts via an intracrine pathway. *Biol. Reprod.* **2009**, *80*, 398–406. [CrossRef] [PubMed]
40. Ren, S.; Nguyen, L.; Wu, S.; Encinas, C.; Adams, J.S.; Hewison, M. Alternative splicing of vitamin D-24-hydroxylase: A novel mechanism for the regulation of extrarenal 1,25-dihydroxyvitamin D synthesis. *J. Biol. Chem.* **2005**, *280*, 20604. [CrossRef] [PubMed]
41. Sadeghi, K.; Wessner, B.; Laggner, U.; Ploder, M.; Tamandl, D.; Friedl, J.; Zügel, U.; Steinmeyer, A.; Pollak, A.; Roth, E.; et al. Vitamin D3 down-regulates monocyte TLR expression and triggers hyporesponsiveness to pathogen-associated molecular patterns. *Eur. J. Immunol.* **2006**, *36*, 361. [CrossRef] [PubMed]
42. Liu, Y.J. IPC: Professional type 1 interferon-producing cells and plasmacytoid dendritic cell precursors. *Annu. Rev. Immunol.* **2005**, *23*, 275–306. [CrossRef] [PubMed]
43. Colonna, M.; Trinchieri, G.; Liu, Y.J. Plasmacytoid dendritic cells in immunity. *Nat. Immunol.* **2004**, *5*, 1219–1226. [CrossRef] [PubMed]
44. Mahnke, K.; Johfnson, T.S.; Ring, S.; Enk, A.H. Tolerogenic dendritic cells and regulatory T cells: A two-way relationship. *J. Dermatol. Sci.* **2007**, *46*, 159–167. [CrossRef] [PubMed]
45. Brennan, A.; Katz, D.R.; Nunn, J.D.; Barker, S.; Hewison, M.; Fraher, L.J.; O'Riordan, J.L. Dendritic cells from human tissues express receptors for the immunoregulatory vitamin D3 metabolite, dihydroxycholecalciferol. *Immunology* **1987**, *61*, 457–461. [PubMed]
46. Dam, T.N.; Moller, B.; Hindkjaer, J.; Kragballe, K. The vitamin D3 analog calcipotriol suppresses the number and antigen-presenting function of Langerhans cells in normal human skin. *J. Investig. Dermatol. Symp. Proc.* **1996**, *1*, 72–77. [PubMed]
47. Gauzzi, M.C.; Purificato, C.; Donato, K.; Jin, Y.; Wang, L.; Daniel, K.C.; Maghazachi, A.A.; Belardelli, F.; Adorini, L.; Gessani, S. Suppressive effect of 1α,25-dihydroxyvitamin D3 on type I IFN-mediated monocyte differentiation into dendritic cells: Impairment of functional activities and chemotaxis. *J. Immunol.* **2005**, *174*, 270–276. [CrossRef] [PubMed]
48. Penna, G.; Adorini, L. 1α,25-dihydroxyvitamin D3 inhibits differentiation, maturation, activation, and survival of dendritic cells leading to impaired alloreactive T cell activation. *J. Immunol.* **2000**, *164*, 2405–2411. [CrossRef] [PubMed]
49. Griffin, M.D.; Lutz, W.H.; Phan, V.A.; Bachman, L.A.; McKean, D.J.; Kumar, R. Potent inhibition of dendritic cell differentiation and maturation by vitamin D analogs. *Biochem. Biophys. Res. Commun.* **2000**, *270*, 701–708. [CrossRef] [PubMed]
50. Gregori, S.; Casorati, M.; Amuchastegui, S.; Smiroldo, S.; Davalli, A.M.; Adorini, L. Regulatory T cells induced by 1α,25-dihydroxyvitamin D3 and mycophenolate mofetil treatment mediate transplantation tolerance. *J. Immunol.* **2001**, *167*, 1945–1953. [CrossRef] [PubMed]
51. Barrat, F.J.; Cua, D.J.; Boonstra, A.; Richards, D.F.; Crain, C.; Savelkoul, H.F.; de Waal-Malefyt, R.; Coffman, R.L.; Hawrylowicz, C.M.; O'Garra, A. In vitro generation of IL-10-producing regulatory CD4+ T cells is induced by immunosuppressive drugs and inhibited by Th1- and Th2-inducing cytokines. *J. Exp. Med.* **2002**, *195*, 603–616. [CrossRef] [PubMed]
52. Steinman, R.M.; Hawiger, D.; Nussenzweig, M.C. Tolerogenic dendritic cells. *Annu. Rev. Immunol.* **2003**, *21*, 685–711. [CrossRef] [PubMed]
53. Lemire, J.M.; Archer, D.C.; Beck, L.; Spiegelberg, H.L. Immunosuppressive actions of 1,25-dihydroxyvitamin D3: Preferential inhibition of Th1 functions. *J. Nutr.* **1995**, *125*, 1704S–1708S. [CrossRef] [PubMed]

54. Boonstra, A.; Barrat, F.J.; Crain, C.; Heath, V.L.; Savelkoul, H.F.; O'Garra, A. 1α,25-Dihydroxyvitamin D3 has a direct effect on naive CD4(+) T cells to enhance the development of Th2 cells. *J. Immunol.* **2001**, *167*, 4974–4980. [CrossRef] [PubMed]
55. Overbergh, L.; Decallonne, B.; Waer, M.; Rutgeerts, O.; Valckx, D.; Casteels, K.M.; Laureys, J.; Bouillon, R.; Mathieu, C. 1α,25-dihydroxyvitamin D3 induces an autoantigenspecific T-helper 1/T-helper 2 immune shift in NOD mice immunized with GAD65 (p524–543). *Diabetes* **2000**, *49*, 1301–1307. [CrossRef] [PubMed]
56. O'Kelly, J.; Hisatake, J.; Hisatake, Y.; Bishop, J.; Norman, A.; Koeffler, H.P. Normal myelopoiesis but abnormal T lymphocyte responses in vitamin D receptor knockout mice. *J. Clin. Investig.* **2002**, *109*, 1091–1099. [CrossRef] [PubMed]
57. Gregori, S.; Giarratana, N.; Smiroldo, S.; Uskokovic, M.; Adorini, L. A 1α,25-dihydroxyvitamin D (3) analog enhances regulatory T-cells and arrests autoimmune diabetes in NOD mice. *Diabetes* **2002**, *51*, 1367–1374. [CrossRef] [PubMed]
58. Weaver, C.T.; Hatton, R.D.; Mangan, P.R.; Harrington, L.E. IL-17 family cytokines and the expanding diversity of effector T cell lineages. *Annu. Rev. Immunol.* **2007**, *25*, 821–852. [CrossRef] [PubMed]
59. Daniel, C.; Sartory, N.A.; Zahn, N.; Radeke, H.H.; Stein, J.M. Immune modulatory treatment of TNBS colitis with calcitriol is associated with a change of a Th1/Th17 to a Th2 and regulatory T cell profile. *J. Pharmacol. Exp. Ther.* **2008**, *324*, 23–33. [CrossRef] [PubMed]
60. Liu, N.; Nguyen, L.; Chun, R.F.; Lagishetty, V.; Ren, S.; Wu, S.; Hollis, B.; DeLuca, H.F.; Adams, J.S.; Hewison, M. Altered endocrine and autocrine metabolism of vitamin D in amouse model of gastrointestinal inflammation. *Endocrinology* **2008**, *149*, 4799–4808. [CrossRef] [PubMed]
61. Adorini, L.; Penna, G.; Giarratana, N.; Roncari, A.; Amuchastegui, S.; Daniel, K.C.; Uskokovic, M. Dendritic cells as key targets for immunomodulation by Vitamin D receptor ligands. *J. Steroid Biochem. Mol. Biol.* **2004**, *89–90*, 437–441. [CrossRef] [PubMed]
62. Dong, X.; Bachman, L.A.; Kumar, R.; Griffin, M.D. Generation of antigen-specific, interleukin-10-producing T cells using dendritic cell stimulation and steroid hormone conditioning. *Transpl. Immunol.* **2003**, *11*, 323–333. [CrossRef]
63. Provvedini, D.M.; Manolagas, S.C. 1α,25-dihydroxyvitamin D3 receptor distribution and effects in subpopulations of normal human T lymphocytes. *J. Clin. Endocrinol. Metab.* **1989**, *68*, 774–779. [CrossRef] [PubMed]
64. Urry, Z.; Xystrakis, E.; Richards, D.F.; McDonald, J.; Sattar, Z.; Cousins, D.J.; Corrigan, C.J.; Hickman, E.; Brown, Z.; Hawrylowicz, C.M. Ligation of TLR9 induced on human IL-10-secreting Tregs by 1α,25-dihydroxyvitamin D3 abrogates regulatory function. *J. Clin. Investig.* **2009**, *119*, 387–398. [CrossRef] [PubMed]
65. Nunn, J.D.; Katz, D.R.; Barker, S.; Fraher, L.J.; Hewison, M.; Hendy, G.N.; O'Riordan, J.L. Regulation of human tonsillar T-cell proliferation by the active metabolite of vitamin D3. *Immunology* **1986**, *59*, 479–484. [PubMed]
66. Vanham, G.; Ceuppens, J.L.; Bouillon, R. T lymphocytes and their CD4 subset are direct targets for the inhibitory effect of calcitriol. *Cell Immunol.* **1989**, *124*, 320–333. [CrossRef]
67. Veldman, C.M.; Cantorna, M.T.; DeLuca, H.F. Expression of 1,25-dihydroxyvitamin D(3) receptor in the immune system. *Arch. Biochem. Biophys.* **2000**, *374*, 334–338. [CrossRef] [PubMed]
68. Willheim, M.; Thien, R.; Schrattbauer, K.; Bajna, E.; Holub, M.; Gruber, R.; Baier, K.; Pietschmann, P.; Reinisch, W.; Scheiner, O.; et al. Regulatory effects of 1α,25-dihydroxyvitamin D3 on the cytokine production of human peripheral blood lymphocytes. *J. Clin. Endocrinol. Metab.* **1999**, *84*, 3739–3744. [CrossRef] [PubMed]
69. Iho, S.; Iwamoto, K.; Kura, F.; Okuno, Y.; Takahashi, T.; Hoshino, T. Mechanism in 1,25(OH)2D3-induced suppression of helper/suppressor function of CD4/CD8 cells to immunoglobulin production in B cells. *Cell Immunol.* **1990**, *127*, 12–25. [CrossRef]
70. Meehan, T.F.; DeLuca, H.F. CD8(+) T cells are not necessary for 1α,25-dihydroxyvitamin D(3) to suppress experimental autoimmune encephalomyelitis in mice. *Proc. Natl. Acad. Sci. USA* **2002**, *99*, 5557–5560. [CrossRef] [PubMed]
71. Topilski, I.; Flaishon, L.; Naveh, Y.; Harmelin, A.; Levo, Y.; Shachar, I. The anti-inflammatory effects of 1,25-dihydroxyvitamin D3 on Th2 cells in vivo are due in part to the control of integrin-mediated T lymphocyte homing. *Eur. J. Immunol.* **2004**, *34*, 1068–1076. [CrossRef] [PubMed]

72. Sigmundsdottir, H.; Pan, J.; Debes, G.F.; Alt, C.; Habtezion, A.; Soler, D.; Butcher, E.C. DCs metabolize sunlight-induced vitamin D3 to "program" T cell attraction to the epidermal chemokine CCL27. *Nat. Immunol.* **2007**, *8*, 285–293. [CrossRef] [PubMed]
73. Yu, S.; Bruce, D.; Froicu, M.; Weaver, V.; Cantorna, M.T. Failure of T cell homing, reduced CD4/CD8αα intraepithelial lymphocytes, and inflammation in the gut of vitamin D receptor KO mice. *Proc. Natl. Acad. Sci. USA* **2008**, *105*, 20834–20839. [CrossRef] [PubMed]
74. Rolf, L.; Muris, H.A.; Hupperts, R.; Damoiseaux, J. Vitamin D effects on B cell function in autoimmunity. *Ann. N.Y. Acad. Sci.* **2014**, *1317*, 84–91. [CrossRef] [PubMed]
75. Baumgarth, N. The double life of a B-1 cell: Selfreactivity selects for protective effector functions. *Nat. Rev. Immunol.* **2011**, *11*, 34–46. [CrossRef] [PubMed]
76. Van Leeuwen, M.; Damoiseaux, J.; Duijvestijn, A.; Tervaert, J.W. The therapeutic potential of targeting B cells and anti-oxLDL antibodies in atherosclerosis. *Autoimmun. Rev.* **2009**, *9*, 53–57. [CrossRef] [PubMed]
77. Kalampokis, I.; Yoshizaki, A.; Tedder, T.F. IL-10-producing regulatory B cells (B10 cells) in autoimmune disease. *Arthritis Res. Ther.* **2013**, *15*, S1. [CrossRef] [PubMed]
78. Mauri, C.; Bosma, A. Immune regulatory function of B cells. *Annu. Rev. Immunol.* **2012**, *30*, 221–241. [CrossRef] [PubMed]
79. Fillatreau, S.; Sweenie, C.H.; McGeachy, M.J.; Gray, D.; Anderton, S.M. B cells regulate autoimmunity by provision of IL-10. *Nat. Immunol.* **2002**, *3*, 944–950. [CrossRef] [PubMed]
80. Mauri, C.; Gray, D.; Mushtaq, N.; Londei, M. Prevention of arthritis by interleukin 10-producing B cells. *J. Exp. Med.* **2003**, *197*, 489–501. [CrossRef] [PubMed]
81. Sun, J.B.; Flach, C.F.; Czerkinsky, C.; Holmgren, J. B lymphocytes promote expansion of regulatory T cells in oral tolerance: Powerful induction by antigen coupled to cholera toxin B subunit. *J. Immunol.* **2008**, *181*, 8278–8287. [CrossRef] [PubMed]
82. Blair, P.A.; Noreña, L.Y.; Flores-Borja, F.; Rawlings, D.J.; Isenberg, D.A.; Ehrenstein, M.R.; Mauri, C. CD19(+) CD24(hi) CD38(hi) B cells exhibit regulatory capacity in healthy individuals but are functionally impaired in systemic lupus erythematosus patients. *Immunity* **2010**, *32*, 129–140. [CrossRef] [PubMed]
83. Duddy, M.; Niino, M.; Adatia, F.; Hebert, S.; Freedman, M.; Atkins, H.; Kim, H.J.; Bar-Or, A. Distinct effector cytokine profiles of memory and naive human B cell subsets and implication in multiple sclerosis. *J. Immunol.* **2007**, *178*, 6092–6099. [CrossRef] [PubMed]
84. Knippenberg, S.; Peelen, E.; Smolders, J.; Thewissen, M.; Menheere, P.; Tervaert, J.W.; Hupperts, R.; Damoiseaux, J. Reduction in IL-10 producing B cells (B reg) in multiple sclerosis is accompanied by a reduced naive/memory B reg ratio during a relapse but not in remission. *J. Neuroimmunol.* **2011**, *239*, 80–86. [CrossRef] [PubMed]
85. Sun, F.; Ladha, S.S.; Yang, L.; Liu, Q.; Shi, S.X.; Su, N.; Bomprezzi, R.; Shi, F.D. Interleukin-10 producing-B cells and their association with responsiveness to rituximab in myasthenia gravis. *Muscle Nerve* **2014**, *49*, 487–494. [CrossRef] [PubMed]
86. Chen, S.; Sims, G.P.; Chen, X.X.; Gu, Y.Y.; Chen, S.; Lipsky, P.E. Modulatory effects of 1,25-dihydroxyvitamin D3 on human B cell differentiation. *J. Immunol.* **2007**, *179*, 1634–1647. [CrossRef] [PubMed]
87. Morgan, J.W.; Kouttab, N.; Ford, D.; Maizel, A.L. Vitamin D-mediated gene regulation in phenotypically defined human B cell subpopulations. *Endocrinology* **2000**, *141*, 3225–3234. [CrossRef] [PubMed]
88. Heine, G.; Niesner, U.; Chang, H.D.; Steinmeyer, A.; Zügel, U.; Zuberbier, T.; Radbruch, A.; Worm, M. 1,25-dihydroxyvitamin D(3) promotes IL-10 production in human B cells. *Eur. J. Immunol.* **2000**, *38*, 2210–2218. [CrossRef] [PubMed]
89. Lugar, P.L.; Love, C.; Grammer, A.C.; Dave, S.S.; Lipsky, P.E. Molecular characterization of circulating plasma cells in patients with active systemic lupus erythematosus. *PLoS ONE* **2012**, *7*, e44362. [CrossRef] [PubMed]
90. Lemire, J.M.; Adams, J.S.; Sakai, R.; Jordan, S.C. 1α,25-dihydroxyvitamin D3 suppresses proliferation and immunoglobulin production by normal human peripheral blood mononuclear cells. *J. Clin. Investig.* **1984**, *74*, 657–661. [CrossRef] [PubMed]
91. Chen, W.C.; Vayuvegula, B.; Gupta, S. 1,25-Dihydroxyvitamin D3-mediated inhibition of human B cell differentiation. *Clin. Exp. Immunol.* **1987**, *69*, 639–646. [PubMed]
92. Provedini, D.M.; Tsoukas, C.D.; Deftos, L.J.; Manolagas, S.C. 1α,25- Dihydroxyvitamin D3-binding macromolecules in human B lymphocytes: Effects on immunoglobulin production. *J. Immunol.* **1986**, *136*, 2734–2740.

93. Iho, S.; Takahashi, T.; Kura, F.; Sugiyama, H.; Hoshino, T. The effect of 1,25-dihydroxyvitamin D3 on in vitro immunoglobulin production in human B cells. *J. Immunol.* **1986**, *136*, 4427–4431. [PubMed]
94. Hartmann, B.; Heine, G.; Babina, M.; Steinmeyer, A.; Zügel, U.; Radbruch, A.; Worm, M. Targeting the vitamin D receptor inhibits the B cell-dependent allergic immune response. *Allergy* **2011**, *66*, 540–548. [CrossRef] [PubMed]
95. Heine, G.; Anton, K.; Henz, B.M.; Worm, M. 1α,25-dihydroxyvitamin D3 inhibits anti-CD40 plus IL-4-mediated IgE production in vitro. *Eur. J. Immunol.* **2002**, *32*, 3395–3404. [CrossRef]
96. Nashold, F.E.; Hoag, K.A.; Goverman, J.; Hayes, C.E. Rag-1-dependent cells are necessary for 1,25-dihydroxyvitamin D(3) prevention of experimental autoimmune encephalomyelitis. *J. Neuroimmunol.* **2001**, *119*, 16–29. [CrossRef]
97. Spach, K.M.; Nashold, F.E.; Dittel, B.N.; Hayes, C.E. IL-10 signaling is essential for 1,25 dihydroxyvitamin D3-mediated inhibition of experimental autoimmune encephalomyelitis. *J. Immunol.* **2006**, *177*, 6030–6037. [CrossRef] [PubMed]
98. Chun, R.F.; Liu, P.T.; Modlin, R.L.; Adams, J.S.; Hewison, M. Impact of vitamin D on immune function: Lessons learned from genome-wide analysis. *Front. Phisiol.* **2014**, *5*, 151. [CrossRef] [PubMed]
99. Kallas, M.; Green, F.; Hewison, M.; White, C.; Kline, G. Rare causes of calcitriol-mediated hypercalcemia: A case report and literature review. *J. Clin. Endocrinol. Metab.* **2010**, *95*, 3111–3117. [CrossRef] [PubMed]
100. Reichel, H.; Koeffler, H.P.; Norman, A.W. Regulation of 25-hydroxyvitamin D3 metabolism in a human promyelocytic leukemia cell line (HL60):1,25-dihydroxyvitamin D3 stimulates the synthesis of 24, 25-dihydroxyvitamin D3. *Arch. Biochem. Biophys.* **1986**, *251*, 222–231. [CrossRef]
101. Akutsu, N.; Lin, R.; Bastien, Y.; Bestawros, A.; Enepekides, D.J.; Black, M.J.; White, J.H. Regulation of gene expression by 1α,25-dihydroxyvitamin D3 and its analog EB1089 undergrowth-inhibitory conditions in squamous carcinoma cells. *Mol. Endocrinol.* **2001**, *15*, 1127–1139. [CrossRef] [PubMed]
102. Lin, R.; Nagai, Y.; Sladek, R.; Bastien, Y.; Ho, J.; Petrecca, K.; Sotiropoulou, G.; Diamandis, E.P.; Hudson, T.J.; White, J.H. Expression profiling in squamous carcinoma cells reveals pleiotropic effects of vitaminD3 analog EB1089 signaling on cell proliferation, differentiation, and immune system regulation. *Mol. Endocrinol.* **2002**, *16*, 1243–1256. [CrossRef] [PubMed]
103. Wang, T.T.; Tavera-Mendoza, L.E.; Laperriere, D.; Libby, E.; MacLeod, N.B.; Nagai, Y.; Bourdeau, V.; Konstorum, A.; Lallemant, B.; Zhang, R.; et al. Large-scale in silico and microarray-based identification of direct1,25-dihydroxyvitamin D3 target genes. *Mol. Endocrinol.* **2005**, *19*, 2685–2695. [CrossRef] [PubMed]
104. Liu, P.T.; Schenk, M.; Walker, V.P.; Dempsey, P.W.; Kanchanapoomi, M.; Wheelwright, M.; Vazirnia, A.; Zhang, X.; Steinmeyer, A.; Zügel, U.; et al. Convergence of IL-1 β and VDR activation pathways in human TLR2/1-induced antimicrobial responses. *PLoS ONE* **2009**, *4*, e5810. [CrossRef] [PubMed]
105. Wang, T.T.; Dabbas, B.; Laperriere, D.; Bitton, A.J.; Soualhine, H.; Tavera-Mendoza, L.E.; Dionne, S.; Servant, M.J.; Bitton, A.; Seidman, E.G.; et al. Direct and indirect induction by 1,25-dihydroxyvitamin D3 of the NOD2/CARD15-defensin β2 innate immune pathway defective in Crohn disease. *J. Biol. Chem.* **2010**, *285*, 2227–2231. [CrossRef] [PubMed]
106. Moresco, E.M.; LaVine, D.; Beutler, B. Toll-like receptors. *Curr. Biol.* **2011**, *21*, R488–R493. [CrossRef] [PubMed]
107. Bacchetta, J.Z.J.; Sea, J.L.; Sea, J.L.; Chun, R.F.; Lisse, T.S.; Zavala, K.; Nayak, A.; Wesseling-Perry, K.; Westerman, M.; Hollis, B.W.; et al. Suppression of iron-regulatory hepcidin by vitamin D. *J. Am. Soc. Nephrol.* **2014**, *25*, 564–572. [CrossRef] [PubMed]
108. Ganz, T. Iron in innate immunity: Starve the invaders. *Curr. Opin. Immunol.* **2009**, *21*, 63–67. [CrossRef] [PubMed]
109. Drakesmith, H.; Prentice, A.M. Hepcidin and the iron-infection axis. *Science* **2012**, *338*, 768–772. [CrossRef] [PubMed]
110. Kim, T.H.; Lee, B.; Kwon, E.; Choi, S.J.; Lee, Y.H.; Song, G.G.; Sohn, J.; Ji, J.D. Regulation of TREM-1 expression by 1,25-dihydroxyvitamin D3 in human monocytes/macrophages. *Immunol. Lett.* **2013**, *154*, 80–85. [CrossRef] [PubMed]
111. Fritsche, J.; Mondal, K.; Ehrnsperger, A.; Andreesen, R.; Kreutz, M. Regulation of 25-hydroxyvitamin D3-1α-hydroxylase and production of 1 α,25-dihydroxyvitamin D3 by human dendritic cells. *Blood* **2003**, *102*, 3314–3316. [CrossRef] [PubMed]

112. Hewison, M.; Freeman, L.; Hughes, S.V.; Evans, K.N.; Bland, R.; Eliopoulos, A.G.; Kilby, M.D.; Moss, P.A.; Chakraverty, R. Differential regulation of vitamin D receptor and its ligand in human monocyte-derived dendritic cells. *J. Immunol.* **2003**, *170*, 5382–5390. [CrossRef] [PubMed]
113. Ooi, J.H.; McDaniel, K.L.; Weaver, V.; Cantorna, M.T. Murine CD8+T cells but not macrophages express the vitamin D1α-hydroxylase. *J. Nutr. Biochem.* **2014**, *25*, 58–65. [CrossRef] [PubMed]
114. Wilkinson, R.J.; Llewelyn, M.; Toossi, Z.; Patel, P.; Pasvol, G.; Lalvani, A.; Wright, D.; Latif, M.; Davidson, R.N. Influence of vitamin D deficiency and vitamin D receptor polymorphisms on tuberculosis among Gujarati Asians in west London: A case-control study. *Lancet* **2000**, *355*, 618–621. [CrossRef]
115. Ustianowski, A.; Shaffer, R.; Collin, S.; Wilkinson, R.J.; Davidson, R.N. Prevalence and associations of vitamin D deficiency in foreign-born persons with tuberculosis in London. *J. Infect.* **2005**, *50*, 432–437. [CrossRef] [PubMed]
116. Williams, B.; Williams, A.J.; Anderson, S.T. Vitamin D deficiency and insufficiency in children with tuberculosis. *Pediatr. Infect. Dis. J.* **2008**, *27*, 941–942. [CrossRef] [PubMed]
117. Wejse, C.; Gomes, V.F.; Rabna, P.; Gustafson, P.; Aaby, P.; Lisse, I.M.; Andersen, P.L.; Glerup, H.; Sodemann, M. Vitamin D as supplementary treatment for tuberculosis: A double-blind, randomized, placebo-controlled trial. *Am. J. Respir. Crit. Care Med.* **2009**, *179*, 843–850. [CrossRef] [PubMed]
118. Martineau, A.R.; Timms, P.M.; Bothamley, G.H.; Hanifa, Y.; Islam, K.; Claxton, A.P.; Packe, G.E.; Moore-Gillon, J.C.; Darmalingam, M.; Davidson, R.N.; et al. High-dose vitamin D (3) during intensive-phase antimicrobial treatment of pulmonary tuberculosis: A double-blind randomised controlled trial. *Lancet* **2011**, *377*, 242–250. [CrossRef]
119. Coussens, A.K.; Wilkinson, R.J.; Hanifa, Y.; Nikolayevskyy, V.; Elkington, P.T.; Islam, K.; Timms, P.M.; Venton, T.R.; Bothamley, G.H.; Packe, G.E.; et al. Vitamin D accelerates resolution of inflammatory responses during tuberculosis treatment. *Proc. Natl. Acad. Sci. USA.* **2012**, *109*, 15449–15454. [CrossRef] [PubMed]
120. Jeng, L.; Yamshchikov, A.V.; Judd, S.E.; Blumberg, H.M.; Martin, G.S.; Ziegler, T.R.; Tangpricha, V. Alterations in vitamin D status and antimicrobial peptide levels in patients in the intensive care unit with sepsis. *J. Transl. Med.* **2009**, *7*, 28. [CrossRef] [PubMed]
121. Cannell, J.J.; Vieth, R.; Umhau, J.C.; Holick, M.F.; Grant, W.B.; Madronich, S.; Garland, C.F.; Giovannucci, E. Epidemic influenza and vitamin D. *Epidemiol. Infect.* **2006**, *134*, 1129–1140. [CrossRef] [PubMed]
122. Hossein-nezhad, A.; Spira, A.; Holick, M.F. Influence of vitamin D status and vitamin D3 supplementation on genome wide expression of white blood cells: A randomized double-blind clinical trial. *PLoS ONE* **2013**, *8*, e58725. [CrossRef] [PubMed]
123. Lagishetty, V.; Misharin, A.V.; Liu, N.Q.; Lisse, T.S.; Chun, R.F.; Ouyang, Y.; McLachlan, S.M.; Adams, J.S.; Hewison, M. Vitamin D deficiency in mice impairs colonic antibacterial activity and predisposes to colitis. *Endocrinology* **2010**, *151*, 2423–2432. [CrossRef] [PubMed]
124. Hooper, L.V.; Stappenbeck, T.S.; Hong, C.V.; Gordon, J.I. Angiogenins: A new class of microbicidal proteins involved in innate immunity. *Nat. Immunol.* **2003**, *4*, 269–273. [CrossRef] [PubMed]
125. Packey, C.D.; Sartor, R.B. Commensal bacteria, traditional and opportunistic pathogens, dysbiosis and bacterial killing in inflammatory bowel diseases. *Curr. Opin. Infect. Dis.* **2009**, *22*, 292–301. [CrossRef] [PubMed]
126. Lauridsen, A.L.; Vestergaard, P.; Hermann, A.P.; Brot, C.; Heickendorff, L.; Mosekilde, L.; Nexo, E. Plasma concentrations of 25-hydroxy-vitamin D and 1,25-dihydroxy-vitamin D are related to the phenotype of Gc (vitamin D-binding protein): Across-sectional study on 595 early postmenopausal women. *Calcif. Tissue Int.* **2005**, *77*, 15–22. [CrossRef] [PubMed]
127. Lauridsen, A.L.; Vestergaard, P.; Nexo, E. Mean serum concentration of vitamin D-binding protein (Gc globulin) is related to the Gc phenotype in women. *Clin. Chem.* **2001**, *47*, 753–756. [PubMed]
128. Wang, T.J.; Zhang, F.; Richards, J.B.; Kestenbaum, B.; Van Meurs, J.B.; Berry, D.; Kiel, D.P.; Streeten, E.A.; Ohlsson, C.; Koller, D.L.; et al. Common genetic determinants of vitamin D insufficiency: A genome-wide association study. *Lancet* **2010**, *376*, 180–188. [CrossRef]
129. Chun, R.F.; Lauridsen, A.L.; Suon, L.; Zella, L.A.; Pike, J.W.; Modlin, R.L.; Martineau, A.R.; Wilkinson, R.J.; Adams, J.; Hewison, M. Vitamin D-binding protein directs monocyte responses to 25-hydroxy-and 1,25-dihydroxy vitamin D. *J. Clin. Endocrinol. Metab.* **2010**, *95*, 3368–3376. [CrossRef] [PubMed]

130. Chun, R.F.; Peercy, B.E.; Adams, J.S.; Hewison, M. Vitamin D binding protein and monocyte response to 25-hydroxy vitamin D and 1,25-dihydroxyvitamin D: Analysis by mathematical modeling. *PLoS ONE* **2012**, *7*, e30773. [CrossRef] [PubMed]
131. Kim, M.; Mirandola, L.; Pandey, A.; Nguyen, D.D.; Jenkins, M.R.; Turcel, M.; Cobos, E.; Chiriva-Internati, M. Application of vitamin D and derivatives in hematological malignancies. *Cancer Lett.* **2012**, *319*, 8–22. [CrossRef] [PubMed]
132. Miyaura, C.; Abe, E.; Kuribayashia, T.; Tanakaa, H.; Konno, K.; Nishii, Y.; Suda, T. 1α,25-dihydroxyvitamin D3 induced differentiation of human myeloid leukemia cells. *Biochem. Biophys. Res. Commun.* **1981**, *105*, 937–943. [CrossRef]
133. Choudhuri, U.; Adams, J.A.; Byrom, N.; McCarthy, D.M.; Barrett, J. 1,25-Dihydroxyvitamin D3 induces normal mononuclear blood cells to differentiate in the direction of monocyte-macrophages. *Haematologia (Budap)* **1990**, *23*, 9–19. [PubMed]
134. Takahashi, T.; Suzuki, A.; Ichiba, S.; Okuno, Y.; Sugiyama, H.; Imura, H.; Nakamura, K.; Iho, S.; Hoshino, T. Effect of 1,25(OH)2D3 on normal human CFU-GM: Target cells of the agent in the suppression of colony formation and induction of macrophage colonies. *Int. J. Hematol.* **1991**, *54*, 57–63. [PubMed]
135. Hall, A.C.; Juckett, M.B. The Role of vitamin D in hematologic disease and stem cell transplantation. *Nutrients* **2013**, *5*, 2206–2221. [CrossRef] [PubMed]
136. Bunce, C.M.; Brown, G.; Hewison, M. Vitamin D and hematopoiesis. *Trends Endocrinol. Metab.* **1997**, *8*, 245–251. [CrossRef]
137. Grande, A.; Montanari, M.; Tagliafico, E.; Manfredini, R.; Zanocco Marani, T.; Siena, M.; Tenedini, E.; Gallinelli, A.; Ferrari, S. Physiological levels of 1α,25 dihydroxyvitamin D3 induce the monocytic commitment of CD34+ hematopoietic progenitors. *J. Leukoc. Biol.* **2002**, *71*, 641–651. [CrossRef] [PubMed]
138. Friedman, A.D. Transcriptional control of granulocyte and monocyte development. *Oncogene* **2007**, *26*, 6816–6828. [CrossRef] [PubMed]
139. Labrecque, J.; Allan, D.; Chambon, P.; Iscove, N.N.; Lohnes, D.; Hoang, T. Impaired granulocytic differentiation in vitro in hematopoietic cells lacking retinoic acid receptors α1 and gamma. *Blood* **1998**, *92*, 607–615. [PubMed]
140. Studzinski, G.P.; Harrison, J.S.; Wang, X.; Sarkar, S.; Kalia, V.; Danilenko, M. Vitamin D control of hematopoietic cell differentiation and leukemia. *J. Cell Biochem.* **2015**, *116*, 1500–1512. [CrossRef] [PubMed]
141. Hughes, P.J.; Marcinkowska, E.; Gocek, E.; Studzinski, G.P.; Brown, G. Vitamin D3-driven signals for myeloid cell differentiation-implications for differentiation therapy. *Leuk. Res.* **2010**, *34*, 553–565. [CrossRef] [PubMed]
142. Kharbanda, S.; Saleem, A.; Emoto, Y.; Stone, R.; Rapp, U.; Kufe, D. Activation of raf-1 and mitogen-activated protein kinases during monocytic differentiation of human myeloid leukemia cells. *J. Biol. Chem.* **1994**, *269*, 872–878. [PubMed]
143. Okazaki, T.; Bielawska, A.; Domae, N.; Bell, R.M.; Hannun, Y.A. Characteristics and partial purification of a novel cytosolic, magnesium-independent, neutral sphingomyelinase activated in the early signal transduction of 1α,25-dihydroxyvitamin D3-induced HL-60 cell differentiation. *J. Biol. Chem.* **1994**, *269*, 4070–4077. [PubMed]
144. Okazaki, T.; Bielawska, A.; Bell, R.M.; Hannun, Y.A. Role of ceramide as a lipid mediator of 1α,25-dihydroxyvitamin D3-induced HL-60 cell differentiation. *J. Biol. Chem.* **1990**, *265*, 15823–15831. [PubMed]
145. Hmama, Z.; Nandan, D.; Sly, L.; Knutson, K.L.; Herrera-Velit, P.; Reiner, N.E. 1α,25-dihydroxyvitamin D3–induced myeloid cell differentiation is regulated by a vitamin D receptor–phosphatidylinositol 3-kinase signaling complex. *J. Exp. Med.* **1999**, *190*, 1583–1594. [CrossRef] [PubMed]
146. Zhang, Y.; Zhang, J.; Studzinski, G.P. AKT pathway is activated by 1, 25-dihydroxyvitamin D3 and participates in its anti-apoptotic effect and cell cycle control in differentiating HL60 cells. *Cell Cycle* **2006**, *5*, 447–451. [CrossRef] [PubMed]
147. Yu, S.; Cantorna, M.T. The vitamin D receptor is required for iNKT cell development. *Proc. Natl. Acad. Sci. USA* **2008**, *105*, 5207–5212. [CrossRef] [PubMed]
148. Weeres, M.A.; Robien, K.; Ahn, Y.O.; Neulen, M.L.; Bergerson, R.; Miller, J.S.; Verneris, M.R. The effects of 1,25-dihydroxyvitamin D3 on in vitro human NK cell development from hematopoietic stem cells. *J. Immunol.* **2014**, *193*, 3456–3462. [CrossRef] [PubMed]

149. Jeanson, N.T.; Scadden, D.T. Vitamin D receptor deletion leads to increased hematopoietic stem and progenitor cells residing in the spleen. *Blood* **2010**, *116*, 4126–4129. [CrossRef] [PubMed]
150. Cortes, M.; Chen, M.J.; Stachura, D.L.; Liu, S.Y.; Kwan, W.; Wright, F.; Vo, L.T.; Theodore, L.N.; Esain, V.; Frost, I.M.; et al. Developmental Vitamin D availability impacts hematopoietic stem cell production. *Cell Rep.* **2016**, *17*, 458–468. [CrossRef] [PubMed]
151. Cortes, M.; Liu, S.Y.; Kwan, W.; Alexa, K.; Goessling, W.; North, T.E. Accumulation of the vitamin D precursor cholecalciferol antagonizes hedgehog signaling to impair hemogenic endothelium formation. *Stem Cell Rep.* **2015**, *5*, 471–479. [CrossRef] [PubMed]
152. Muto, A.; Kizaki, M.; Yamato, K.; Kawai, Y.; Kamata-Matsushita, M.; Ueno, H.; Ohguchi, M.; Nishihara, T.; Koeffler, H.P.; Ikeda, Y. 1,25-dihydroxyvitamin D3 induces differentiation of a retinoic acid–resistant acute promyelocytic leukemia cell line (UF-1) associated with expression of p21 WAF1/CIP1 and p27 KIP1. *Blood* **1999**, *93*, 2225–2233. [PubMed]
153. Bastie, J.N.; Balitrand, N.; Guidez, F.; Guillemot, I.; Larghero, J.; Calabresse, C.; Chomienne, C.; Delva, L. 1α,25-dihydroxyvitamin D3 transrepresses retinoic acid transcriptional activity via vitamin D receptor in myeloid cells. *Mol. Endocrinol.* **2004**, *18*, 2685–2699. [CrossRef] [PubMed]
154. Ji, Y.; Studzinski, G.P. Retinoblastoma protein and CCAAT/enhancer-binding protein β are required for 1,25-dihydroxyvitamin D3-induced monocytic differentiation of HL60 cells. *Cancer Res.* **2004**, *64*, 370–377. [CrossRef] [PubMed]
155. Gemelli, C.; Orlandi, C.; Zanocco Marani, T.; Martello, A.; Vignudelli, T.; Ferrari, F.; Montanari, M.; Parenti, S.; Testa, A.; Grande, A.; et al. The vitamin D3/Hox-A10 pathway supports MafB function during the monocyte differentiation of human CD34+ hemopoietic progenitors. *J. Immunol.* **2008**, *181*, 5660–5672. [CrossRef] [PubMed]
156. Koeffler, H.P.; Hirji, K.; Itri, L. 1,25-dihydroxyvitamin D3: In vivo and in vitro effects on human preleukemic and leukemic cells. *Cancer Treat. Rep.* **1985**, *69*, 1399–1407. [PubMed]
157. Callens, C.; Coulon, S.; Naudin, J.; Radford-Weiss, I.; Boissel, N.; Raffoux, E.; Wang, P.H.; Agarwal, S.; Tamouza, H.; Paubelle, E.; et al. Targeting iron homeostasis induces cellular differentiation and synergizes with differentiating agents in acute myeloid leukemia. *J. Exp. Med.* **2010**, *207*, 731–750. [CrossRef] [PubMed]
158. Molnar, I.; Stark, N.; Lovato, J.; Powell, B.L.; Cruz, J.; Hurd, D.D.; Mathieu, J.S.; Chen, T.C.; Holick, M.F.; Cambra, S.; et al. Treatment of low-risk myelodysplastic syndromes with high-dose daily oral cholecalciferol (2000–4000 IU vitamin D(3)). *Leukemia* **2007**, *21*, 1089–1092. [CrossRef] [PubMed]
159. Motomura, S.; Kanamori, H.; Maruta, A.; Kodama, F.; Ohkubo, T. The effect of 1-hydroxyvitamin D3 for prolongation of leukemic transformation-free survival in myelodysplastic syndromes. *Am. J. Hematol.* **1991**, *38*, 67–68. [CrossRef] [PubMed]
160. Siitonen, T.; Timonen, T.; Juvonen, E.; Terävä, V.; Kutila, A.; Honkanen, T.; Mikkola, M.; Hallman, H.; Kauppila, M.; Nyländen, P.; et al. Valproic acid combined with 13-*cis* retinoic acid and 1,25-dihydroxyvitamin D3 in the treatment of patients with myelodysplastic syndromes. *Haematologica* **2007**, *92*, 1119–1122. [CrossRef] [PubMed]
161. Ferrero, D.; Darbesio, A.; Giai, V.; Genuardi, M.; Dellacasa, C.M.; Sorasio, R.; Bertini, M.; Boccadoro, M. Efficacy of a combination of human recombinant erythropoietin + 13-*cis*-retinoic acid and dihydroxylated vitamin D3 to improve moderate to severe anaemia in low/intermediate risk myelodysplastic syndromes. *Br. J. Haematol.* **2009**, *144*, 342–349. [CrossRef] [PubMed]
162. Slapak, C.A.; Desforges, J.F.; Fogaren, T.; Miller, K.B. Treatment of acute myeloid leukemia in the elderly with low-dose cytarabine, hydroxyurea, and calcitriol. *Am. J. Hematol.* **1992**, *41*, 178–183. [CrossRef] [PubMed]
163. Harrison, J.S.; Bershadskiy, A. Clinical experience using vitamin D and analogs in the treatment of myelodysplasia and acute myeloid leukemia: A review of the literature. *Leuk. Res. Treat.* **2012**, *2012*, 125814. [CrossRef] [PubMed]
164. Ferrero, D.; Campa, E.; Dellacasa, C.; Campana, S.; Foli, C.; Boccadoro, M. Differentiating agents + low-dose chemotherapy in the management of old/poor prognosis patients with acute myeloid leukemia or myelodysplastic syndrome. *Haematologica* **2004**, *89*, 619–620. [PubMed]

165. Hellstrom, E.; Robert, K.H.; Hellström, E.; Robèrt, K.H.; Samuelsson, J.; Lindemalm, C.; Grimfors, G.; Kimby, E.; Oberg, G.; Winqvist, I.; et al. Treatment of myelodysplastic syndromes with retinoic acid and 1α-hydroxy-vitamin D3 in combination with low-dose ara-c is not superior to ara-c alone. Results from a randomized study. The scandinavian myelodysplasia group (SMG). *Eur. J. Haematol.* **1990**, *45*, 255–261. [CrossRef] [PubMed]
166. Radujkovic, A.; Schnitzler, P.; Ho, A.D.; Dreger, P.; Luft, T. Low serum vitamin D levels are associated with shorter survival after first-line azacitidine treatment in patients with myelodysplastic syndrome and secondary oligoblastic acute myeloid leukemia. *Clin. Nutr.* **2017**, *36*, 542–551. [CrossRef] [PubMed]
167. Lee, H.J.; Muindi, J.R.; Tan, W.; Hu, Q.; Wang, D.; Liu, S.; Wilding, G.E.; Ford, L.A.; Sait, S.N.; Block, A.W.; et al. Low 25(OH) vitamin D3 levels are associated with adverse outcome in newly diagnosed, intensively treated adult acute myeloid leukemia. *Cancer* **2014**, *120*, 521–529. [CrossRef] [PubMed]
168. Pardanani, A.; Drake, M.T.; Finke, C.; Lasho, T.L.; Rozell, S.A.; Jimma, T.; Tefferi, A. Vitamin D insufficiency in myeloproliferative neoplasms and myelodysplastic syndromes: Clinical correlates and prognostic studies. *Am. J. Hematol.* **2011**, *86*, 1013–1016. [CrossRef] [PubMed]
169. Park, W.H.; Seol, J.G.; Kim, E.S.; Binderup, L.; Koeffler, H.P.; Kim, B.K.; Lee, Y.Y. The induction of apoptosis by a combined 1,25(OH)2D3 analog, EB1089 and TGF-β1 in NCI-H929 multiple myeloma cells. *Int. J. Oncol.* **2002**, *20*, 533–542. [CrossRef] [PubMed]
170. Park, W.H.; Seol, J.G.; Kim, E.S.; Jung, C.W.; Lee, C.C.; Binderup, L.; Koeffler, H.P.; Kim, B.K.; Lee, Y.Y. Cell cycle arrest induced by the vitamin D(3) analog EB1089 in NCI-H929 myeloma cells is associated with induction of the cyclin-dependent kinase inhibitor p27. *Exp. Cell Res.* **2000**, *254*, 279–286. [CrossRef] [PubMed]
171. Puthier, D.; Bataille, R.; Barillé, S.; Mellerin, M.P.; Harousseau, J.L.; Ponzio, A.; Robillard, N.; Wijdenes, J.; Amiot, M. Myeloma cell growth arrest, apoptosis, and interleukin-6 receptor modulation induced by EB1089, a vitamin D3 derivative, alone or in association with dexamethasone. *Blood* **1996**, *88*, 4659–4666. [PubMed]
172. Luong, Q.T.; Koeffler, H.P. Vitamin D compounds in leukemia. *J. Steroid Biochem. Mol. Biol.* **2005**, *97*, 195–202. [CrossRef] [PubMed]
173. Kelly, J.L.; Salles, G.; Goldman, B.; Fisher, R.I.; Brice, P.; Press, O.; Casasnovas, O.; Maloney, D.G.; Soubeyran, P.; Rimsza, L.; et al. Low serum vitamin D levels are associated with inferior survival in follicular lymphoma: A prospective evaluation in swog and lysa studies. *J. Clin. Oncol.* **2015**, *33*, 1482–1490. [CrossRef] [PubMed]
174. Tracy, S.I.; Maurer, M.J.; Witzig, T.E.; Drake, M.T.; Ansell, S.M.; Nowakowski, G.S.; Thompson, C.A.; Inwards, D.J.; Johnston, P.B.; Micallef, I.N.; et al. Vitamin D insufficiency is associated with an increased risk of early clinical failure in follicular lymphoma. *Blood Cancer J.* **2017**, *7*, e595. [CrossRef] [PubMed]
175. Hohaus, S.; Tisi, M.C.; Bellesi, S.; Maiolo, E.; Alma, E.; Tartaglia, G.; Corrente, F.; Cuccaro, A.; D'Alo', F.; Basile, U.; et al. Vitamin D deficiency and supplementation in patients with aggressive B-cell lymphomas treated with immunochemotherapy. *Cancer Med.* **2018**, *7*, 270–281. [CrossRef] [PubMed]
176. Cho, H.J.; Shin, D.Y.; Kim, J.H.; Bae, J.Y.; Lee, K.H.; See, C.J.; Kim, N.; Park, E.K.; Ra, E.K.; Lee, J.E.; et al. Impact of vitamin D receptor gene polymorphisms on clinical outcomes of HLA-matched sibling hematopoietic stem cell transplantation. *Clin. Transplant.* **2012**, *26*, 476–483. [CrossRef] [PubMed]
177. Bogunia-Kubik, K.; Middleton, P.; Norden, J.; Dickinson, A.; Lange, A. Association of vitamin D receptor polymorphisms with the outcome of allogeneic haematopoietic stem cell transplantation. *Int. J. Immunogenet.* **2008**, *35*, 207–213. [CrossRef] [PubMed]
178. Middleton, P.G.; Cullup, H.; Dickinson, A.M.; Norden, J.; Jackson, G.H.; Taylor, P.R.; Cavet, J. Vitamin D receptor gene polymorphism associates with graft-versus-host disease and survival in HLA-matched sibling allogeneic bone marrow transplantation. *Bone Marrow Transplant.* **2002**, *30*, 223–228. [CrossRef] [PubMed]
179. Glotzbecker, B.; Ho, V.T.; Aldridge, J.; Kim, H.T.; Horowitz, G.; Ritz, J.; Soiffer, R.; Avigan, D.; Rosenblatt, J. Low levels of 25-hydroxyvitamin D before allogeneic hematopoietic SCT correlate with the development of chronic GVHD. *Bone Marrow Transplant.* **2013**, *48*, 593–597. [CrossRef] [PubMed]
180. Von Bahr, L.; Blennow, O.; Alm, J.; Björklund, A.; Malmberg, K.J.; Mougiakakos, D.; Le Blanc, A.; Oefner, P.J.; Labopin, M.; Ljungman, P.; et al. Increased incidence of chronic GvHD and CMV disease in patients with vitamin D deficiency before allogeneic stem cell transplantation. *Bone Marrow Transplant.* **2015**, *50*, 1217–1223. [CrossRef] [PubMed]

181. Hansson, M.E.; Norlin, A.C.; Omazic, B.; Wikström, A.C.; Bergman, P.; Winiarski, J.; Remberger, M.; Sundin, M. Vitamin D levels affect outcome in pediatric hematopoietic stem cell transplantation. *Biol. Blood Marrow Transplant.* **2014**, *20*, 1537–1543. [CrossRef] [PubMed]
182. Caballero-Velázquez, T.; Montero, I.; Sánchez-Guijo, F.; Parody, R.; Saldaña, R.; Valcarcel, D.; López-Godino, O.; Ferra, I.; Coll, C.; Cuesta, M.; et al. Immunomodulatory effect of vitamin D after allogeneic stem cell transplantation: Results of a prospective multicenter clinical trial. *Clin. Cancer Res.* **2016**, *22*, 5673–5681. [CrossRef] [PubMed]

International Journal of
*Molecular Sciences*

MDPI

*Review*

# Vitamin D and Influenza—Prevention or Therapy?

**Beata M. Gruber-Bzura**

Department of Biochemistry and Biopharmaceuticals, National Medicines Institute, Str. Chelmska 30/34, 00-725 Warsaw, Poland; b.gruber@nil.gov.pl; Tel.: +22-841-39-91; Fax: +22-841-06-52

Received: 28 June 2018; Accepted: 1 August 2018; Published: 16 August 2018

**Abstract:** Vitamin D generates many extraskeletal effects due to the vitamin D receptor (VDR) which is present in most tissues throughout the body. The possible role of vitamin D in infections is implied from its impact on the innate and adaptive immune responses. A significant effect is also the suppression of inflammatory processes. Because vitamin D could be acknowledged as a "seasonal stimulus", as defined by R. Edgar Hope-Simpson, it would be crucial to prove it from a potential easy and cheap prophylaxis or therapy support perspective as far as influenza infections are concerned. The survey of the literature data generates some controversies and doubts about the possible role of vitamin D in the prevention of influenza virus. The most important point is to realise that the broad spectrum of this vitamin's activity does not exclude such a possibility. According to most of the authors, more randomized controlled trials with effective, large populations are needed to explore the preventive effect of vitamin D supplementation on viral influenza infections.

**Keywords:** vitamin D; influenza; respiratory tract infections

## 1. Introduction

The popularity of vitamin D as a vitamin with a broad spectrum of activity is still growing. There are many papers published each year about its properties, including ensuring our bone health. Now, we know that vitamin D is associated with cancer, diabetes, cardiac, and gastrointestinal diseases, and, most interestingly, even with events of unknown etiology, such as inflammatory bowel disease [1]. The phenomenon of its multidirectional activity is possibly due to the presence of the VDR in most nonskeletal human cells. New perspectives on vitamin D activity and targets are very important from, of course, the scientific point of view, but, and this is crucial, because of increasing awareness of the deficiency of this vitamin in different populations across the world and the need for its supplementation.

Through the VDR, vitamin D can also modulate the innate and acquired immune system. This has prompted the idea of exploring the impact of vitamin D on the efficacy of our immune system in fighting off difficult-to-treat viral infections, such as influenza, especially because reaching an optimal level of this "medicine" is cheap and easy for everybody.

The aim of this paper was to present recent data on the possible role of vitamin D in modulating the immune response in influenza virus infection and its potential preventive and therapeutic potential in that disease.

This review is based on an electronic search of articles in the PubMed database, including papers published mostly in the last eight years up until 2018 in that field. The relevant papers are also included. All research articles were found with a combination of the following keywords: vitamin D and influenza, vitamin D and respiratory illnesses, vitamin D and influenza vaccines, and vitamin D and infections. Published articles included in this meta-analysis were selected on the basis of the following criteria: they must have been published in English or Polish, concerning the association between the immune response and vitamin D serum concentration or supplementation, and defining the type of trial, the number of participants, outcome measure, and including statistical analysis.

## 2. The Metabolism, Action, and Guidance Serum Concentration of Vitamin D

### 2.1. Metabolism

Vitamin D comes from two sources: skin synthesis from the precursor—7-dehydrocholesterol—to cholecalciferol upon UVB radiation, and from the diet as cholecalciferol ($D_3$) or ergocalciferol ($D_2$). Metabolic pathways (Figure 1), common for both forms, include: 25-hydroxylation to calcidiol (25(OH)D), which is carried by the liver enzymes CYP2R1 and CYP27A1 (cytochrome P450-associated 25-hydroxylases), followed by 1α-hydroxylation to the active metabolite 1α,25–dihydroxyvitamin $D_3$ (calcitriol, 1α,25$(OH)_2$D), catalysed by cytochrome P450-associated 25(OH)D(3)-1α-hydroxylase (CYP27B1), the enzyme present in the kidney but also in other extrarenal tissues, including immune cells [2–4]. Due to a developed feedback loop system, the metabolic activation of chole- and ergocalciferol and catabolic reactions are strictly regulated. The positive regulators of 1α,25$(OH)_2$D production are parathormone (PTH), secreted by parathyroid glands, and calcium level, and the negative ones are phosphate level and fibroblast growth factor-23 (FGF-23). All of them affect the activity of 1α-hydroxylase [5].

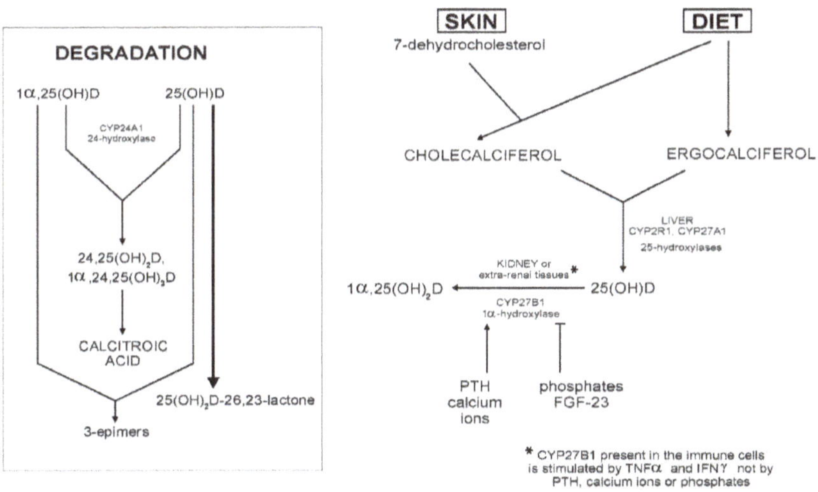


**Figure 1.** Metabolic pathways of vitamin D. Abbreviations: CYP24A1 (cytochrome P450-associated 24-hydroxylase); CYP2R1 and CYP27A1 (cytochrome P450-associated 25-hydroxylases); CYP27B1 (cytochrome P450-associated 25(OH)D3-1α-hydroxylase); PTH (parathormone); FGF-23 (fibroblast growth factor-23); TNFα (tumour necrosis factor α); IFNγ (interferon γ).

Unlike the renal form, CYP27B1 present in the immune cells is not regulated by PTH, FGF-23, calcium, or phosphate signaling, but is stimulated by cytokines such as tumour necrosis factor α (TNF α) and interferon (IFNγ) [4,6]. In turn, CYP27B1 in keratinocytes is upregulated in response to injury and Toll-like receptor (TLR) activation [4]. Extrarenal expression of that enzyme may also be promoted by alternate pathogen recognition receptors (PRRs) [7]. Regulation of extrarenal 1α-hydroxylase is strongly dependent on the concentration of the circulating 25(OH)D [5].

Both metabolites, 25(OH)D and 1α,25(OH)D, are transported in the bloodstream by the carrier vitamin D binding protein (DBP), with higher affinity of 25(OH)D to that transporter [5]. Degradation of both metabolites is catalysed by CYP24A1, a multicatalytic enzyme that results in 24-hydroxylation and the formation of 24,25$(OH)_2$D, and 1α,24,25$(OH)_2$D, which are subsequently converted to calcitroic acid [2,5]. The circulating calcidiol may be also converted by CYP24A1 to the inactive 25$(OH)_2$D-26,23-lactone and 24,25$(OH)_2$D [2].

Some metabolites, such as 3-epimers, which are formed by epimerization of the C-3 in ring A of 25(OH)D, 1α,25$(OH)_2$D, and 24,25$(OH)_2$D, have only slightly weaker biological activity than 1α,25$(OH)_2$D. Such epimers were first reported in human keratinocytes in 1994 [8].

### 2.2. Nongenomic and Genomic Action

Vitamin D affects cellular metabolism by genomic and nongenomic pathways. Mostly, it acts through binding with VDR, and then, after forming a heterodimer with the retinoid X receptor (RXR) and translocating to the nucleus, binds to vitamin D responsive elements (VDRE) in DNA and controls gene transcription.

The nongenomic pathway includes rapid reactions at the cell membranes and has been reported for thyroid hormone, estrogen, and corticosteroids [5,9–11]. Probably via a membrane-bound receptor protein (1α,25$(OH)_2$D membrane-associated rapid response steroid-binding protein, also known as endoplasmic reticulum stress protein 57), vitamin D regulates rapid cellular calcium efflux and calcium-activated chloride channel activity [11].

Nongenomic responses to the active vitamin D metabolite continue via the messenger systems of phospholipase C, protein kinase C, and phosphatidylinositol-3′-kinase (PI3K), initiating the opening of calcium channels and Ras/MAPK signal transduction [7].

Beyond the role in calcium homeostasis and bone metabolism, vitamin D generates many extraskeletal effects through the VDR, which is present in most tissues throughout the body [11,12]. Affinity to the VDR is much higher for 1α,25$(OH)_2$D ($K_a$ = $10^{-10}$ M) than for 25(OH)D ($K_a$ = $10^{-8}$ M) [12]. Activation of that receptor is the basis for the regulation of approximately 3% of the human genome, affecting immune and neurological functions and playing a role in skin, cardiovascular, and gastrointestinal diseases, cancer inhibition, or inhibition of autoimmune diseases [5,10].

As was found in recent years, polymorphism in some enzymes and proteins related to vitamin D, such as DBP, CYP28B1, CYP2R1, CYP24A1, or VDR (especially polymorphs FokI, TaqI, ApaI, and BsmI), can affect the individual's response to anti-infectious treatment, such as interferon/ribavirin therapy in chronic hepatitis C [13,14], susceptibility of the individuals to cancer, tuberculosis, ulcerative colitis, and Crohn's disease, or the increased risk of type 1 diabetes, as was noted in European people [15–18].

### 2.3. Guidance Serum Concentrations

The most valuable indicator of the body's vitamin D status is the serum levels of 25(OH)D, because of the relatively high affinity of that metabolite to DBP and its long serum half-life, ca. 25 days. Measurement of 1α,25$(OH)_2$D to define vitamin D level in the organism is not recommended, as its serum half-life is only a few hours (ca. 7 h) [9,12]. Guidance serum concentration of 25(OH)D, which indicates an efficient level of vitamin D in the organism, is 30–80 ng/mL (ca. 75–200 nM/L). A severe vitamin D deficiency is reflected in serum concentrations below 10 ng/mL [12,19–23].

## 3. Vitamin D as an Anti-Infective Agent

The clear functions of vitamin D in the immune system are difficult to define because the immune response is not a static process and depends on the stage of infection.

The VDR, which has also been detected in immunological cells, suggests that vitamin D can regulate some processes related to immunity. As was shown in vitro, activated human T and B cells and also the endothelial cells lining the upper and lower respiratory tract can transform inactive metabolite 25(OH)D into active 1α,25$(OH)_2$D. This compound acts on immune cells in an autocrine, paracrine, or intracrine way (i.e., throughout the pathways inside the cells) [24–26].

The possible role of vitamin D in infectious diseases is implied by its impact on the innate and adaptive immune responses (Figure 2):

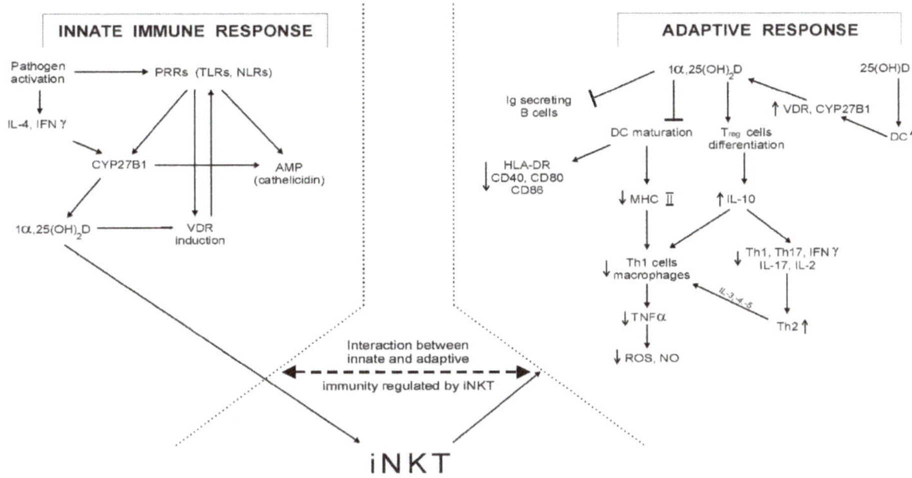


**Figure 2.** The role of vitamin D in the immune response. Abbreviations: PRRs (pathogen recognition receptors); TLRs (Toll-like receptors); NLRs (nucleotide-binding oligomerization domain (NOD)-like receptors); IL (interleukin); IFNγ (interferon γ); CYP27B1 (cytochrome P450-associated 25(OH)D(3)-1α-hydroxylase); AMP (antimicrobial peptides); VDR (vitamin D receptor); iNKT (invariant NK T cells); Ig (immunoglobulin); DC (dendritic cells); Treg cells (regulatory T cells); HLA-DR (human leukocyte antigens); CD (costimulatory molecules); MHC (major histocompatibility complex); Th cells (T-helper cells); IFNγ (interferon γ); ROS (reactive oxygen species); NO (nitric oxide).

### *3.1. Vitamin D and the Innate Response*

#### 3.1.1. Pathogen Recognition Receptors—PRRs

The innate immune response can be defined, generally, as nonspecific, although it proves to be the first line of defense against infective agents and initiates antigen presentation [27,28].

The crucial points for the innate immune response are the Toll-like receptors (TLRs), being a subgroup of various intracellular innate PRRs which is present in macrophages, polymorphonuclear cells, monocytes, and epithelial cells. TLRs recognize molecules related to the pathogen; for example, the lipopolysaccharides of bacteria or viral nucleic acids and proteins. Such activated TLRs release cytokines which induce reactive oxygen species and antimicrobial peptides (AMPs), cathelicidins, and defensins [4,6,7,24,25]. Several TLRs affect or are affected by VDR induction. For example, expression of the coreceptor for TLR4, costimulatory molecule CD-14, is induced by 1α,25(OH)$_2$D in monocytes and epidermal keratinocytes. In turn, the increased expression of CYP27B1 in macrophages is the indirect result of AMPs, which stimulates TLR2 [6]. As shown by Greiller and Martineau [7], ligation of the TLR2/1 heterodimer in macrophages has been demonstrated to upregulate CYP27B1, similarly to the ligation of TLR8 by CL097 or TLR4 by lipopolysaccharide (LPS).

The exact mechanism of TLR ligation-induced CYP27B1 production is not fully understood, but it is possible that other TLRs or alternate PRRs may enhance extrarenal activation of that enzyme, allowing calcitriol to have more extensive effects on the immune response [7]. Upon viral infection, pathogen-associated molecular patterns (PAMPs) can also be recognized by other PRRs, such as retinoic-acid-inducible gene-I (RIG-I)-like receptors and nucleotide binding-oligomerisation domain (NOD)-like receptors (NLRs). In myeloid and epithelial cells, the intracellular receptor NOD2 is induced by 1α,25(OH)$_2$D via two VDREs in the NOD2 gene. The addition of lysosomal breakdown products of bacterial peptidoglycan to calcitriol-induced NOD2 enhanced NFκB signalling and AMP such as beta defensin 2 expression [7,29].

The inflammatory cytokines TNF$\alpha$ and interleukins (IL) IL-1$\beta$, -6, and -12 are produced at an early stage of the innate immune response. These cytokines, among others, induce synthesis of acute phase proteins and contribute to the recruitment and activation of cells of the adaptive immune response. PRR signalling also results in the production of chemokine ligands (CXCLs), such as CXCL8–CXCL10 and IL-15, which generates neutrophils and natural killer cells (NK). These cells have an immune role, especially in bacterial innate immunity [7].

Upregulation of TLRs can also involve other mechanisms. Neutrophils, unlike macrophages, express the VDR, but they do not show an active 1$\alpha$-hydroxylase. Thus, in these types of cells, it is not possible to induce transformation of 25(OH)D into the active metabolite upon TLR stimulation. In such cases, the surface proteins, such as the triggering receptor on myeloid cells-1 (TREM-1) or transforming growth factor $\beta$ (TGF$\beta$), present on the neutrophils or epithelial keratinocytes, respectively, may participate in the cell response to circulating 1$\alpha$,25$(OH)_2$D, including TLR signalling, via TREM-1, or by stimulation of CYP27B1 expression, via TGF$\beta$ [29]. TGF$\beta$ can cooperate with calcitriol to induce 5-lipooxygenase (5-LO), which catalyzes the synthesis of leukotrienes, compounds which participate, among others, in the phagocytosis of bacteria [29].

One of the features of the antibacterial innate response is the destruction of the pathogens by autophagy [26]. According to Chun et al. [29], recent data suggest that this process is important for the antibacterial response induced by vitamin D against *Mycobacterium tuberculosis* infection.

TLR-released AMPs have a broad spectrum of activity, not only microbial but also antiviral, and have been shown to inactivate the influenza virus [24]. The antiviral effects of AMPs are the result of, among other effects, the destruction of envelope proteins done by cathelicidins. Regarding antibacterial activity, AMPs induce, among other effects, membrane disruption. In humans, the active antimicrobial cathelicidin 37-residue, the amphipathic, helical peptide LL-37, is cleaved from the human cathelicidin propeptide (hCAP18). The majority of cathelicidin is stored in neutrophil granules, but also the other types of immune cells, as monocytes and NK and B lymphocytes can express hCAP18 [4]. Production of cathelicidins in human macrophages in vitro is stimulated by the active metabolite of vitamin D, 1$\alpha$,25$(OH)_2$D, via increased expression of the VDR [25]. According to Sundaram and Coleman [25], VDRE upregulation by TLRs leads to the transcription of cathelicidin, which kills intracellular *Mycobacterium tuberculosis*. As was shown by Szymczak and Pawliczak [26], hCAP18 also has activity against viruses and other bacteria. The lung epithelial cells, during viral infection, are capable of converting calcidiol into the active metabolite calcitriol, leading to increased hCAP18 production. As shown by Beard et al. [4], cathelicidin expression in macrophages and keratinocytes is induced by CYP27B1, and if there is no 25(OH)D, VDR, or CYP27B1, the ability of these cells to produce cathelicidins is significantly impaired. Following Szymczak and Pawliczak [26], not only TLR signaling, but also cytokines such as IL-4 and IFN$\gamma$ may affect the CYP27B1 expression. The presence of IFN$\gamma$ stimulates macrophage CYP27B1. It is also interesting that 1$\alpha$,25$(OH)_2$D participates in the negative feedback mechanism that self-inhibits the hyperactivation of TLRs [26].

It is worth noting that the impact of 1$\alpha$,25$(OH)_2$D on viral pattern recognition receptor-driven cytokine production varies between pathogens. As shown by Fitch et al. [30], in viral responses, it failed to modify TLR7/8- or respiratory syncytial virus (RSV)-stimulated innate cytokine production, even in supraphysiologic concentrations.

#### 3.1.2. Antimicrobial Peptides—AMPs

Vitamin D also regulates the other type of AMPs: defensins. Human beta defensin 2 is modestly stimulated by 1$\alpha$,25$(OH)_2$D and its antiviral effects arise from chemoattractive properties for neutrophils and monocytes [4,25]. However, serum 25(OH)D concentration was not associated with levels of serum AMPs in patients with community-acquired pneumonia [25].

Genome analysis revealed VDRE within the promoters of cathelicidin and beta defensin 2. Surprisingly, only cathelicidin appeared to be transcriptionally induced by 1$\alpha$,25$(OH)_2$D in monocytes [29]. As shown by the authors, the increased level of beta defensin 2 resulted from the decreased expression of 1α,25(OH)$_2$D and interleukin-1 (IL-1) from monocytes, and that AMPs, for gene induction, require the cooperation of VDRE in the promoter and nuclear factor κB (NFκB) as the other transcription factor.

### 3.2. Vitamin D and the Adaptive Response

#### 3.2.1. T Lymphocytes

The basis of the adaptive response is: antigen presentation to B and T cells, and the antigen-stimulated production of antibodies and a wide spectrum of cytokines, chemokines, enzymes, and hormones. The initial observation related to the role of vitamin D in the immune system was the presence of the VDR in the activated lymphocytes [29].

T lymphocytes (T cells) include a few types of cells (called subgroups): CD8$^+$ T cells, expressing relatively high levels VDR and vitamin D-activating 1α-hydroxylase; CD4$^+$ T cells; NK cells; and memory cells. Activated CD8$^+$ T cells can be differentiated into cytotoxic lymphocytes (CTLs), crucial for control against intracellular pathogens and cancer. Activated CD4$^+$ T cells can be differentiated into T helper cells (Th cells), such as regulatory T cells (Treg) (called suppressor T cells), γδT cells, and Th17, Th9, Th1, and Th2 cells, which produce different profiles of cytokines. So, Th1 cells produce IL-2, TNFα, and interferon γ (IFNγ), and Th2 cells produce IL-3, -4, -5, -10, and -13. CD4$^+$ T-induced cells, via cytokine secretion, provide support to other immune cells, such as CTLs, and to the serum antibody response, via CD40:CD40 ligand costimulation of antigen-specific B cells [6,25–27,29,31]. Besides cytokine production, the role of Th cells includes: the support of immunoglobulin production, macrophage activation, and production of eosinophiles and mastocytes (Th2) [27]. The Th1-stimulated response is the key for many bacterial and viral infections, but if it is an uncontrolled process, it leads to autoimmunity [28].

What is the role of vitamin D in that type of immune response? It acts as a modulator of Th cell proliferation and cytokine production, but also through promoting Treg cells, which are responsible for anti-infectious action, for suppressing immune responses, and for limiting inflammatory processes [28]. The exact mechanism is not well-known. One of the reports focused on the ability of 25(OH)D to stimulate Treg cells through induction of the antigen-presenting dendritic cells (DC) expressing VDR and CYP27B1 [32]. It is interesting to note that, as reported by Jeffrey et al. [32], the form of vitamin D significant for the generation of Treg cells was non-DBP-bound 25(OH)D. Bruce et al. [28] mentioned that 1α,25(OH)$_2$D regulates invariant NK T cells (iNKT), which can act as regulatory cells and participate in the interaction between innate and adaptive immunity. Induction of iNKT has been shown to be protective, including against autoimmune diseases. In turn, Sigmundsdottir et al. [33] indicate that 1α,25(OH)$_2$D-stimulated chemokine receptor 10 (CCR10) expression on T cells enables communication with other immune-competent cells, as it recognizes CCL27 secreted by keratinocytes throughout the organism. In vitro, 1α,25(OH)$_2$D inhibits the expression of Th1 cytokines and stimulates Th2 cytokines. As shown previously, the other subgroup of Th cells, i.e., Th17 cells, secrete IL-17, playing a role in autoimmune processes [6,29]. Inhibition of Th1 cells was also noted in vivo, in mouse DC [25].

Inverse associations between vitamin D concentrations and disease activity in patients with inflammatory bowel disease, type 1 diabetes, multiple sclerosis, rheumatoid arthritis, or autoimmune thyroiditis have been noted [34].

1α,25(OH)$_2$D appeared to limit autoimmune reactions, among other means, by inhibiting Th17 cell activity [35]. Apart from that, 1α,25(OH)$_2$D can decrease the ability to present antigens by the inhibition of DC maturation. The result of the decrease is the diminished expression of the human leukocyte antigen HLA-DR and costimulatory molecules such as CD40, CD80, and CD86. 1α,25(OH)$_2$D induces the differentiation of Treg cells, which induces IL-10 production. This cytokine is thought to inhibit IL-12. 1α,25(OH)$_2$D-treated DC expressed less costimulatory MHC (major histocompatibility complex) II than intact cells [28]. Accordingly, Th1 cell and macrophage production is diminished, although the ability to induce Treg cells is maintained [6,25,28]. The induced IL-10 also suppresses Th1 and Th17 cells and thus production of IFNγ, IL-17, and IL-2, leading to immune tolerance [6]. The above events make Th2 cells predominant. As a result, the enhanced secretion of IL-4, -5, and -13 further suppresses Th1 cells. In human monocytes in vitro, 1α,25(OH)$_2$D was shown as the inhibitor of Th1 cell-mediated cytokines and tumour necrosis factor α (TNFα), and in vivo in mice as the suppressor of the secretion and production of Th17 cells by downregulation of IL-23 and -6 [6,25,28]. As shown by Bruce et al. [28], treatment of naïve CD4$^+$ Th cells during Th17 cells priming with 1α,25(OH)$_2$D inhibits IL-17 production. The other mechanism of immunomodulation related to vitamin D is the impact on DC gene expression, which is independent of the differentiation of these cells [29]. As shown by Chun et al. [29], DC gene expression can be regulated by two major metabolites of vitamin D: 25(OH)D and 1α,25(OH)$_2$D.

On the basis of animal studies, it was shown that the promotion of Th2 cells may have adverse effects on allergic diseases such as asthma atopic dermatitis through induction of the inflammatory processes [6]. The increased production of Th2 cytokines (IL-4, -5, -13), noted in the acute phase of atopic dermatitis, suppresses cathelicidin and increases susceptibility to infection. In the chronic phase of the disease, Th1 cells were predominant [6].

By inhibiting IFNγ, 1α,25(OH)$_2$D inhibits the stimulation of reactive oxygen species and nitric oxide production. These effects, along with the suppression of IL-17 secretion, are responsible for the reduction of resistance to pathogens such as *Toxoplasma* and *Citrobacter* [6,36,37]. Ehrchen et al. [36] showed that VDR-knockout mice developed an altered Th1 response in *Leishmania major* infection, as indicated by the normal production of IFNγ by CD4$^+$ and CD8$^+$ Th cells. In turn, Rajapakse et al. [37] observed reduced IFNγ and IL-12 levels in *Toxoplasma gondii*-infected mice, indicating inhibition of Th1 cell activity. Following 1α,25(OH)$_2$D treatment, reduced counts of CD4$^+$ Th cells and splenocytes and the marked induction of apoptosis were also noted. Ryz et al. [38] showed the reduction of Th17 cells in *Citrobacter rodentium*-infected mice treated with 1α,25(OH)$_2$D. According to the impaired Th17 response, a defect in the production of antimicrobial peptide REG3γ was reported.

1α,25(OH)$_2$D impairs the antigen-presenting and T cell-stimulatory capacity of monocytes and macrophages, with a decrease in MHC II and CD40, -80, and -86. Suppression of IL-12 and IL-23, which are involved in Th1 differentiation, is due to the 1α,25(OH)$_2$D-mediated NFκB activation [7,28].

It is worth noting that human T cell responses are regulated in the presence of DC and the inactive metabolite of vitamin D, i.e., 25(OH)D. As described by Jeffery et al. [32], CYP27B1 is induced in DC upon maturation with LPS or upon T cell contact, which results in the synthesis and release of 1α,25(OH)$_2$D, which, in turn, affects T cell responses.

#### 3.2.2. B Lymphocytes

1α,25(OH)$_2$D inhibits proliferation and acts as a proapoptotic agent in activated human B cells in vitro. Although it does not act on the production of these cells, it is thought to inhibit their differentiation [25]. As shown by Fang et al. [39] in mice, the immune protection induced by the influenza virus primary infection significantly relies on the presence of B lymphocytes.

Some suppressive effects of 1α,25(OH)$_2$D were noted with reference to immunoglobulin (Ig)-secreting B cells. 1α,25(OH)$_2$D was specifically able to inhibit the development of them after mitogenic stimulation [40]. The impact on Ig is not the only mechanism of B cell–vitamin D interaction. Other reports have also shown the regulation of B cells by 1α,25(OH)$_2$D through IL-10 and CCR-10 [41]. As reported by Heine et al. [42], human B cells, upon activation by the receptor CD40 and IL-4 signals, show increased expression of the gene for the 25(OH)D 1α-hydroxylase CYP1α, followed by the production of significant amounts of 1α,25(OH)$_2$D. 1α,25(OH)$_2$D enhances IL-10 expression in B cells by the transcriptional activity of VDR or through modulation of calcium signalling. On the basis of these studies, it can be suggested that the inactive metabolite of vitamin D, 25(OH)D, can also modulate the immune response.

### 3.3. Anti-Inflammatory Action of Vitamin D

As reported by Penna et al. [43] in studies with use of the VDR agonist elocalcitol, this compound inhibited IL-17 and proinflammatory cytokine IFN$\gamma$ secretion in prostate-draining lymph node T cells from elocalcitol-treated nonobese diabetic mice [43,44]. These results focus on the anti-inflammatory properties of vitamin D.

As such, this vitamin acts through the promotion of the feedback loop to prevent the overactivity of antibacterial processes and inflammatory events. Such action includes the downregulation of TLR2 and TLR4 on monocytes [25,29]. $1\alpha$,25(OH)$_2$D has been reported to decrease proinflammatory chemokine production in human respiratory epithelial cells and downregulate proinflammatory cytokines such as IL-6, -8, and TNF$\alpha$ in many different cells in vitro [25]. As shown in human lymphocytes, the anti-inflammatory effect of vitamin D can be carried out in part through NF$\kappa$B inhibition. This transcription factor regulates expression of the genes encoding the inflammatory proteins produced during infection, such as cytokines, chemokines, acute phase proteins, or inducible effector enzymes [6,25].

Vitamin D-modulated T-cell proliferation is a part of the mechanisms leading to anti-inflammatory responses via the increased proliferation of the subgroup of T cells, CD8$\alpha$. These types of T cells, unlike CD8$^+$ T cells, are not cytotoxic, but may play a role in suppressing gastroinflammation [29].

### 3.4. Time-Dependence of Immunomodulatory Effects of Vitamin D

The immunomodulatory effect of vitamin D is related to early or late phases of infection. The correlation between serum 25(OH)D levels and the levels of $\alpha_1$-antichymotrypsin, an acute-phase protein in patients with tuberculosis, was associated with the illness, but not with the initial, acute-phase response to infection [45]. Also, the different interactions of $1\alpha$,25(OH)$_2$D with different mechanisms were reported to depend on time in human leukemia cells. Tse et al. [46] reported the time-dependent biphasic regulation of NF$\kappa$B in HL-60 cells. After exposure of HL-60 cells to $1\alpha$,25(OH)$_2$D, there was an early, ca. 4h suppression and a late, 8–72h-prolonged reactivation of that transcription factor. Alongside that stimulation, there was an upregulation of inflammatory and anti-apoptotic genes such as TNF$\alpha$, IL-1$\beta$, and Bcl-xL [46]. Such an effect can indirectly impact the immune regulation shown by vitamin D because the inflammatory processes are strictly related to the immune response of the cells to pathogens. According to Sundaram and Coleman [25], the results of the study in vivo on the immune response to allergens in mice suggested that vitamin D supplementation (100 ng of $1\alpha$,25(OH)$_2$D given as injection) given after the initial period of sensitization prevented high levels of eosinophils and the local inflammatory response in bronchoalveolar lavage fluid and lung tissues. Surprisingly, the lack of this effect was reported during supplementation every day during the study.

## 4. Vitamin D versus Influenza—Prevention or Therapy?

### 4.1. Anti-Infective Mechanisms of Vitamin D

The following ascertainment very accurately announces the possible relations between the main seasonal infection, i.e., influenza and vitamin D status in the body:

""Whoever wishes to investigate medicine properly should proceed thus: in the first place to consider the seasons of the year ... " (Hippocrates, ca. 400 BC) [24]."

Vitamin D could be acknowledged as a "seasonal stimulus", following R. Edgar Hope-Simpson, the British practitioner and self-educated epidemiologist. After documentation that influenza A epidemics in temperate latitudes are most intense in the months following the winter solstice, he hypothesized that solar radiation produces a "seasonal stimulus" that affects the pathogenesis of influenza A. He theorized that there is a seasonal steroid hormone system with an impact on the human immune system whose substrate levels are low during the influenza season, but peak when influenza is rare [24,47].

There are a few arguments which support vitamin D as a likely candidate for the abovementioned "seasonal stimulus". In summary, most important is the 1$\alpha$,25(OH)$_2$D-stimulated production of AMPs, such as defensin and cathelicidin. As mentioned, these endogenous antibiotics act directly, destroying not only microbial pathogens, but also viruses, including the influenza virus [24,48,49]. The production of cathelicidin is dose-dependent on the serum level of 1$\alpha$,25(OH)$_2$D. As shown by Lang and Samaras [49], 30 ng/mL is necessary for the optimal induction of cathelicidin mRNA, but a higher level of 40 ng/mL was not more efficient. The next argument which supports the antiviral activity of vitamin D is the modulation of the inflammatory response as mentioned above. The release of proinflammatory cytokines by the influenza virus appeared to correlate with the severity of illness [24]. Khare et al. [50] noted that treatment of human lung A549 epithelial cells with 100 or 30 nM of 1$\alpha$,25(OH)$_2$D prior to or post-H1N1 (influenza A virus) exposure significantly decreased the levels of infection-induced TNF$\alpha$, IFN$\beta$, and IFN-stimulated gene-15 (ISG15) and downregulated IL-8 and IL-6 RNA levels. Following Helming et al. [51], 1$\alpha$,25(OH)$_2$D also potentially diminishes the proinflammatory cytokine production by the modulation of macrophages, which prevents them from the secretion of too many cyto- and chemokines. On the basis of the obtained data, the authors suggested that 1$\alpha$,25(OH)$_2$D participates in a negative feedback loop in which IFN$\gamma$-activated macrophages induce the release of 1$\alpha$,25(OH)$_2$D. When this metabolite is accumulated at an efficient concentration, VDR expression is synergistically induced and it translocates into the nucleus. Then, it can suppress the genes making the proinflammatory proteins, such as CCL5, CXCL16, IFI203, FCGR1, FCGR3, TLR2, IRF2, CXCL10, and CXCL9. Tight control of IFN$\gamma$ responses is crucial for the consequences of granulomatous diseases such as tuberculosis and sarcoidosis [51].

### 4.2. *Human Studies on Influenza Prevention by Vitamin D*

A survey of the literature data generates some controversies and doubts about the possible role of vitamin D for the prevention of influenza infections. However, there are data obtained in vitro or in vivo which denote the antiviral activity of vitamin D in the case of influenza. Nowadays, the final conclusion is that its significance as an anti-influenza agent remains unresolved, but it does not mean that these considerations are senseless. It is most important to realise that the broad spectrum of vitamin D activity does not exclude such a role.

Some of the following studies do not strictly concern influenza infection, but also the influenza-like respiratory illnesses of respiratory tract infections (RIs) and pneumonia. However, the time period of the studies, i.e., October–March, or the winter months, does not exclude influenza infections, most common in the autumn and winter. According to Cannell et al. [24], if vitamin D is a "seasonal stimulus", as has already been mentioned in this paper, then vitamin D deficiency should predispose patients to respiratory infections.

#### 4.2.1. Beneficial Effects

Moan et al. [52] compared the seasonality of deaths from influenza and pneumonia in Norway with vitamin D serum levels. The time period of the studies was 1980–2000. The final conclusion of these studies was that the high numbers of winter influenza and pneumonia deaths in Norway were related to low vitamin D levels in this season. The data support the hypothesis that vitamin D acts as a protector against influenza and pneumonia, although it is not clear if it requires any help, or which mechanism dominates in the battle against viral infections.

Laaksi et al. [53] conducted a placebo-controlled double-blind study (October–March) which involved 164 volunteering young Finnish men (18–28 years of age) undergoing military training. The subjects were randomly assigned to the intervention group ($n$ = 80), which received 400 IU of vitamin D per day, or the placebo group ($n$ = 84). After six months of the study, the supplemented group showed a mean serum concentration of 25(OH)D ($\pm$SD) of 71.6 $\pm$ 22.9 nM/L ($n$ = 58) and the placebo group showed 51.3 $\pm$ 15.5 nM/L of 25(OH)D ($n$ = 50) ($p$ < 0.001). The main outcome considered was the number of days absent from duty due to respiratory infection. The proportion of men remaining healthy throughout the six-month study period was greater in the supplemented group (51.3%) as compared to the placebo group (35.5%) ($p = 0.045$). The above results provided some evidence for the preventive effect of vitamin D supplementation against respiratory tract infection, and according to the Cox regression analysis, the authors noted that the hazard ratio for absence from duty due to the respiratory tract infection was lower in the supplemented group as compared to the placebo group. As shown by the authors, randomized controlled trials with higher doses and larger populations are needed to explore the preventive effect of vitamin D supplementation on acute respiratory tract infection.

Promising and encouraging results on supplementation with vitamin D to prevent influenza were presented by Urashima et al. [54], who conducted a randomized, double-blind, placebo-controlled trial comparing vitamin $D_3$ supplements (1200 IU/day) with placebo in schoolchildren from December to March. The outcome was the incidence of influenza A diagnosed with presence of the influenza antigen. As reported, influenza A occurred in 18 out of 167 (10.8%) children in the vitamin group compared with 31 out of 167 (18.6%) children in the placebo group (relative risk (RR) = 0.58; 95% CI: 0.34, 0.99; $p = 0.04$). The reduction of influenza incidence was greater in children who, among other factors, had not been taking other vitamin D supplements. It is interesting to note that in children with a previous diagnosis of asthma, asthma attacks as a secondary outcome occurred in two children receiving vitamin D compared with twelve children receiving placebo (RR = 0.17; 95% CI: 0.04, 0.73; $p = 0.06$). The above study, conducted up to 2012, was reported by Jorde et al. [55] as one properly performed randomized clinical trial on influenza prevention.

The positive role of vitamin D in respiratory infections and lung function was confirmed by Berry et al. [56], who used cross-sectional data from 6789 participants in a nationwide 1958 British birth cohort. The authors measured 25(OH)D, lung function, forced vital capacity, and respiratory infections from the age of 45 years. They showed a linear association between vitamin D status and seasonal infections and lung function. Each 10 nM/L increase in 25(OH)D was associated with a 7% lower risk of infection. Similar results were also obtained in other studies. Aregbesola et al. [57] investigated the risk of hospitalized pneumonia in an ageing general population in eastern Finland. On the basis of the study, the authors suggested an inverse effect of the serum 25(OH)D concentration on the risk of developing pneumonia. Jones et al. [58] examined 46 residual blood samples from adults and children, some of whom experienced influenza virus infections of the respiratory tract. Assays were performed for retinol binding protein (RBP), vitamin D, and antibody isotypes. Results showed that 44 samples exhibited RBP and/or vitamin D insufficiencies or deficiencies. Besides, vitamin D correlated with blood IgM and IgG3, while RBP correlated with IgG4 and IgA. It is known that vitamins A and D are critical for healthy immune responses at mucosal surfaces in mice. Especially, IgA is a first line of defense against mucosal pathogens. So, according to the authors, the results suggested that also in humans, there is a correlation between vitamin A and D levels and antibody profile. The authors suggest that vitamins may support the dendritic cell development necessary for antigen presentation; T-cell activation and homing; B-cell activation, division, and maturation; and/or the stabilization of differentiated antibody-producing cells. According to Jones et al. [58], these results recognize the need for further studies on the correction of vitamin supplementation, particularly at the time of respiratory virus vaccination, to improve vaccine efficacy and for protection against respiratory tract diseases.

Intriguing effects were presented by Mamani et al. [59] and Brance et al. [60] in their studies. Both studies showed an inverse correlation between the level of 25(OH)D and the severity of CAP (community-acquired pneumonia), which was defined as the CURB65 score (confusion, uremia, respiratory rate, low blood pressure, $\geq$65 years). In addition, according to Brance et al. [60], higher 25(OH)D concentrations were found to be correlated with lower CCI (Charlson comorbidity index). Nanri et al. [61], in a nested case-control study in a cohort of workers in four companies in Japan during the winter season, found lower influenza risk to be associated with vitamin D sufficiency ($\geq$30 ng/mL), but only among unvaccinated participants. In a subgroup vaccinated earlier against influenza, serum 25(OH)D concentration did not correlate significantly with the incidence of physician-diagnosed influenza.

#### 4.2.2. Studies Showing No Relevant Effects

However, there is still one fly in the ointment. There are in vitro and in vivo data, as well as data resulting from human studies, which do not prove any significance of vitamin D supplementation in viral respiratory infections.

Gui et al. [62] showed the negative impact of $1\alpha$,25(OH)$_2$D treatment on the innate immune response generated by the H9N2 infection in mice, especially at the later stage of the disease. Although it decreased the influenza M gene (encoding the M protein related to inflammatory response and virus replication), IL-6, and IFN$\beta$ in A549 cells prior to and post-infection with H9N2 influenza, the authors found that it did not affect virus replication in vitro and in vivo. Besides, the effect of $1\alpha$,25(OH)$_2$D treatment was dependent on the stage of the illness. As shown in vivo, $1\alpha$,25(OH)$_2$D downregulated pulmonary inflammation in mice two days post-infection, but increased the inflammatory response 4 to 6 days post-infection. Simultaneously, the expression of the antiviral cytokine IFN$\beta$ was significantly higher at two days post-infection and lower on days 4 and 8. These effects were consistent with the period of maximum body weight loss and the lung damage in calcitriol-treated mice. The reason for the positive anti-inflammatory activity of $1\alpha$,25(OH)$_2$D noted in A549 cells and the opposite effect reported in mice during the later stage of infection is not clear. The authors indicate two possible explanations. The first is that the activity of vitamin D in vivo is complex and has an impact on so many pathways and mechanisms that it can affect one component of this system, but not the other. Secondly, avian influenza viruses such as H9N2 induce different mechanisms in mice and in humans. This is in agreement with the conclusion given by Grant and Giovanucci [48], who discussed, regarding the data showing that suppressing proinflammatory cytokines by vitamin D did not reduce the risk of death in mice infected with H5N1 viruses, that such an effect should not be applied to the H1N1 infections in humans because of the differences in immune response. The anti-inflammatory activity of $1\alpha$,25(OH)$_2$D on A549 infected with H1N1 was also shown by Khare et al. [50]. The authors noted that $1\alpha$,25(OH)$_2$D treatment prior to or post-infection downregulated IL-6 and IL-8 RNA levels and decreased the levels of infection-induced TNF$\alpha$, IFN$\beta$, and ISG15. $1\alpha$,25(OH)$_2$D did not affect viral clearance, similarly to findings reported by Gui et al. [62], but reduced autophagy and restored increased apoptosis seen in the H1N1 infection back to its constitutive level.

In the studies presented by Jorde et al. [55], vitamin D appeared to make influenza infection a significantly more prolonged disease than in the patients receiving placebo. In the study, 569 subjects from 10 different clinical trials were included. Of the subjects, 289 were randomized to receive vitamin D (1111–6800 IU/day) and 280 received placebo. Influenza-like disease was reported in 38 subjects in the vitamin D group and 42 in the placebo group. In these groups, 25 and 26 subjects, respectively, showed clinical symptoms of influenza according to the defined criteria. In the vitamin D group, the duration of the illness was significantly longer than in the placebo group (2–60 days versus 2–18 days; $p = 0.007$). One of the weaknesses of this study, as emphasized by the authors, was that the study was retrospective and relied on self-reported symptoms; thus, there was not a definite diagnosis of influenza.

Some conflicting results were presented by Urashima et al. [63]. As was given by the authors, a randomized controlled trial on the effects of vitamin D supplementation on influenza illness during the 2009 H1N1 pandemic revealed that influenza A or B occurred less in the vitamin D group than in the placebo group only during the first half of the study. During the second month, the vitamin D group's results were similar to those of the placebo group. The authors observed similar effects, i.e., preventive action of vitamin D supplementation only in the initial part of the studies, in the studies conducted among the students who received 2000 IU of vitamin D per day for two months. As shown by post-hoc analysis, influenza A occurred significantly less in the vitamin group (2/148, 1.4%) compared with the placebo group (8/99, 8.1%), but only in the first month of the study. The initial benefit was lost during the second month. These results could have been related with longer supplementation with vitamin D and are indirectly consistent with the results of Urashima et al. [54], who did not note the impact of vitamin D on influenza A incidence in children who had been taking more than one vitamin D supplement.

The lack of any correlation between respiratory infections and vitamin D supplementation was showed by Li-Ng et al. [64], who described a randomized controlled trial for the prevention of symptomatic upper respiratory tract infections, conducted during the winter. In total, 162 adults received 2000 IU of vitamin D per day for 12 weeks. There was no difference in the incidence of infections and in the duration or severity of respiratory tract infection symptoms between the supplemented and the placebo groups (48 vs. 50 cases, respectively, $p = 0.57$, and 5.4 ± 4.8 days vs. 5.3 ± 3.1 days, respectively, $p = 0.86$). It is worth noting that after 12 weeks, the mean serum concentration of 25(OH)D in the supplemented group was 88.5 ± 23.2 nM/L and in the placebo group was 63.0 ± 25.8 nM/L. As a matter of fact, the serum concentration of 25(OH)D in the supplemented group was not too efficient; the value 88.5 ± 23.2 nM/L is placed near the defined lower limit of the 25(OH)D level reported as the guidance level (75–200 nM/L) [12]. In studies conducted in 2007, the same authors found a significant reduction in colds and influenza in women taking 800 or 2000 IU/day of vitamin D. Surprisingly, in the final year of the study, women supplemented with 2000 IU were still vitamin D-deficient. [65]. According to Aloia and Li-Ng [65], the trial should use enough cholecalciferol to raise 25(OH)D levels to those achieved by natural skin synthesis in the summer, i.e., ca. 50 ng/mL. A similar lack of protective effects of vitamin D was found by Lappe et al. [66] in double-blind, placebo-controlled, population-based, randomized clinical trials. Among healthy postmenopausal older women with a mean baseline serum 25(OH)D level of 32.8 ng/mL, supplementation with vitamin D and calcium compared with placebo did not result in a significantly lower risk of all types of cancer after four years.

#### 4.2.3. Critical Views

##### Studies on the Role of Vitamin D Supplementation

A systematic review and meta-analysis done by Martineau et al. [67] pointed to the issue which is most crucial for studies based on vitamin D supplementation. The authors presenting the results of 26 eligible randomized controlled trials showed that vitamin D supplementation significantly reduced the risk of acute respiratory tract infection (RI), but the protective effects were observed in those receiving daily or weekly vitamin D without additional bolus, and not in those receiving one or more bolus doses. The beneficial effects were negatively correlated with baseline 25(OH)D levels < 25 nM/L. In addition, the authors found that baseline vitamin D status and dosing frequency independently modified the effect of vitamin D supplementation on the risk of acute RI. On the basis of the results, the authors suggested that high doses of vitamin D supplemented as bolus firstly can evoke some adverse effects in circulating vitamin D metabolites. Secondly, the high concentrations after bolus may dysregulate the activity of the enzymes responsible for the synthesis and degradation of 1$\alpha$,25(OH)D, which results in decreased concentrations of it in extrarenal tissues. In turn, the correlation is such: the stronger the effect, the lower the vitamin D level may be, according to the authors, based on the principle that people who are most deficient in a micronutrient will be the most likely to respond to its replacement [67]. This may explain the results obtained by Aloia and Li-Ng in 2007, described above, that showed that supplementation in women with a vitamin D-deficit was more protective against colds and influenza than in the case of vitamin D-sufficient women [65].

On the basis of the studies mentioned above, in which supplementation with vitamin D and the analogous higher serum concentration of this vitamin did not give the expected effect, Grant et al. [68] proposed a new approach to vitamin D randomized controlled trials to provide true confirmation of the role of its vitamin dosing being solely to achieve the targets set for achieved 25(OH)D concentrations.

As shown by the authors, the assumption that the vitamin D dose–response relationship is linear is not true. The supplemented vitamin D has no direct health effect, because the conversion to 25(OH)D varies from individual to individual regarding genetic polymorphisms of CYP enzymes, intestinal absorption, or body mass. Grant et al. [68] proposed a design strategy which targets supplementation to the chosen baseline status while ensuring achievement of the desired status. This could be done through checking 25(OH)D concentrations periodically during the trial, as well as the baseline, and through recruiting nonreplete subjects. With such an approach, randomized controlled trials would have an increased potential for detecting causality.

Seasonality of Vitamin D Levels and Influenza Rate

As shown by Shaman et al. [69], seasonal variation in the serum concentration of 25(OH)D, which contributes to immune function, has been hypothesized to be the underlying source of observed influenza seasonality in temperate regions. So, the authors studied whether 25(OH)D levels could be used to simulate influenza infection rates. The studies were done in two regions of the United States. On the basis of best-fitting simulations which could reproduce the observed seasonal cycle of influenza, the authors concluded that it is unlikely that seasonal variations in vitamin D levels determine the seasonality of influenza in temperate regions. These results suggest that influenza transmission may be dependent on different factors. On the basis of a reanalysis of laboratory experiments, Shaman et al. [70] revealed that absolute humidity strongly modulates the airborne survival and transmission of the influenza virus. Also, Koep et al. [71] studied the impact of absolute humidity in the indoor school environment on virus survival. As noticed, classroom humidification may be a feasible approach to increase indoor absolute humidity to levels that may decrease influenza virus survival and transmission. In turn, Yang et al. [72] studied other meteorological factors which influence the dynamics of influenza in tropical Africa. They computed the monthly viral positive rate for three circulating influenza subtypes: A/H1N1, A/H3N2, and B, among patients presenting influenza-like illness or severe acute RI in two Ugandan cities. The impact of temperature, relative and absolute humidity, and precipitation, as well as interactions among the above influenza subtypes on the epidemic dynamics of each influenza subtype were studied. As observed, the associations with weather variables differed by influenza subtype. The models showed that precipitation and temperature were negatively correlated with A/H1N1 activity. A mutually negative association between A/H3N2 and B activity was identified.

Kroll et al. [73] mentioned the other factor conjugated with the level of 25(OH)D and parathyroid hormone (PTH) concentrations, which are known to have a reciprocal seasonal relationship with 25(OH)D. These two compounds vary in sinusoidal pattern throughout the year, even in ergocalciferol-treated patients. This means that 25(OH)D is higher in the summer and lower in winter, while PTH shows the reverse pattern. According to the authors, in these observations, held across three latitudinal regions, both genders and multiple years are applicable for patient care.

Independently of the results of the conducted studies, which support or do not support the positive role of vitamin D serum levels and supplementation in the prevention against influenza infections with influenza-related diseases, such as RI and pneumonia, most of the authors agree that this topic deserves further study. The questions: "What dose of vitamin D ensures more resistance to influenza infections?" and "Does vitamin D-deficiency mean more susceptibility to influenza infections?" remain open and require more clinical trials.

A summary of the results obtained in the human studies on the role of vitamin D in the upper respiratory tract infections and defined influenza can be found in Table 1.

**Table 1.** The selective human studies on the effect of vitamin D supplementation on influenza, upper respiratory tract infections (URI or RI), and pneumonia incidence published between the years 2009 and 2018.

| Author, Year | Trial/Duration/Location | Sample Size/Participants | Vitamin D Dose Supplemented per Day | Outcome Measure | Result | Statistical Significance |
|---|---|---|---|---|---|---|
| Li-Ng et al., 2009 [64] | Randomized controlled study/winter season/USA | 162/adults | 2000 IU | Primary outcome: incidence of URI symptoms on the basis of biweekly questionnaire | 48 cases of URI symptoms in the supplemented group vs. 50 cases of URI symptoms in the placebo group | $p = 0.57$ |
| | | | | Secondary outcome: duration and severity of URI symptoms on the basis of biweekly questionnaire | $5.4 \pm 4.8$ days in the vitamin group vs. $5.3 \pm 3.1$ days in the placebo group | 95% CI: −1.8 to 2.1 $p = 0.86$ |
| Laaksi et al., 2010 [53] | Randomized, double-blind, placebo-controlled study/Oct–March/Finland | 164/men 18–28 years old, undergoing military training | 400 IU | The number of days absent from duty due to URI | 51.3% of men remaining healthy in the intervention group vs 35.5% in the placebo group | $p = 0.045$ |
| Urashima et al., 2010 [54] | Multicenter, randomized, double-blind, placebo-controlled parallel-group study/Dec–March/Japan | 167/schoolchildren | 1200 IU | Primary outcome: influenza A antigen vs incidence of influenza A* | 18/167 (10.8%) children in the suppl. group vs 31/167 (18.6%) children in the placebo group | RR = 0.58; 95% CI: 0.34, 0.99; $p = 0.04$ |
| | | | | Secondary outcome: asthma attacks in the case of a previous diagnosis of asthma | Asthma attack incidence in 2 children in the suppl. group vs 12 children in the placebo group | RR = 0.17; 95% CI: 0.04, 0.73; $p = 0.06$ |
| Berry et al., 2011 [57] | Cross-sectional study/Great Britain | 6789/1958 British birth cohort aged above 45 years | - | 25(OH)D serum concentration vs. RI | Each 10 nM/L increase in 25(OH)D was associated with a 7% lower risk of RI | 95% CI: 3.11% |
| | | | | 25(OH)D serum concentration/lung function vs forced expiratory volume in 1 s (FEV1) and forced vital capacity (FVC) | Each 10 nM/L increase in 25(OH)D was associated with 8 mL higher volume of FEV1 and 13 mL higher volume of FVC | 95% CI: 3, 13 for FEV1 95% CI: 7, 20 for FVC |
| Jorde et al., 2012 [55] | Retrospective study in subjects participating in randomized, double-blind, placebo-controlled studies/fall–winter 2010/Norway | 569/men and females aged 32–84 years from 10 different clinical trials | 1111–6800 IU | Primary outcome: influenza-like diseases and influenza reported on the basis of the questionnaire | 25/289 subjects receiving vitamin D and 26/280 subjects receiving placebo were reported as infected with defined influenza | $p = 0.064$ |
| | | | | Secondary outcome: Duration and severity of the illness | The median duration of the illness in 24/25 subjects infected with influenza was 7 days and in the placebo group was 4 days. | $p = 0.007$ |

**Table 1.** *Cont.*

| Author, Year | Trial/Duration/Location | Sample Size/Participants | Vitamin D Dose Supplemented per Day | Outcome Measure | Result | Statistical Significance |
|---|---|---|---|---|---|---|
| Aregbesola et al., 2013 [56] | Prospective study/ 9.8 years/Finland | 1421/723 men and 698 women, 53–73 years old | - | 25(OH)D serum concentration vs the risk of incident hospitalized pneumonia | The subjects in the lowest 25(OH)D serum conc. tertile had a 2.6-fold higher risk of developing pneumonia with the subjects in the highest tertile | 95% CI: 1.4 to 5.0; $p = 0.05$ |
| Jones et al., 2015 [58] | Laboratory study/USA | 46 blood samples/adults and children | - | Retinol binding protein (RBP) vs. antibody isotypes | 44 samples exhibited the correlated RBP and/or vitamin D insufficiency or deficiency as follows: RBP <22,000 ng/mL, 25(OH)D <30 ng/mL RBP correlated with: | $R_S = 0.31$, $p = 0.04$ |
| | | | | | IgA | $R_S = 0.47$, $p = 0.0009$ |
| | | | | | IgA/IgM | $R_S = 0.33$, $p = 0.03$ |
| | | | | | IgA/IgG1 | $R_S = 0.51$, $p = 0.0003$ |
| | | | | | IgG4 | $R_S = 0.39$, $p = 0.0082$ |
| | | | | 25(OH)D concentration vs. antibody isotypes | 25(OH)D concentration correlated with: | |
| | | | | | IgM | $R_S = 0.36$, $p = 0.01$ |
| | | | | | IgG3 | $R_S = 0.32$, $p = 0.03$ |
| | | | | | IgG3/IgG1 | $R_S = 0.36$, $p = 0.01$ |
| | | | | | IgG3/IgA | $R_S = 0.31$, $p = 0.04$ |
| Mamani et al., 2017 [59] | Case-control study/Iran | 73 patients with CAP, 76 healthy controls | - | 25(OH)D serum concentration vs. the risk of the incidence of CAP | The risk of pneumonia among the subjects with 25(OH)D <10 ng/mL was 3.69 | 95% CI: 1.46, 9.31; $p = 0.006$ |
| Brance et al., 2018 [60] | Observational study/July 2015–June 2016/Argentina | 167 patients with CAP, 59% women, 57.4 ± 19.6 years old | - | 25(OH)D serum concentration vs. CURB65 score and CCI | 25(OH)D serum concentration inversely correlated with the severity of | |
| | | | | | CAP (CURB65 score) | $p = 0.049$ |
| | | | | | and with CCI | $p = 0.07$ |

RR—Relative risk; $R_S$—Spearman's rank correlation coefficient; CI—Confidence interval; CAP—community-acquired pneumonia; CCI—Charlson comorbidity index.

## 5. Does Vitamin D Affect the Immunogenicity of Influenza Vaccines?

### *5.1. Vitamin D as an Adjuvant*

According to annual reports, the influenza virus causes 3–5 million severe cases which result in 250,000 to 350,000 deaths worldwide [49]. Vaccination is the most important preventive strategy against influenza, with the crucial expected outcome being antibody production, although it is still not sufficiently effective to reduce the morbidity and mortality associated with influenza in humans [74]. As shown in this paper, vitamin D undoubtedly plays the role of an immunomodulating agent. Its significance in innate and adaptive immunity prompted scientists to check if there is any influence of vitamin D on the response of the patients to immunization by vaccination with live or inactive attenuated influenza virus, especially because some studies on influenza prevention provide a negative correlation between enhanced post-vaccine immunization and obese patients, whereas obesity implies vitamin D deficits [75,76]. These observations indirectly point to the possible role of vitamin D improvement of the immune response to influenza vaccines. Principi et al. [77] mentioned animal studies in which the results encourage regarding vitamin D as a kind of adjuvant agent for better influenza vaccine efficacy. The studies showed that the addition of vitamin D or $1\alpha$,25(OH)$_2$D to a variety of vaccine preparations could increase induced immunity to the herpes simplex virus, diphtheria toxoid, tetanus toxoid, hepatitis B surface antigen, poliovirus, or HIVgp160. Following Sadarangani et al. [75,78], mature adult mice that were immunized subcutaneously or intramuscularly with inactivated polio vaccine, *Haemophilus influenzae* type b oligosaccharide conjugated to diphtheria toxoid vaccine, and hepatitis B surface antigen (HBsAg), coadministered with $1\alpha$,25(OH)$_2$D, demonstrated production of antigen-specific mucosal immunity (IgA and IgG antibodies) as well as enhanced systemic immune response. As demonstrated by Lang and Samaras [49], vitamin D coadministered with trivalent influenza vaccine (TIV) in mice was shown to enhance the anti-hemagglutinin antibody response and mucosal immunity.

Even though vitamin D has profound effects on immunity, there are insufficient data to show the real relationship between vitamin D status and influenza vaccine immunization [45]. Sadarangani et al. [75,78] suggest that studying the impact of VDR-related polymorphism can provide further insight into the complex interactions in the context of vaccine-induced immune response.

As mentioned earlier, according to Jones et al. [58], the vitamins which support vaccine efficacy should be supplemented, particularly at the time of respiratory virus vaccination, to improve antigen presentation and antibody profile, and this warrants further studies.

### *5.2. Human Studies on the Impact of Vitamin D on Immunization*

#### 5.2.1. Beneficial Effects

The general target groups for influenza vaccination are children and older adults. Following Principi et al. [77], influenza is common in pediatrics. Children are the main cause of the spread of the infection, and because of the immaturity of the immune system at ages less than five years, they can develop severe disease. The other target group, older persons, have impaired immune responses to influenza vaccination secondary to immunosenescence, inflammaging, and impaired T-cell responses or loss of dendritic cell function [25,75]. Additionally, this group of the population is more often exposed to vitamin D deficit because of decreased skin synthesis and decreased renal production of the active metabolite, $1\alpha$,25(OH)$_2$D [75].

Literature data provide evidence both for and against the beneficial effect of vitamin D supplementation on antibody production, and the possible reasons for the negative results obtained are broadly discussed.

The most promising results were obtained by Chadha et al. [79] in prostate cancer patients. Response to the trivalent influenza vaccine, Fluzone, for the period 2006–2007 in 28 of 35 participants was defined as the increase in antibody titer at three months against all three strains of the virus positively correlated with 25(OH)D level, which was tested as a continuous variable in relation to serological response ($p = 0.0446$). The median baseline 25(OH)D level was 44.88 ng/mL (range: 9.16–71.98 ng/mL). This outcome was also evaluated in studies on adults aged above 50 years by Sundaram et al. [80], but the results obtained by the authors were opposite. Prospective cohort studies conducted during two influenza seasons (2008–2009) showed that more than 25% of the 1103 participants were vitamin D-deficient (<25 ng/mL), and the deficiency was associated with a greater frequency of post-vaccine seroprotection for the H1N1 virus, but only in the first year of the study. It was not related to seroprotection or seroconversion for any other strain in either year. Therefore, no consistent association was found between vitamin D levels and serological response to influenza vaccination in older adults.

Sadarangani et al. [75] reported a weak positive correlation between 25(OH)D levels and change in influenza-specific granzyme B response on day 75 post-vaccination ($p = 0.04$) in a cohort of 159 healthy subjects (50–74 years old) vaccinated with one dose of trivalent 2010–2011 influenza vaccine containing A/California/H1N1-like virus. The range of 25(OH)D defined in the studied group was from 36.6 to 52.2 ng/mL. These results do not exclude the role of vitamin D in improving immunization. Granzyme B is a serine protease produced by NK cells, dendritic cells, and cytotoxic T cells. It is known to induce the cytotoxic T-cell-mediated apoptosis of virus-infected host cells and has been evaluated as a promising marker of immunity response to the influenza vaccine. It is worth noting is that granzyme B response negatively correlates with increasing age [75]. In the same study, no correlation between 25(OH)D levels and humoral immune response was detected at any time point after vaccination.

#### 5.2.2. Studies Showing No Relevant Effects

No association between vitamin D and post-vaccine serological response was shown in HIV-infected patients vaccinated with monovalent influenza A (H1N1) by Crum-Cianflone et al. [81]. During a prospective cohort study of 124 participants (64 HIV-infected and 64 uninfected), seroconversion measured as >4-fold increase in antibody titer was achieved on day 28 post-vaccination in 56% of the HIV-infected persons versus 74% of those HIV-uninfected, but vitamin D deficiency was insignificantly different between both groups.

Studies on the immunization of children vaccinated with live attenuated or inactivated influenza vaccines in dependence of the 25(OH)D serum concentration conducted by Lin et al. [82] also did not bring an answer to the question: "Can vitamin D act as an adjuvant in influenza vaccines?". Vitamin D levels in serum and influenza antibody titers were measured prior to and 21 days post-vaccination with live attenuated or inactivated influenza vaccine. Surprisingly, low vitamin D levels were associated with higher antibody titer against live attenuated virus, and this effect was strain-specific ($p < 0.05$). Similar results were obtained by Principi et al. [77] four years earlier. On the basis of a prospective, randomized, single-blind, placebo-controlled study with 116 children, the authors indicated that daily supplementation with 1000 IU of vitamin D for four months starting from the injection of the first dose of the trivalent influenza vaccine Fluarix did not significantly modify the antibody response. Similarly, Lee et al. [74] observed no correlation between 25(OH)D levels and post-vaccination antibody titer in a retrospective observational study conducted among 437 young healthy members of the military. Only 224 of them (51.3%) demonstrated an increase in anti-influenza post-vaccination titer, which was not associated with 25(OH)D levels.

A summary of the results obtained by some authors in vitro or in vivo or in clinical controlled studies on the impact of vitamin D on the serological response to anti-influenza vaccines can be found in Table 2.

**Table 2.** Results of human studies on the association between vitamin D serum level and the serological response to anti-influenza vaccines published between the years 2011 and 2018.

| Author, Year | Trial/Location | Sample Size/participants | Anti-Influenza Vaccine/Vitamin D Supplemented per Day | Outcome Measure | Time Points | Results | Statistical Significance |
|---|---|---|---|---|---|---|---|
| Chadha et al., 2011 [79] | Prospective study/USA | 35/prostate cancer patients | Fluzone 2006–2007 trivalent vaccine containing: A/New Caledonia/20/99(H1N1), A/Wisconsin/67/2005(H3N2), B/Malaysia/2506/2004 viruses/- | Serum 25(OH)D conc. vs. antibody titer with HAI | 3 months post-vaccination * | 28 of 35 subjects showed 4-fold antibody titer increase against all three strains correlated with serum 25(OH)D conc. in the range: 9.16–71.98 ng/mL | $p = 0.0446$ |
| Sundaram et al., 2013 [80] | Prospective, cohort study/USA | 1103/adult volunteers aged ≥50 years | Trivalent vaccines containing: | Serum 25(OH)D conc. vs | Pre-vaccination and 21–28 days post-vaccination in two seasons: fall 2008–spring 2009 and fall 2009–spring 2010 | ≥4-fold rise in HAI to post-vaccination against H1N1 strain associated with vitamin D deficiency, i.e., | OR = 1.68, 95% CI = 1.13–2.49 |
| | | | Season 1: A/Brisbane/59/2009-like(H1N1), A/Brisbane/10/2007-like(H3N2), B/Florida/4/2006-like viruses | antibody titer with HAI | | serum 25(OH)D concentration <25 ng during season 1 | |
| | | | Season 2: A/Brisbane/59/2009-like(H1N1), A/Brisbane/10/2007-like(H3N2), B/Brisbane/60/2008-like, influenza A(H1N1)pdm09-like[A(H1N1)pdm09]/- | | | | |
| Principi et al., 2013 [77] | Prospective, randomized, single-blinded, placebo-controlled study/Italy | 116/children with a history of recurrent acute otitis media previously unvaccinated | Fluarix, trivalent vaccine containing: A/California/7/2009(H1N1)-like, A/Perth/16/2009(H3N2)-like, B/Brisbane/60/2008(B)-like viruses/1000 IU for 4 months | Antibody titer with HAI vs. vitamin D supplementation | 3 months post-vaccination | No significant correlation | No significance |
| Sadarangani et al., 2016 [75] | Retrospective, cohort study/USA | 159/healthy subjects aged 50–74 | Fluarix 2010–2011, trivalent vaccine containing: A/H1N1/California/2009-like virus/- | Antibody titer with VNA or HAI vs. serum 25(OH)D conc. | 0, 28, and 75 days post-vaccination | No significant correlation | No significance |

**Table 2.** *Cont.*

| Author, Year | Trial/Location | Sample Size/participants | Anti-Influenza Vaccine/Vitamin D Supplemented per Day | Outcome Measure | Time Points | Results | Statistical Significance |
|---|---|---|---|---|---|---|---|
| Crum-Cianflone et al., 2016 [81] | Prospective cohort study/USA | 128/adults, 64 HIV-infected and 64 HIV-uninfected | Monovalent 2009 influenza A (H1N1) vaccine containing A/California/7/2009(H1N1)/ 2009–2010/- | Seroconversion vs. HIV infection | 0 days, 28 days, and 6 months post-vaccination | Seroconversion at 28 days post-vaccination in 56% of HIV-infected patients vs. 74% HIV-uninfected persons. | $p = 0.03$ |
| | | | | Serum 25(OH)D conc. vs. HIV infection | | Vitamin D deficiency was not significantly prevalent in HIV-infected patients (25%) compared to HIV-uninfected persons (17%) | $p = 0.39$ |
| | | | | Serum 25(OH)D conc. vs. seroconversion | | No associations between serum 25(OH)D conc. and antibody responses at 28 days and after 6 months | No significance |
| Lin et al., 2017 [82] | Prospective cohort study/USA | 135/children aged 3–17 years | Live attenuated influenza vaccine, LAIV B lineages (B Brisbane and B Massachusetts) and LAIV A strains (A/H1N1 and A/H3N2) or inactivated influenza vaccine, IIV/- | Serum 25(OH)D conc. vs. antibody titer with HAI | Pre-vaccination and 21 days post-vaccination | Serum 25(OH)D conc. were >20 ng/mL in 55% of persons. | $p < 0.05$ |
| | | | | | | Post-vaccination antibody titers for LAIV B were higher among those with lower 25(OH)D levels and among younger participants; | $p < 0.05$ |
| | | | | | | no associations between 25(OH)D conc. and responses to LAIV A strains or to any IIV strains | |
| Lee et al., 2018 [74] | Retrospective, cross-sectional observational study/USA | 437/young healthy military members | Monovalent influenza A (H1N1) vaccine 2009 (strain A/California/7/2009/H1N1) | Immunogenicity vs. serum 25(OH)D conc. (seroprotection was defined as antibody titer ≥ 1:40 with MN | At least 30 days post-vaccination | 34.8% of participants were vitamin D-deficient, 38.2% were insufficient, and 27.0% were normal in regard to serum 25(OH)D conc.; 51.4% of total participants showed seroprotection | Geometric mean titer: insufficient vs. normal: OR 1.25 (0.78–2.01); deficient vs normal: OR 1.10 (0.68–1.78); continuous 25(OH)D conc.: OR 0.98 (0.84–1.15) |
| | | | | | | No associations between antibody response and any baseline characteristic | 95% CI, $p > 0.05$ |

HAI—hemagglutination antibody inhibition assay; OR—odd ratios; VNA—viral neutralization assay; MN—microneutralization test.

### 5.2.3. Critical Views

The authors have a critical approach to the clinical studies which do not prove a beneficial effect of vitamin D on the immune system or indicate the lack of any effect of this vitamin on post-vaccine immunization. The main potential reasons for this outcome which are most often considered are: too short a time between vaccination and antibody assays, which may be not sufficient to reveal immunostimulation; or too low a dose of vitamin D, giving too low a concentration level of 25(OH)D to mediate in immunization processes [49]. For example, Lee et al. [74], in their studies, used the clinical cutoffs for insufficiency of 20–30 ng/mL and for deficiency of <20 ng/mL, while as demonstrated by the authors, some experts have shown that a higher level is required (≥40 ng/mL). Some authors also admit that in their studies, T cells were not included. Therefore, any potential impact of vitamin D on this part of the immune response has not been explored [74,80]. The next potential reason is too-small samples being used in the studies or the profiles of the immune system in the patients with comorbid diseases being different than in the healthy subjects, which can potentially determine the effect of vitamin D on the response to the influenza antigen, as observed by Chadha et al. [79]. Besides, studies such as the above should be conducted on a group which is representative for other populations [74]. As Lang and Samaras [49] mentioned, for example, the antibody level present in the preimmunized serum can mask the potential immunomodulatory effect of vitamin D on the immune response post-vaccination because of the inverse relationship between antibody levels prior to and antibody increase post-vaccination. Few studies have included subjects already on vitamin D supplementation, or they were conducted in selected populations [49,74,75,78].

The work described and done by Surman et al. [83] raised the very important question of the interaction of different factors resulting in the immune response. The authors reported in mice immunized with an attenuated influenza virus vaccine that double deficiencies for vitamin A and D reduced antibody response in the respiratory tract to a greater extent than deficiency for one of these vitamins. Although supplementation with vitamin A had a greater corrective effect than vitamin D for the restitution of seroprotection (IgG and IgA responses), the best results were with the two vitamins combined and administered at the time of the vaccination of the animals. These studies, although they are animal-based, give some useful information for designing human clinical trials for the improvement of influenza vaccine efficacy and show that the approach to the significance and the potential role of supplementation with micronutrients, such as vitamins, to reach this aim should regard their interaction and synergism of action [74,83].

Lee et al. [74] suggested the real role of vitamin D in immunization against influenza viruses. It is known that vitamin D supports anti-inflammatory processes through its impact on T cells. Hence, according to the authors, a measure of increased immunity may not be the mechanism of action by which vitamin D functions. It is possible that its higher level reduces the severity of the inflammatory response brought on by infections.

So, the presented results are not too enthusiastic regarding the improvement of influenza vaccine efficacy by vitamin D. However, in light of its fantastic properties, it is undoubtedly the only vitamin characterized by such a broad spectrum of activity in the immune system.

## 6. Summary

The survey of the literature concerning the role of vitamin D in the immune system and immunization, especially against influenza viruses, does not give an unequivocal and one-word answer of "yes" or "no" to the questions: "Does vitamin D supplementation enhance the host's resistance to influenza?" and "Does vitamin D supplementation play a role in the therapy of viral infection diseases?". The authors, as presented, despite the lack of the expected results, do not exclude the significance of this vitamin for the anti-influenza battle waged in human organisms. Some of them suggest that the effects of 1α,25(OH)$_2$D on chemokine expression and secretion vary between pathogens [7]. The results of the presented studies also do not exclude the premises to including vitamin D as an adjuvant in directing new influenza vaccine design. Following Wiwanitkit [84], in spite of the unconfirmed usefulness of vitamin D as an adjuvant for influenza vaccination, giving it simultaneously still poses some clinical advantages. According to Grant and Giovanucci [48], vitamin D supplements or fortified foods should be evaluated further as a possibly useful component of a programme to reduce influenza mortality rates, especially in elderly persons.

The studies clearly show that vitamin D is, undoubtedly, part of the complex factors which affect the immune response. So, assessing vitamin D status and maintaining optimal serum levels should be considered in all ageing adults and children, and micronutrients should be regarded as one of the essential factors which improve our health condition overall and also support our fight against diseases.

**Funding:** This research received no external funding.

**Conflicts of Interest:** The author declares no conflict of interest.

## References

1. Gruber, B.M. The Phenomenon of Vitamin D. *Eur. PMC* **2015**, *69*, 127–139. Available online: http://www.phmd.pl/fulltxt.php?ICID=1137498 (accessed on 23 January 2015).
2. Jones, G. Vitamin D safety: Its mechanisms and application. *Stand. Med. Pediatr.* **2012**, *9*, 605–609.
3. Kim, D. The role of vitamin D in thyroid diseases. *Int. J. Mol. Sci.* **2017**, 18. [CrossRef] [PubMed]
4. Beard, J.A.; Bearden, A.; Striker, R. Vitamin D and the anti-viral state. *J. Clin. Virol.* **2011**, *50*, 194–200. [CrossRef] [PubMed]
5. Kienreich, K.; Grübler, M.; Tomaschitz, A.; Schmid, J.; Verheyen, N.; Rutters, F.; Dekker, J.M.; Pilz, S. Vitamin D, arterial hypertension & cerebrovascular disease. *Indian J. Med. Res.* **2013**, *137*, 669–679. Available online: http://www.ijmr.org.in (accessed on 23 September 2013). [PubMed]
6. Bikle, D.D. Extraskeletal actions of vitamin D. *Ann. N. Y. Acad. Sci.* **2016**, *1376*, 29–51. [CrossRef] [PubMed]
7. Greiller, C.L.; Martineau, A.R. Modulation of the immune response to respiratory viruses by vitamin D. *Nutrients* **2015**, *7*, 4240–4270. [CrossRef] [PubMed]
8. Odroważ-Sypniewska, G.; Karczmarewicz, E.; Paprotny, Ł.; Płudowski, P. 3-epi-25(OH)D—A new metabolite, potential biological function, interference in laboratory assays. *Stand. Med. Pediatr.* **2012**, *9*, 680–686.
9. McCullough, P.; Amend, J. Results of daily oral dosing with up to 60,000 international units (iu) of vitamin D3 for 2 to 6 years in 3 adult males. *J. Steroid Biochem. Mol. Biol.* **2017**, *173*, 308–312. [CrossRef] [PubMed]
10. Christakos, S.; Hewison, M.; Gardner, D.G.; Wagner, C.L.; Sergeev, I.N.; Rutten, E.; Pittas, A.G.; Boland, R.; Ferrucci, L.; Bikle, D.D. Vitamin D: Beyond bone. *Ann. N. Y. Acad. Sci.* **2013**, *1287*, 45–58. [CrossRef] [PubMed]
11. Moukayed, M.; Grant, W.B. Molecular link between vitamin D and cancer prevention. *Nutrients* **2013**, *5*, 3993–4021. [CrossRef] [PubMed]
12. Lorenc, R.S.; Karczmarewicz, E.; Kryśkiewicz, E.; Płudowski, P. Vitamin D provision and supplementation standards. *Stand. Med. Pediatr.* **2012**, *9*, 595–604.
13. Abdelsalam, A.; Rashed, L.; Salman, T.; Hammad, L.; Sabry, D. Molecular assessment of vitamin D receptor polymorphism as a valid predictor to the response of interferon/Ribavirin based therapy in Egyptian patients with Chronic Hepatitis C. *J. Dig. Dis.* **2016**, *17*, 547–553. [CrossRef] [PubMed]
14. Lange, C.M.; Bojunga, J.; Ramos-Lopez, E.; von Wagner, M.; Hassler, A.; Vermehren, J.; Herrmann, E.; Badenhoop, K.; Zeuzem, S.; Sarrazin, C. Vitamin D deficiency and a CYP27B1-1260 promoter polymorphism are associated with chronic hepatitis C and poor response to interferon-alfa based therapy. *J. Hepatol.* **2011**, *54*, 887–893. [CrossRef] [PubMed]
15. Xue, L.N.; Xu, K.Q.; Zhang, W.; Wang, Q.; Wu, J.; Wang, X.Y. Associations between vitamin D receptor polymorphisms and susceptibility to ulcerative colitis and Crohn's disease: A meta-analysis. *Inflamm. Bowel Dis.* **2013**, *19*, 54–60. [CrossRef] [PubMed]
16. Misiorowski, W. Vitamin D in type 1 and type 2 diabetes in adulthood. *Stand. Med. Pediatr.* **2012**, *9*, 639–644.
17. Holick, C.N.; Stanford, J.L.; Kwon, E.M.; Ostrander, E.A.; Nejentsev, S.; Peters, U. Comprehensive association analysis of the vitamin D pathway genes, VDR, CYP28B1, and CYP24A1, in prostate cancer. *Cancer Epidemiol. Biomark. Prev.* **2007**, *16*, 1990–1999. [CrossRef] [PubMed]

18. Areeshi, M.Y.; Mandal, R.K.; Akhter, N.; Panda, A.K.; Haque, S. Evaluating the association between TaqI variant of vitamin D receptor gene and susceptibility to tuberculosis: A meta-analysis. *Toxicol. Int.* **2014**, *21*, 140–147. [CrossRef] [PubMed]
19. Pittas, A.G.; Laskowski, U.; Kos, L.; Saltzman, E. The role of vitamin D In adults requiring nutrition therapy. *J. Parenter. Enteral Nutr.* **2010**, *34*, 70–78. [CrossRef] [PubMed]
20. Bischoff-Ferrari, H.A.; Shao, A.; Dawson-Hughes, B.; Hathcock, J.; Giovanucci, E.; Willet, W.C. Benefit-risk assessment of vitamin D supplementation. *Osteoporos. Int.* **2010**, *21*, 1121–1132. [CrossRef] [PubMed]
21. Heaney, R.P. Vitamin D in health and disease. *Clin. J. Am. Soc. Nephrol.* **2008**, *3*, 1535–1541. [CrossRef] [PubMed]
22. Dougherty, K.A.; Schall, J.J.; Zemel, B.S.; Tuluc, F.; Hou, X.; Ritstein, R.M.; Stallings, V.A. Safety and efficacy of high-dose daily vitamin $D_3$ supplementation in children and young adult infected with human immunodeficiency virus. *J. Pediatr. Infect. Dis. Soc.* **2014**, *3*, 294–303. [CrossRef] [PubMed]
23. Cashman, K.D.; Ritz, C.; Kely, M. ODIN Collaborators. Improved dietary guidelines for vitamin D: Application of individual participant data (IPD)-level meta-regression analyses. *Nutrients* **2017**, *9*, 469. [CrossRef] [PubMed]
24. Cannell, J.J.; Vieth, R.; Umhau, J.C.; Holick, M.F.; Grant, W.B.; Madronich, S.; Garland, C.F.; Giovannucci, E. Epidemic influenza and vitamin D. *Epidemiol. Infect.* **2006**, *134*, 1129–1140. [CrossRef] [PubMed]
25. Sundaram, M.E.; Coleman, L.A. Vitamin D and influenza. *Adv. Nutr.* **2012**, *3*, 517–525. [CrossRef] [PubMed]
26. Szymczak, I.; Pawliczak, R. The active metabolite of vitamin $D_3$ as a potential immunomodulator. *Scand. J. Immunol.* **2015**, *83*, 83–91. [CrossRef] [PubMed]
27. Jakóbisiak, M. Główne komponenty i zasadnicze cechy odpowiedzi immunologicznej. In *Immunologia*, 2nd ed.; Wydawnictwo Naukowe PWN: Warsaw, Poland, 1995; pp. 28–36.
28. Bruce, D.; Ooi, J.H.; Yu, S.; Cantorna, M.T. Vitamin D and host resistance to infection? Putting the cart in front of the horse. *Exp. Biol. Med.* **2010**, *235*, 921–927. [CrossRef] [PubMed]
29. Chun, R.F.; Liu, P.T.; Modlin, R.L.; Adams, J.S.; Hewison, M. Impact of vitamin D on immune function: Lessons learned from genome-wide analysis. *Front. Physiol.* **2014**, *5*, 1–15. [CrossRef] [PubMed]
30. Fitch, N.; Becker, A.B.; HayGlass, K.T. Vitamin D[1,25[OH]2D3] differentially regulates human innate cytokine responses to bacterial versus viral pattern recognition receptor stimuli. *J. Immunol.* **2016**, *196*, 2965–2972. [CrossRef] [PubMed]
31. Jasińska, J. The role of receptor CD40-ligand CD40 (cd40/D40L) system in inflammatory processes. *Alergia* **2015**, *4*, 39–42.
32. Jeffery, L.E.; Wood, A.M.; Qureshi, O.S.; Hou, T.Z.; Gardner, D.; Briggs, Z.; Kaur, S.; Raza, K.; Sansom, D.M. Availability of 25-hydroxyvitamin D(3) to APCs controls the balance between regulatory and inflammatory T cell responses. *J. Immunol.* **2012**, *189*, 5155–5164. [CrossRef] [PubMed]
33. Sigmundsdottir, H.; Pan, J.; Debes, G.F.; Alt, C.; Habtezion, A.; Soler, D.; Butcher, E.C. DCs metabolize sunlight-induced vitamin D3 to 'program' T cell attraction to the epidermal chemokine CCL27. *Nat. Immunol.* **2007**, *8*, 285–293. [CrossRef] [PubMed]
34. Amital, H.; Shoenfeld, Y. Disease associations of vitamin D in autoimmune disorders-prevention and therapy. *Stand. Med. Pediatr.* **2012**, *9*, 620–622.
35. Tang, J.; Zhou, R.; Luger, D.; Zhu, W.; Silver, P.B.; Grajewski, R.S.; Su, S.B.; Chan, C.C.; Adorini, L.; Caspi, R.R. Calcitriol suppresses antirenal autoimmunity through inhibitory effects on the Th17 effector response. *J. Immunol.* **2009**, *182*, 4624–4632. [CrossRef] [PubMed]
36. Ehrchen, J.; Helming, L.; Varga, G.; Pasche, B.; Loser, K.; Gunzer, M.; Sunderkötter, C.; Sorg, C.; Roth, J.; Lengeling, A. Vitamin D receptor signaling contributes to susceptibility to infection with Leishmania major. *FASEB J.* **2007**, *21*, 3208–3218. [CrossRef] [PubMed]
37. Rajapakse, R.; Mousli, M.; Pfaff, A.W.; Uring-Lambert, B.; Marcellin, L.; Bronner, C.; Jeanblanc, M.; Villard, O.; Letscher-Bru, V.; Klein, J.P.; et al. 1,25-dihydroxyvitamin D3 induces splenocyte apoptosis and enhances BALB/c mice sensitivity to toxoplasmosis. *J. Steroid Biochem. Mol. Biol.* **2005**, *96*, 179–185. [CrossRef] [PubMed]
38. Ryz, N.R.; Patterson, S.J.; Zhang, Y.; Ma, C.; Huang, T.; Bhinder, G.; Wu, X.; Chan, J.; Glesby, J.; Sham, H.P.; et al. Active vitamin D (1,25-dihydroxyvitamin D3) increases host susceptibility to Citrobacter rodentium by suppressing mucosal Th17 responses. *Am. J. Ohysiol. Gastrointest. Liver Physiol.* **2012**, *303*, G1299–G1311. [CrossRef] [PubMed]

39. Fang, Y.; Banner, D.; Kelvin, A.A.; Huang, S.S.; Paige, C.J.; Corfe, S.A.; Kane, K.P.; Bleackley, R.C.; Rowe, T.; Leon, A.J.; et al. Seasonal H1N1 influenza virus infection induces cross-protective pandemic H1N1 virus immunity through a CD8-independent, B cell dependent mechanism. *J. Virol.* **2012**, *86*, 2229–2238. [CrossRef] [PubMed]
40. Shiozawa, K.; Shiozawa, S.; Shimizu, S.; Fujita, T. 1α,25-dihydroxyvitamin $D_3$ inhibits pokeweed mitogen-stimulated human B-cell activation: An analysis using serum-free culture conditions. *Immunology* **1985**, *56*, 161–167. [PubMed]
41. Shirakawa, A.-K.; Nagakubo, D.; Hieshima, K.; Nakayama, T.; Jin, Z.; Yoshie, O. 1,25-Dihydroxyvitamin D3 Induces CCR10 Expression in Terminally Differentiating Human B Cells. *J. Immunol.* **2008**, *180*, 2786–2795. [CrossRef]
42. Heine, G.; Niesner, U.; Chang, H.D.; Steinmeyer, A.; Zügel, U.; Zuberbier, T.; Radbruch, A.; Worm, M. 1,25-dihydroxyvitamin D3promotes IL-10 production in human B cells. *Eur. J. Immunol.* **2008**, *38*, 2210–2218. [CrossRef]
43. Penna, G.; Amuchastegui, S.; Cossetti, C.; Aquilano, F.; Mariani, R.; Sanvito, F.; Doglioni, C.; Adorini, L. Treatment of experimental autoimmune prostatitis in nonobese diabetic mice by the vitamin D receptor agonist elocalcitol. *J. Immunol.* **2006**, *177*, 8504–8511. [CrossRef] [PubMed]
44. Langrish, L.L.; Chen, Y.; Blumenschein, W.M.; Mattson, J.; Basham, B.; Sedgwick, J.D.; McClanahan, T.; Kastelein, R.A.; Cua, D.J. IL-23 drives a pathogenic T cell population that induces autoimmune inflammation. *J. Exp. Med.* **2005**, *201*, 233–240. [CrossRef] [PubMed]
45. Friis, H.; Range, N.; Pedersen, M.L.; Mølgaard, C.; Changalucha, J.; Krarup, H.; Magnussen, P.; Søborg, C.; Andersen, A.B. Hypovitaminosis D is common among pulmonary tuberculosis patients in Tanzania but is not explained by the acute phase response. *J. Nutr.* **2008**, *138*, 2474–2480. [CrossRef] [PubMed]
46. Tse, A.K.; Wan, C.K.; Shen, X.L.; Zhu, G.Y.; Cheung, H.Y.; Yang, M.; Fong, W.F. 1,25-dihydroxyvitamin D3 induces biphasic NF-kappaB responses during HL-60 leukemia cells differentiation through protein induction and PI3K/Akt-dependent phosphorylation/degradation of IκB. *Exp. Cell Res.* **2007**, *313*, 1722–1734. [CrossRef] [PubMed]
47. Hope-Simpson, R. The role of season in the epidemiology of influenza. *Epidemiol. Infect.* **1981**, *86*, 35–47.
48. Grant, W.B.; Giovannucci, E. The possible roles of solar ultraviolet-B radiation and vitamin D in reducing case-fatality rates from the 1918–1919 influenza pandemic in the United States. *Dermato-Endocrinol.* **2009**, *1*, 215–219. [CrossRef]
49. Lang, P.O.; Samaras, D. Aging adults and seasonal influenza: Does the vitamin D status (H)arm the body? *J. Aging Res.* **2011**, *2012*. [CrossRef] [PubMed]
50. Khare, D.; Godbole, N.M.; Pawar, S.D.; Mohan, V.; Pandey, G.; Gupta, S.; Kumar, D.; Dhole, T.N.; Godbole, M.M. Calcitriol [1,25[OH]2D3] pre-and post-treatment suppresses inflammatory response to influenza A (H1N1) infection in human lung A549 epithelial cells. *Eur. J. Nutr.* **2013**, *52*, 1405–1415. [CrossRef] [PubMed]
51. Helming, L.; Böse, J.; Ehrchen, J.; Schiebe, S.; Frahm, T.; Geffers, R.; Probst-Kepper, M.; Balling, R.; Lengeling, A. 1α,25-dihydroxyvitamin $D_3$ is a potent suppressor of interferon γ-mediated macrophage activation. *Blood* **2005**, *106*, 4351–4358. [CrossRef] [PubMed]
52. Moan, J.; Dahlback, A.; Ma, L.W.; Juzeniene, A. Influenza, solar radiation and vitamin D. *Dermato-Endocrinol.* **2009**, *1*, 307–309. [CrossRef] [PubMed]
53. Laaksi, I.; Ruohola, J.-P.; Mattila, V.; Auvinen, A.; Ylikomi, T.; Pihlajamäki, H. Vitamin D supplementation for the prevention of acute respiratory tract infection: A randomized, double-blinded trial among young Finnish men. *J. Infect. Dis.* **2010**, *202*, 809–814. [CrossRef] [PubMed]
54. Urashima, M.; Segawa, T.; Okazaki, M.; Kurihara, M.; Wada, Y.; Ida, H. Randomized trial of vitamin D supplementation to prezent seasonal influenza A in schoolchildren. *Am. J. Clin. Nutr.* **2010**, *91*, 1255–1260. [CrossRef] [PubMed]
55. Jorde, R.; Witham, M.; Janssens, W.; Rolighed, L.; Borchhardt, K.; De Boer, I.H.; Grimnes, G.; Hutchinson, M.S. Vitamin D supplementation did not prevent influenza-like illness as diagnosed retrospectively by questionnaires in subjects participating in randomized clinical trials. *Scand. J. Infect. Dis.* **2012**, *44*, 126–132. [CrossRef] [PubMed]
56. Berry, D.J.; Hesketh, K.; Power, C.; Hyppönen, E. Vitamin D status has a linear association with seasonal infections and lung function in British adults. *Br. J. Nutr.* **2011**, *106*, 1433–1440. [CrossRef] [PubMed]

57. Aregbesola, A.; Voutilainen, S.; Nurmi, T.; Virtanen, J.K.; Ronkainen, K.; Tuomainen, T.P. Serum 25-hydroxyvitamin D3 and the risk of pneumonia in an ageing general population. *J. Epidemiol. Community Health* **2013**, *67*, 533–536. [CrossRef] [PubMed]
58. Jones, B.G.; Oshansky, C.M.; Bajracharya, R.; Tang, L.; Sun, Y.; Wong, S.S.; Webby, R.; Thomas, P.G.; Hurwitz, J.L. Retinol binding protein and vitamin D associations with serum antibody isotypes, serum influenza virus-specific neutralizing activities and airway cytokine profiles. *Clin. Exp. Immunol.* **2015**, *183*, 239–247. [CrossRef] [PubMed]
59. Mamani, M.; Muceli, N.; Ghasemi Basir, H.R.; Vashegbani, M.; Poorolajal, J. Association between serum concentration of 25-hydroxyvitamin D and community-acquired pneumonia: A case-control study. *Int. J. Gen. Med.* **2017**, *13*, 423–429. [CrossRef] [PubMed]
60. Brance, M.L.; Miljevic, J.N.; Tizziani, R.; Taberna, M.E.; Grossi, G.P.; Toni, P.; Valentini, E.; Trepat, A.; Zaccardi, J.; Moro, J.; et al. Serum 25-hydroxyvitamin D levels in hospitalized adults with community-acquired pneumonia. *Clin. Respir. J.* **2018**, *12*, 2220–2227. [CrossRef] [PubMed]
61. Nanri, A.; Nakamoto, K.; Sakamoto, N.; Imai, T.; Akter, S.; Nonaka, D.; Mizoue, T. Association of serum 25-hydroxyvitamin D with influenza in case-control study nested in a cohort of Japanese employees. *Clin. Nutr.* **2017**, *36*, 1288–1293. [CrossRef] [PubMed]
62. Gui, B.; Chen, Q.; Hu, C.; Zhu, C.; He, G. Effects of calcitriol (1,25-dihydroxy-vitamin D3) on the inflammatory response induced by H9N2 influenza virus infection in human lung A549 epithelial cells and in mice. *Virol. J.* **2017**, *14*, 10–20. [CrossRef] [PubMed]
63. Urashima, M.; Mezawa, H.; Noya, M.; Camargo, C.A. Jr. Effects of witamin D supplements on influenza A illness during the 2009 H1N1 pandemic: A randomized controlled trial. *Food Funct.* **2014**, *5*, 2365–2370. [CrossRef] [PubMed]
64. Li-Ng, M.; Aloia, J.F.; Pollack, S.; Cunha, B.A.; Mikhail, M.; Yeh, J.; Berbari, N. A randomized controlled. trial of vitamin D3 supplementation for the prevention of symptomatic upper respiratory tract infections. *Epidemiol. Infect.* **2009**, *137*, 1396–1404. [CrossRef] [PubMed]
65. Aloia, J.F.; Li-Ng, M. Re: Epidemic influenza and vitamin D. *Epidemiol. Infect.* **2007**, *137*. [CrossRef]
66. Lappe, J.; Watson, P.; Travers-Gustafson, D.; Recker, R.; Garland, C.; Gorham, E.; Baggerly, K.; McDonnell, S.L. Effect of vitamin D and calcium supplementation on cancer incidence in older women: A randomized clinical trial. *JAMA Netw.* **2017**, *317*, 1234–1243. [CrossRef] [PubMed]
67. Martineau, R.; Jolliffe, D.A.; Hooper, R.L.; Greenberg, L.; Aloi, J.F.; Bergman, P.; Dubnov-Raz, G.; Esposito, S.; Ganmaa, D.; Ginde, A.A.; et al. Vitamin D supplementation to prevent cute respiratory tract infections: Systematic review and meta-analysis of individual participant data. *BMJ* **2017**, *356*. [CrossRef]
68. Grant, W.B.; Boucher, B.J.; Bhattoa, H.P.; Lahore, H. Why vitamin D clinical trials should be based on 25-hydroxyvitamin D concentrations. *J. Steroid Biochem. Mol. Biol.* **2018**, *177*, 266–269. [CrossRef] [PubMed]
69. Shaman, J.; Jeon, C.Y.; Giovannucci, E.; Lipsitch, M. Shortcomings of vitamin D-based model simulations of seasonal influenza. *PLoS ONE* **2011**, *6*, e20743. [CrossRef] [PubMed]
70. Shaman, J.; Pitzer, V.E.; Viboud, C.; Grenfell, B.T.; Lipsitch, M. Absolute humidity and the seasonal onset of influenza in the continental United States. *PLoS Biol.* **2010**, *8*, e1000316. [CrossRef]
71. Koep, T.H.; Enders, F.T.; Pierret, C.; Ekker, S.C.; Krageschmidt, D.; Neff, K.L.; Lipsitch, M.; Shaman, J.; Huskins, W.C. Predictors of indoor absolute humidity and estimated effects on influenza virus survival in grade schools. *BMC Infect. Dis.* **2013**, *13*. [CrossRef] [PubMed]
72. Yang, W.; Cummings, M.J.; Bakamutumaho, B.; Kayiwa, J.; Owor, N.; Namagambo, B.; Byaruhanga, T.; Lutwama, J.; O'Donnell, M.R.; Shaman, J. Dynamics of influenza in tropical Africa: Temperature, humidity, and co-circulating (sub)types. *Influenza Other Respir. Viruses* **2018**, *12*, 446–456. [CrossRef] [PubMed]
73. Kroll, M.H.; Bi, C.; Garber, C.C.; Kaufman, H.W.; Liu, D.; Caston-Balderrama, A.; Zhang, K.; Clarke, N.; Xie, M.; Reitz, R.E.; et al. Temporal relationship betwen vitamin D status and parathyroid hormone in the United States. *PLoS ONE* **2015**. [CrossRef] [PubMed]
74. Lee, R.U.; Won, S.H.; Hansen, C.; Crum-Cianflone, N.F. 25-hydroxyvitamin D, influenza vaccine response and healthcare encounters among a young adult population. *PLoS ONE* **2018**, *13*, e0192479. [CrossRef]
75. Sadarangani, S.P.; Ovsyannikova, I.G.; Goergen, K.; Grill, D.E.; Poland, G.A. Vitamin D, leptin and impact on immune response to seasonal influenza A/H1N1 vaccine in older persons. *Hum. Vaccines Immunother.* **2016**, *12*, 691–698. [CrossRef] [PubMed]

76. Galant, K.; Barg, E.; Kazanowska, B. Vitamin D and metabolic, autoimmunologic and neoplasm diseases. *Pediatr. Endocrinol. Diabetes Metab.* **2016**, *24*, 31–37. [CrossRef] [PubMed]
77. Principi, N.; Marchisio, P.; Terranova, L.; Zampiero, A.; Baggi, E.; Daleno, C.; Tirelli, S.; Pelucchi, C.; Esposito, S. Impact of vitamin D administration on immunogenicity of trivalent inactivated influenza vaccine in previously unvaccinated children. *Hum. Vaccines Immunother.* **2013**, *9*, 969–974. [CrossRef] [PubMed]
78. Sadarangani, S.P.; Whitaker, J.A.; Poland, G.A. "Let there be light": The role of vitamin D in the immune response to vaccines. *Expert. Rev. Vaccines* **2015**, *14*, 1427–1440. [CrossRef] [PubMed]
79. Chadha, M.K.; Fakih, M.; Muindi, J.; Tian, L.; Mashtare, T.; Johnson, C.S.; Trump, D. Effect of 25-hydroxyvitamin D status on serological response to influenza vaccine in prostate cancer patients. *Prostate* **2011**, *71*, 368–372. [CrossRef] [PubMed]
80. Sundaram, M.E.; Talbot, H.K.; Zhu, Y.; Griffin, M.R.; Spencer, S.; Shay, D.K.; Coleman, L.A. Vitamin D is not associated with serologic response to influenza vaccine in adults over 50 years old. *Vaccine* **2013**, *31*, 2057–2061. [CrossRef] [PubMed]
81. Crum-Cianflone, N.F.; Won, S.; Lee, R.; Lalani, T.; Ganesan, A.; Burgess, T.; Agan, B.K. Vitamin D levels and influenza vaccine immunogenicity among HIV-infected and HIV-uninfected adults. *Vaccine* **2016**, *34*, 5040–5046. [CrossRef] [PubMed]
82. Lin, C.J.; Martin, J.M.; Cole, K.S.; Zimmerman, R.K.; Susick, M.; Moehling, K.K.; Levine, M.Z.; Spencer, S.; Flannery, B.; Nowalk, M.P. Are children's vitamin D levels and BMI associated with antibody titers produced in response to 2014-2015 influenza vaccine? *Hum. Vaccines Immunother.* **2017**, *13*, 1661–1665. [CrossRef] [PubMed]
83. Surman, S.L.; Penkert, R.R.; Jones, B.G.; Sealy, R.E.; Hurwitz, J.L. Vitamin supplementation at the time of immunization with a cold-adapted influenza virus vaccine corrects poor mucosal antibody responses in mice deficient for vitamins A and D. *Clin. Vaccine Immunol.* **2016**, *23*. [CrossRef] [PubMed]
84. Wiwanitkit, V. Vitamin D and influenza vaccination. *Hum. Vaccines Immunother.* **2013**, *9*. [CrossRef] [PubMed]

International Journal of
*Molecular Sciences*

MDPI

*Review*

# The Impact of Vitamin D on the Immunopathophysiology, Disease Activity, and Extra-Musculoskeletal Manifestations of Systemic Lupus Erythematosus

**Anselm Mak** [1,2]

1 Department of Medicine, National University of Singapore, Singapore 119228, Singapore; mdcam@nus.edu.sg
2 Division of Rheumatology, University Medicine Cluster, National University Health System, Singapore 119228, Singapore

Received: 27 June 2018; Accepted: 8 August 2018; Published: 10 August 2018

**Abstract:** Over the past two decades it has been increasingly recognized that vitamin D, aside from its crucial involvement in calcium and phosphate homeostasis and the dynamics of the musculoskeletal system, exerts its influential impact on the immune system. The mechanistic roles that vitamin D plays regarding immune activation for combating infection, as well as pathologically and mediating autoimmune conditions, have been progressively unraveled. In vitro and in vivo models have demonstrated that the action of vitamin D on various immunocytes is not unidirectional. Rather, how vitamin D affects immunocyte functions depends on the context of the immune response, in the way that its suppressive or stimulatory action offers physiologically appropriate and immunologically advantageous outcomes. In this review, the relationship between various aspects of vitamin D, starting from its adequacy in circulation to its immunological functions, as well as its autoimmune conditions, in particular systemic lupus erythematosus (SLE), a prototype autoimmune condition characterized by immune-complex mediated inflammation, will be discussed. Concurring with other groups of investigators, our group found that vitamin D deficiency is highly prevalent in patients with SLE. Furthermore, the circulating vitamin D levels appear to be correlated with a higher disease activity of SLE as well as extra-musculoskeletal complications of SLE such as fatigue, cardiovascular risk, and cognitive impairment.

**Keywords:** lupus; systemic lupus erythematosus; vitamin D; disease damage; disease activity

## 1. Vitamin D, Its Nature and Impact on Various Body Systems

### 1.1. The Physiology and Health Impact of Vitamin D

Vitamin D is one of the steroid-based vitamins that chiefly regulates the absorption of calcium from the intestine, facilitates calcium reabsorption in the kidneys, and mobilizes calcium and phosphate from bones by activating the osteoclasts [1,2]. Humans obtain vitamin D from two sources, a small amount from diet, while the majority is synthesized in the epidermal layer of the skin by converting 7-dehydrocholesterol in the epidermis to cholecalciferol (vitamin $D_3$) via the action of ultraviolet B between the wavelengths of 280 and 315 nm [3]. By binding to the vitamin D-binding protein in the bloodstream, cholecalciferol is transported to the liver, where it is hydroxylated to 25-hydroxycholecalciferol [25-(OH)$D_3$] under the action of hepatic 25$\alpha$-hydroxylase [4]. Subsequently, in the final processing step, 25-(OH)$D_3$ is converted to 1-25$\alpha$ dihydroxycholecalciferol (1,25-[OH]$_2D_3$) or calcitriol, through the action of 1$\alpha$-hydroxylase in the renal cortices [4]. Unlike the 1$\alpha$-hydroxylation in the liver, the 25$\alpha$-hydroxylation in the kidneys is a tightly regulated process under the meticulous control of the parathyroid hormone (PTH) and serum calcium and phosphate levels [5]. Furthermore, as a checking mechanism to prevent the overproduction of 1,25-$(OH)_2D_3$, which can otherwise induce hypercalcaemia and hyperphosphataemia, 1,25-$(OH)_2D_3$ is capable of stimulating the 24α-hydroxylation of 25-$(OH)D_3$, leading to the formation of the physiologically-inactive 24, 25-$(OH)_2D_3$ [6]. Being an active form of vitamin D, 1,25-$(OH)_2D_3$ initiates the downstream signaling pathways with its association to the vitamin D receptors (VDR), a group of nuclear receptors that heterodimerize with the retinoid X receptor (RXR) and bind to target DNA sequences, named the vitamin D responsive elements (VDRE). The VDRE is situated physically and/or functionally to the promotor region of the target genes that functions to modulate the cell growth, proliferation, and apoptosis, amongst other physiological functions [7].

Besides maintaining calcium and phosphate homeostasis, vitamin D has been demonstrated to diversely impact several body systems, as evidenced by the observation that its deficiency can lead to general health issues [3,8]. For instance, a low vitamin D level has been shown to be related to an increased prevalence of hypertension; cardiovascular disease; respiratory infections; periodontal disease; certain malignancies including breast, colon, and prostate cancers [3,9–11]; and autoimmune conditions such as multiple sclerosis (MS), type 1 diabetes mellitus (DM), rheumatoid arthritis (RA), and systemic lupus erythematosus (SLE) [12–14]. Such observations have led to the postulation that vitamin D supplementation may be a favorable strategy to combat certain health issues, apart from those related to the musculoskeletal system and calcium-phosphate homeostasis. In fact, it is evident that adequate vitamin D levels help reduce the fall risk and the prevalence of certain types of cancers including breast cancer [15], as well as that of autoimmune conditions, such as RA and MS, in a dose-dependent fashion [16–18]. Besides the potential reduction in disease prevalence, adequate vitamin D also exerts an impact on survival. A systematic review and meta-analysis revealed that an increase in the serum 25-$(OH)D_3$ level from 54 to 110 nmol/L was associated with an estimated reduction of mortality by 20% in patients with vitamin-D sensitive diseases, including cardiovascular disease, cancers, respiratory infections, DM, Alzheimer's disease, falls, and MS [19]. In addition, an increment of the serum vitamin D level to 105 nmol/L was shown to reduce the incidence of cardiovascular disease and cancers, including colorectal and breast malignancy, by 15% and 30%, respectively [3]. Nevertheless, the results of the studies addressing the potential relationship between the amount of vitamin D intake or the vitamin D levels and the prevalence of various vitamin D-related disease conditions should be interpreted with caution, because issues such as the difference in study design; insufficient sample size and statistical power; heterogeneity in meta-analyses; and intangible confounding factors, including assessment of sun exposure, skin pigmentation, drug interactions, and alcohol use, may confound accurate interpretation.

### *1.2. Vitamin D and the Immune Cells*

#### A Brief Outlook

Besides the kidneys, immune cells, including the dendritic cells (DC), macrophages, monocytes, and lymphocytes (B and T cells), are capable of converting 25-$(OH)D_3$ to 1,25-$(OH)_2D_3$ [20–23]. Such a local regulation of the intracellular active vitamin D level theoretically endows the immunocytes with a higher level of vitamin D-related immune response control in the inflamed sites, and it also directly demonstrates the important role that vitamin D contributes to the immune system. VDR is also expressed in immunocytes, and hence, the immune-physiological effects of 1,25-$(OH)_2D_3$ in various immunocytes can be implemented via binding with the VDR, which mediates the gene transcription via association with the VDRE [7]. Another piece of evidence that suggests a potential role of vitamin D in the immune system and the mediation of autoimmune diseases is that a few well-described VDR polymorphisms, namely, *ApaI*, *FokI*, *BsmI*, and *TaqI*, were demonstrated to be associated with the risk of the development of autoimmunity [24,25]. Nevertheless, data regarding the mechanism as to how VDRE affects the downstream functions of immunocytes are limited. With the use of the chromatin precipitation technique, followed by deep sequencing datasets of VDR-binding sites from calcitriol-treated human B cells and monocytes, the transcriptome data identified a small set of VDR targeted genes that were upregulated [26]. The upregulated calcitriol-responsive VDR target genes demonstrated significant relationships with leukocyte transendothelial migration, Fc$\gamma$ receptor-mediated phagocytosis, and transcriptional regulation by VDR [26], illustrating a pivotal role of the VDR target genes in immune regulation [26]. In addition, in the FoxP3$^{+}$ regulatory T cells (Treg), 1,25-$(OH)_2D_3$-associated VDR was shown to target the VDRE, which comprises an intronic conserved noncoding sequence region of the human *FoxP3* gene, leading to the enhancement of the FoxP3 promotor activity as well as the resultant suppressive activity of Treg, secondary to the vitamin D treatment [27].

### 1.3. Action of Vitamin D in Immunocytes of the Innate Immune System

#### 1.3.1. Dendritic Cells

Immunocytes, ranging from the DC to lymphocytes, have been shown to be influenced by vitamin D because of their expression of VDR [7,23]. During the early differentiation stage, the DC that was differentiated from monocytes in the presence of 1,25-$(OH)_2D_3$ remained in a tolerogenic state, as characterized by a reduction in interleukin (IL)-12 and an increase in IL-10 production, leading to the subsequent reduction in allogenic T-cell activation and enhanced Treg differentiation [28]. However, the effect of 1,25-$(OH)_2D_3$ on inflammatory DCs, which are differentiated and matured inflammatory, are much less substantial [28]. While the role of vitamin D in lupus DC has not been well investigated, being the main driver of interferon (IFN)$\alpha$ production and antigenic presentation, by plasmacytoid DC and myeloid DC, respectfully, the potential role of vitamin D in lupus DC warrants further investigation.

#### 1.3.2. Macrophages

Unlike the unidirectional action of 1,25-$(OH)_2D_3$ in DC, 1,25-$(OH)_2D_3$ has dual effects in the macrophages, depending on their activation status. Taking bacterial infection as an example, during the initial phase of infection, 1,25-$(OH)_2D_3$ enhances the differentiation of monocytes to macrophages [29]. In tuberculosis infection, where the macrophage IFN$\gamma$ receptors are stimulated, the resultant activation of Cyp27Bq potentiates the conversion of 25-$(OH)D_3$ to 1,25-$(OH)_2D_3$ by 1$\alpha$-hydroxylation in the macrophages [30]. 1,25-$(OH)_2D_3$ induces the production of IL-1$\beta$ that enhances the production of the antimicrobial cathelicidin, attempting to facilitate the clearance of the pathogens by the macrophages [31]. At the later phase of infection, where the macrophages are sufficiently or over activated, 1,25-$(OH)_2D_3$ acts to dampen the proinflammatory response by reducing the production of IL-1$\beta$, IL-6, tumor necrosis factor (TNF)$\alpha$, nuclear factor kappa-B ligand (RANKL), nitric oxide, and co-oxygenase-2 (COX-2), and increasing the production of IL-10, which is an anti-inflammatory cytokine [32,33]. While data regarding the action of vitamin D in lupus macrophages are scant, such dual effects impacted by vitamin D on macrophages would be potentially operative in the macrophages from patients with SLE, as lupus macrophages have also been documented to possess dual inflammatory and anti-inflammatory properties [34,35].

### 1.4. Action of Vitamin D in Immune Cells of Adaptive Immune System

Upon activation, different T cell subsets, including the CD4$^{+}$, CD8$^{+}$, and TCR$\gamma\delta^{+}$ T cells, express VDR [36]. The subsequent T-cell activation, which involves the propagation of T-cell receptor (TCR) signaling, has been shown to require an activation of VDR via its association with 1,25-$(OH)_2D_3$ [37]. Compared with DC and the B cells (which will be discussed in a subsequent section), the actions of vitamin D on T cells are more complicated, because of the differential actions of vitamin D in different well-described T cell subsets, including CD4$^{+}$ (Th1, Th2, Th17, and Treg) and CD8$^{+}$ T cells. As a result, the impact of vitamin D on T cell physiology is subset-dependent.

#### 1.4.1. CD4$^+$ T cells—The Th1, Th2, Th17, and Treg Subtypes

The pathogenic role of the CD4$^+$ Th1 T cells in mediating the pro-inflammatory response in SLE has been well described in the literature and extensively reviewed elsewhere [38]. Although insufficient data have yet been shown regarding how vitamin D affects the lupus CD4$^+$ Th1 cells, the data thus far have shown that the VDR expression in non-lupus CD4$^+$ T cells is not as substantial as other T cell subtypes. Nevertheless, during the initial phase of CD4$^+$ T cell differentiation, 1,25-(OH)$_2$D$_3$ treatment is capable of inhibiting the IFN$\gamma$ production chiefly via the downregulation of IL-2 production in Th1 cells [39]. However, the role played by vitamin D in inflammatory response and autoimmunity appears to be more substantial in Th2 and Th17 than in Th1 cells, partly due to the low expression of VDR in the latter [40,41]. Th2 cells are able to suppress experimental autoimmune encephalitis (EAE), a murine model of MS that is predominantly Th17 mediated. Upon activation by 1,25-(OH)$_2$D$_3$, the Th2 cells suppress the Th17-induced inflammatory response in EAE through the transcription factor, GATA3, and the signal transducer and activator of the transcription protein (STAT)-6 activation, although the presence of IL-4 is essential [42]. Similar to MS, Th17 also plays a pivotal role in the pathophysiology of RA. The culture of the peripheral blood mononuclear cells (PBMC), from RA patients with 1,25-(OH)$_2$D$_3$, has been shown to restore IL-4 levels and Th2 polarization [43], potentially leading to less severe arthritis. Notably, in conditions where IL-4 is abundant, 1,25-(OH)$_2$D$_3$ does not appear to increase the IL-4 production further, and suppresses other Th subtypes [43]. Collectively, these data demonstrate the potential of the role of vitamin D in Th2 cells for the suppression of Th17-mediated autoimmune disease via GATA3 and STAT-6 activation and IL-4 production, at least in the Th17-driven EAE model.

The Th17 cells are increasingly recognized as playing essential roles in driving a number of autoimmune conditions, including SLE, RA, and MS [43–45]. In a murine model of retinal autoimmunity, 1,25-(OH)$_2$D$_3$ was shown to inhibit Th17 activity and reduce the expression of IL-17 driven cytokines, namely IL-17 and IL-22/23, in CD4$^+$ memory and CCR6$^+$ T cells [46]. Besides the inhibition of Th17 activity, 1,25-(OH)$_2$D$_3$ also inhibits Th17 differentiation, as evidenced by the observation that when naïve T cells were given conditions towards Th17 polarization, the presence of 1,25-(OH)$_2$D$_3$ inhibits the Th17-related cytokines and intracellular transcription factors, including RORC and CCR6 [47]. In a six-month prospective study of 20 lupus patients with vitamin D deficiency, the adequate supplementation of vitamin D (cholecalciferol) led to reductions in the frequencies of Th1 and Th17 CD4$^+$ T cells, while the frequency of the Treg cells increased [48]. The SLE disease activity in these patients remained stable, without the necessity to escalate the immunosuppressive therapy [48]. While the mechanism as to how the differentiation and activity of Th17 are affected by 1,25-(OH)$_2$D$_3$ is not completely understood, current data suggest that the regulation of IL-17A can be mediated by the binding of VDR to the *IL-17A* promotor region, leading to competition with nuclear factor of activated T cells (NFAT) binding in the same promotor site, and the subsequent recruitment of histone deacetylase (HDAC) and RUNX1, which depress the expression of the *IL-17A* gene [49].

One interesting phenomenon with regard to Th17 cells is their plasticity. Upon stimulation with IL-12 and TNF$\alpha$, Th17 cells manifest Th1 properties by expressing Tbet and IFN$\gamma$, and these non-classic Th17 cells appear to be more pathogenic than the classic Th17 cells in driving autoimmunity [50]. Indeed, 1,25-(OH)$_2$D$_3$ inhibits the proportion of non-classic Th17 cells that express IFN$\gamma$ and IL-17 [50]. In our preliminary knockout mouse study, we were able to demonstrate the plasticity between the Th1 and Th17 cells, and the superior pathogenic potential of Th17 cells in SLE [51] (which will be discussed in a subsequent section).

Treg cells play a crucial role in dampening the proinflammatory responses in many autoimmune conditions, including SLE, type I DM, RA, and MS [52,53]. In the EAE model, 1,25-(OH)$_2$D$_3$ induces the expression of FoxP3 in the lymphoid organs, which is IL-10 signaling dependent [54]. The in vitro treatment of Treg with 1,25-(OH)$_2$D$_3$ induces the production of the IL-10 and expressions of co-inhibitory molecules, including PD1 and CTLA4, which dampen the excessive pro-inflammatory T cell response [55]. Similar to the action of vitamin D in the promotor regions of the Th17 cells, VDR binds to three VDRE regions in the non-coding sequence of the FoxP3 promotor that controls FoxP3 expression in the Treg population [56]. Furthermore, 1,25-(OH)$_2$D$_3$ enhances the Treg suppressive activity and number by inducing the indoleamine 2,3-dioxygenase (IDO) expression, and yet, low vitamin D levels in the SLE patients dampen the Treg migratory ability [56]. Summarizing the current data described, 1,25-$(OH)_2D_3$ furnishes the suppressive activity on the $CD4^+$ T cells, particularly in the Th17 and Th1 subsets that predominantly produce the pro-inflammatory IFNγ and TNFα. In contrast, 1,25-$(OH)_2D_3$ enhances the activity of Th2 and Treg, which chiefly express IL-4 and IL-10 as well as co-inhibitory molecules, including cytotoxic T-lymphocyte–associated antigen 4 (CTLA-4) and programmed cell death-1 (PD1), which dampen the pro-inflammatory responses. As such, 1,25-$(OH)_2D_3$ likely possesses a strong potential to manipulate the immune system against autoimmunity.

#### 1.4.2. $CD8^+$ T Cells

The exact pathophysiological function of the $CD8^+$ T cells in SLE has not been fully characterized, although studies have demonstrated its potential suppressive action in SLE [38]. While limited data has been shown in lupus $CD8^+$ T cells in relation to vitamin D, the current data have demonstrated that $CD8^+$ T cells express a higher VDR level than $CD4^+$ T cells in the non-SLE setting [36]. An adoptive transfer of $VDR^{-/-}$ $CD8^+$ T cells in $Rag^{-/-}$ mice led to severe colitis (predominantly $CD8^+$ T-cell mediated), with an increase in the IFNγ and IL-17 expression in the intestine, particularly when IL-10 is absent [57]. Upon 1,25-$(OH)_2D_3$ treatment, the number of hyperactivated $CD8^+$ cells were substantially reduced [57]. As for the clinical studies, 1,25-$(OH)_2D_3$ was found to inhibit $CD8^+$ activated IFNγ and TNFα secretion in the patients with MS. In the patients with psoriasis, which is a $CD8^+$ dominant T-cell mediated autoimmune condition that manifests as chronic cutaneous and joint inflammation, topical vitamin D (calcipotriol) is one of the therapeutic options of the condition [58].

#### 1.4.3. B Cells

B cells are pathologically important in SLE as they produce lupus-related autoantibodies and function as antigen presenting cells (APC) [59]. The role of vitamin D that is played specifically by lupus B cells has only been very scarcely reported. In a cross-sectional study of 32 patients with SLE, the patients with a high B cell activation status were shown to have a lower mean 25(OH)-vitamin D levels compared to those with a low B cell activation [60]. Nevertheless, the involvement of vitamin D in B cells regarding the physiological control of inflammatory responses is evidenced by the knowledge that the VDR associates with VDRE in lymphoblastoid B cell lines [61]. Thus far, the data have shown that the action of 1,25-$(OH)_2D_3$ appears to be dual, depending on the stage of differentiation of the B cells. Studies indicated that 1,25-$(OH)_2D_3$ decreased the proliferation of B cells and immunoglobulin (Ig) class switching, and induced apoptosis [62]. Vitamin D inhibits B cell differentiation by interfering nuclear NF-κB translocation and CD40 co-stimulation [63]. At the other end of the B-cell development spectrum, 1,25-$(OH)_2D_3$ stimulates the development of plasma cells from terminally differentiated B cells, and enhances the migration of plasma cells towards inflammatory mucosal surfaces via the induction of CCR10 receptors [64]. While vitamin D has been shown to reduce the production of lupus-related autoantibodies, such as anti-nuclear antibody (ANA), independent of its impact on B cell differentiation [65,66], VDR binds to the IL-10 promotor and enhances IL-10 production, potentially alleviating autoimmunity [67]. The role that vitamin D exerts on B cells as APC is not clear, although there is one study that demonstrated that CD86 expression was reduced on B cells primed with 1,25-$(OH)_2D_3$, reducing the potency of allogenic T cell stimulation and the subsequent T-cell mediated inflammatory response [68].

## 2. Hypovitaminosis D and Its Impact on Disease Activity and Major Extra-Skeletal Manifestations in Patients with Systemic Lupus Erythematosus

### *2.1. Pathology Related to T Cell Subsets in SLE—A Brief Discussion*

SLE is a multi-systemic autoimmune condition characterized by immune-complex induced inflammation as a result of the association between autoantigens and autoantibodies [59]. Upstream to auto-antibody formation, the SLE have been found to be substantially T-cell driven, particularly the $CD4^+$ subset [38]. The Th1 and Th17 $CD4^+$ T cells are chiefly responsible for driving the pro-inflammatory response in SLE [38]. Using a CD137 ligand (CD137L) knockout B6.MRL.lpr$^{-/-}$ spontaneous murine SLE model, our group has recently demonstrated that while the $CD4^+Tbet^+$ (Th1) subset is increased in the B6.MRL.lpr$^{-/-}$ mice, the absence of the co-stimulatory molecule CD137L increases the severity of SLE via two pathways [50]. Firstly, the absence of a costimulatory function shifts the plastic Th1 subset into the more pathogenic $CD4^+ROR\gamma t^+$ (Th17) T cells. Secondly, the lack of CD137L signaling leads to a reduced serum level and intracellular IL-10 expression secondary to the significant reduction of the splenic $CD11b^+$ population [51]. As discussed previously, as 1,25-$(OH)_2D_3$ dampens the proinflammatory response of Th17, it is theoretically possible to suppress the SLE disease with vitamin D treatment. Indeed, at least three clinical studies have demonstrated the beneficial effects of vitamin D in terms of the reduction in the frequency of, as well as the IL-17 expression of, the Th17 cells [46,69,70]. Although these studies were probably not powered enough to demonstrate the clinical improvement of SLE, further experiments with larger sample sizes will be necessary to explore the potential clinical benefit of vitamin D supplementation in Th17 activity.

### 2.2. Low Vitamin D Level and SLE Disease Activity

As exposure to the UV light is one of the most potent contributors to SLE flare, clinicians frequently advise SLE patients to minimize sunlight exposure and use sunscreen. The effect of drugs such as glucocorticoids, calcineurin inhibitors, and anticonvulsants, as well as compromised 1$\alpha$-hydroxylation of 25-$(OH)D_3$ due to renal insufficiency, further contribute to low vitamin D in patients with SLE [71–73].

Since four decades ago, the relationship between vitamin D and SLE has been described, with the first report published in the late 1970s, which described that 7 out of 12 paediatric SLE patients had low serum 1, 25-$(OH)_2D_3$ levels [74]. Thereafter, a number of larger-scale case-control studies confirmed the relationship between hypovitaminosis D and the prevalence of SLE [75–78], particularly in patients with lupus nephritis [75]. In our recent age- and gender-matched case-control study of 61 SLE patients and 61 healthy controls, we found that a deficiency of the total 25$(OH)D_3$ level, as measured by liquid chromatography-tandem mass spectrometry, was significantly more prevalent in the patient group (19.7% vs. 3.3%, $p = 0.003$) [79]. Recently, a meta-analysis of 18 studies consisting of 1083 patients with SLE and 1273 healthy controls showed that the SLE patients had a significantly lower level of vitamin D compared with the healthy controls [76]. In addition, vitamin D deficiency was significantly more prevalent in the SLE patients when compared with the healthy controls, with a relative risk of 2.3 ($p = 0.002$) [80].

To further to the observation that hypovitaminosis D is more prevalent in SLE patients, low vitamin D levels also appear to be associated with a high disease activity of SLE [81–85]. In an observational study comprising 142 patients with 25-$(OH)D_3$ level less than 30 ng/mL, the serum level of 25-$(OH)D_3$ was found to be significantly higher in those patients with low disease activity (SLE disease activity index [SLEDAI] 1–5), and lower in those with high disease activity [81]. In addition, higher anti-dsDNA levels were found to be associated with low vitamin D levels in SLE patients [60,83]. Furthermore, the relationship between hypovitaminosis D and the SLE disease activity does not spare the paediatric population. In a case-control study of 35 paediatic-onset SLE patients in Taiwan, low serum 25-$(OH)D_3$ levels were significantly associated with active SLE disease activity compared to patients with inactive lupus [86]. Moreover, patients with active lupus nephritis had a significantly lower 25-$(OH)D_3$ level compared with those without nephritis [86].

## 3. Vitamin D and Its Major Extra-Musculoskeletal Impacts

### 3.1. Cardiovascular Risk and Its Biomarkers

The patients with low vitamin D levels or with a vitamin D deficiency were found to have a high prevalence of cardiovascular risk factors such as dyslipidaemia [87], hypertension [88], fasting glucose and insulin resistance [89], metabolic syndrome [90], positive antiphospholipid antibodies [87], and a high increased hsCRP level [91]. In a prospective study of 890 patients with SLE in a large international inception cohort, multiple logistic regression analyses revealed that patients at the high quantiles of 25-(OH)D were less likely to possess cardiovascular risk factors, including hypertension and hyperlipidaemia, while a non-significant trend of reduction of the hazard ratio of cardiovascular events was noted across successively higher quantiles of 25-(OH)D levels [88].

Recently, biophysical markers of cardiovascular disease such as carotid plaques, carotid intima-media thickness, endothelium-dependent dilation, and arterial stiffness have emerged as potential biomarkers of cardiovascular disease [92]. A few studies have addressed the potential relationship between hypovitaminosis D and the unfavorable alterations of these biophysical cardiovascular risk markers. By measuring the total plaque area (TPA) with the use of carotid ultrasound in a matched case-control study, the 25-(OH)D level was inversely associated with age-adjusted TPA [93]. In addition, a logistic regression model consisting of ACE inhibitor non-use, the 25-(OH)D level and LDL-c levels had a diagnostic accuracy of 84% in predicting the accelerated atherosclerosis [93]. Again, because of the difference in research methodology, the patient characteristics and the operator-dependent nature of the measurement of biophysical markers, not all of these studies detected a significant relationship between vitamin D deficiency and the presence cardiovascular risk factor [94].

### 3.2. Fatigue and Sleep

Fatigue is very prevalent in patients with SLE. Up to 80% of patients with SLE report symptoms of fatigue during their disease course [95]. Fatigue in SLE patients has been found to be related to low serum vitamin D levels [95–99]. A cross-sectional study of 90 patients with SLE found that those with vitamin D deficiency (25-(OH)$D_3$ $\leq$10 ng/mL) reported higher fatigue scores that those with adequate vitamin D levels [100]. In an observational study of 60 patients who took vitamin $D_3$, an inverse relationship between the 25-$(OH)_3$ level and fatigue score was identified [101]. Apart from fatigue, the vitamin D level was found to be related to sleep quality in the patients with SLE [102]. With the use of the Pittsburgh Sleep Quality Index, a retrospective study of 60 SLE patients revealed a significant relationship between a low serum vitamin D level and poor sleep quality [102].

### 3.3. Cognitive Impairment

While the association between cognitive impairment and low vitamin D has been extensively reported in the general population, especially amongst the elderly, the relationship between cognition and vitamin D in SLE has been scarcely reported. In our recent age- and gender-matched case-control study of 122 subjects, we found that the 25(OH)$D_3$ level was significantly lower in the patients with SLE compared with that of healthy individuals [79]. Specifically, the deficiency of 25(OH)$D_3$ was associated with a poorer subclinical cognitive function in terms of the total throughput score with the use of the Automated Neuropsychological Assessment Matrix, even after adjusting for age, education, duration of SLE, cumulative steroid dose, SLE disease activity, and SLE-related damaged and anxiety level in patients with SLE [79].

## 4. Vitamin D Supplementation in SLE

Over the past 10 years or so, a few studies have been designed to study the effect of vitamin D supplementation in patients with SLE [7,101,103–105]. While some studies demonstrated a reduction of SLE disease activity with vitamin D supplementation coupled with reduction in autoantibody levels [106,107], some did not [100,105]. For instance, in a recently published randomized control trial of vitamin D supplementation (50,000 U weekly for three weeks, then monthly for three months, versus placebo) in 90 patients with SLE, no change was detected in SLE disease activity between the two groups, using the SLEDAI [108]. In one of the largest single-arm prospective observational studies of 1006 patients with SLE with a low 25-(OH)D level (<40 ng/mL) who received vitamin D2 treatment (50,000 units weekly), a 20-unit increase in serum 25-(OH)D level was shown to be associated with a mean reduction of 0.22 points in the SLEDAI scale, corresponding to a reduction of 21% in the odds of active SLE with SLEDAI $\geq$5 [109]. In the most updated meta-analysis of the three studies, which involved 233 patients in the vitamin D treatment group and 128 in the placebo group, vitamin D supplementation was found to be significantly associated with a reduction in anti-dsDNA positivity ($p = 0.005$), without statistically significant heterogeneity amongst the studies [110]. Furthermore, the vitamin D supplement led to a reduction in fatigue [98,104], decrease in the Th1/Th17 and memory B cells that would otherwise enhance lupus-related inflammation, and increase in Treg cells that dampen lupus-related proinflammatory response [48,104]. While the presence of the type 1-IFN signature has recently been implicated in active SLE, vitamin D supplementation however failed to demonstrate a significant change in the IFN signature after 12-weeks of treatment [105]. As for cardiovascular risk, cholecalciferol supplementation in the patients deficient of vitamin D appears to improve the endothelial function, and calcitriol enhances the endothelial nitric oxide expression in human endothelial cells [111]. Nevertheless, at the time of writing of this review, there is yet to be any published study that demonstrates the clinical benefit of vitamin D supplementation in reducing cardiovascular events and mortality in SLE patients.

## 5. Conclusions

Besides its important involvement in calcium and phosphate homeostasis and the dynamics of the musculoskeletal system, vitamin D has an influential impact on the immune system. Current data have shown that while vitamin D generally suppresses the proinflammatory properties of APC, Th1 and Th17 $CD4^+$ T cells, and B cells, it enhances the anti-inflammatory characteristics of Treg and Th2 cells in many autoimmune conditions. While how vitamin D impacts SLE immunocytes needs to be answered by further research, the potentially relevant effects of vitamin D in individual immunocytes, and their cross-talks in the settling of lupus has been proposed (Figure 1). The heightened prevalence of extra-musculoskeletal complications, such as cardiovascular risk, fatigue, sleep disturbance, and cognitive impairment, highlights the potential of vitamin D supplementation as an adjunct therapeutic option for patients with SLE. Although not all clinical trials demonstrated clinically significant benefits of vitamin D supplementation in patients with SLE, in terms of an improvement of SLE disease activity and its associated complications, the theoretical clinical advantage of an adequate vitamin D level, the positive signals detected in various meta-analyses, and the favorable tolerability of vitamin D supplementation in nearly all of these clinical trials warrant sufficient vitamin D intake, supplementation, and monitoring in all patients with SLE in current clinical practice.

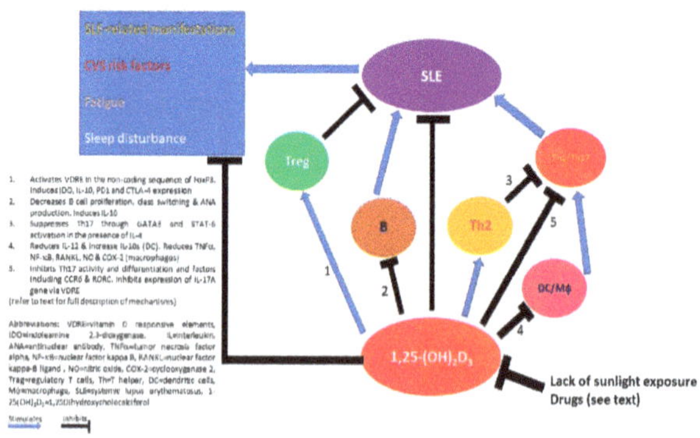


**Figure 1.** Potential mechanisms of how 1,25-$(OH)_2D_3$ interacts with the environment and immune cells in mediating the clinical manifestations of systemic lupus erythematosus (SLE).

**Funding:** This research received no external funding.

**Conflicts of Interest:** The author declares no conflict of interest.

## References


1. Pike, J.W.; Christakos, S. Biology and Mechanisms of Action of the Vitamin D Hormone. *Endocrinol. Metab. Clin. N. Am.* **2017**, *46*, 815–843. [CrossRef] [PubMed]
2. Masuyama, R. Role of local vitamin D signaling and cellular calcium transport system in bone homeostasis. *J. Bone Miner. Metab.* **2014**, *32*, 1–9. [CrossRef] [PubMed]
3. Wimalawansa, S.J. Non-musculoskeletal benefits of vitamin D. *J. Steroid Biochem. Mol. Biol.* **2018**, *175*, 60–81. [CrossRef] [PubMed]
4. Wintermeyer, E.; Ihle, C.; Ehnert, S.; Stöckle, U.; Ochs, G.; de Zwart, P.; Flesch, I.; Bahrs, C.; Nussler, A.K. Crucial Role of Vitamin D in the Musculoskeletal System. *Nutrients* **2016**, *8*, 319. [CrossRef] [PubMed]
5. Steingrimsdottir, L.; Gunnarsson, O.; Indridason, O.S.; Franzson, L.; Sigurdsson, G. Relationship between serum parathyroid hormone levels, vitamin D sufficiency, and calcium intake. *JAMA* **2005**, *294*, 2336–2341. [CrossRef] [PubMed]
6. Christakos, S.; Ajibade, D.V.; Dhawan, P.; Fechner, A.J.; Mady, L.J. Vitamin D: Metabolism. *Endocrinol. Metab. Clin. N. Am.* **2010**, *39*, 243–253. [CrossRef] [PubMed]
7. Holick, M.F. Vitamin D: A millenium perspective. *J. Cell Biochem.* **2003**, *88*, 296–307. [CrossRef] [PubMed]
8. Caccamo, D.; Ricca, S.; Currò, M.; Ientile, R. Health Risks of Hypovitaminosis D: A Review of New Molecular Insights. *Int. J. Mol. Sci.* **2018**, *19*, 892. [CrossRef] [PubMed]
9. Semba, R.D.; Houston, D.K.; Bandinelli, S.; Sun, K.; Cherubini, A.; Cappola, A.R.; Guralnik, J.M.; Ferrucci, L. Relationship of 25-hydroxyvitamin D with all-cause and cardiovascular disease mortality in older community-dwelling adults. *Eur. J. Clin. Nutr.* **2010**, *64*, 203–209. [CrossRef] [PubMed]
10. Edvardsen, K.; Veierød, M.B.; Brustad, M.; Braaten, T.; Engelsen, O.; Lund, E. Vitamin D-effective solar UV radiation, dietary vitamin D and breast cancer risk. *Int. J. Cancer* **2011**, *128*, 1425–1433. [CrossRef] [PubMed]
11. Meeker, S.; Seamons, A.; Maggio-Price, L.; Paik, J. Protective links between vitamin D, inflammatory bowel disease and colon cancer. *World J. Gastroenterol.* **2016**, *22*, 933–948. [CrossRef] [PubMed]
12. Holick, M.F. Vitamin D deficiency. *N. Engl. J. Med.* **2007**, *357*, 266–281. [CrossRef] [PubMed]
13. Cutolo, M.; Otsa, K. Review: Vitamin D, immunity and lupus. *Lupus* **2008**, *17*, 6–10. [CrossRef] [PubMed]
14. Bragazzi, N.L.; Watad, A.; Neumann, S.G.; Simon, M.; Brown, S.B.; Abu Much, A.; Harari, A.; Tiosano, S.; Amital, H.; Shoenfeld, Y. Vitamin D and rheumatoid arthritis: An ongoing mystery. *Curr. Opin. Rheumatol.* **2017**, *29*, 378–388. [CrossRef] [PubMed]
15. Lappe, J.M.; Travers-Gustafson, D.; Davies, K.M.; Recker, R.R.; Heaney, R.P. Vitamin D and calcium supplementation reduces cancer risk: Results of a randomized trial. *Am. J. Clin. Nutr.* **2007**, *85*, 1586–1591. [CrossRef] [PubMed]
16. Dankers, W.; Colin, E.M.; van Hamburg, J.P.; Lubberts, E. Vitamin D in Autoimmunity: Molecular Mechanisms and Therapeutic Potential. *Front. Immunol.* **2017**, *7*, 697. [CrossRef] [PubMed]
17. Munger, K.L.; Zhang, S.M.; O'Reilly, E.; Hernán, M.A.; Olek, M.J.; Willett, W.C.; Ascherio, A. Vitamin D intake and incidence of multiple sclerosis. *Neurology* **2004**, *13*, 60–65. [CrossRef]
18. Merlino, L.A.; Curtis, J.; Mikuls, T.R.; Cerhan, J.R.; Criswell, L.A.; Saag, K.G.; Iowa Women's Health Study. Vitamin D intake is inversely associated with rheumatoid arthritis: Results from the Iowa Women's Health Study. *Arthritis Rheum.* **2004**, *50*, 72–77. [CrossRef] [PubMed]
19. Grant, W.B. An estimate of the global reduction in mortality rates through doubling vitamin D levels. *Eur. J. Clin. Nutr.* **2011**, *65*, 1016–1026. [CrossRef] [PubMed]
20. Bartels, L.E.; Hvas, C.L.; Agnholt, J.; Dahlerup, J.F.; Agger, R. Human dendritic cell antigen presentation and chemotaxis are inhibited by intrinsic 25-hydroxy vitamin D activation. *Int. Immunopharmacol.* **2010**, *10*, 922–928. [CrossRef] [PubMed]
21. Tung, Y.C.; Ou, T.T.; Tsai, W.C. Elevated 1- $\alpha$ hydroxylase activity in monocytes from patients with active tuberculosis. *Clin. Dev. Immunol.* **2013**, *2013*, 928138. [CrossRef] [PubMed]
22. Koeffler, H.P.; Reichel, H.; Bishop, J.E.; Norman, A.W. Gamma-Interferon stimulates production of 1,25-dihydroxyvitamin $D_3$ by normal human macrophages. *Biochem. Biophys. Res. Commun.* **1985**, *127*, 596–603. [CrossRef]
23. Baeke, F.; Takiishi, T.; Korf, H.; Gysemans, C.; Mathieu, C. Vitamin D: Modulator of the immune system. *Curr. Opin. Pharmacol.* **2010**, *10*, 482–496. [CrossRef] [PubMed]

24. Mahto, H.; Tripathy, R.; Das, B.K.; Panda, A.K. Association between vitamin D receptor polymorphisms and systemic lupus erythematosus in an Indian cohort. *Int. J. Rheum. Dis.* **2018**, *21*, 468–476. [CrossRef] [PubMed]
25. Mao, S.; Huang, S. Association between vitamin D receptor gene BsmI, FokI, ApaI and TaqI polymorphisms and the risk of systemic lupus erythematosus: A meta-analysis. *Rheumatol. Int.* **2014**, *34*, 381–388. [CrossRef] [PubMed]
26. Satoh, J.; Tabunoki, H. Molecular network of chromatin immunoprecipitation followed by deep sequencing-based vitamin D receptor target genes. *Mult. Scler.* **2013**, *19*, 1035–1045. [CrossRef] [PubMed]
27. Kang, S.W.; Kim, S.H.; Lee, N.; Lee, W.W.; Hwang, K.A.; Shin, M.S.; Lee, S.H.; Kim, W.U.; Kang, I. 1,25-Dihyroxyvitamin $D_3$ promotes FOXP3 expression via binding to vitamin D response elements in its conserved noncoding sequence region. *J. Immunol.* **2012**, *188*, 5276–5282. [CrossRef] [PubMed]
28. Penna, G.; Adorini, L. 1 Alpha,25-dihydroxyvitamin $D_3$ inhibits differentiation, maturation, activation, and survival of dendritic cells leading to impaired alloreactive T cell activation. *J. Immunol.* **2000**, *164*, 2405–2411. [CrossRef] [PubMed]
29. Rook, G.A.; Steele, J.; Fraher, L.; Barker, S.; Karmali, R.; O'Riordan, J.; Stanford, J. Vitamin $D_3$, gamma interferon, and control of proliferation of Mycobacterium tuberculosis by human monocytes. *Immunology* **1986**, *57*, 159–163. [PubMed]
30. Fabri, M.; Stenger, S.; Shin, D.M.; Yuk, J.M.; Liu, P.T.; Realegeno, S.; Lee, H.M.; Krutzik, S.R.; Schenk, M.; Sieling, P.A.; et al. Vitamin D is required for IFN-$\gamma$-mediated antimicrobial activity of human macrophages. *Sci. Transl. Med.* **2011**, *3*, 104ra102. [CrossRef] [PubMed]
31. Liu, P.T.; Stenger, S.; Tang, D.H.; Modlin, R.L. Cutting edge: Vitamin D-mediated human antimicrobial activity against Mycobacterium tuberculosis is dependent on the induction of cathelicidin. *J. Immunol.* **2007**, *179*, 2060–2063. [CrossRef] [PubMed]
32. Wang, Q.; He, Y.; Shen, Y.; Zhang, Q.; Chen, D.; Zuo, C.; Qin, J.; Wang, H.; Wang, J.; Yu, Y. Vitamin D inhibits COX-2 expression and inflammatory response by targeting thioesterase superfamily member 4. *J. Biol. Chem.* **2014**, *289*, 11681–11694. [CrossRef] [PubMed]
33. Korf, H.; Wenes, M.; Stijlemans, B.; Takiishi, T.; Robert, S.; Miani, M.; Eizirik, D.L.; Gysemans, C.; Mathieu, C. 1,25-Dihydroxyvitamin $D_3$ curtails the inflammatory and T cell stimulatory capacity of macrophages through an IL-10-dependent mechanism. *Immunobiology* **2012**, *217*, 1292–1300. [CrossRef] [PubMed]
34. Li, J.; Yu, Y.F.; Liu, C.H.; Wang, C.M. Significance of M2 macrophages in glomerulonephritis with crescents. *Pathol. Res. Pract.* **2017**, *213*, 1215–1220. [CrossRef] [PubMed]
35. Maria, N.I.; Davidson, A. Renal Macrophages and Dendritic Cells in SLE Nephritis. *Curr. Rheumatol. Rep.* **2017**, *19*, 81. [CrossRef] [PubMed]
36. Veldman, C.M.; Cantorna, M.T.; DeLuca, H.F. Expression of 1,25-dihydroxyvitamin D(3) receptor in the immune system. *Arch. Biochem. Biophys.* **2000**, *374*, 334–338. [CrossRef] [PubMed]
37. Von Essen, M.R.; Kongsbak, M.; Schjerling, P.; Olgaard, K.; Odum, N.; Geisler, C. Vitamin D controls T cell antigen receptor signaling and activation of human T cells. *Nat. Immunol.* **2010**, *11*, 344–349. [CrossRef] [PubMed]
38. Mak, A.; Kow, N.Y. The pathology of T cells in systemic lupus erythematosus. *J. Immunol. Res.* **2014**, *2014*, 419029. [CrossRef] [PubMed]
39. Staeva-Vieira, T.P.; Freedman, L.P. 1,25-dihydroxyvitamin $D_3$ inhibits IFN-$\gamma$ and IL-4 levels during in vitro polarization of primary murine $CD4^+$ T cells. *J. Immunol.* **2002**, *168*, 1181–1189. [CrossRef] [PubMed]
40. Palmer, M.T.; Lee, Y.K.; Maynard, C.L.; Oliver, J.R.; Bikle, D.D.; Jetten, A.M.; Weaver, C.T. Lineage-specific effects of 1,25-dihydroxyvitamin D(3) on the development of effector CD4 T cells. *J. Biol. Chem.* **2011**, *286*, 997–1004. [CrossRef] [PubMed]
41. Chang, S.H.; Chung, Y.; Dong, C. Vitamin D suppresses Th17 cytokine production by inducing C/EBP homologous protein (CHOP) expression. *J. Biol. Chem.* **2010**, *285*, 38751–38755. [CrossRef] [PubMed]
42. Waddell, A.; Zhao, J.; Cantorna, M.T. NKT cells can help mediate the protective effects of 1,25-dihydroxyvitamin $D_3$ in experimental autoimmune encephalomyelitis in mice. *Int. Immunol.* **2015**, *27*, 237–244. [CrossRef] [PubMed]

43. Colin, E.M.; Asmawidjaja, P.S.; van Hamburg, J.P.; Mus, A.M.; van Driel, M.; Hazes, J.M.; van Leeuwen, J.P.; Lubberts, E. 1,25-dihydroxyvitamin $D_3$ modulates Th17 polarization and interleukin-22 expression by memory T cells from patients with early rheumatoid arthritis. *Arthritis Rheum.* **2010**, *62*, 132–142. [CrossRef] [PubMed]
44. Tahmasebinia, F.; Pourgholaminejad, A. The role of Th17 cells in auto-inflammatory neurological disorders. *Prog. Neuropsychopharmacol. Biol. Psychiatry* **2017**, *79*, 408–416. [CrossRef] [PubMed]
45. Suárez-Fueyo, A.; Bradley, S.J.; Tsokos, G.C. T cells in Systemic Lupus Erythematosus. *Curr. Opin. Immunol.* **2016**, *43*, 32–38. [CrossRef] [PubMed]
46. Reihani, H.; Rastin, M.; Mahmoudi, M.; Ghoryani, M.; Abdollahi, N.; Tabasi, N.S.; Zamani Taghizadeh Rabe, S.; Sahebari, M. Influence of 1 Alpha, 25-Dihydroxyvitamin $D_3$ on T Helper 17 Cells and Related Cytokines in Systemic Lupus Erythematosus. *Iran. J. Immunol.* **2015**, *12*, 82–93. [PubMed]
47. Joshi, S.; Pantalena, L.C.; Liu, X.K.; Gaffen, S.L.; Liu, H.; Rohowsky-Kochan, C.; Ichiyama, K.; Yoshimura, A.; Steinman, L.; Christakos, S.; et al. 1,25-dihydroxyvitamin D(3) ameliorates Th17 autoimmunity via transcriptional modulation of interleukin-17A. *Mol. Cell Biol.* **2011**, *31*, 3653–3669. [CrossRef] [PubMed]
48. Terrier, B.; Derian, N.; Schoindre, Y.; Chaara, W.; Geri, G.; Zahr, N.; Mariampillai, K.; Rosenzwajg, M.; Carpentier, W.; Musset, L.; et al. Restoration of regulatory and effector T cell balance and B cell homeostasis in systemic lupus erythematosus patients through vitamin D supplementation. *Arthritis Res. Ther.* **2012**, *14*, R221. [CrossRef] [PubMed]
49. Mukasa, R.; Balasubramani, A.; Lee, Y.K.; Whitley, S.K.; Weaver, B.T.; Shibata, Y.; Crawford, G.E.; Hatton, R.D.; Weaver, C.T. Epigenetic instability of cytokine and transcription factor gene loci underlies plasticity of the T helper 17 cell lineage. *Immunity* **2010**, *32*, 616–627. [CrossRef] [PubMed]
50. Maggi, L.; Santarlasci, V.; Capone, M.; Rossi, M.C.; Querci, V.; Mazzoni, A.; Cimaz, R.; De Palma, R.; Liotta, F.; Maggi, E.; et al. Distinctive features of classic and nonclassic (Th17 derived) human Th1 cells. *Eur. J. Immunol.* **2012**, *42*, 3180–3188. [CrossRef] [PubMed]
51. Mak, A.; Dharmadhikari, B.; Schwarz, H. Lack of CD137-CD137 Ligand Signalling Aggravates Glomerulonephritis and Reduces the Survival of Lupus-Prone B6.MRLlpr Mice [abstract]. *Arthritis Rheumatol.* **2017**, *69* (Suppl. 10), 2573.
52. Mizui, M.; Tsokos, G.C. Targeting regulatory T Cells to treat patients with systemic lupus erythematosus. *Front. Immunol.* **2018**, *9*, 786. [CrossRef] [PubMed]
53. Dominguez-Villar, M.; Hafler, D.A. Regulatory T cells in autoimmune disease. *Nat. Immunol.* **2018**. [CrossRef] [PubMed]
54. Spach, K.M.; Nashold, F.E.; Dittel, B.N.; Hayes, C.E. IL-10 signaling is essential for 1,25-dihydroxyvitamin $D_3$-mediated inhibition of experimental autoimmune encephalomyelitis. *J. Immunol.* **2006**, *177*, 6030–6037. [CrossRef] [PubMed]
55. Urry, Z.; Chambers, E.S.; Xystrakis, E.; Dimeloe, S.; Richards, D.F.; Gabryšová, L.; Christensen, J.; Gupta, A.; Saglani, S.; Bush, A.; et al. The role of 1$\alpha$,25-dihydroxyvitamin $D_3$ and cytokines in the promotion of distinct Foxp3$^+$ and IL-10$^+$ CD4$^+$ T cells. *Eur. J. Immunol.* **2012**, *42*, 2697–2708. [CrossRef] [PubMed]
56. Handono, K.; Marisa, D.; Kalim, H. Association between the low levels of vitamin D and Treg function in systemic lupus erythematosus patients. *Acta Med. Indones.* **2013**, *45*, 26–31. [PubMed]
57. Chen, J.; Bruce, D.; Cantorna, M.T. Vitamin D receptor expression controls proliferation of naïve CD8$^+$ T cells and development of CD8 mediated gastrointestinal inflammation. *BMC Immunol.* **2014**, *15*, 6. [CrossRef] [PubMed]
58. Reich, K.; Zschocke, I.; Bachelez, H.; de Jong, E.M.G.J.; Gisondi, P.; Puig, L.; Warren, R.B.; Ortland, C.; Mrowietz, U.; PSO-TOP Study Group. A Topical Treatment Optimization Programme (TTOP) improves clinical outcome for calcipotriol/betamethasone gel in psoriasis: Results of a 64-week multinational randomized phase IV study in 1790 patients (PSO-TOP). *Br. J. Dermatol.* **2017**, *177*, 197–205. [CrossRef] [PubMed]
59. Kow, N.Y.; Mak, A. Costimulatory pathways: Physiology and potential therapeutic manipulation in systemic lupus erythematosus. *Clin. Dev. Immunol.* **2013**, *2013*, 245928–245940. [CrossRef] [PubMed]
60. Ritterhouse, L.L.; Crowe, S.R.; Niewold, T.B.; Kamen, D.L.; Macwana, S.R.; Roberts, V.C. Vitamin D deficiency is associated with an increased autoimmune response in healthy individuals and in patients with systemic lupus erythematosus. *Ann. Rheum. Dis.* **2011**, *70*, 1569–1574. [CrossRef] [PubMed]

61. Ramagopalan, S.V.; Heger, A.; Berlanga, A.J.; Maugeri, N.J.; Lincoln, M.R.; Burrell, A.; Handunnetthi, L.; Handel, A.E.; Disanto, G.; Orton, S.M.; et al. A ChIP-seq defined genome-wide map of vitamin D receptor binding: Associations with disease and evolution. *Genome Res.* **2010**, *20*, 1352–1360. [CrossRef] [PubMed]
62. Lemire, J.M.; Adams, J.S.; Sakai, R.; Jordan, S.C. 1 alpha,25-dihydroxyvitamin $D_3$ suppresses proliferation and immunoglobulin production by normal human peripheral blood mononuclear cells. *J. Clin. Investig.* **1984**, *74*, 657–661. [CrossRef] [PubMed]
63. Geldmeyer-Hilt, K.; Heine, G.; Hartmann, B.; Baumgrass, R.; Radbruch, A.; Worm, M. 1,25-dihydroxyvitamin $D_3$ impairs NF-κB activation in human naïve B cells. *Biochem. Biophys. Res. Commun.* **2011**, *407*, 699–702. [CrossRef] [PubMed]
64. Shirakawa, A.K.; Nagakubo, D.; Hieshima, K.; Nakayama, T.; Jin, Z.; Yoshie, O. 1,25-dihydroxyvitamin $D_3$ induces CCR10 expression in terminally differentiating human B cells. *J. Immunol.* **2008**, *180*, 2786–2795. [CrossRef] [PubMed]
65. Chen, S.; Sims, G.P.; Chen, X.X.; Gu, Y.Y.; Chen, S.; Lipsky, P.E. Modulatory effects of 1,25-dihydroxyvitamin $D_3$ on human B cell differentiation. *J. Immunol.* **2007**, *179*, 1634–1647. [CrossRef] [PubMed]
66. Iho, S.; Takahashi, T.; Kura, F.; Sugiyama, H.; Hoshino, T. The effect of 1,25-dihydroxyvitamin $D_3$ on in vitro immunoglobulin production in human B cells. *J. Immunol.* **1986**, *136*, 4427–4431. [PubMed]
67. Heine, G.; Niesner, U.; Chang, H.D.; Steinmeyer, A.; Zügel, U.; Zuberbier, T.; Radbruch, A.; Worm, M. 1,25-dihydroxyvitamin D(3) promotes IL-10 production in human B cells. *Eur. J. Immunol.* **2008**, *38*, 2210–2218. [CrossRef] [PubMed]
68. Drozdenko, G.; Scheel, T.; Heine, G.; Baumgrass, R.; Worm, M. Impaired T cell activation and cytokine production by calcitriol-primed human B cells. *Clin. Exp. Immunol.* **2014**, *178*, 364–372. [CrossRef] [PubMed]
69. Drozdenko, G.; Heine, G.; Worm, M. Oral vitamin D increases the frequencies of $CD38^+$ human B cells and ameliorates IL-17-producing T cells. *Exp. Dermatol.* **2014**, *23*, 107–112. [CrossRef] [PubMed]
70. Wahono, C.S.; Rusmini, H.; Soelistyoningsih, D.; Hakim, R.; Handono, K.; Endharti, A.T.; Kalim, H.; Widjajanto, E. Effects of $1,25(OH)_2D_3$ in immune response regulation of systemic lupus erythematosus (SLE) patient with hypovitamin D. *Int. J. Clin. Exp. Med.* **2014**, *7*, 22–31. [PubMed]
71. Sumethkul, K.; Boonyaratavej, S.; Kitumnuaypong, T.; Angthararuk, S.; Cheewasat, P.; Manadee, N.; Sumethkul, V. The predictive factors of low serum 25-hydroxyvitamin D and vitamin D deficiency in patients with systemic lupus erythematosus. *Rheumatol. Int.* **2013**, *33*, 1461–1467. [CrossRef] [PubMed]
72. Chaiamnuay, S.; Chailurkit, L.O.; Narongroeknawin, P.; Asavatanabodee, P.; Laohajaroensombat, S.; Chaiamnuay, P. Current daily glucocorticoid use and serum creatinine levels are associated with lower 25(OH) vitamin D levels in Thai patients with systemic lupus erythematosus. *J. Clin. Rheumatol.* **2013**, *19*, 121–125. [CrossRef] [PubMed]
73. Mok, C.C. Vitamin D and systemic lupus erythematosus: An update. *Expert. Rev. Clin. Immunol.* **2013**, *9*, 453–463. [CrossRef] [PubMed]
74. O'Regan, S.; Chesney, R.W.; Hamstra, A.; Eisman, J.A.; O'Gorman, A.M.; Deluca, H.F. Reduced serum 1,25-(OH)2 vitamin $D_3$ levels in prednisone-treated adolescents with systemic lupus erythematosus. *Acta Paediatr. Scand.* **1979**, *68*, 109–111. [CrossRef] [PubMed]
75. Kamen, D.L.; Cooper, G.S.; Bouali, H.; Shaftman, S.R.; Hollis, B.W.; Gilkeson, G.S. Vitamin D deficiency in systemic lupus erythematosus. *Autoimmun. Rev.* **2006**, *5*, 114–117. [CrossRef] [PubMed]
76. Garf, K.E.; Marzouk, H.; Farag, Y.; Rasheed, L.; Garf, A.E. Vitamin D status in Egyptian patients with juvenile-onset systemic lupus erythematosus. *Rheumatol. Int.* **2015**, *35*, 1535–1540. [CrossRef] [PubMed]
77. Shahin, D.; El-Farahaty, R.M.; Houssen, M.E.; Machaly, S.A.; Sallam, M.; ElSaid, T.O.; Neseem, N.O. Serum 25-OH vitamin D level in treatment-naïve systemic lupus erythematosus patients: Relation to disease activity, IL-23 and IL-17. *Lupus* **2017**, *26*, 917–926. [CrossRef] [PubMed]
78. Abaza, N.M.; El-Mallah, R.M.; Shaaban, A.; Mobasher, S.A.; Al-Hassanein, K.F.; Abdel Zaher, A.A.; El-Kabarity, R.H. Vitamin D Deficiency in Egyptian Systemic Lupus Erythematosus Patients: How Prevalent and Does It Impact Disease Activity? *Integr. Med. Insights* **2016**, *11*, 27–33. [CrossRef] [PubMed]
79. Tay, S.H.; Ho, C.S.; Ho, R.C.; Mak, A. 25-Hydroxyvitamin $D_3$ Deficiency Independently Predicts Cognitive Impairment in Patients with Systemic Lupus Erythematosus. *PLoS ONE* **2015**, *10*, e0144149. [CrossRef] [PubMed]
80. Bae, S.C.; Lee, Y.H. Association between Vitamin D level and/or deficiency, and systemic lupus erythematosus: A meta-analysis. *Cell. Mol. Biol.* **2018**, *64*, 7–13. [CrossRef] [PubMed]

81. Eloi, M.; Horvath, D.V.; Ortega, J.C.; Prado, M.S.; Andrade, L.E.; Szejnfeld, V.L.; de Moura Castro, C.H. 25-Hydroxivitamin D Serum Concentration, Not Free and Bioavailable Vitamin D, Is Associated with Disease Activity in Systemic Lupus Erythematosus Patients. *PLoS ONE* **2017**, *12*, e0170323. [CrossRef] [PubMed]
82. Borba, V.Z.; Vieira, J.G.; Kasamatsu, T.; Radominski, S.C.; Sato, E.I.; Lazaretti-Castro, M. Vitamin D deficiency in patients with active systemic lupus erythematosus. *Osteoporos. Int.* **2009**, *20*, 427–433. [CrossRef] [PubMed]
83. Mok, C.C.; Birmingham, D.J.; Ho, L.Y.; Hebert, L.A.; Song, H.; Rovin, B.H. Vitamin D deficiency as marker for disease activity and damage in systemic lupus erythematosus: A comparison with anti-dsDNA and anti-C1q. *Lupus* **2012**, *21*, 36–42. [CrossRef] [PubMed]
84. Sahebari, M.; Nabavi, N.; Salehi, M. Correlation between serum 25(OH)D values and lupus disease activity: An original article and a systematic review with meta-analysis focusing on serum VitD confounders. *Lupus* **2014**, *23*, 1164–1177. [CrossRef] [PubMed]
85. Yap, K.S.; Northcott, M.; Hoi, A.B.; Morand, E.F.; Nikpour, M. Association of low vitamin D with high disease activity in an Australian systemic lupus erythematosus cohort. *Lupus Sci. Med.* **2015**, *2*, e000064. [CrossRef] [PubMed]
86. Lin, T.C.; Wu, J.Y.; Kuo, M.L.; Ou, L.S.; Yeh, K.W.; Huang, J.L. Correlation between disease activity of pediatric-onset systemic lupus erythematosus and level of vitamin D in Taiwan: A case-cohort study. *J. Microbiol. Immunol. Infect.* **2018**, *51*, 110–114. [CrossRef] [PubMed]
87. Mok, C.C.; Birmingham, D.J.; Leung, H.W.; Hebert, L.A.; Song, H.; Rovin, B.H. Vitamin D levels in Chinese patients with systemic lupus erythematosus: Relationship with disease activity, vascular risk factors and atherosclerosis. *Rheumatology* **2012**, *51*, 644–652. [CrossRef] [PubMed]
88. Lertratanakul, A.; Wu, P.; Dyer, A.; Urowitz, M.; Gladman, D.; Fortin, P.; Bae, S.C.; Gordon, C.; Clarke, A.; Bernatsky, S.; et al. 25-hydroxyvitamin D and cardiovascular disease in patients with systemic lupus erythematosus: Data from a large international inception cohort. *Arthritis Care Res.* **2014**, *66*, 1167–1176. [CrossRef] [PubMed]
89. Sabio, J.M.; Vargas-Hitos, J.A.; Martinez-Bordonado, J.; Navarrete-Navarrete, N.; Díaz-Chamorro, A.; Olvera-Porcel, C.; Zamora, M.; Jiménez-Alonso, J. Association between low 25-hydroxyvitamin D, insulin resistance and arterial stiffness in nondiabetic women with systemic lupus erythematosus. *Lupus* **2015**, *24*, 155–163. [CrossRef] [PubMed]
90. Wang, L.M.; Zheng, Z.H.; Li, T.F.; Han, L.S.; He, Y.J.; Zhang, Y.L.; Zeng, H.L.; Liu, S.Y. 25-hydroxyvitamin D is associated with metabolic syndrome among premenopausal women with systemic lupus erythematosus in China. *Lupus* **2017**, *26*, 403–409. [CrossRef] [PubMed]
91. Robinson, A.B.; Tangpricha, V.; Yow, E.; Gurion, R.; McComsey, G.A.; Schanberg, L.E.; APPLE Investigators. Vitamin D deficiency is common and associated with increased C-reactive protein in children and young adults with lupus: An Atherosclerosis Prevention in Pediatric Lupus Erythematosus substudy. *Lupus Sci. Med.* **2014**, *1*, e000011. [CrossRef] [PubMed]
92. Mak, A.; Kow, N.Y.; Schwarz, H.; Gong, L.; Tay, S.H.; Ling, L.H. Endothelial dysfunction in systemic lupus erythematosus—A case-control study and an updated meta-analysis and meta-regression. *Sci. Rep.* **2017**, *7*, 7320. [CrossRef] [PubMed]
93. Ravenell, R.L.; Kamen, D.L.; Spence, J.D.; Hollis, B.W.; Fleury, T.J.; Janech, M.G.; Almeida, J.S.; Shaftman, S.R.; Oates, J.C. Premature atherosclerosis is associated with hypovitaminosis D and angiotensin-converting enzyme inhibitor non-use in lupus patients. *Am. J. Med. Sci.* **2012**, *344*, 268–273. [CrossRef] [PubMed]
94. Jung, J.Y.; Koh, B.R.; Bae, C.B.; Kim, H.A.; Suh, C.H. Carotid subclinical atherosclerosis is associated with disease activity but not vitamin D in Korean systemic lupus erythematosus. *Lupus* **2014**, *23*, 1517–1522. [CrossRef] [PubMed]
95. Mahieu, M.A.; Ramsey-Goldman, R. Candidate Biomarkers for Fatigue in Systemic Lupus Erythematosus: A Critical Review. *Curr. Rheumatol. Rev.* **2017**, *13*, 103–112. [CrossRef] [PubMed]
96. Ahn, G.E.; Ramsey-Goldman, R. Fatigue in systemic lupus erythematosus. *Int. J. Clin. Rheumtol.* **2012**, *7*, 217–227. [CrossRef] [PubMed]
97. Yuen, H.K.; Cunningham, M.A. Optimal management of fatigue in patients with systemic lupus erythematosus: A systematic review. *Ther. Clin. Risk Manag.* **2014**, *10*, 775–786. [CrossRef] [PubMed]

98. Salman-Monte, T.C.; Torrente-Segarra, V.; Almirall, M.; Corzo, P.; Mojal, S.; Carbonell-Abelló, J. Prevalence and predictors of vitamin D insufficiency in supplemented and non-supplemented women with systemic lupus erythematosus in the Mediterranean region. *Rheumatol. Int.* **2016**, *36*, 975–985. [CrossRef] [PubMed]
99. Carrión-Barberà, I.; Salman-Monte, T.C.; Castell, S.; Castro, F.; Ojeda, F.; Carbonell, J. Prevalence and factors associated with fatigue in female patients with systemic lupus erythematosus. *Med. Clin.* **2018**. [CrossRef]
100. Ruiz-Irastorza, G.; Egurbide, M.V.; Olivares, N.; Martinez-Berriotxoa, A.; Aguirre, C. Vitamin D deficiency in systemic lupus erythematosus: Prevalence, predictors and clinical consequences. *Rheumatology* **2008**, *47*, 920–923. [CrossRef] [PubMed]
101. Ruiz-Irastorza, G.; Gordo, S.; Olivares, N.; Egurbide, M.V.; Aguirre, C. Changes in vitamin D levels in patients with systemic lupus erythematosus: Effects on fatigue, disease activity, and damage. *Arthritis Care Res.* **2010**, *62*, 1160–1165. [CrossRef] [PubMed]
102. Gholamrezaei, A.; Bonakdar, Z.S.; Mirbagher, L.; Hosseini, N. Sleep disorders in systemic lupus erythematosus. Does vitamin D play a role? *Lupus* **2014**, *23*, 1054–1058. [CrossRef] [PubMed]
103. Andreoli, L.; Dall'Ara, F.; Piantoni, S.; Zanola, A.; Piva, N.; Cutolo, M.; Tincani, A. A 24-month prospective study on the efficacy and safety of two different monthly regimens of vitamin D supplementation in pre-menopausal women with systemic lupus erythematosus. *Lupus* **2015**, *24*, 499–506. [CrossRef] [PubMed]
104. Piantoni, S.; Andreoli, L.; Scarsi, M.; Zanola, A.; Dall'Ara, F.; Pizzorni, C.; Cutolo, M.; Airò, P.; Tincani, A. Phenotype modifications of T-cells and their shift toward a Th2 response in patients with systemic lupus erythematosus supplemented with different monthly regimens of vitamin D. *Lupus* **2015**, *24*, 490–498. [CrossRef] [PubMed]
105. Aranow, C.; Kamen, D.L.; Dall'Era, M.; Massarotti, E.M.; Mackay, M.C.; Koumpouras, F.; Coca, A.; Chatham, W.W.; Clowse, M.E.; Criscione-Schreiber, L.G.; et al. Randomized, Double-Blind, Placebo-Controlled Trial of the Effect of Vitamin $D_3$ on the Interferon Signature in Patients With Systemic Lupus Erythematosus. *Arthritis Rheumatol.* **2015**, *67*, 1848–1857. [CrossRef] [PubMed]
106. Abou-Raya, A.; Abou-Raya, S.; Helmii, M. The effect of vitamin D supplementation on inflammatory and hemostatic markers and disease activity in patients with systemic lupus erythematosus: A randomized placebo-controlled trial. *J. Rheumatol.* **2013**, *40*, 265–272. [CrossRef] [PubMed]
107. Lima, G.L.; Paupitz, J.; Aikawa, N.E.; Takayama, L.; Bonfa, E.; Pereira, R.M. Vitamin D Supplementation in Adolescents and Young Adults With Juvenile Systemic Lupus Erythematosus for Improvement in Disease Activity and Fatigue Scores: A Randomized, Double-Blind, Placebo-Controlled Trial. *Arthritis Care Res.* **2016**, *68*, 91–98. [CrossRef] [PubMed]
108. Karimzadeh, H.; Shirzadi, M.; Karimifar, M. The effect of Vitamin D supplementation in disease activity of systemic lupus erythematosus patients with Vitamin D deficiency: A randomized clinical trial. *J. Res. Med. Sci.* **2017**, *22*, 4. [CrossRef] [PubMed]
109. Petri, M.; Bello, K.J.; Fang, H.; Magder, L.S. Vitamin D in systemic lupus erythematosus: Modest association with disease activity and the urine protein-to-creatinine ratio. *Arthritis Rheum.* **2013**, *65*, 1865–1871. [CrossRef] [PubMed]
110. Franco, A.S.; Freitas, T.Q.; Bernardo, W.M.; Pereira, R.M.R. Vitamin D supplementation and disease activity in patients with immune-mediated rheumatic diseases: A systematic review and meta-analysis. *Medicine* **2017**, *96*, e7024. [CrossRef] [PubMed]
111. Reynolds, J.A.; Haque, S.; Williamson, K.; Ray, D.W.; Alexander, M.Y.; Bruce, I.N. Vitamin D improves endothelial dysfunction and restores myeloid angiogenic cell function via reduced CXCL-10 expression in systemic lupus erythematosus. *Sci. Rep.* **2016**, *6*, 22341. [CrossRef] [PubMed]

International Journal of
*Molecular Sciences*

MDPI

*Review*

# The Anti-Inflammatory Effects of Vitamin D in Tumorigenesis

**Wei Liu [1,2], Lei Zhang [1,2], Hui-Jing Xu [1,2], Yan Li [1,2], Chuan-Min Hu [2], Jing-Yan Yang [3] and Mei-Yan Sun [1,2,*]**

1 College of Laboratory Medicine, Jilin Medical University, Jilin 132013, China; liuweijldx@sina.com (W.L.); zhanglei0358@sina.com (L.Z.); renyunqingzheng@126.com (H.-J.X.); jlliyang2004@163.com (Y.L.)
2 Laboratory of Antibody Engineering, Jilin Medical University, Jilin 132013, China; chuminhu@163.com
3 Department of Physical Education, Jilin Medical University, Jilin 132013, China; yjy8899@126.com
* Correspondence: sunmy990@163.com; Tel.: +86-432-6456-0321

Received: 18 July 2018; Accepted: 8 September 2018; Published: 13 September 2018

**Abstract:** In conjunction with the classical functions of regulating intestinal, bone, and kidney calcium and phosphorus absorption, as well as bone mineralization of vitamin D, the population-based association between low vitamin D status and increased cancer risk is now generally accepted. Inflammation is causally related to oncogenesis. It is widely thought that vitamin D plays an important role in the modulation of the inflammation system by regulating the production of inflammatory cytokines and immune cells, which are crucial for the pathogenesis of many immune-related diseases. Mechanistic studies have shown that vitamin D influences inflammatory processes involved in cancer progression, including cytokines, prostaglandins, MAP kinase phosphatase 5 (MKP5), the nuclear factor kappa B (NF-κB) pathway, and immune cells. Multiple studies have shown that vitamin D has the potential to inhibit tumor development by interfering with the inflammation system. The present review summarizes recent studies of the mechanisms of vitamin D on regulating the inflammation system, which contributes to its potential for cancer prevention and therapy. This review helps answer whether inflammation mediates a causal relationship between vitamin D and tumorigenesis.

**Keywords:** vitamin D; inflammation; tumorigenesis; cytokines; immune cells

---

## 1. Introduction

Vitamin D is a fat-soluble steroid derivative, which plays an important role in calcium homeostasis and bone metabolism through its actions in intestine, bone, kidney, and the parathyroid glands [1–3]. Moreover, preclinical and clinical studies strongly suggest that vitamin D deficiency increases the risk of developing multiple malignancies. Inflammation is appropriately added as one of the 10 hallmarks of cancer [4]. Cytokines and immune cells in the inflammatory microenvironment essentially serve as direct growth and migratory factors for cancer cells [5,6]. Studies have shown that tissues with chronic inflammation generally exhibit high cancer incidence [7–9]. Vitamin D can modulate the innate and adaptive immune responses. Studies of tumor cells revealed that vitamin D exerts important regulatory effects on some of the key molecular pathways involved in inflammation. Studies have shown the role of vitamin D in the inflammatory microenvironment, but the evidence linking vitamin D and immune response in the context of cancer is still scarce. In the current review, we aim to provide an overview of anti-inflammatory actions and vitamin D in tumorigenesis; the results suggest the beneficial effects of vitamin D supplementation in decreasing the risk and adverse outcomes of cancer, although the precise effect remains to be elucidated in large clinical trials.

## 2. Vitamin D and Epidemiology

Epidemiological data showed that vitamin D deficiency is significantly associated with high risk of multiple tumors and poor prognosis [10–13]. Garland et al. first determined that low levels of calcitriol may be associated with high incidence of colorectal cancer [14]. This hypothesis was subsequently verified in a large number of meta-analyses, including breast, prostate, colon, lung, and other cancers [15]. The results showed that high serum 25(OH)D levels are negatively correlated with the incidence or mortality of multiple cancers [16]. A meta-analysis of 17,332 cancer patients indicated that a 10 nmol/L increase in the blood 25(OH)D level confers a 4% reduction in overall mortality of cancer patients and high vitamin D levels are significantly associated with tumor prognostic indicators [17]. In a large case-cohort study within a Japan Public Health Center-based prospective study, the results showed that higher vitamin D concentrations are associated with lower risk of total cancer. These findings support the hypothesis that vitamin D has protective effects against cancer [18].

The correlation between vitamin D and the risk of gastrointestinal cancer were mostly reported. Gorham et al. found that daily intake of vitamin D above 1000 IU reduces the risk of colorectal cancer by about 50% for people who receive less than 100 IU daily vitamin D intake [19]. Ma et al. found that high-dose vitamin D intake and high levels of serum 25(OH)D reduce the risk of colorectal cancer by 12% and 33%, respectively [20]. The vitamin D intake of the respondents was estimated by the Health Professionals Follow-Up Study and the Nurses' Health Study based on 120,000 men and women, including 365 confirmed cases of pancreatic cancer, during a 16-year follow-up. The results of the study showed that respondents who consumed more vitamin D in their diet had a lower incidence of pancreatic cancer than those with a lower vitamin D intake [21]. Biochemical evidence clearly indicated that hepatic carcinoma cells are responsive to the inhibitory effect of vitamin D. Severe 25(OH)D deficiency identifies a poor prognosis in patients with hepatocellular carcinoma [22].

Vitamin D has also been implicated in the development and progression of other cancers [23]. Several studies suggested that decreased serum levels of 25(OH)D are correlated with increased risk of prostate cancer (PCa) [24]. Studies have shown that, for a 20 ng/mL increase in serum 25(OH)D levels, the risk of breast cancer is reduced by 26%, suggesting that the risk of breast cancer is significantly reduced with the increase in serum vitamin D levels [25]. Kim et al. performed a meta-analysis of vitamin D intake, serum 25(OH)D concentrations, and the incidence and prognosis of breast cancer. A dose-response analysis of 13 studies showed that every 100 IU/day increase in vitamin D intake decreases the incidence of breast cancer by 2% [26].

## 3. Vitamin D and Metabolism

Vitamin D3 ($D_3$) is the main form of vitamin D in the human body, which can be either synthesized from 7-dehydrocholesterol when skin is exposed to ultraviolet-B (UVB) light or obtained from the diet (Figure 1). In the liver, vitamin D is metabolized by vitamin D 25-hydroxylase (*Cytochrome P450 2R1, CYP2R1*) and sterol 27-hydroxylase (*Cytochrome P450 27A1, CYP27A1*) to 25(OH)D, which is the major circulating form of vitamin D in serum. 25(OH)D is further metabolized by 25(OH)D 1$\alpha$-hydroxylase (*Cytochrome P450 27R1, CYP27B1*) in the mitochondria of the kidney epithelial cells to 1,25-dihydroxy vitamin D3 (1,25$(OH)_2D_3$), also known as calcitriol [27]. In some cancers, such as parathyroid carcinomas, the expression levels and activity of *CYP27B1* in the cancer cells are lower than in normal cells [28]. Hsu et al. compared CYP27B1 activity in samples from normal prostate epithelial cells, cancer-derived prostate epithelial cells, PCa cells lines, and samples of benign prostatic hyperplasia. *CYP27B1* expression is significantly reduced in the benign prostatic hyperplasia cells and is reduced further in the cancer-derived cells and cell lines. Decreased expression of *CYP27B1* is correlated with a decrease in growth inhibition in response to 25(OH)D [29]. In order to maintain the homeostasis of the organism, especially regarding the levels of calcium and phosphate, the enzyme 1,25 dihydroxyvitamin D3 24 hydroxylase (*Cytochrome P450 24R1, CYP24R1*) plays a key role in converting calcitriol to biologically inactive metabolites [30]. Hobaus et al. found that 77 (60%) out of 127 colorectal tumors show increased *CYP24A1* gene copy-number and that more than six copies of *CYP24A1* are positively correlated with *CYP24A1* mRNA expression suggestive of a causal relationship [31]. Albertson et al. proved that the amplification of the *CYP24A1* gene in human breast tumors and analysis of the datasets from The Cancer Genome Atlas confirms that a subset of human breast cancers (10–13%) exhibit alterations in the *CYP24* gene, with the most frequent changes being amplifications and up-regulation at the mRNA level [32]. Borkowski et al. found that high *CYP24A1* expression is significantly correlated with poor patient outcome in multiple lung cancer cohorts [33]. Better understanding the regulation of vitamin D hydroxylases in tumorigenesis may provide targeting strategies in the future.

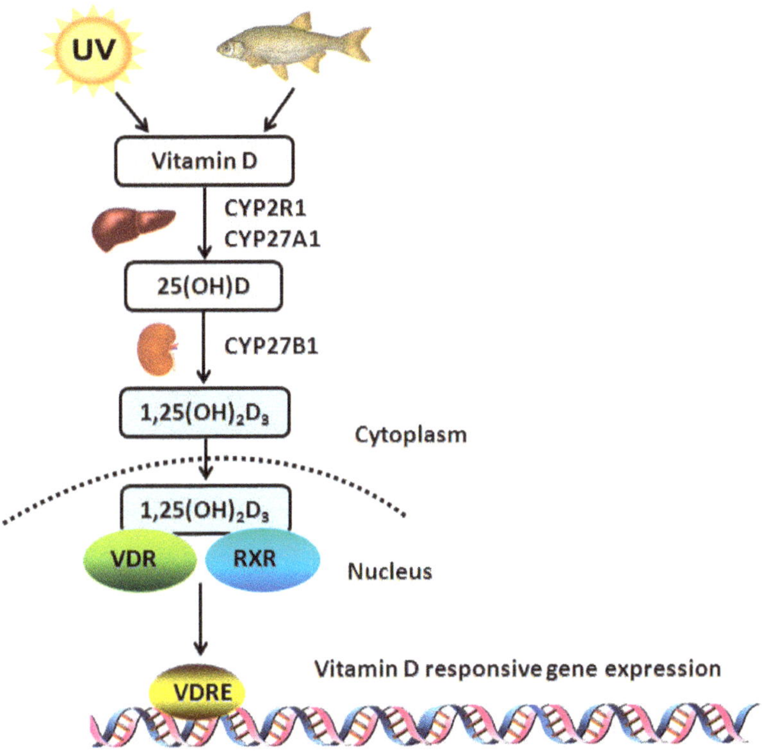


**Figure 1.** Vitamin metabolism. Vitamin D is synthesized by UVB radiation in the epidermis of skin or obtained from diet, hydroxylated to form 25(OH)D by CYP2R1 and CYP27A1 in liver, further metabolized by CYP27B1 in kidney to 1,25(OH)2D3 (calcitriol), and binds to the vitamin D receptor (VDR) which then heterodimerizes with retinoid X receptor (RXR). The $1,25(OH)_2D_3$-VDR-RXR complex binds to promoter regions (VDREs) of vitamin D-responsive genes to modulate gene expression.

Calcitriol attaches to a member of the ligand-activated transcription factor steroid hormone receptor superfamily, known as the VDR, which displays a typical domain structure of a nuclear receptor with a highly conserved DNA-binding region and ligand-binding domain. Binding between the ligand-binding domain and calcitriol (or its analogs) induces heterodimerization of VDR and the RXR to form the calcitriol-VDR-RXR complex, binding the complex to vitamin D response elements (VDREs) in multiple regulatory regions, located at promoters and distal sites of target genes, which, along with the recruitment of co-modulators, plays an important role in regulating cell proliferation and differentiation [34,35]. The presence of the VDR in tumor cells is a prerequisite for the antineoplastic effects of calcitriol. The cell growth evaluated by MTT assay is greatly increased in CYP24-induced and VDR-diminished cells than non-responding cells by 25(OH)$D_3$ activity ($p < 0.01$). In addition, 23 cases of low *VDR* expression have a poorer prognosis than 19 cases of high *VDR* expression [36]. Lopes et al. investigated the immunohistochemical expression of the *VDR* in situ in a range of benign lesions and carcinomas of the mammary gland. The percentage of positive cases for the VDR is higher in benign lesions than in invasive tumors (93.5% and 56.2%, respectively) [37].

## 4. Vitamin D and Inflammation

In the context of cancer, the role of the immune system is not straightforward. Interactions can occur between cancer cells and host immune cells in the tumor microenvironment to create an immunosuppressive network that promotes tumor growth and protects the tumor from immune attack [38]. Population-based studies, as well as molecular studies, have demonstrated that vitamin D is implicated in many immune-related diseases, such as asthma, atherosclerosis, type 2 diabetes, and autoimmune diseases [39,40]. Recent studies indicated that a persistent inflammatory microenvironment can induce tumor production [41–46]. In the course of the development of the tumor, some immune cells often change from the "protector" of self-organizations to the "accomplice" of tumor cells, and nourish the cancer [47]. By mediating complex pathways, the tumor inflammatory microenvironment induces the expression of a variety of pro-inflammatory cytokines, promotes angiogenesis and tumor growth, invasion and metastasis, and accelerates the development of tumors [48–51]. Several mechanisms for how vitamin D affects inflammatory microenvironment in cancers have been explored, including regulating the interaction between immune and tumor cells to regulate the levels of cytokines, inhibiting NF-κB signaling pathway, up-regulating MKP5, and inhibiting the prostaglandins pathway and immune cells (macrophages, DCs, B cells, and T cells) [52,53] (Figure 2). Non-steroidal anti-inflammatory drugs (NSAIDs) are generally used to relieve acute pain and treat chronic inflammation, such as arthritis, among which the most commonly used are aspirin, ibuprofen, and piroxicam. A recent study has confirmed that NSAIDs have a certain degree of anti-tumor effect, which further confirms the close relationship between inflammation and cancer [54].

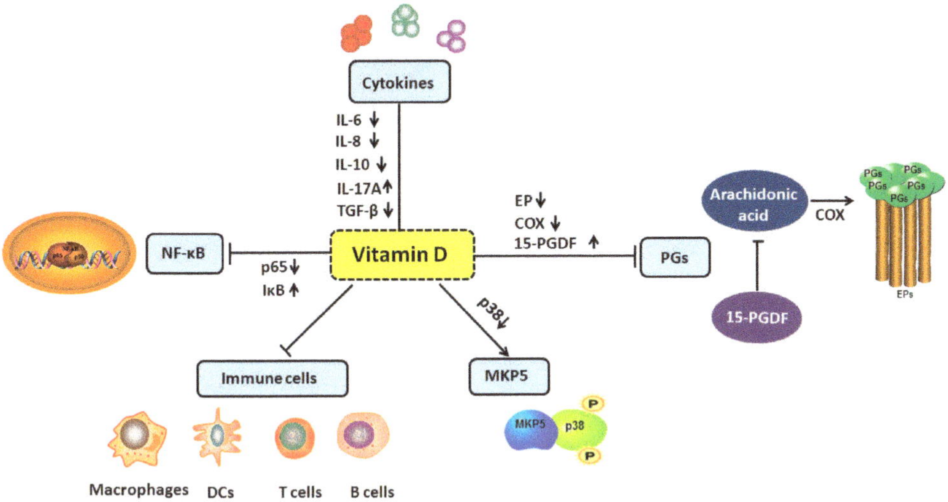


**Figure 2.** The effects of vitamin D of inflammation in tumorigenesis. These include the following. (1) Regulate the levels of cytokines including *IL-6, IL-8, IL-17A, IL-10 and TGF-β*; (2) inhibit NF-κB signaling pathway; (3) up-regulate the expression of MAP kinase phosphatase 5; (4) inhibit the prostaglandins pathway via reducing PG receptors (EPs), decreasing *COX-2* expression and increasing *15-PGDH* expression; (5) inhibit the immune cells via VDR including macrophages, DCs, B cells and T cells. Upward arrows indicate activation, downward arrows indicate inhibition.

### 4.1. Vitamin D and Cytokines

Cytokines produced in the inflammatory microenvironment have been correlated with cancer pathogenesis. Several cytokines are found to be higher in the blood of colorectal cancer patients than in healthy controls [55]. By microarray-based methods, Powell et al. found that several cytokine genes (*IL-1β, IL-6, IL-8, and TGF-β1*) are over-expressed in PCa patients compared to European American men [56]. Interestingly, studies revealed that vitamin D exerts critical regulatory effects on some cytokines' effects on natural and acquired immunity. Barrat et al. found that vitamin D leads to enhanced *IL-10* gene expression and inhibition of the *Th1*- and *Th2*-specific transcription factors [57]. Dauletbaev et al. demonstrated that high concentrations of 25(OH)$D_3$, 1,25$(OH)_2D_3$ and the synthetic analogue paricalcitol moderately down-regulate *IL-8* in hyperinflammatory macrophages from cystic fibrosis patients [58].

IL-6 is a pro-inflammatory cytokine with pro-tumorigenic capacity and is a crucial key effector in PCa and colorectal cancer progression [59]. For example, IL-6 up-regulates *Mcl-1* gene transcription through p38-MAPK and JAK-STAT pathways, inhibiting tumor necrosis factor-relatedapoptosis inducing ligand (TRAIL)-induced tumor cells apoptosis [60]. The results showed that IL-6, as the effect factor of COX-2, can promote angiogenesis and tumor growth [61]. Cathcart et al. demonstrated that IL-6 down-regulates p53 protein levels and results in a concomitant increase in *MMP-14* expression, leading to enhanced cancer cell invasion and metastasis [62]. Several reports have shown that vitamin D may influence the regulation of IL-6 synthesis. 25(OH)D pretreatment inhibits both UV- and tumor necrosis factor alpha (TNF-α)-stimulated IL-6 production in normal cells via p38 MAPK inhibition, demonstrating its significant anti-inflammatory effects in cancer cells [63].

IL-8 has been found to be an important angiogenic factor. It can promote the initiation, migration, and invasion of tumors by promoting the proliferation of neoplastic cells and expression of *MMP-2* and *MMP-9*, increasing collagenase activity and inhibiting tumor cell apoptosis [64,65]. In the study by Singh et al., IL-8 activity was observed to play a crucial role in breast cancer, being regulated by HER-2 positive cancers through chemokine receptor 1/2 (CXCR1/2) ligands. The *IL-8* levels influence breast cancer stem cell activity by enhancing or down-regulating tumorigenesis [66]. Bao et. al found that calcitriol affects the stability of *IL-8* mRNA through the ATTTA motif in the 3′-flankingregion or other post-transcriptional regulations, thereby delaying PCa cell-induced human umbilical vein endothelial cell migration and tube formation [67]. Another study in colon cancer cells showed that calcitriol reduces the production of IL-8, and the anti-inflammatory effect is associated with depression of IL-8 production due to increased release of the soluble form CD14 (sCD14) depending on ERK1/2 [68].

IL-10 is an important anti-inflammatory cytokine due to its effects in down-regulating *Th1* cytokine (*IL-2* and *IFN-γ*) and *Th2* cytokine (*TNF-α, IL-6,* and *IL-8*) production, and its potential to promote resolution of inflammation [69,70]. *IL-10* normally plays an important role in controlling the development of chronic inflammatory processes and its absence results in chronic inflammatory responses (particularly in the gut), which are probably induced by normal enteric antigens [71]. IL-10 could inhibit the production of pro-inflammatory cytokines, which in turn stimulates tumor growth [72]. In a mouse model of colitis anti-CTLA-4, treatment induces IL-10$^+$ Tregs that express the cell surface receptor inducible co-stimulator ligand (ICOS) with potent indoleamine 2,3 dioxygenase (IDO)-dependent anti-inflammatory properties [73]. These observations are particularly interesting as expression of *ICOS* is an immunologic marker correlated with clinical responses in PCa patients, who receive anti-CTLA-4 therapy [74]. It has been observed that the expression of *TLR9* by human adaptive IL-10-Treg populations is regulated by calcitriol [75,76].

TGF-β promotes tumor invasiveness through MMP-2, MMP-9, MT-MMP1 and urokinase-like plasminogen activator is up-regulated in both pancreatic ductal adenocarcinoma and hepatocarcinoma [77]. TGF-β1 induces integrin $\alpha v\beta 3$ expression and the enhanced expression of $\alpha v\beta 3$ potentiates TGF-β1-induced responses, which enhance epithelial-mesenchymal transition (EMT) in mammary epithelial cells in lung fibroblasts [78]. Artaza et al. found that vitamin D suppress TGF-β-mediated fibrosis through modulating multiple pro-fibrotic proteins, for instance, lowering collagen I and III expression and raising expression levels of *MMP-8* in mesenchymal multipotent cells [79].

### 4.2. Vitamin D and Prostaglandins

Prostaglandins (PGs) are pro-inflammatory molecules that bind to specific receptors and play a key role in mediating a series of cellular activities, such as cell proliferation, differentiation, and apoptosis, and promote tumorigenesis and cancer growth [80–84]. PG-related metabolic enzymes and EPs in the inflammatory microenvironment suggests that PGs are closely related to the occurrence and development of these tumors [85].

PGs mediate different signal transduction pathways through the interaction with EPs [86]. More importantly, the interactions between prostaglandin E2 (PGE2) and EP2 can produce positive feedback signals to increase TNF-α, IL-6, CXCL1, and COX-2 levels of neutrophils and tumor-associated fibroblasts [87], thereby promoting tumorigenesis by amplifying inflammation and shaping the inflammatory microenvironment [88,89]. Bazzani et al. indicated PGE2 induces epidermal growth factor receptor (EGFR) nuclear translocation and growth through the release of EGFR ligands in lung adenocarcinoma cells [90]. Moreno et al. further confirmed the mechanisms of suppressing the biological activity of PGs by vitamin D. Calcitriol decreases the mRNA expression of the *PGE2*, inhibits the oncogene *COX-2*, and induces the expression of the putative tumor suppressor *15-PGDH* in prostate cells, suggesting that calcitriol may play an important role in the chemoprevention of PCa [91].

COX-2 is a key enzyme in the synthesis of PGs. Clinical trials and epidemiological studies have suggested that COX-2 is involved in tumorigenesis and its inhibition can reduce the risk of cancer [92–94]. Studies have shown that COX-2 can up-regulate the expression of vascular endothelial growth factor, such as *VEGF-D* (lymphangiogenic factor) and *VEGFR-3* (VEGF-D receptor), which promote tumor angiogenesis in inflammatory mammary carcinomas [95]. High expression of *COX-2* in lung adenocarcinoma cells has also been found to activate tyrosine kinase receptor activity and induce over-expression of the *VEGF-C* gene, which in turn promotes tumor angiogenesis [96]. *COX-2* over-expression in tumor cells also stimulates production of VEGF-A, which causes blood vascular endothelial cell migration and tube formation [97]. On the other hand, over-expression of *COX-2* has a significant inhibitory effect on the apoptosis of tumor cells. Recent studies have shown that COX-2 can inhibit the apoptosis of tumor cells by activating the *Bcl2* gene pathway. The selective COX-2 inhibitor NS-398 can significantly inhibit the function of tumor cells to secrete PGE2 and maintain the cells in G1 phase [98,99]. In PCa cells, calcitriol reduces the expression levels of *COX-2* [100]. Moreno et al. found that calcitriol abolishes c-fos induction and growth stimulation by arachidonic acid, which reflects the effect of calcitriol in decreasing endogenous synthesis of PGs due to COX-2 suppression in PCa cells [91]. In addition, calcitriol interferes with the COX-2/PGE2 pathway, inhibits the activity levels of the phosphorylation of *ERK* and *ERα*, then downregulates the expression of *CYP1B1*, consequently inhibiting the proliferation of breast cancer cells [101].

15-hydroxyprostaglandin dehydrogenase (15-PGDH) is an antagonist of COX-2 and exerts a strong inhibitory effect on the development of cancer [102,103]. Yan et al. suggested that the TGF-β-mediated induction of 15-PGDH and crosstalk between the TGF-β and prostaglandin pathways may represent a significant additional effector of TGF-β-mediated suppression of cancer in the gastrointestinal tract [104]. Studies have shown that the absence or reduction of *15-PGDH* promotes the development of tumors in breast cancer, medullary thyroid carcinoma, PCa, and bladder cancer [105–108], thus supporting a tumor-suppressor role for 15-PGDH in cancer. Several studies highlighted the calcitriol-mediated suppression of the oncogene *COX-2*, an increase in the expression of the putative tumor suppressor *15-PGDH* in prostate cells, a decrease in the levels of biologically active PGs, thereby reducing the development of cancer [91]. The up-regulation of the *IGFBP-3* gene has been shown to be crucial in calcitriol-mediated inhibition of LNCaP cell growth. Calcitriol regulation of androgen-responsive genes, as well as genes involved in androgen catabolism, suggests that there are interactions between calcitriol and androgen signaling pathways in LNCaP cells [109].

In conclusion, vitamin D inhibits the synthesis and biological actions of pro-inflammatory PGs by three mechanisms: reducing PG receptors, decreasing *COX-2* expression, and increasing *15-PGDH* expression.

### 4.3. Vitamin D and MAP Kinase Phosphatase 5

MKP5 is another novel calcitriol-responsive gene. MKP5 preferentially binds and inhibits the activation of p38 MAPK, which is a family of serine/threonine-directed kinases classified as stress-activated kinases. A consequence of p38 activation is an increase in the production of pro-inflammatory cytokines that sustain and amplify the inflammatory response [110,111]. Calcitriol up-regulates *MKP5* expression, leading to subsequent inhibition of the production of pro-inflammatory cytokines, such as *IL-6*, by interfering with the signaling of pleiotropic inflammatory cytokines, such as *TNFα*, supporting a role for calcitriol in the prevention and/or early treatment of PCa [63].

### 4.4. Vitamin D and Nuclear Factor Kappa B Signal Pathway

NF-κB transcription factor is a heterodimer consisting of p50 and p65. It binds to its inhibitory protein inhibitor of NF-κB (IκB) and exists as an inactive hetero-oligomer in the cytoplasm. NF-κB is vital for the regulation of genes that control various responses in eukaryotic cells [112]. Studies revealed that calcitriol inhibits inflammatory responses by modulating the NF-κB signal pathway. Calcitriol inhibits lipopolysaccharide (LPS)-induced p38 phosphorylation and TNF-α and IL-6 production through increased binding of the VDR and histone H4 acetylation at the identified VDRS of the murine and human MKP1 promoters [63,113]. Calcitriol can also increase IκB levels by increasing mRNA stability and decreasing its phosphorylation, thereby reducing the nuclear translocation of NF-κB [114]. Calcitriol suppresses activation of the p65 subunit of the NF-κB complex in colon cancer cells, thereby preventing binding of NF-κB to DNA [115,116]. Thus, calcitriol may serve as an effective inhibitor for cancer via suppression of the NF-κB signal pathway.

## 4.5. Vitamin D and Immune Cells

As the VDR is expressed on immune cells (B cells, T cells, and antigen presenting cells, such as macrophages and DCs), vitamin D could modulate the innate and adaptive immune response [117]. There is increasing epidemiologic evidence linking vitamin D deficiency and autoimmune diseases, including multiple sclerosis (MS), rheumatoid arthritis (RA), diabetes mellitus (DM), inflammatory bowel disease, and systemic lupus erythematosus (SLE) [118–120]. At present, the role of vitamin D in reducing inflammation-dependent carcinogenesis via regulating immune cells is not straight forward, but it is generally recognized that the interactions occurring between tumor and host immune cells in the tumor microenvironment could promote tumor growth and protect the tumor from immune attack. We addressed several mechanisms, through which vitamin D affects different immune cells.

The relationships between B cells and vitamin D have been less intensively studied, but current research suggests that vitamin D can be used as a modulator of allergic immune responses by affecting B cells. Chen et al. found that it inhibits the ongoing proliferation of activated B cells and induces their apoptosis by up-regulating the expression of *p27*, whereas initial cell division is unimpeded. These results indicate that calcitriol may play an important role in the maintenance of B cell homeostasis [121]. Vitamin D deficiency impairs rituximab-mediated cellular cytotoxicity and the outcome of patients with diffuse large B cells lymphoma [122].

A large variety of immune cells express the VDR. Notably, expression of *VDR* increases on T cells upon antigenic stimulation. Jeffery et al. observed that stimulation of $CD4^+$ $CD25^-$ T cells in the presence of calcitriol inhibits production of pro-inflammatory cytokines, including IFN-$\gamma$, IL-17, and IL-21, induced high levels of *CTLA-4* and *FoxP3*, but does not substantially affect T cell division [123]. T-cell cytokines also control vitamin D metabolism in macrophages. For example, IFN-$\gamma$, a Th1 cytokine, up-regulates the macrophage CYP27B1, leading to enhanced bioconversion of $25(OH)D_3$ to its active metabolite-calcitriol. In contrast, the Th2 cytokine IL-4 induces catabolism of $25(OH)D_3$ to the inactive metabolite $24,25(OH)_2D_3$ [124]. Sheikh et al. observed that the stimulation of $CD4^+$ T cells with vitamin D suppresses proliferation capacity; enhanced the expression of *PD1*, *PD-L1*, and *CTLA-4* inhibitory markers on $CD4^+$ T cells; and diminished the percentage of pro-inflammatory cytokines, including IFN-$\gamma$, IL-17, and IL-22, except IL-4 in $CD4^+$ T cells [125]. Conditional targeting experiments showed that VDR function in T cells is necessary. Neither calcitriol nor T-cell-specific VDR targeting influences $CD4^+$ $Foxp3^+$ T-cell proportions in the periphery or the central nervous system (CNS) [126]. Palmer et al. suggested that vitamin D deficiency may promote autoimmunity by favoring the inordinate production of Th17 and Th9 cells at the expense of regulatory IL-10-producing T cells [127]. Calcitriol suppresses the inflammatory infiltrates and inhibits the expression of *P65*, *ROR$\gamma$t*, and *IL-17* in the spleen tissues of model mice [128]. These results suggest a potential mechanism, through which vitamin D metabolism links the T-cell-mediated immune responses to the adaptive immune responses.

The innate immune system interacts with vitamin D in several interesting ways. First, the primary role of macrophages is to engulf and kill bacteria. Toll-like receptors (TLRs) activation of human macrophages up-regulates expression of the *VDR* and the vitamin *CYP27B1* genes, leading to induction of the antimicrobial peptide cathelicidin and killing of intracellular *Mycobacterium tuberculosis* [129]. Chen et al. found a novel regulatory mechanism for vitamin D to control innate immunity: calcitriol promotes negative feedback regulation of TLR signaling via targeting microRNA-155-SOCS1 in macrophages [130]. Moreover, calcitriol or its analogs have been shown to initiate the differentiation of myeloid progenitors into macrophages, and to reduce *MCP-1* and *IL-6* expression via inhibiting the activation of NF-κB in macrophages [131]. Helming et al. found that calcitriol can selectively suppress the key effector functions of IFN-$\gamma$-activated macrophages. The deactivation of IFN-$\gamma$-stimulated macrophages is dependent on a functional VDR and calcitriol acts specifically on IFN-$\gamma$-activated macrophages, whereas the steroid has no effects on resting macrophages [132].

Dendritic cells (DCs) are the most potent antigen-presenting cells. DCs modulate T cell development and they can be not only immunogenic but also tolerogenic, both intrathymically and in the periphery [133]. A number of studies demonstrated that calcitriol inhibits the differentiation, maturation, and immunostimulatory capacity of human DCs. Studies have consistently reported that in vitro treatment of DCs with VDR agonists leads to down-regulated expression of the costimulatory molecules *CD40*, *CD80*, and *CD86*, and to decreased *IL-12* and enhanced *IL-10* production, resulting in decreased T cells activation [134,135]. Szeles et al. found that monocyte-derived DCs are able to turn on calcitriol sensitive genes in the early phases of differentiation if the precursor is present. Their data collectively suggest that exogenously- or endogenously-generated calcitriol autonomously regulates a large set of its targets not via inhibition of differentiation and maturation, leading to the previously characterized tolerogenic state [136]. Takeda et al. observed a significant increase in Foxp3 (+) regulatory T cells and a decrease in $CD80^{+}CD86^{+}$DCs in the mesenteric lymph nodes, spleen, and atherosclerotic lesions in oral calcitriol-treated mice in association with increased *IL-10* and decreased *IL-12* mRNA expression. $CD11c^{+}$ DCs from the calcitriol group showed reduce proliferative activity of T lymphocytes, suggesting the suppression of DC maturation [137]. Cytokines secreted by vitamin D-treated DCs are significantly more potent in driving differentiation of IL-22-producing T cells, as compared to secreted cytokines of not-vitamin D-treated DCs [138].

## 5. Conclusions

A large number of epidemiological surveys have shown that vitamin D deficiency is related to the high incidence of many types of tumors. Experimental data have also confirmed that vitamin D can inhibit the occurrence and metastasis of tumors through many different processes. Among the above, factors related to the inflammatory microenvironment are attracting considerable attention. The treatment of tumor inflammatory microenvironment components as a new target can overcome the limitations of many current traditional treatment methods. Although multiple molecular pathways of vitamin D action in cancer cells have been identified, these pathways provide a mechanistic basis for its potential efficacy via anti-inflammatory actions in cancer. The mechanism of vitamin D in inhibiting tumors by affecting the tumor microenvironment needs to be further explored and studied. Vitamin D and its analogues as anti-cancer drug candidates still need additional prospective intervention experiments to provide direct evidence.

**Author Contributions:** M.-Y.S. conceived the idea, W.L. fully wrote the manuscript, and Y.L. and C.-M.H. contributed to collate the data. L.Z., H.-J.X. and J.-Y.Y. contributed to critical editing the manuscript. All authors approved the final version before submitting.

**Funding:** This work was supported by the National Natural Science Youth Fund of the People's Republic of China (No. 81601234), the National Natural Science Fund of the People's Republic of China (No. 81771304) and the Science and Technology Innovation Development Fund of Jilin City (No. 201750246).

**Conflicts of Interest:** The authors declare no conflicts of interest.

## References

1. Bikle, D.D. Vitamin D metabolism, mechanism of action, and clinical applications. *Chem. Biol.* **2014**, *21*, 319–329. [CrossRef] [PubMed]
2. Jones, G. Metabolism and biomarkers of vitamin D. *Scand. J. Clin. Lab. Investig. Suppl.* **2012**, *243*, 7–13.
3. Christakos, S.; Dhawan, P.; Liu, Y.; Peng, X.; Porta, A. New insights into the mechanisms of vitamin D action. *J. Cell Biochem.* **2003**, *88*, 695–705. [CrossRef] [PubMed]
4. Hanahan, D.; Weinberg, R.A. Hallmarks of cancer: The next generation. *Cell* **2011**, *144*, 646–674. [CrossRef] [PubMed]
5. Chen, F.; Zhuang, X.; Lin, L.; Yu, P.; Wang, Y.; Shi, Y.; Hu, G.; Sun, Y. New horizons in tumor microenvironment biology challenges and opportunities. *BMC Med.* **2015**, *13*, 45. [CrossRef] [PubMed]
6. Ariztia, E.V.; Lee, C.J.; Gogoi, R.; Fishman, D.A. The tumor microenvironment: Key to early detection. *Crit. Rev. Clin. Lab. Sci.* **2006**, *43*, 393–425. [CrossRef] [PubMed]
7. Grivennikov, S.I.; Greten, F.R.; Karin, M. Immunity, inflammation, and cancer. *Cell* **2010**, *140*, 883–899. [CrossRef] [PubMed]

8. Gajewski, T.F.; Schreiber, H.; Fu, Y.X. Innate and adaptive immune cells in the tumor microenvironment. *Nat. Immunol.* **2013**, *14*, 1014–1022. [CrossRef] [PubMed]
9. Zamarron, B.F.; Chen, W. Dual roles of immune cells and their factors in cancer development and progression. *Int. J. Biol. Sci.* **2011**, *7*, 651–658. [CrossRef] [PubMed]
10. Tretli, S.; Schwartz, G.G.; Torjesen, P.A.; Robsahm, T.E. Serum levels of 25-hydroxyvitamin D and survival in Norwegian patients with cancer of breast, colon, lung, and lymphoma: A population-based study. *Cancer Causes Control* **2012**, *23*, 363–370. [CrossRef] [PubMed]
11. Tagliabue, E.; Raimondi, S.; Gandini, S. Vitamin D, Cancer Risk, and Mortality. *Adv. Food Nutr. Res.* **2015**, *75*, 1–52. [PubMed]
12. Ordonez Mena, J.M.; Brenner, H. Vitamin D and cancer: An overview on epidemiological studies. *Adv. Exp. Med. Biol.* **2014**, *810*, 17–32. [PubMed]
13. Pilz, S.; Tomaschitz, A.; Obermayer-Pietsch, B.; Dobnig, H.; Pieber, T.R. Epidemiology of vitamin D insufficiency and cancer mortality. *Anticancer Res.* **2009**, *29*, 3699–3704. [PubMed]
14. Garland, C.F.; Garland, F.C. Do sunlight and vitamin D reduce the likelihood of colon cancer? *Int. J. Epidemiol.* **1980**, *9*, 227–231. [CrossRef] [PubMed]
15. Cho, M.; Peddi, P.F.; Ding, K.; Chen, L.; Thomas, D.; Wang, J.; Lockhart, A.C.; Tan, B.; Wang-Gillam, A. Vitamin D deficiency and prognostics among patients with pancreatic adenocarcinoma. *J. Transl. Med.* **2013**, *11*, 206. [CrossRef] [PubMed]
16. Budhathoki, S.; Hidaka, A.; Yamaji, T.; Sawada, N.; Tanaka-Mizuno, S.; Kuchiba, A.; Charvat, H.; Goto, A.; Kojima, S.; Sudo, N.; et al. Plasma 25-hydroxyvitamin D concentration and subsequent risk of total and site specific cancers in Japanese population: Large case-cohort study within Japan Public Health Center-based Prospective Study cohort. *BMJ* **2018**, *360*, k671. [CrossRef] [PubMed]
17. Li, M.; Chen, P.; Li, J.; Chu, R.; Xie, D.; Wang, H. Review: The impacts of circulating 25-hydroxyvitamin D levels on cancer patient outcomes: A systematic review and meta-analysis. *J. Clin. Endocrinol. Metab.* **2014**, *99*, 2327–2336. [CrossRef] [PubMed]
18. Colston, K.W.; Hansen, C.M. Mechanisms implicated in the growth regulatory effects of vitamin D in breast cancer. *Endocr. Relat. Cancer* **2002**, *9*, 45–59. [CrossRef] [PubMed]
19. Gorham, E.D.; Garland, C.F.; Garland, F.C.; Grant, W.B.; Mohr, S.B.; Lipkin, M.; Newmark, H.L.; Giovannucci, E.; Wei, M.; Holick, M.F. Vitamin D and prevention of colorectal cancer. *J. Steroid. Biochem. Mol. Biol.* **2005**, *97*, 179–194. [CrossRef] [PubMed]
20. Ma, Y.; Zhang, P.; Wang, F.; Yang, J.; Liu, Z.; Qin, H. Association between vitamin D and risk of colorectal cancer: A systematic review of prospective studies. *J. Clin. Oncol.* **2011**, *29*, 3775–3782. [CrossRef] [PubMed]
21. Skinner, H.G.; Michaud, D.S.; Giovannucci, E.; Willett, W.C.; Colditz, G.A.; Fuchs, C.S. Vitamin D intake and the risk for pancreatic cancer in two cohort studies. *Cancer Epidemiol. Biomarkers Prev.* **2006**, *15*, 1688–1695. [CrossRef] [PubMed]
22. Krishnan, A.V.; Trump, D.L.; Johnson, C.S.; Feldman, D. The role of vitamin D in cancer prevention and treatment. *Endocrinol. Metab. Clin. N. Am.* **2010**, *39*, 401–418. [CrossRef] [PubMed]
23. Moukayed, M.; Grant, W.B. The roles of UVB and vitamin D in reducing risk of cancer incidence and mortality: A review of the epidemiology, clinical trials, and mechanisms. *Rev. Endocr. Metab. Disord.* **2017**, *18*, 167–182. [CrossRef] [PubMed]
24. Giovannucci, E. The epidemiology of vitamin D and cancer incidence and mortality: A review (United States). *Cancer Causes Control* **2005**, *16*, 83–95. [CrossRef] [PubMed]
25. Yin, L.; Grandi, N.; Raum, E.; Haug, U.; Arndt, V.; Brenner, H. Meta-analysis: Serum vitamin D and breast cancer risk. *Eur. J. Cancer* **2010**, *46*, 2196–2205. [CrossRef] [PubMed]
26. Kim, Y.; Je, Y. Vitamin D intake, blood 25(OH)D levels, and breast cancer risk or mortality: A meta-analysis. *Br. J. Cancer* **2014**, *110*, 2772–2784. [CrossRef] [PubMed]
27. Schuster, I. Cytochromes P450 are essential players in the vitamin D signaling system. *Biochim. Biophys. Acta* **2011**, *1814*, 186–199. [CrossRef] [PubMed]
28. Segersten, U.; Correa, P.; Hewison, M.; Hellman, P.; Dralle, H.; Carling, T.; Akerstrom, G.; Westin, G. 25-hydroxyvitamin $D_3$-1$\alpha$-hydroxylase expression in normal and pathological parathyroid glands. *J. Clin. Endocrinol. Metab.* **2002**, *87*, 2967–2972. [CrossRef] [PubMed]

29. Hsu, J.Y.; Feldman, D.; McNeal, J.E.; Peehl, D.M. Reduced 1α-hydroxylase activity in human prostate cancer cells correlates with decreased susceptibility to 25-hydroxyvitamin $D_3$-induced growth inhibition. *Cancer Res.* **2001**, *61*, 2852–2856. [PubMed]
30. Haussler, M.R.; Whitfield, G.K.; Haussler, C.A.; Hsieh, J.C.; Thompson, P.D.; Selznick, S.H.; Dominguez, C.E.; Jurutka, P.W. The nuclear vitamin D receptor: Biological and molecular regulatory properties revealed. *J. Bone Miner. Res.* **1998**, *13*, 325–349. [CrossRef] [PubMed]
31. Hobaus, J.; Hummel, D.M.; Thiem, U.; Fetahu, I.S.; Aggarwal, A.; Mullauer, L.; Heller, G.; Egger, G.; Mesteri, I.; Baumgartner-Parzer, S.; et al. Increased copy-number and not DNA hypomethylation causes overexpression of the candidate proto-oncogene CYP24A1 in colorectal cancer. *Int. J. Cancer* **2013**, *133*, 1380–1388. [CrossRef] [PubMed]
32. Albertson, D.G.; Ylstra, B.; Segraves, R.; Collins, C.; Dairkee, S.H.; Kowbel, D.; Kuo, W.L.; Gray, J.W.; Pinkel, D. Quantitative mapping of amplicon structure by array CGH identifies *CYP24* as a candidate oncogene. *Nat. Genet.* **2000**, *25*, 144–146. [CrossRef] [PubMed]
33. Borkowski, R.; Du, L.; Zhao, Z.; McMillan, E.; Kosti, A.; Yang, C.R.; Suraokar, M.; Wistuba, I.I.; Gazdar, A.F.; Minna, J.D.; et al. Genetic mutation of p53 and suppression of the miR-17 approximately 92 cluster are synthetic lethal in non-small cell lung cancer due to upregulation of vitamin D Signaling. *Cancer Res.* **2015**, *75*, 666–675. [CrossRef] [PubMed]
34. Sharan, D.; Ajeesh, P.S.; Rameshkumar, R.; Mathankumar, M.; Paulina, R.J.; Manjula, M. Virtual reality based therapy for post operative rehabilitation of children with cerebral palsy. *Work* **2012**, *41*, 3612–3615. [PubMed]
35. Deeb, K.K.; Trump, D.L.; Johnson, C.S. Vitamin D signalling pathways in cancer: Potential for anticancer therapeutics. *Nat. Rev. Cancer* **2007**, *7*, 684–700. [CrossRef] [PubMed]
36. Mimori, K.; Tanaka, Y.; Yoshinaga, K.; Masuda, T.; Yamashita, K.; Okamoto, M.; Inoue, H.; Mori, M. Clinical significance of the overexpression of the candidate oncogene *CYP24* in esophageal cancer. *Ann. Oncol.* **2004**, *15*, 236–241. [CrossRef] [PubMed]
37. Lopes, N.; Sousa, B.; Martins, D.; Gomes, M.; Vieira, D.; Veronese, L.A.; Milanezi, F.; Paredes, J.; Costa, J.L.; Schmitt, F. Alterations in Vitamin D signalling and metabolic pathways in breast cancer progression: A study of VDR, CYP27B1 and CYP24A1 expression in benign and malignant breast lesions. *BMC Cancer* **2010**, *10*, 483. [CrossRef] [PubMed]
38. De Souza, A.P.; Bonorino, C. Tumor immunosuppressive environment: Effects on tumor-specific and nontumor antigen immune responses. *Expert. Rev. Anticancer Ther.* **2009**, *9*, 1317–1332. [CrossRef] [PubMed]
39. Wacker, M.; Holick, M.F. Vitamin D—effects on skeletal and extraskeletal health and the need for supplementation. *Nutrients* **2013**, *5*, 111–148. [CrossRef] [PubMed]
40. Wobke, T.K.; Sorg, B.L.; Steinhilber, D. Vitamin D in inflammatory diseases. *Front. Physiol.* **2014**, *5*, 244. [PubMed]
41. Bu, P.; Wang, L.; Chen, K.Y.; Srinivasan, T.; Murthy, P.K.; Tung, K.L.; Varanko, A.K.; Chen, H.J.; Ai, Y.; King, S.; et al. A miR-34a-Numb Feedforward Loop Triggered by Inflammation Regulates Asymmetric Stem Cell Division in Intestine and Colon Cancer. *Cell Stem Cell* **2016**, *18*, 189–202. [CrossRef] [PubMed]
42. Mishra, A.; Liu, S.; Sams, G.H.; Curphey, D.P.; Santhanam, R.; Rush, L.J.; Schaefer, D.; Falkenberg, L.G.; Sullivan, L.; Jaroncyk, L.; et al. Aberrant overexpression of IL-15 initiates large granular lymphocyte leukemia through chromosomal instability and DNA hypermethylation. *Cancer Cell* **2012**, *22*, 645–655. [CrossRef] [PubMed]
43. Tili, E.; Michaille, J.J.; Wernicke, D.; Alder, H.; Costinean, S.; Volinia, S.; Croce, C.M. Mutator activity induced by microRNA-155 (miR-155) links inflammation and cancer. *Proc. Natl. Acad. Sci. USA* **2011**, *108*, 4908–4913. [CrossRef] [PubMed]
44. El Rayes, T.; Catena, R.; Lee, S.; Stawowczyk, M.; Joshi, N.; Fischbach, C.; Powell, C.A.; Dannenberg, A.J.; Altorki, N.K.; Gao, D.; et al. Lung inflammation promotes metastasis through neutrophil protease-mediated degradation of Tsp-1. *Proc. Natl. Acad. Sci. USA* **2015**, *112*, 16000–16005. [CrossRef] [PubMed]
45. Katoh, H.; Wang, D.; Daikoku, T.; Sun, H.; Dey, S.K.; Dubois, R.N. CXCR2-expressing myeloid-derived suppressor cells are essential to promote colitis-associated tumorigenesis. *Cancer Cell* **2013**, *24*, 631–644. [CrossRef] [PubMed]
46. Li, W.; Xiao, J.; Zhou, X.; Xu, M.; Hu, C.; Xu, X.; Lu, Y.; Liu, C.; Xue, S.; Nie, L.; et al. STK4 regulates TLR pathways and protects against chronic inflammation-related hepatocellular carcinoma. *J. Clin. Investig.* **2015**, *125*, 4239–4254. [CrossRef] [PubMed]

47. Lu, H.; Ouyang, W.; Huang, C. Inflammation, a key event in cancer development. *Mol. Cancer Res.* **2006**, *4*, 221–233. [CrossRef] [PubMed]
48. Balkwill, F.; Mantovani, A. Inflammation and cancer: Back to Virchow? *Lancet* **2001**, *357*, 539–545. [CrossRef]
49. Cai, G.H.; Li, M.X.; Lu, L.; LokYi Chan, R.; Wang, J.H.; Cho, C.H. The current role and therapeutic targets of vitamin D in gastrointestinal inflammation and cancer. *Curr. Pharm. Des.* **2015**, *21*, 2917–2923. [CrossRef] [PubMed]
50. Wang, K.; Karin, M. Tumor-Elicited Inflammation and Colorectal Cancer. *Adv. Cancer Res.* **2015**, *128*, 173–196. [PubMed]
51. Candido, J.; Hagemann, T. Cancer-related inflammation. *J. Clin. Immunol.* **2013**, *33*, S79–S84. [CrossRef] [PubMed]
52. Raymond, E.; Dalgleish, A.; Damber, J.E.; Smith, M.; Pili, R. Mechanisms of action of tasquinimod on the tumour microenvironment. *Cancer Chemother. Pharmacol.* **2014**, *73*, 1–8. [CrossRef] [PubMed]
53. Han, Y.; Zhang, Y.; Jia, T.; Sun, Y. Molecular mechanism underlying the tumor-promoting functions of carcinoma-associated fibroblasts. *Tumour Biol.* **2015**, *36*, 1385–1394. [CrossRef] [PubMed]
54. Piazuelo, E.; Lanas, A. NSAIDS and gastrointestinal cancer. *Prostaglandins Other Lipid Mediat.* **2015**, *120*, 91–96. [CrossRef] [PubMed]
55. Groblewska, M.; Mroczko, B.; Wereszczynska-Siemiatkowska, U.; Kedra, B.; Lukaszewicz, M.; Baniukiewicz, A.; Szmitkowski, M. Serum interleukin 6 (IL-6) and C-reactive protein (CRP) levels in colorectal adenoma and cancer patients. *Clin. Chem. Lab. Med.* **2008**, *46*, 1423–1428. [CrossRef] [PubMed]
56. Powell, I.J.; Dyson, G.; Land, S.; Ruterbusch, J.; Bock, C.H.; Lenk, S.; Herawi, M.; Everson, R.; Giroux, C.N.; Schwartz, A.G.; et al. Genes associated with prostate cancer are differentially expressed in African American and European American men. *Cancer Epidemiol. Biomark. Prev.* **2013**, *22*, 891–897. [CrossRef] [PubMed]
57. Barrat, F.J.; Cua, D.J.; Boonstra, A.; Richards, D.F.; Crain, C.; Savelkoul, H.F.; de Waal-Malefyt, R.; Coffman, R.L.; Hawrylowicz, C.M.; O'Garra, A. In vitro generation of interleukin 10-producing regulatory $CD4^+$ T cells is induced by immunosuppressive drugs and inhibited by T helper type 1 (Th1)- and Th2-inducing cytokines. *J. Exp. Med.* **2002**, *195*, 603–616. [CrossRef] [PubMed]
58. Dauletbaev, N.; Herscovitch, K.; Das, M.; Chen, H.; Bernier, J.; Matouk, E.; Berube, J.; Rousseau, S.; Lands, L.C. Down-regulation of IL-8 by high-dose vitamin D is specific to hyperinflammatory macrophages and involves mechanisms beyond up-regulation of DUSP1. *Br. J. Pharmacol.* **2015**, *172*, 4757–4771. [CrossRef] [PubMed]
59. Nguyen, D.P.; Li, J.; Tewari, A.K. Inflammation and prostate cancer: The role of interleukin 6 (IL-6). *BJU Int.* **2014**, *113*, 986–992. [CrossRef] [PubMed]
60. Lippitz, B.E. Cytokine patterns in patients with cancer: A systematic review. *Lancet Oncol.* **2013**, *14*, e218–228. [CrossRef]
61. Dalwadi, H.; Krysan, K.; Heuze-Vourc'h, N.; Dohadwala, M.; Elashoff, D.; Sharma, S.; Cacalano, N.; Lichtenstein, A.; Dubinett, S. Cyclooxygenase-2-dependent activation of signal transducer and activator of transcription 3 by interleukin-6 in non-small cell lung cancer. *Clin. Cancer Res.* **2005**, *11*, 7674–7682. [CrossRef] [PubMed]
62. Cathcart, J.M.; Banach, A.; Liu, A.; Chen, J.; Goligorsky, M.; Cao, J. Interleukin-6 increases matrix metalloproteinase-14 (MMP-14) levels via down-regulation of p53 to drive cancer progression. *Oncotarget* **2016**, *7*, 61107–61120. [CrossRef] [PubMed]
63. Nonn, L.; Peng, L.; Feldman, D.; Peehl, D.M. Inhibition of p38 by vitamin D reduces interleukin-6 production in normal prostate cells via mitogen-activated protein kinase phosphatase 5: Implications for prostate cancer prevention by vitamin D. *Cancer Res.* **2006**, *66*, 4516–4524. [CrossRef] [PubMed]
64. Cheng, D.; Kong, H.; Li, Y. Prognostic value of interleukin-8 and MMP-9 in nasopharyngeal carcinoma. *Eur. Arch. Otorhinolaryngol.* **2014**, *271*, 503–509. [CrossRef] [PubMed]
65. Haraguchi, M.; Komuta, K.; Akashi, A.; Matsuzaki, S.; Furui, J.; Kanematsu, T. Elevated IL-8 levels in the drainage vein of resectable Dukes' C colorectal cancer indicate high risk for developing hepatic metastasis. *Oncol. Rep.* **2002**, *9*, 159–165. [CrossRef] [PubMed]
66. Singh, J.K.; Simoes, B.M.; Howell, S.J.; Farnie, G.; Clarke, R.B. Recent advances reveal IL-8 signaling as a potential key to targeting breast cancer stem cells. *Breast Cancer Res.* **2013**, *15*, 210. [CrossRef] [PubMed]
67. Bao, B.Y.; Yao, J.; Lee, Y.F. 1$\alpha$,25-dihydroxyvitamin $D_3$ suppresses interleukin-8-mediated prostate cancer cell angiogenesis. *Carcinogenesis* **2006**, *27*, 1883–1893. [CrossRef] [PubMed]

68. Hidaka, M.; Wakabayashi, I.; Takeda, Y.; Fukuzawa, K. Vitamin $D_3$ derivatives increase soluble CD14 release through ERK1/2 activation and decrease IL-8 production in intestinal epithelial cells. *Eur. J. Pharmacol.* **2013**, *721*, 305–312. [CrossRef] [PubMed]
69. John, M.; Au, B.T.; Jose, P.J.; Lim, S.; Saunders, M.; Barnes, P.J.; Mitchell, J.A.; Belvisi, M.G.; Chung, K.F. Expression and release of interleukin-8 by human airway smooth muscle cells: Inhibition by Th-2 cytokines and corticosteroids. *Am. J. Respir. Cell Mol. Biol.* **1998**, *18*, 84–90. [CrossRef] [PubMed]
70. Schandene, L.; Alonso-Vega, C.; Willems, F.; Gerard, C.; Delvaux, A.; Velu, T.; Devos, R.; de Boer, M.; Goldman, M. B7/CD28-dependent IL-5 production by human resting T cells is inhibited by IL-10. *J. Immunol.* **1994**, *152*, 4368–4374. [PubMed]
71. de Vries, J.E. Immunosuppressive and anti-inflammatory properties of interleukin 10. *Ann. Med.* **1995**, *27*, 537–541. [CrossRef] [PubMed]
72. Dennis, K.L.; Blatner, N.R.; Gounari, F.; Khazaie, K. Current status of interleukin-10 and regulatory T-cells in cancer. *Curr. Opin. Oncol.* **2013**, *25*, 637–645. [CrossRef] [PubMed]
73. Coquerelle, C.; Oldenhove, G.; Acolty, V.; Denoeud, J.; Vansanten, G.; Verdebout, J.M.; Mellor, A.; Bluestone, J.A.; Moser, M. Anti-CTLA-4 treatment induces IL-10-producing $ICOS^+$ regulatory T cells displaying IDO-dependent anti-inflammatory properties in a mouse model of colitis. *Gut* **2009**, *58*, 1363–1373. [CrossRef] [PubMed]
74. Chen, H.; Liakou, C.I.; Kamat, A.; Pettaway, C.; Ward, J.F.; Tang, D.N.; Sun, J.; Jungbluth, A.A.; Troncoso, P.; Logothetis, C.; et al. Anti-CTLA-4 therapy results in higher $CD4^+ICOS^{hi}$ T cell frequency and IFN-$\gamma$ levels in both nonmalignant and malignant prostate tissues. *Proc. Natl. Acad. Sci. USA* **2009**, *106*, 2729–2734. [CrossRef] [PubMed]
75. Sharma, S.; Stolina, M.; Lin, Y.; Gardner, B.; Miller, P.W.; Kronenberg, M.; Dubinett, S.M. T cell-derived IL-10 promotes lung cancer growth by suppressing both T cell and APC function. *J. Immunol.* **1999**, *163*, 5020–5028. [PubMed]
76. Urry, Z.; Xystrakis, E.; Richards, D.F.; McDonald, J.; Sattar, Z.; Cousins, D.J.; Corrigan, C.J.; Hickman, E.; Brown, Z.; Hawrylowicz, C.M. Ligation of TLR9 induced on human IL-10-secreting Tregs by 1$\alpha$,25-dihydroxyvitamin $D_3$ abrogates regulatory function. *J. Clin. Investig.* **2009**, *119*, 387–398. [CrossRef] [PubMed]
77. Sannino, G.; Armbruster, N.; Bodenhofer, M.; Haerle, U.; Behrens, D.; Buchholz, M.; Rothbauer, U.; Sipos, B.; Schmees, C. Role of BCL9L in transforming growth factor-$\beta$ (TGF-$\beta$)-induced epithelial-to-mesenchymal-transition (EMT) and metastasis of pancreatic cancer. *Oncotarget* **2016**, *7*, 73725–73738. [CrossRef] [PubMed]
78. Mori, S.; Kodaira, M.; Ito, A.; Okazaki, M.; Kawaguchi, N.; Hamada, Y.; Takada, Y.; Matsuura, N. Enhanced Expression of Integrin $\alpha v\beta 3$ Induced by TGF-$\beta$ Is Required for the Enhancing Effect of Fibroblast Growth Factor 1 (FGF1) in TGF-$\beta$-Induced Epithelial-Mesenchymal Transition (EMT) in Mammary Epithelial Cells. *PLoS ONE* **2015**, *10*, e0137486. [CrossRef] [PubMed]
79. Artaza, J.N.; Norris, K.C. Vitamin D reduces the expression of collagen and key profibrotic factors by inducing an antifibrotic phenotype in mesenchymal multipotent cells. *J. Endocrinol.* **2009**, *200*, 207–221. [CrossRef] [PubMed]
80. Telliez, A.; Furman, C.; Pommery, N.; Henichart, J.P. Mechanisms leading to COX-2 expression and COX-2 induced tumorigenesis: Topical therapeutic strategies targeting COX-2 expression and activity. *Anticancer Agents Med. Chem.* **2006**, *6*, 187–208. [CrossRef] [PubMed]
81. Bonventre, J.V.; Huang, Z.; Taheri, M.R.; O'Leary, E.; Li, E.; Moskowitz, M.A.; Sapirstein, A. Reduced fertility and postischaemic brain injury in mice deficient in cytosolic phospholipase $A_2$. *Nature* **1997**, *390*, 622–625. [CrossRef] [PubMed]
82. Jiang, J.; Qiu, J.; Li, Q.; Shi, Z. Prostaglandin E2 Signaling: Alternative Target for Glioblastoma? *Trend. Cancer* **2017**, *3*, 75–78. [CrossRef] [PubMed]
83. Funk, C.D.; FitzGerald, G.A. COX-2 inhibitors and cardiovascular risk. *J. Cardiovasc. Pharmacol.* **2007**, *50*, 470–479. [CrossRef] [PubMed]
84. FitzGerald, G.A.; Loll, P. COX in a crystal ball: Current status and future promise of prostaglandin research. *J. Clin. Investig.* **2001**, *107*, 1335–1337. [CrossRef] [PubMed]
85. Ricciotti, E.; FitzGerald, G.A. Prostaglandins and inflammation. *Arterioscler. Thromb. Vasc. Biol.* **2011**, *31*, 986–1000. [CrossRef] [PubMed]

86. Harizi, H.; Corcuff, J.B.; Gualde, N. Arachidonic-acid-derived eicosanoids: Roles in biology and immunopathology. *Trends Mol. Med.* **2008**, *14*, 461–469. [CrossRef] [PubMed]
87. Chen, Y.; Hughes-Fulford, M. Prostaglandin E2 and the protein kinase A pathway mediate arachidonic acid induction of c-fos in human prostate cancer cells. *Br. J. Cancer* **2000**, *82*, 2000–2006. [PubMed]
88. Kawahara, K.; Hohjoh, H.; Inazumi, T.; Tsuchiya, S.; Sugimoto, Y. Prostaglandin E2-induced inflammation: Relevance of prostaglandin E receptors. *Biochim. Biophys. Acta* **2015**, *1851*, 414–421. [CrossRef] [PubMed]
89. Ma, X.; Aoki, T.; Tsuruyama, T.; Narumiya, S. Definition of Prostaglandin $E_2$-EP2 Signals in the Colon Tumor Microenvironment That Amplify Inflammation and Tumor Growth. *Cancer Res.* **2015**, *75*, 2822–2832. [CrossRef] [PubMed]
90. Bazzani, L.; Donnini, S.; Finetti, F.; Christofori, G.; Ziche, M. PGE2/EP3/SRC signaling induces EGFR nuclear translocation and growth through EGFR ligands release in lung adenocarcinoma cells. *Oncotarget* **2017**, *8*, 31270–31287. [CrossRef] [PubMed]
91. Moreno, J.; Krishnan, A.V.; Swami, S.; Nonn, L.; Peehl, D.M.; Feldman, D. Regulation of prostaglandin metabolism by calcitriol attenuates growth stimulation in prostate cancer cells. *Cancer Res.* **2005**, *65*, 7917–7925. [CrossRef] [PubMed]
92. Rothwell, P.M.; Fowkes, F.G.; Belch, J.F.; Ogawa, H.; Warlow, C.P.; Meade, T.W. Effect of daily aspirin on long-term risk of death due to cancer: Analysis of individual patient data from randomised trials. *Lancet* **2011**, *377*, 31–41. [CrossRef]
93. Bos, P.D.; Zhang, X.H.; Nadal, C.; Shu, W.; Gomis, R.R.; Nguyen, D.X.; Minn, A.J.; van de Vijver, M.J.; Gerald, W.L.; Foekens, J.A.; et al. Genes that mediate breast cancer metastasis to the brain. *Nature* **2009**, *459*, 1005–1009. [CrossRef] [PubMed]
94. Regulski, M.; Regulska, K.; Prukala, W.; Piotrowska, H.; Stanisz, B.; Murias, M. COX-2 inhibitors: A novel strategy in the management of breast cancer. *Drug Discov. Today* **2016**, *21*, 598–615. [CrossRef] [PubMed]
95. Clemente, M.; Sanchez-Archidona, A.R.; Sardon, D.; Diez, L.; Martin-Ruiz, A.; Caceres, S.; Sassi, F.; Dolores Perez-Alenza, M.; Illera, J.C.; Dunner, S.; et al. Different role of COX-2 and angiogenesis in canine inflammatory and non-inflammatory mammary cancer. *Vet. J.* **2013**, *197*, 427–432. [CrossRef] [PubMed]
96. Su, J.L.; Shih, J.Y.; Yen, M.L.; Jeng, Y.M.; Chang, C.C.; Hsieh, C.Y.; Wei, L.H.; Yang, P.C.; Kuo, M.L. Cyclooxygenase-2 induces EP1- and HER-2/Neu-dependent vascular endothelial growth factor-C up-regulation: A novel mechanism of lymphangiogenesis in lung adenocarcinoma. *Cancer Res.* **2004**, *64*, 554–564. [CrossRef] [PubMed]
97. Sawaoka, H.; Kawano, S.; Tsuji, S.; Tsujii, M.; Murata, H.; Hori, M. Effects of NSAIDs on proliferation of gastric cancer cells in vitro: Possible implication of cyclooxygenase-2 in cancer development. *J. Clin. Gastroenterol.* **1998**, *27*, S47–S52. [CrossRef] [PubMed]
98. Nzeako, U.C.; Guicciardi, M.E.; Yoon, J.H.; Bronk, S.F.; Gores, G.J. COX-2 inhibits Fas-mediated apoptosis in cholangiocarcinoma cells. *Hepatology* **2002**, *35*, 552–559. [CrossRef] [PubMed]
99. Hung, W.C.; Chang, H.C.; Pan, M.R.; Lee, T.H.; Chuang, L.Y. Induction of p27(KIP1) as a mechanism underlying NS398-induced growth inhibition in human lung cancer cells. *Mol. Pharmacol.* **2000**, *58*, 1398–1403. [CrossRef] [PubMed]
100. Krishnan, A.V.; Feldman, D. Molecular pathways mediating the anti-inflammatory effects of calcitriol: Implications for prostate cancer chemoprevention and treatment. *Endocr. Relat. Cancer* **2010**, *17*, R19–38. [CrossRef] [PubMed]
101. Yuan, L.; Jiang, R.; Yang, Y.; Ding, S.; Deng, H. 1,25-Dihydroxyvitamin $D_3$ inhibits growth of the breast cancer cell line MCF-7 and downregulates cytochrome P4501B1 through the COX-2/PGE2 pathway. *Oncol. Rep.* **2012**, *28*, 2131–2137. [CrossRef] [PubMed]
102. Choi, S.H.; Kim, B.G.; Robinson, J.; Fink, S.; Yan, M.; Sporn, M.B.; Markowitz, S.D.; Letterio, J.J. Synthetic triterpenoid induces 15-PGDH expression and suppresses inflammation-driven colon carcinogenesis. *J. Clin. Investig.* **2014**, *124*, 2472–2482. [CrossRef] [PubMed]
103. Hazra, S.; Batra, R.K.; Tai, H.H.; Sharma, S.; Cui, X.; Dubinett, S.M. Pioglitazone and rosiglitazone decrease prostaglandin E2 in non-small-cell lung cancer cells by up-regulating 15-hydroxyprostaglandin dehydrogenase. *Mol. Pharmacol.* **2007**, *71*, 1715–1720. [CrossRef] [PubMed]
104. Yan, M.; Rerko, R.M.; Platzer, P.; Dawson, D.; Willis, J.; Tong, M.; Lawrence, E.; Lutterbaugh, J.; Lu, S.; Willson, J.K.; et al. 15-Hydroxyprostaglandin dehydrogenase, a COX-2 oncogene antagonist, is

a TGF-β-induced suppressor of human gastrointestinal cancers. *Proc. Natl. Acad. Sci. USA* **2004**, *101*, 17468–17473. [CrossRef] [PubMed]

105. Wolf, I.; O'Kelly, J.; Rubinek, T.; Tong, M.; Nguyen, A.; Lin, B.T.; Tai, H.H.; Karlan, B.Y.; Koeffler, H.P. 15-hydroxyprostaglandin dehydrogenase is a tumor suppressor of human breast cancer. *Cancer Res.* **2006**, *66*, 7818–7823. [CrossRef] [PubMed]
106. Quidville, V.; Segond, N.; Lausson, S.; Frenkian, M.; Cohen, R.; Jullienne, A. 15-Hydroxyprostaglandin-dehydrogenase is involved in anti-proliferative effect of non-steroidal anti-inflammatory drugs COX-1 inhibitors on a human medullary thyroid carcinoma cell line. *Prostaglandins Other Lipid Mediat.* **2006**, *81*, 14–30. [CrossRef] [PubMed]
107. Swami, S.; Krishnan, A.V.; Moreno, J.; Bhattacharyya, R.B.; Peehl, D.M.; Feldman, D. Calcitriol and genistein actions to inhibit the prostaglandin pathway: Potential combination therapy to treat prostate cancer. *J. Nutr.* **2007**, *137*, 205S–210S. [CrossRef] [PubMed]
108. Gee, J.R.; Montoya, R.G.; Khaled, H.M.; Sabichi, A.L.; Grossman, H.B. Cytokeratin 20, AN43, PGDH, and COX-2 expression in transitional and squamous cell carcinoma of the bladder. *Urol. Oncol.* **2003**, *21*, 266–270. [CrossRef]
109. Krishnan, A.V.; Shinghal, R.; Raghavachari, N.; Brooks, J.D.; Peehl, D.M.; Feldman, D. Analysis of vitamin D-regulated gene expression in LNCaP human prostate cancer cells using cDNA microarrays. *Prostate* **2004**, *59*, 243–251. [CrossRef] [PubMed]
110. Park, J.I.; Lee, M.G.; Cho, K.; Park, B.J.; Chae, K.S.; Byun, D.S.; Ryu, B.K.; Park, Y.K.; Chi, S.G. Transforming growth factor-β1 activates interleukin-6 expression in prostate cancer cells through the synergistic collaboration of the Smad2, p38-NF-κB, JNK, and Ras signaling pathways. *Oncogene* **2003**, *22*, 4314–4332. [CrossRef] [PubMed]
111. Kim, E.K.; Choi, E.J. Pathological roles of MAPK signaling pathways in human diseases. *Biochim. Biophys. Acta* **2010**, *1802*, 396–405. [CrossRef] [PubMed]
112. Xia, Y.; Shen, S.; Verma, I.M. NF-κB, an active player in human cancers. *Cancer Immunol. Res.* **2014**, *2*, 823–830. [CrossRef] [PubMed]
113. Zhang, Y.; Leung, D.Y.; Richers, B.N.; Liu, Y.; Remigio, L.K.; Riches, D.W.; Goleva, E. Vitamin D inhibits monocyte/macrophage proinflammatory cytokine production by targeting MAPK phosphatase-1. *J. Immunol.* **2012**, *188*, 2127–2135. [CrossRef] [PubMed]
114. Cohen-Lahav, M.; Shany, S.; Tobvin, D.; Chaimovitz, C.; Douvdevani, A. Vitamin D decreases NF-κB activity by increasing IκBα levels. *Nephrol. Dial. Transplant.* **2006**, *21*, 889–897. [CrossRef] [PubMed]
115. Sun, J.; Mustafi, R.; Cerda, S.; Chumsangsri, A.; Xia, Y.R.; Li, Y.C.; Bissonnette, M. Lithocholic acid down-regulation of NF-κB activity through vitamin D receptor in colonic cancer cells. *J. Steroid. Biochem. Mol. Biol.* **2008**, *111*, 37–40. [CrossRef] [PubMed]
116. Tse, A.K.; Zhu, G.Y.; Wan, C.K.; Shen, X.L.; Yu, Z.L.; Fong, W.F. 1α,25-Dihydroxyvitamin $D_3$ inhibits transcriptional potential of nuclear factor κB in breast cancer cells. *Mol Immunol* **2010**, *47*, 1728–1738. [CrossRef] [PubMed]
117. Holick, M.F. Vitamin D: A millenium perspective. *J. Cell Biochem.* **2003**, *88*, 296–307. [CrossRef] [PubMed]
118. Munger, K.L.; Levin, L.I.; Hollis, B.W.; Howard, N.S.; Ascherio, A. Serum 25-hydroxyvitamin D levels and risk of multiple sclerosis. *JAMA* **2006**, *296*, 2832–2838. [CrossRef] [PubMed]
119. Littorin, B.; Blom, P.; Scholin, A.; Arnqvist, H.J.; Blohme, G.; Bolinder, J.; Ekbom-Schnell, A.; Eriksson, J.W.; Gudbjornsdottir, S.; Nystrom, L.; et al. Lower levels of plasma 25-hydroxyvitamin D among young adults at diagnosis of autoimmune type 1 diabetes compared with control subjects: Results from the nationwide Diabetes Incidence Study in Sweden (DISS). *Diabetologia* **2006**, *49*, 2847–2852. [CrossRef] [PubMed]
120. Merlino, L.A.; Curtis, J.; Mikuls, T.R.; Cerhan, J.R.; Criswell, L.A.; Saag, K.G. Vitamin D intake is inversely associated with rheumatoid arthritis: Results from the Iowa Women's Health Study. *Arthritis. Rheum.* **2004**, *50*, 72–77. [CrossRef] [PubMed]
121. Chen, S.; Sims, G.P.; Chen, X.X.; Gu, Y.Y.; Lipsky, P.E. Modulatory effects of 1,25-dihydroxyvitamin $D_3$ on human B cell differentiation. *J. Immunol.* **2007**, *179*, 1634–1647. [CrossRef] [PubMed]
122. Bittenbring, J.T.; Neumann, F.; Altmann, B.; Achenbach, M.; Reichrath, J.; Ziepert, M.; Geisel, J.; Regitz, E.; Held, G.; Pfreundschuh, M. Vitamin D deficiency impairs rituximab-mediated cellular cytotoxicity and outcome of patients with diffuse large B-cell lymphoma treated with but not without rituximab. *J. Clin. Oncol.* **2014**, *32*, 3242–3248. [CrossRef] [PubMed]

123. Jeffery, L.E.; Burke, F.; Mura, M.; Zheng, Y.; Qureshi, O.S.; Hewison, M.; Walker, L.S.; Lammas, D.A.; Raza, K.; Sansom, D.M. 1,25-Dihydroxyvitamin $D_3$ and IL-2 combine to inhibit T cell production of inflammatory cytokines and promote development of regulatory T cells expressing CTLA-4 and FoxP3. *J. Immunol.* **2009**, *183*, 5458–5467. [CrossRef] [PubMed]
124. Edfeldt, K.; Liu, P.T.; Chun, R.; Fabri, M.; Schenk, M.; Wheelwright, M.; Keegan, C.; Krutzik, S.R.; Adams, J.S.; Hewison, M.; et al. T-cell cytokines differentially control human monocyte antimicrobial responses by regulating vitamin D. metabolism. *Proc. Natl. Acad. Sci. USA* **2010**, *107*, 22593–22598. [CrossRef] [PubMed]
125. Sheikh, V.; Kasapoglu, P.; Zamani, A.; Basiri, Z.; Tahamoli-Roudsari, A.; Alahgholi-Hajibehzad, M. Vitamin D3 inhibits the proliferation of T helper cells, downregulate $CD4^+$ T cell cytokines and upregulate inhibitory markers. *Hum. Immunol.* **2018**, *79*, 439–445. [CrossRef] [PubMed]
126. Mayne, C.G.; Spanier, J.A.; Relland, L.M.; Williams, C.B.; Hayes, C.E. 1,25-Dihydroxyvitamin $D_3$ acts directly on the T lymphocyte vitamin D receptor to inhibit experimental autoimmune encephalomyelitis. *Eur. J. Immunol.* **2011**, *41*, 822–832. [CrossRef] [PubMed]
127. Palmer, M.T.; Lee, Y.K.; Maynard, C.L.; Oliver, J.R.; Bikle, D.D.; Jetten, A.M.; Weaver, C.T. Lineage-specific effects of 1,25-dihydroxyvitamin $D_3$ on the development of effector CD4 T cells. *J. Biol. Chem.* **2011**, *286*, 997–1004. [CrossRef] [PubMed]
128. Sun, D.; Luo, F.; Xing, J.C.; Zhang, F.; Xu, J.Z.; Zhang, Z.H. 1,25$(OH)_2D_3$ inhibited Th17 cells differentiation via regulating the NF-κB activity and expression of IL-17. *Cell Prolif.* **2018**, e12461. [CrossRef] [PubMed]
129. Liu, P.T.; Stenger, S.; Li, H.; Wenzel, L.; Tan, B.H.; Krutzik, S.R.; Ochoa, M.T.; Schauber, J.; Wu, K.; Meinken, C.; et al. Toll-like receptor triggering of a vitamin D-mediated human antimicrobial response. *Science* **2006**, *311*, 1770–1773. [CrossRef] [PubMed]
130. Chen, Y.; Liu, W.; Sun, T.; Huang, Y.; Wang, Y.; Deb, D.K.; Yoon, D.; Kong, J.; Thadhani, R.; Li, Y.C. 1,25-Dihydroxyvitamin D promotes negative feedback regulation of TLR signaling via targeting microRNA-155-SOCS1 in macrophages. *J. Immunol.* **2013**, *190*, 3687–3695. [CrossRef] [PubMed]
131. Sanchez-Nino, M.D.; Bozic, M.; Cordoba-Lanus, E.; Valcheva, P.; Gracia, O.; Ibarz, M.; Fernandez, E.; Navarro-Gonzalez, J.F.; Ortiz, A.; Valdivielso, J.M. Beyond proteinuria: VDR activation reduces renal inflammation in experimental diabetic nephropathy. *Am. J. Physiol. Renal. Physiol.* **2012**, *302*, F647–F657. [CrossRef] [PubMed]
132. Helming, L.; Bose, J.; Ehrchen, J.; Schiebe, S.; Frahm, T.; Geffers, R.; Probst-Kepper, M.; Balling, R.; Lengeling, A. 1α,25-Dihydroxyvitamin $D_3$ is a potent suppressor of interferon γ-mediated macrophage activation. *Blood* **2005**, *106*, 4351–4358. [CrossRef] [PubMed]
133. Steinman, R.M.; Banchereau, J. Taking dendritic cells into medicine. *Nature* **2007**, *449*, 419–426. [CrossRef] [PubMed]
134. Berer, A.; Stockl, J.; Majdic, O.; Wagner, T.; Kollars, M.; Lechner, K.; Geissler, K.; Oehler, L. 1,25-Dihydroxyvitamin $D_3$ inhibits dendritic cell differentiation and maturation in vitro. *Exp. Hematol.* **2000**, *28*, 575–583. [CrossRef]
135. Canning, M.O.; Grotenhuis, K.; de Wit, H.; Ruwhof, C.; Drexhage, H.A. 1-α,25-Dihydroxyvitamin $D_3$ (1,25$(OH)_2D_3$) hampers the maturation of fully active immature dendritic cells from monocytes. *Eur. J. Endocrinol.* **2001**, *145*, 351–357. [CrossRef] [PubMed]
136. Szeles, L.; Keresztes, G.; Torocsik, D.; Balajthy, Z.; Krenacs, L.; Poliska, S.; Steinmeyer, A.; Zuegel, U.; Pruenster, M.; Rot, A.; et al. 1,25-dihydroxyvitamin $D_3$ is an autonomous regulator of the transcriptional changes leading to a tolerogenic dendritic cell phenotype. *J. Immunol.* **2009**, *182*, 2074–2083. [CrossRef] [PubMed]
137. Takeda, M.; Yamashita, T.; Sasaki, N.; Nakajima, K.; Kita, T.; Shinohara, M.; Ishida, T.; Hirata, K. Oral administration of an active form of vitamin $D_3$ (calcitriol) decreases atherosclerosis in mice by inducing regulatory T cells and immature dendritic cells with tolerogenic functions. *Arterioscler. Thromb. Vasc. Biol.* **2010**, *30*, 2495–2503. [CrossRef] [PubMed]
138. Sommer, A.; Fabri, M. Vitamin D regulates cytokine patterns secreted by dendritic cells to promote differentiation of IL-22-producing T cells. *PLoS ONE* **2015**, *10*, e0130395. [CrossRef] [PubMed]

International Journal of
*Molecular Sciences*

MDPI

*Article*

# Calcitriol and Its Analogs Establish the Immunosuppressive Microenvironment That Drives Metastasis in 4T1 Mouse Mammary Gland Cancer

**Agata Pawlik [1], Artur Anisiewicz [1], Beata Filip-Psurska [1], Marcin Nowak [2], Eliza Turlej [1], Justyna Trynda [1], Joanna Banach [1], Paweł Gretkierewicz [1] and Joanna Wietrzyk [1,*]**

1 Department of Experimental Oncology, Hirszfeld Institute of Immunology and Experimental Therapy, Polish Academy of Sciences, 53-114 Wroclaw, Poland; agata.pawlik@iitd.pan.wroc.pl (A.P.); artur.anisiewicz@iitd.pan.wroc.pl (A.A.); filip@iitd.pan.wroc.pl (B.F.-P.); turlej@iitd.pan.wroc.pl (E.T.); justynatrynda@wp.pl (J.T.); joanna.banach@iitd.pan.wroc.pl (J.B.); pgretkierewicz@gmail.com (P.G.)

2 Faculty of Veterinary Medicine, Wroclaw University of Environmental and Life Sciences, 50-3754 Wroclaw, Poland; marcin.nowak@upwr.edu.pl

* Correspondence: wietrzyk@iitd.pan.wroc.pl; Tel.: +48-71-370-99-92

Received: 22 June 2018; Accepted: 18 July 2018; Published: 20 July 2018

**Abstract:** In our previous study, calcitriol and its analogs PRI-2191 and PRI-2205 stimulated 4T1 mouse mammary gland cancer metastasis. Therefore, we aimed to analyze the inflammatory response in 4T1-bearing mice treated with these compounds. Gene expression analysis of the splenocytes and regional lymph nodes demonstrated prevalence of the T helper lymphocytes (Th2) response with an increased activity of regulatory T (Treg) lymphocytes in mice treated with these compounds. We also observed an increased number of mature granulocytes and B lymphocytes and a decreased number of $TCD4^+$, $TCD4^+CD25^+$, and $TCD8^+$, as well as natural killer (NK) $CD335^+$, cells in the blood of mice treated with calcitriol and its analogs. Among the splenocytes, we observed a significant decrease in NK $CD335^+$ cells and an increase in $TCD8^+$ cells. Calcitriol and its analogs decreased the levels of interleukin (IL)-1β and IL-10 and increased the level of interferon gamma (IFN-γ) in the plasma. In the tumor tissue, they caused an increase in the level of IL-10. Gene expression analysis of lung tissue demonstrated an increased level of osteopontin (*Spp1*) and transforming growth factor β (TGF-β) mRNA. The expression of *Spp1* was also elevated in lymph nodes. Calcitriol and its analogs caused prevalence of tumor-conducive changes in the immune system of 4T1 tumor-bearing mice, despite the induction of some tumor-disadvantageous effects.

**Keywords:** calcitriol; vitamin D analogs; breast cancer; metastasis; immunosuppression; osteopontin; TGF-β

## 1. Introduction

Metastasis is a complex and multistage process of primary malignant tumor cells spreading to the secondary organs. To invade the surrounding stroma and disrupt the vascular endothelium, tumor cells initially acquire an invasive phenotype. Within the blood stream, the tumor cells must survive and evade physical damage, as well as evade attacks from the immune system [1,2]. Moreover, malignant cells interact with various stromal host cells at primary and secondary tumor locations. Prolonged inflammation is generally accepted as a potentiating factor in the progression of primary tumor and in metastasis [3–6].

Calcitriol, (1,25-dihydroxyvitamin $D_3$; $1,25(OH)_2D_3$), is a hormonally active vitamin $D_3$ metabolite and an essential factor in maintaining the normal functioning of the immune system [7]. This is because of the fact that almost all immune cells express vitamin D receptor (VDR) and calcitriol is able to modulate its functions [8]. For example, the *Vdr* mRNA is expressed in subsets of the T helper lymphocytes (Th). Although Th1 or naïve T cells express low levels of *Vdr*, in differentiated Th17 and Th2 cells, and to a lesser extent in regulatory T (Treg) cells, the expression of *Vdr* has been found to be increased [9]. In addition, cells of the immune system with the 1-hydroxylase enzyme encoded by *CYP27B1* can generate calcitriol independently, regardless of the normal *CYP27B1* regulation in kidney via parathormone (PTH) [10,11]. Although many of the studies have focused on analyzing the effect of vitamin D on the immune system using in vitro and in vivo methods, there are only a few reports on the effect of vitamin D on the immune system during progression and metastasis of solid tumors. It is important to investigate the effects of vitamin D on immune response in solid tumors in the light of the evidence that calcitriol shows an immunosuppressive effect, which is conflicting with the expected properties of an effective anticancer agent in the case of solid tumors [12]. Therefore, researchers are proposing the selection of clinical indications in which systemic immunomodulatory effects of calcitriol could be minimized during treatment with vitamin D and its analogs, for example, in case of superficial bladder cancer [12]. Intravesicular combined treatment of bladder cancer with bacillus Calmette–Guerin (BCG) and calcitriol improved the therapeutic efficacy of the anticancer immunotherapeutic agent, increasing the production of interleukin (IL)-8, thereby enhancing the innate immune cell recruitment [13]. However, Guo et al., using the model of hepatocellular carcinoma, have shown that by increasing the cyclin dependent kinase inhibitor $p27^{kip1}$, calcitriol reduces the secretion of pro-inflammatory cytokines and consequently inhibits signal transducer and activator of transcription 3 (STAT3) signaling activation, which eventually suppresses tumor development in mice [14]. Recently, in the model of the 4T1 mouse mammary gland cancer transplanted subcutaneously, Cao et al. have shown, that vitamin D stimulated the growth of primary tumors and decreased survival time of mice. The authors correlated this unfavorable effect with decreased Th1 response and increased recruitment of myeloid-derived suppressor cells [15].

Calcitriol, as well as other VDR agonists, are well-known inducers of apoptosis and cell cycle arrest and are inhibitors of metastatic invasion of breast cancer cells, as determined by in vitro methods. Despite the expression of VDR in breast cancer cells, vitamin D deficiency is very common in patients with breast cancer. Some authors suggest that a "correction of vitamin D deficiency or provision of supplemental vitamin D in women living with breast cancer would be predicted to delay recurrence and extend survival" [16–18].

The vitamin D deficiency may promote breast cancer metastasis [19]. Furthermore, the antimetastatic effect of VDR agonists in immune-deficient mice has been previously observed [18]. These findings motivated us to evaluate the antimetastatic activity of calcitriol and its analogs using the 4T1 mouse mammary gland cancer model transplanted orthotopically into immunocompetent mice. Using this model, we performed a series of experiments and reported that calcitriol and its low-toxic analogs PRI-2191 [20] and PRI-2205 [21] stimulated metastasis in 4T1 tumor-bearing mice [22]. The increased metastatic potential of 4T1 cells was accompanied with increased perfusion of blood in tumor tissue. During our previous aforementioned investigations, one of the cytokines, the transforming growth factor-β (TGF-β), was found to be significantly elevated in the plasma of mice treated with vitamin D and its analogs [22]. Therefore, in this study, we aimed to analyze the general inflammatory response of the 4T1 tumor-bearing mice treated with calcitriol, PRI-2191, and PRI-2205 to evaluate the role of immunosuppression as one of the mechanisms of pro-metastatic action of vitamin D and its analogs.

## 2. Results

### 2.1. Phenotypical Analyses of Cells Harvested from Blood

#### Phenotype of Peripheral Blood Lymphocytes

After treatment with calcitriol and its analogs, the $CD4^+$ T lymphocytes were found to be increased on days 14 and 21 (on day 14, $p < 0.05$ for calcitriol; $p = 0.0539$ for PRI-2191). However, further treatment with study compounds caused a significant decrease in the population of CD4$^+$ cells (Figure 1d). We further analyzed the subpopulation of the regulatory T cells (Tregs; CD25$^+$). Beginning from day 21, the percentage of CD25$^+$ cells was found to be decreased in mice treated with calcitriol and its analogs ($p < 0.05$ on days 21 and 28 for PRI-2191 and PRI-2205 or only for PRI-2205, respectively) (Figure 1e).

After treatment with the study compounds, we observed a significant decrease in the percentage of natural killer (NK) cells on day 21 and of T CD8$^+$ cells on day 28 (Figure 1c,f).

On day 14, calcitriol (but not its analogs) significantly decreased the percentage of B cells. All compounds tended to increase the percentage of CD19$^+$ cells on days 21, 28, and 33 ($p < 0.05$ on day 28 for both analogs and on day 33 for PRI-2205) (Figure 1b).

The blood morphological analysis is presented in the Supplementary Material (Figures S1 and S2).

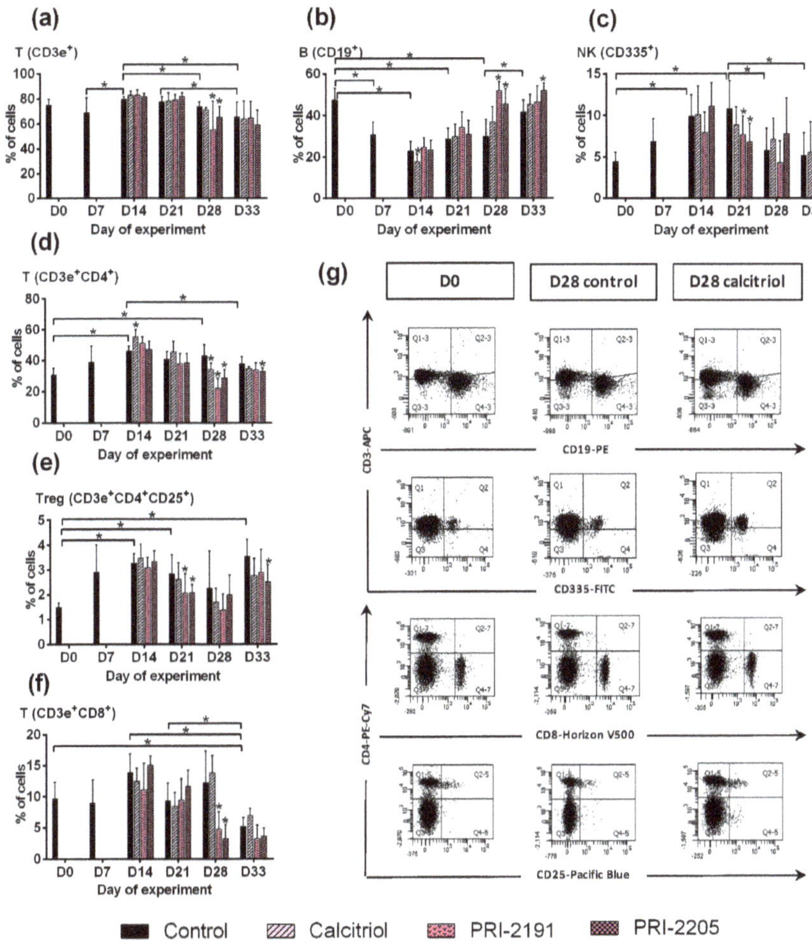


**Figure 1.** The phenotypes of peripheral blood lymphocytes in mice bearing 4T1 mammary gland tumors and treated with calcitriol, PRI-2191, and PRI-2205. (**a**) T lymphocytes CD3e$^+$; (**b**) B lymphocytes CD19$^+$; (**c**) natural killer (NK) cells CD335$^+$; (**d**) TCD4$^+$ lymphocytes; (**e**) TCD4$^+$CD25$^+$ lymphocytes; (**f**) TCD8$^+$ lymphocytes; (**g**) representative dot plots of selected analysis performed on day 28. Data for calcitriol are shown as an example. The number of samples analyzed were 5–7 per group. Data were analyzed using the FACS Diva software. Data are presented as mean ± SD. Statistical analysis: Kruskal–Wallis multiple comparison test. * $p < 0.05$.

## 2.2. Characteristics of Spleen Lymphocytes

### 2.2.1. Phenotype RT[2] Profiler PCR Array ("Mouse T Helper Cell Differentiation") Analysis of Gene Expression

Calcitriol and its analogs significantly modified the expression of genes in splenocytes in our studies. Therefore, we decided to describe only those genes that were upregulated more than 10 times or that were downregulated more than 5 times (Figure 2a).

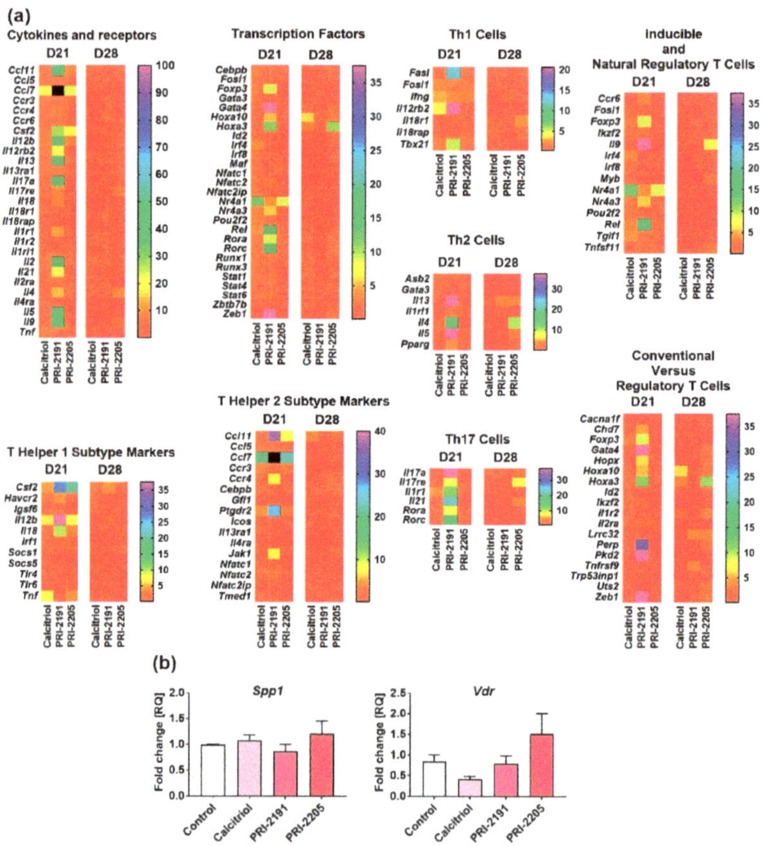


**Figure 2.** The gene expression profile in isolated mononuclear splenocytes from 4T1 tumor-bearing mice treated with calcitriol and its analogs. (**a**) Heat maps presenting mean relative quantification (RQ) values from duplicates. The genes were grouped according to the analysis provided by manufacturer: cytokines and receptors, transcription factors, T helper lymphocytes (Th1) and Th2 subtype markers, and epigenetically regulated genes: conventional versus regulatory T cells, inducible and natural regulatory T cells, Th1, Th2, and Th17 cells. Values outside the scale are marked in black. Detailed data can be found in the Table S1. Fold-change (RQ) of target cDNA was determined using ΔΔCt analysis with reference to beta-actin (*Actb*) and beta-2 microglobulin (*B2m*) and adjusted to the values obtained for the vehicle treated control mice for each treatment group; (**b**) the expression of vitamin D receptor (*Vdr*) and osteopontin (*Spp1*) mRNA levels (day 28). Fold-change (RQ) of target genes was determined using ΔΔCt analysis compared to beta-2 microglobulin (*B2m*) and adjusted to the values obtained for the vehicle treated control mice for each treatment group. Data presentation: mean with standard error of mean. Number of samples analyzed was 2–3 per group. Statistical analysis: Kruskal–Wallis multiple comparison test. * $p < 0.05$.

On day 21, expression of the following genes was found to be upregulated: genes related to Th1, such as colony stimulating factor 2 (granulocyte-macrophage) (*Csf2*) and interleukin 2 (*Il2*); and those related to Th2 cells, such as chemokine (C-C motif) ligand 7 (*Ccl7*), *Ccl11*, prostaglandin D2 receptor 2 (*Ptgdr2*), *Il13*, *Il4*, *Il5*, and *Il9*. Expression of the genes related to Treg cells was also found to be upregulated on day 21: homeobox A3 (*Hoxa3*) and reticuloendotheliosis oncogene, NF-κB subunit (*Rel*). On day 21, the following Th17-related genes were also upregulated (but not as much as the aforementioned genes): interleukin 17a (*Il17a*), interleukin 17 receptor E (*Il17re*), interleukin 1 receptor type 1 (*Il1r1*), *Il21*, RAR-related orphan receptor alpha (*Rora*), and RAR-related orphan receptor gamma (*Rorc*).

On the 28th day of the experiment, the interferon regulatory factors 4 and 8 (*Irf4* and *Irf8*), both related to Treg cells, were the most downregulated genes.

The following groups of genes were upregulated on day 21 and were downregulated on day 28, related to Th1 cells: interleukin 12b (*Il12b*), interleukin 12 receptor, beta 2 (*Il12rb2*), tumor necrosis factor (*Tnf*); related to Th2 cells: ankyrin repeat and SOCS box containing 2 (*Asb2*); and related to Treg cells: nuclear receptor subfamily 4 group A members 1 and 3 (*Nr4a1* and *Nr4a3*), transforming growth factor β (TGF-β)-induced factor homeobox 1 (*Tgif1*) (Figure 2a). On day 21, the expression of *Foxp3* mRNA was found to be upregulated by calcitriol (~3-fold), PRI-2191 (~9-fold), and PRI-2205 (~2-fold), whereas on day 28, calcitriol or its analogs did not change the expression of *Foxp3* (Figure 2a). Table S1 lists all fold change values.

#### 2.2.2. *Vdr* and *Spp1* mRNA Level in Splenocytes

Calcitriol downregulated and PRI-2205 upregulated the expression of *Vdr*. However, there were no statistically significant changes (Figure 2b).

#### 2.2.3. Phenotype of Spleen Lymphocytes

From day 14 to 33, PRI-2205 decreased the percentage of T lymphocytes. Calcitriol, and especially PRI-2191, tended to increase the percentage of these cells (Figure 3a). Among the T lymphocytes, the percentage of $CD4^+$ cells was found to be decreased on day 21 after treatment (Figure 3d). Calcitriol and PRI-2205 decreased the percentage of $CD4^+CD25^+$ cells significantly on day 33 (Figure 3e). All tested compounds decreased the percentage of NK $CD335^+$, which was found to be greater on days 28 and 33 (Figure 3c). A significant increase in $CD8^+$ cells was observed in mice treated with all compounds, especially on day 28 (Figure 3f).

### *2.3. Cytokine Analysis in 4T1 Tumor-Bearing Mice Treated with Calcitriol, PRI-2191, and PRI-2205*

#### 2.3.1. Cytokine Arrays Analysis of Plasma and Supernatants from Lipopolysaccharide (LPS)- or Concanavalin A (ConA)-Stimulated Splenocytes

Plasma cytokines or cytokines secreted into the culture medium by stimulated splenocytes were screened. Spleens were harvested from healthy mice (indicated as D0) and from tumor-bearing mice on days 7, 21, and 33 after transplantation with cancer cells (D7).

On day 21, almost all cytokines were found to be downregulated in plasma by calcitriol and its analogs. However, on day 33, some of the cytokines were found to be upregulated, such as eotaxin, IL-1α, IL-10, IL-13, IL-12p70, IL-23, monocyte chemotactic protein 5 (MCP-5), macrophage-induced gene (MIG), and macrophage inflammatory protein (MIP)-1α (Figure 4a and Table S2A–E).

Calcitriol and its analogs modulated the levels of cytokines secreted from LPS-stimulated splenocytes. On day 21, granulocyte-colony stimulating factor (G-CSF), granulocyte-macrophage (GM)-CSF, IL-4, IL-6, and IL-10 were found to be increased by calcitriol and its analogs, whereas on day 33, their level was found to be decreased as compared with control tumor-bearing mice. However, IL-2 was found to be decreased on both days 21 and 33 by calcitriol and its analogs (Figure 4b and Table S3A–E).

In ConA-stimulated splenocytes (day 21), calcitriol and its analogs stimulated the secretion of GM-CSF, soluble intercellular adhesion molecule-1 (sICAM-1), IFN-$\gamma$, IL-1$\alpha$ and $\beta$, IL-3, IL-4, IL-7, IL-17, IL-27, macrophage (M)-CSF, stromal cell-derived factor 1 (SDF-1), tissue inhibitor of metalloproteinases (TIMP)-1, tumor necrosis factor (TNF$\alpha$), and triggering receptor expressed on myeloid cells (TREM)-1 into the culture media. The secretion of IL-12p70, MIP-1$\beta$, and MIP-2 was downregulated by all study compounds, whereas other cytokines remain unchanged. On day 33, almost all cytokines were found to be decreased in ConA-stimulated splenocytes in case of mice treated with calcitriol and its analogs (Figure 4b and Table S4A–E).

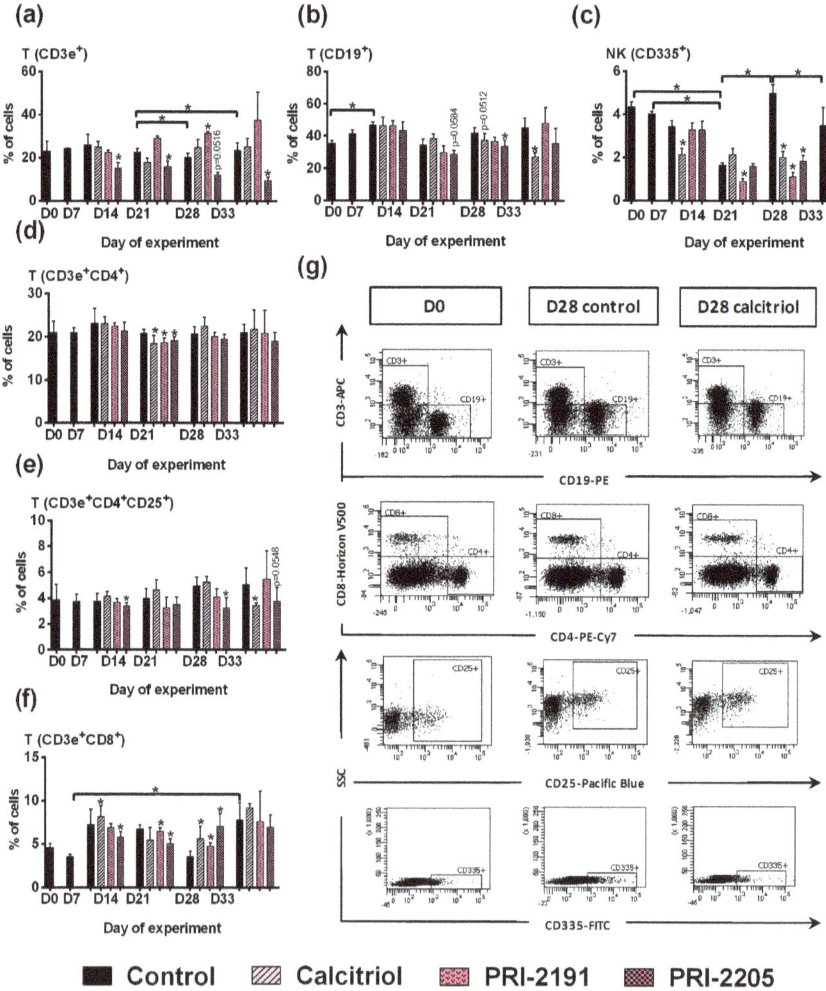


**Figure 3.** Spleen lymphocyte phenotypes in mice bearing 4T1 mammary gland tumors treated with calcitriol, PRI-2191, and PRI-2205. (**a**) T lymphocytes CD3e$^+$; (**b**) B lymphocytes CD19$^+$; (**c**) NK cells CD335$^+$; (**d**) TCD4$^+$ lymphocytes; (**e**) TCD4$^+$CD25$^+$ lymphocytes; (**f**) TCD8$^+$ lymphocytes; (**g**) representative dot plots of selected analysis performed on day 28. Data for calcitriol are shown as an example. Number of samples analyzed was six per group with the following exception: D0 = 2 and D7 = 3. Data were analyzed using the FACS Diva software. Data are presented as mean ± SD. Statistical analysis: Kruskal–Wallis multiple comparison test. * $p < 0.05$.

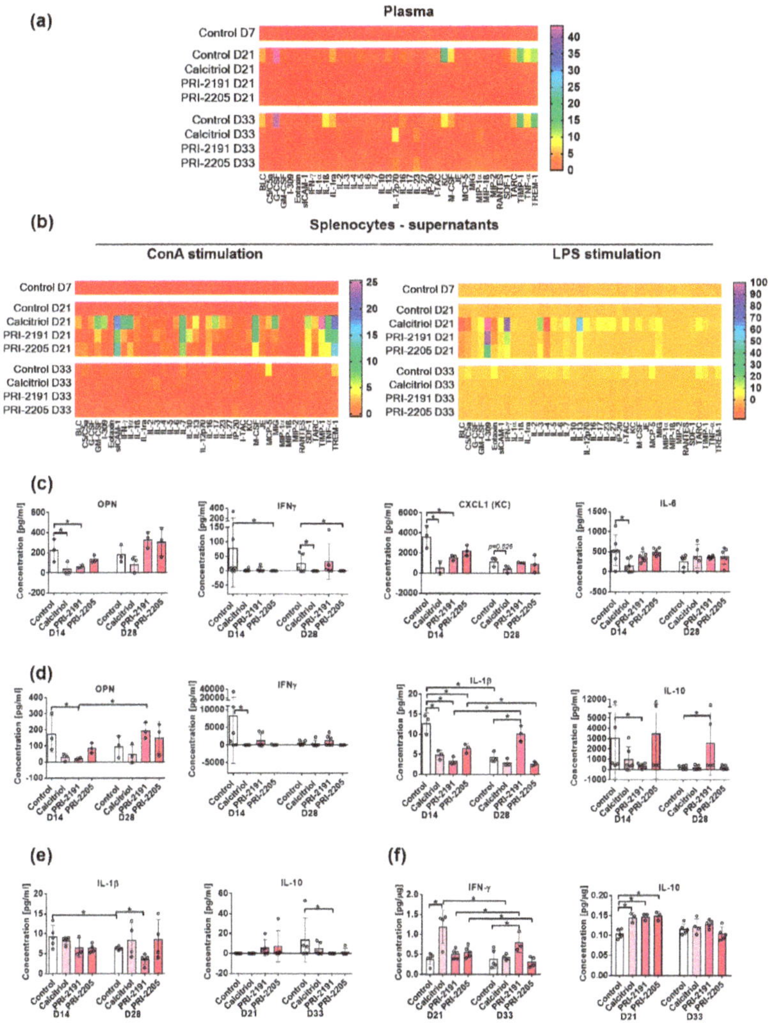


**Figure 4.** Analysis of cytokine arrays of plasma or supernatants from lipopolysaccharide (LPS)-/Concanavalin A (ConA)-stimulated splenocytes and expression of selected cytokines in tumor tissue. (**a**,**e**) Plasma; (**b**–**d**) supernatants obtained from spleen cells stimulated with LPS and ConA; (**f**) tumor tissue lysates. (**a**,**b**)—samples analyzed with proteome profiler array. Samples from three randomly selected mice were pooled in each tested group. Data are presented as fold change as compared with healthy (D0) mice samples; (**c**,**d**) supernatants obtained from spleen cells (harvested on days 14 and 28) stimulated with (**c**) LPS and (**d**) ConA; (**e**) plasma from mice harvested on days 14 or 21 and 28 or 33; (**f**) tumors harvested on days 21 and 33, homogenized and lysed. (**c**–**f**) samples were analyzed with ELISA tests. Number of samples analyzed were 2–6 per group. Data are presented as mean ± SD and individual sample results. Statistical analysis: Kruskal–Wallis multiple comparison test. * $p < 0.05$. OPN—osteopontin; IFN-γ—interferon gamma; IL—interleukin; CXCL-1—chemokine (C–X–C motif) ligand 1; G-CSF—granulocyte-colony stimulating factor; GM—granulocyte-macrophage; sICAM-1—soluble intercellular adhesion molecule-1; M-CSF—macrophage colony-stimulating factor; MCP-5—monocyte chemotactic protein 5; MIG—macrophage-induced gene; MIP—macrophage inflammatory protein; SDF-1—stromal cell-derived factor 1; TIMP-1—stromal cell-derived factor 1; TNF—tumor necrosis factor; TREM—triggering receptor expressed on myeloid cells.

#### 2.3.2. Plasma Levels of Selected Cytokines Measured by ELISA Test

The plasma concentration of IL-1β was found to be decreased by calcitriol and its analogs, especially by PRI-2191 ($p < 0.05$ on day 28). The plasma concentration of IL-10 in control mice was found to be elevated from a non-detectable level on day 21 to a detectable level on day 33 in three out of five mice. All study compounds tended to decrease the level of IL-10 on day 33 ($p < 0.05$ for PRI-2191-treated mice), whereas both analogs tended to increase its level on day 21 (Figure 4e). On days 21 and 28, IL-2 and IL-4 were found to be undetectable in almost all mice with the exception of control animals on day 28 (IL-2) and day 21 (IL-4). Calcitriol and its analogs did not significantly affect the plasma level of IFN-γ and IL-6 (Figure S3C).

#### 2.3.3. Levels of Selected Cytokines in Supernatants from Stimulated Splenocytes

LPS- or ConA-stimulated splenocytes produced low level of TGF-β (Figure S3). CXCL-1 chemokine secretion was inhibited by calcitriol and its analogs in LPS- and ConA-stimulated splenocytes (Figure 4c,d). The level of OPN in the supernatants was found to be decreased by calcitriol and its analogs (both in LPS- and ConA-stimulated splenocytes harvested from mice on day 14). On day 28, only calcitriol tended to decrease the level of osteopontin (OPN), but both analogs tended to increase its level (Figure 4c,d). IL-1β secretion was found to be decreased in LPS- and ConA-stimulated splenocytes collected on day 14 ($p < 0.05$), whereas in mice with more advanced tumors (day 28), the secretion of IL-1β did not change in mice treated with calcitriol or PRI-2205 as compared with control mice, but was found to be increased by PRI-2191 (Figure 4d and Figure S3A). IL-2 and IL-4 were not found to be significantly affected by calcitriol and its analogs (Figure S3A,B). The level of IL-6 from LPS-stimulated splenocytes remained unchanged by PRI-2191 and PRI-2205; however, calcitriol decreased its level on day 14 ($p < 0.05$). Calcitriol showed a similar tendency of decreasing the level of IL-6 on day 14 after ConA stimulation (Figure 4c and Figure S3B). On day 14, the level of IL-10 was found to be decreased by calcitriol and its analogs ($p < 0.05$ for PRI-2191 ConA stimulation). On day 28, PRI-2191 significantly increased the secretion of this cytokine in ConA-stimulated splenocytes (Figure 4d and Figure S3A). In ConA-stimulated splenocytes, IFN-γ secretion diminished in mice treated with calcitriol and its analog, especially on day 14, but in LPS-stimulated splenocytes, the same tendency was observed on day 28 (Figure 4c,d).

### *2.4. Immunological Response in Lymph Nodes during 4T1 Tumor Progression in Mice Treated with Calcitriol, PRI-2191, and PRI-2205*

#### $RT^2$ Profiler PCR Array ("Mouse T Helper Cell Differentiation") Analysis of Gene Expression

Calcitriol and its analogs upregulated the expression of some genes in lymph nodes, such as genes related to Th2 cells by two- to eight-fold: ankyrin repeat and SOCS box containing 2 (*Asb2*), chemokine (C-C motif) ligand 7 (*Ccl7*), CCAAT/enhancer binding protein (C/EBP), beta (*Cebpb*), *Il13*, and *Il4*. The expression of the following genes related to Treg cells was also elevated: calcium voltage-gated channel subunit alpha1 F (*Cacna1f*), FOS like 1 (*Fosl*), GATA binding protein 4 (*Gata4*), urotensin 2 (*Uts2*), and zinc finger E-box binding homeobox 1 (*Zeb1*). An approximately two-fold increase in the expression of *Foxp3* was also observed.

On day 14, we identified some of the genes that were downregulated by calcitriol and its analogs. The following genes related to Th1 cells were downregulated: tumor necrosis factor (*Tnf*) and T-box 21 (*Tbx21*), and those related to Treg cells, such as nuclear receptor subfamily 4 group A member 3 (*Nr4a3*). Moreover, a single gene, chemokine (C–C motif) receptor 3 (*Ccr3*), was found to be downregulated only on day 28 (related to Th2). On days 14 and 28, a single gene, *Il17a*, was found to be downregulated by calcitriol and its analogs (related to Th17). *Hoxa10*, *Csf2*, and *Il9* were found to be downregulated on day 14 and upregulated on day 28 (Figure 5a). Table S5 shows all fold change values.

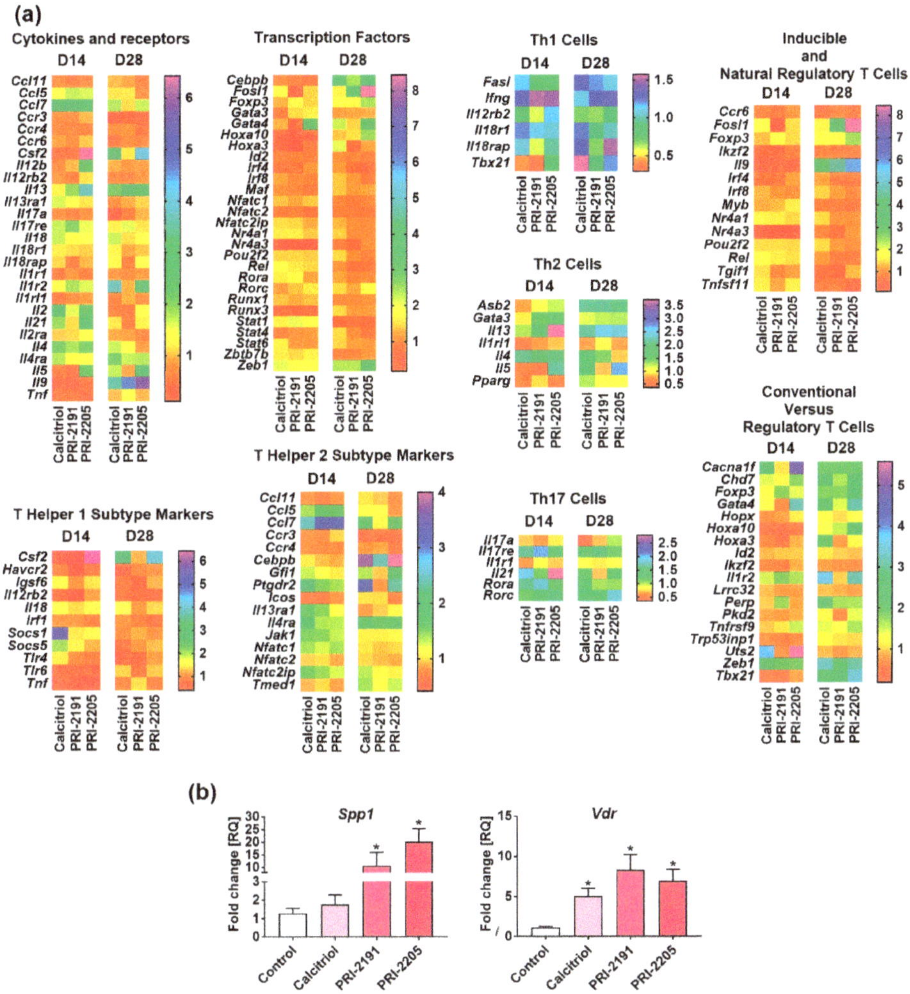


**Figure 5.** The gene expression in regional lymph nodes from 4T1 tumor bearing mice treated with calcitriol and its analogs. (**a**) Heat maps presenting mean relative quantification (RQ) values from duplicates. The genes were grouped according to the analysis provided by the manufacturer: cytokines and receptors, transcription factors, Th1 and Th2 subtype markers, and epigenetically regulated genes: conventional versus regulatory T cells, inducible and natural regulatory T cells, Th1, Th2, and Th17 cells. Lymph node specimens were collected on the days 21 and 28 (after inoculation with 4T1 cells). Fold-change (RQ) of target cDNA was determined using $\Delta\Delta$Ct analysis with reference to actin, beta (*Actb*), and beta-2 microglobulin (*B2m*) and adjusted to the values obtained for the vehicle treated control mice for each treatment group; (**b**) the expression of vitamin D receptor (*Vdr*) and osteopontin (*Spp1*) mRNA levels (day 28). Data presentation: mean with standard error of mean. Fold-change (RQ) of target genes was determined using $\Delta\Delta$Ct method compared with beta-2 microglobulin (*B2m*) and adjusted to the values obtained for the vehicle treated control mice for each treatment group. Number of samples analyzed were 4–6 per group. Statistical analysis: Kruskal–Wallis multiple comparison test. * $p < 0.05$.

### 2.5. Vitamin D Receptor (Vdr) and Osteopontin (Spp1) mRNA Levels in Lymph Nodes

We found a statistically significant increase of *Vdr* in mice treated with all the study compounds. PRI-2191 and PRI-2205 significantly increased the level of *Spp1* in lymph nodes. Calcitriol showed a similar tendency, but the increase was not significant (Figure 5b).

### 2.6. Immune Response of Tumor and Lung Tissue from 4T1 Mammary Gland Tumor-Bearing Mice Treated with Calcitriol, PRI-2191, and PRI-2205

#### 2.6.1. Level of Selected Cytokines in Tumor Tissue

Calcitriol (day 21) and PRI-2191 (day 33) significantly increased the level of INF-γ (Figure 4f). All study compounds decreased the level of IL-1β on day 28, but only PRI-2191 decreased its level significantly (Figure S4a). IL-5 and IL-6 were not found to be affected significantly by the study compounds (Figure S4b,c). However, on day 21, IL-10 was found to be significantly elevated by calcitriol and its analogs (Figure 4g).

#### 2.6.2. Gene Expression Profile in Lung Tissue

Calcitriol and its analogs upregulated the expression of the following genes on day 14 and downregulated their expression on day 28: angiopoietin 1 (*Angpt1*), vascular endothelial growth factor receptor 1 (*Flt1*), transforming growth factor β1 (*Tgfb1*), connective tissue growth factor (*Ctgf*), vascular endothelial growth factor A (*Vegfa*), and platelet-derived growth factor α (*Pdgfa*) (Figure 6a). Calcitriol and PRI-2191 showed a similar profile in the regulation of gene expression for most of the analyzed mRNAs. PRI-2205 also showed a similar tendency, but not as clear as the other two compounds. The expression of some genes differed from the above mentioned pattern. For example, upregulation of the following mRNA levels was observed on both days of analysis (day 14 and 28): cystatin F (leukocystatin; *Cst7*), glucose phosphate isomerase (*Gpi*), and secreted phosphoprotein 1 (osteopontin (OPN), *Spp1*). Moreover, only calcitriol upregulated neurofibromin 2 (merlin; *Nf2*) and TIMP metallopeptidase inhibitor 2 (*Timp2*), whereas only PRI-2205 upregulated methionyl aminopeptidase 2 (*Metap2*) expression (Figure 6a and Table S6).

We performed a real-time PCR analysis of *Spp1* and *Tgfb* in samples of lung tissue. The expression of *Spp1* and *Tgfb* was found to be significantly upregulated on day 28 by study compounds (Figure 6b). On day 14, calcitriol decreased the expression of *Spp1*, whereas PRI-2205 increased the expression of *Tgfb1* (Figure 6b).

#### 2.6.3. Granulocytes Lung Tissue Infiltration

Granulocytes were found to infiltrate the lung tissue of 4T1-tumor bearing mice. Figure 6c shows a photograph of a lung tissue with an evident metastatic lesion (higher magnification of the tissue from control mice: Figure 6e(i)). On the border of the metastatic foci, the nuclei of the cancer cells have been marked with red arrows and the nuclei of the granulocytes with yellow arrows. Regardless of the treatment, infiltration of granulocytes was similar in mice in all groups (Figure 6d,e).

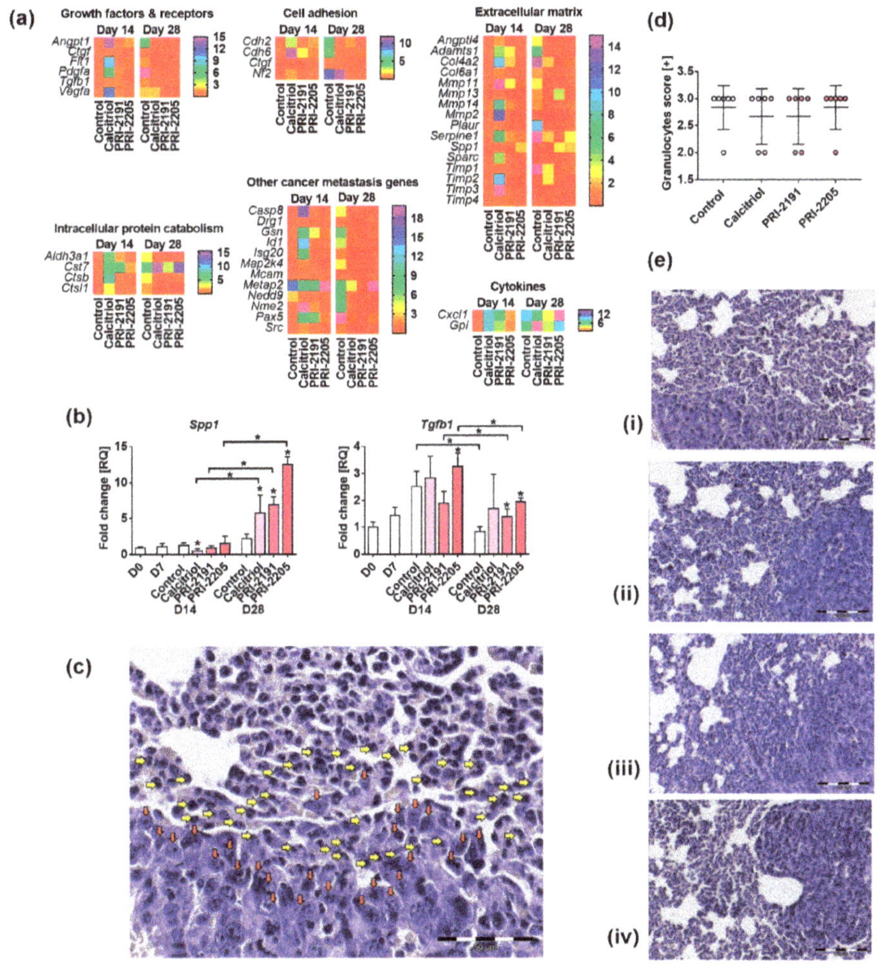


**Figure 6.** Changes in mRNA expression and granulocyte infiltration in lung from mice treated with calcitriol and its analogs. (**a**) Screening of genes correlated with tumor invasion and metastasis using real-time PCR; (**b**) real-time PCR analysis of two selected genes; (**c**–**e**) lung infiltration by granulocytes: (**c**) representative picture of lung tissue with metastatic focus (magnification of 400×, scale bar 50 μm) infiltrated by granulocytes. On the margin of tumor tissue: red arrows—nuclei of tumor cells, yellow arrows—nuclei of granulocytes; (**d**) scoring of granulocytes infiltration in lung tissue; (**e**) representative pictures of lung tissue sections (magnification of 200×, scale bar 100 μm): (**i**) control, (**ii**) calcitriol, (**iii**) PRI-2191, and (**iv**) PRI-2205. Number of mice evaluated were 3–7 per group. Data presentation: (**a**) heat maps presenting mean relative quantification (RQ) values from duplicates. Fold-change (RQ) of target cDNA was determined by calculating the differences in ΔΔCT values with reference to ribosomal protein L13A (*Rpl13a*) and adjusted to the values obtained for the untreated mice (named D0) for each treatment group; (**b**,**d**) mean ± SD. Statistical analysis: Kruskal–Wallis multiple comparison test, * $p < 0.05$.

## 3. Discussion

The hormonally active form of vitamin $D_3$ is known to regulate calcium and phosphorus homeostasis. However, it might also cause toxicity because of the application of potentially active, hyper-physiological doses in cancer treatment, which motivated the researchers to synthesize new analogs, with a split calcemic and antiproliferative activity [23]. One of such compounds synthesized is the vitamin $D_3$ metabolite, (24*R*)-1,24-dihydroxyvitamin $D_3$ (tacalcitol, 1,24$(OH)_2D_3$, PRI-2191). It is well documented that PRI-2191 inhibits cancer cell proliferation by inducing the differentiation of epidermal mouse and human keratinocytes in the case of psoriasis, and leukemic cells in the case of acute myeloid leukemia [24,25]. PRI-2191 also binds to the VDR with an affinity greater than or similar to that of calcitriol, but shows weaker induction of calcemia than that of calcitriol [20,26]. The analogs of vitamin D with a reversed (5*E*,7*E*) geometry of the triene system have been previously reported by our team [21,27] and by another group [28] to show the enhanced antitumor activity as compared with the natural (5*Z*,7*E*) vitamin D compounds. An example of such analog is the PRI-2205, deprived of calcemic activity [21]. The mechanism of action of PRI-2191 in cancer cells has been shown to be similar to calcitriol and is related to the induction of cancer cell differentiation, for example, by increasing the expression of E-cadherin in colon cancer cells [26,29]. However, the mechanism of action of PRI-2205 seems to be different; it induces apoptosis in leukemic cells [21] and enhances the antitumor and antimetastatic activity of 5-fluorouracil (5-FU) and capecitabine in colon cancer in vivo via mechanisms that are not related to the process of cell differentiation [29]. In our previous study, PRI-2205, similar to calcitriol and PRI-2191, induced metastatic spread of the 4T1 mammary gland cancer transplanted orthotopically (the number of metastases increased by 58%, 50%, and 54% over control, respectively, on day 33, Table S7) [22], but the activity of PRI-2205 toward immune response presented herein seems to be different in some aspects. Therefore, in the subsequent paragraphs, we commonly discuss the results obtained for calcitriol, PRI-2191, and PRI-2205, only when the direction of their activity is the same, without analysis of different results for PRI-2205.

According to the literature, calcitriol and its analogs slow down the growth of human [30–32] or mouse [32,33] breast cancer. Moreover, a vitamin D-deficient diet accelerated tumor growth [19] and bone metastasis [34] of transplanted mouse MMTV-*Wnt1* or human MDA-MB-231 breast cancer cells, respectively. However, the exogenous 25-hydroxyvitamin D (25(OH)D) delays spontaneous mammary gland neoplasia [35]. In our previous studies using the model of human and mouse colon cancer, we observed that PRI-2191, when used with 5-fluorouracil, decreased primary tumor growth and the number of lymph node metastases [26,29]. Furthermore, we have previously observed decreased tumor growth in the lung A549 xenograft when PRI-2191 was combined with cytostatic and tyrosine kinases inhibitors [36,37]. However, the cancer cells used in the aforementioned studies, such as MDA-MB-231 [34], MCF-7 [32], MMTV-*PyMT* [35], MMTV-*Wnt1* [19], and HT-29 [26], were sensitive to the direct action of vitamin D and/or immune deficient models were used [26,34,36,37]. Although 4T1 cells express VDR, they are not sensitive to calcitriol in vitro nor in vivo; cell proliferation and primary tumor growth was not affected upon the treatment [22]. Zhang et al. reported contrasting results compared with ours using the same experimental model (4T1); they reported a reduction in the number of lung metastases in 4T1-tumor bearing mice after treatment with calcitriol [38]. On the other hand, Cao et al., using 4T1 cells transplantated subcutaneously (s.c.), report on stimulation of primary tumor growth and reduced survival of mice upon vitamin D treatment. It seems that such divergent observations may be the result of a different scheme of the experiment. For example, Zhang et al. administered calcitriol intraperitoneally one day prior to the inoculation of cancer cells. In case of our study or the one conducted by Cao et al., administration of calcitriol was started 7 or 17 days, respectively, after mice inoculation with the 4T1 cells. However, similar to our results, Zhang et al. also did not notice any significant influence of calcitriol on in situ tumor growth. In our previous studies, we have shown that PRI-2205 did not affect the growth of the primary tumor. However, it increased the number of lung metastases in the 4T1 model, whereas in combination with cyclophosphamide, it significantly increased its anti-cancer properties [39].

Based on our research and the results of other aforementioned studies, we can conclude that it is important to study the effect of calcitriol and its analogs, not only in mice bearing cancers that are sensitive to direct action of these compounds, but also in non-sensitive tumors, where the influence of these compounds may play an important role on the entire host organism, including tumor stroma and the immune system. Interestingly, the modulating role of calcitriol and its analogs may play different roles at different stages of tumor progression. Thus, we can assume that the use of calcitriol or its analogs before tumor implantation [38] or after the administration of immunosuppressive cytostatic cyclophosphamide [39]—therefore, in the body in which the inflammatory process accompanying the development of 4T1 did not start or was suppressed—will have a different effect than the results reported here. The results of this study have shown that calcitriol and its analogs stimulated inflammatory response (increased mRNA level of acute phase proteins in the liver, Figure S5), which is characteristic during the progression of 4T1 tumor [40]. In addition, 4T1 tumor development is accompanied by gradually progressing leukocytosis correlated with splenomegalia [40,41]. Calcitriol and its analogs increased the percentage of lymphocytes and monocytes, whereas they decreased the percentage of granulocytes (Figure S1). This encouraged us to conduct further analysis with respect to the immune modulation in 4T1 tumor-bearing mice after the administration of calcitriol and its analogs.

To evaluate the local and whole body immune response in 4T1 tumor-bearing mice after administration of calcitriol and its analogs, we performed the selected mRNA profiling of splenocytes and regional lymph nodes, as well as cytokine arrays of plasma and ex vivo-stimulated splenocytes. The most upregulated genes by calcitriol and its analogs were the typical ones or those related to Th2 and Treg cells. It is known that these cells are engaged in tumor-conducive immunosuppressive response [42]. In particular, in splenocytes and/or lymph nodes, we have observed an increased expression of mRNA of Th2 signature cytokines [43], such as *Il4*, *Il5*, *Il9*, or *Il13*. However, in splenocytes alone, calcitriol increased the level of *Tnf*, *Il12b*, and *Il12rb2* at an early stage of tumor progression and decreased their levels during the advanced phase of the experiment. Moreover, calcitriol stably increased the expression of *Il2* in splenocytes. The Treg development from naïve periphery $CD4^+$ T lymphocytes (pTreg) is triggered by the combination of cytokines, namely IL-2 and TGF-β (in addition to the antigen stimulation), but signaling of both cytokines is important in pTreg, as well as thymus derived Treg (tTreg), differentiation [44]. In the splenocytes, this activity of calcitriol and its analogs lead to the upregulation of *Rel*, an important gene in Treg differentiation. It is one of the subunits of NF-kB, and it has been shown to be essential in the differentiation and activation of Treg, as well as during cancer progression [45]. In addition, it has been demonstrated that c-Rel (protein encoded by *Rel*) is responsible for the development of tTreg cells, but not for peripherally induced pTregs [46,47]. Other transcription factors responsible for the development of tTregs are members of the NR4a family [44]. Interestingly, our results have shown that mRNA levels of both *Nr4a1* and *Nr4a3* were increased in spleen during the early stage of tumor progression and were decreased during advanced stages. These transcription factors are known as strong inducers of *Foxp3* [48]. In our studies, the expression of *Foxp3* was found to be elevated (up to 8-fold) in splenocytes from mice treated with calcitriol and its analogs, but only during the early stage of tumor growth (day 21). On day 28, the expression level of *Foxp3* was found to be similar to control tumor-bearing mice. Interestingly, transient increase in the levels of *Il12b* and *Il12rb2* in splenocytes from mice treated with calcitriol and its analogs can contribute to the lack of effect on the *Foxp3* expression level during the later stage of tumor progression. The IL-12, with its receptors, can participate in the conversion of Tregs into the Th1-like cells, as well as in maintaining Th1/Th2 hybrids. This process is known as the T cells plasticity [49,50]. However, the metastatic dissemination process in 4T1 tumor-bearing mice started about 1–2 weeks after the transplantation of tumor cells [41] and immunosuppressive cells can promote metastasis during all stages of metastatic cascade [51]. As it has been proven, cancer immunoediting is a dynamic process (consisting of the following phases: elimination, equilibrium, and escape) and during particular stages of carcinogenesis, the immune system can either destroy or promote tumor growth [52]. Therefore, different response of the immune cells to calcitriol and its analogs at the beginning and at the end of the experiment may be a reflection of the influence of growing tumors (not inhibited by the treatment) altering the inflammatory response.

Previous studies have revealed that the secretion of IFN-$\gamma$ and IL-2 by T cells decreases during exposure to calcitriol, whereas the production of IL-4, IL-5, and IL-10 increases, thereby resulting in a shift toward Th2 cells' response [53–55]. Although our gene profiling studies clearly indicated the shift in the immune response toward tumor promotion, mainly at earlier stages of the tumor progression, the cytokine profile in plasma have shown the downregulation of almost all cytokines by calcitriol and its analogs. However, on day 21, the stimulated splenocytes from mice treated with calcitriol and its analogs produced higher levels of cytokines. On day 33, the production of these cytokines was found to be decreased compared with control 4T1 tumor-bearing mice. The influence of calcitriol toward the secretion of cytokines, for example IL-10, may differ in various experimental conditions. Several studies have shown that the expression of IL-10 is induced by calcitriol in various immune cells, and in this way, the known immunosuppressive effect of calcitriol is mediated [54,56]. However, the inhibitory effect of calcitriol on the IL-10 production in vitro, as well as in vivo, has also been reported [57]. This contradictory result might be due to the different cell types studied. Moreover, we hypothesize that the upregulation of IL-10 is related to the duration of the treatment with calcitriol. Indeed, the treatment with calcitriol for several days consistently showed an upregulation of IL-10, whereas there was no regulation or inhibition of IL-10 during shorter incubation times used [58]. In this study, we observed a downregulation of IL-10 secretion by splenocytes during the early stages of tumor progression (day 14), which subsequently increased. In addition, in tumor tissue, IL-10 was found to be upregulated by calcitriol and its analogs during the early stages of tumor growth. Moreover, when we calculated the IL-10/IL-12p70 ratio, which can represent the balance between Th1/Th2 responses [59], it turned out that on day 33, splenocytes from mice treated with calcitriol or its analogs, and stimulated with LPS, produced 18–21 times higher levels of IL-10 over IL-12p17. Although the elevated level of IL-10 over IL-12 suggests the predomination of Th2 response during the treatment, further studies are necessary to exactly determine the type of cells that adopts such a behavior in response to treatment with calcitriol and its analogs during tumor progression.

The growth of 4T1 cells is accompanied by the elevated secretion of cytokines such as G-CSF and chemokine (C–X–C motif) ligand 1 (CXCL1, KC) [40,41], but calcitriol and its analogs only significantly diminished CXCL1 during the early stage of tumor progression, not affecting the secretion of G-CSF in a significant manner. High amounts of G-CSF, the cytokine that is able to switch the T cell cytokine secretion profile to Th2 responses, promote the regulatory T cell and modulate cytokine production [60], and may veil the activity of calcitriol in terms of secretion of various other cytokines. Moreover, the G-CSF enhances proliferation and mobilization of $Ly6G^{+}Ly6C^{+}$ granulocytes and facilitates their homing during metastasis of the target organs even before the onset of cancer cells, which facilitate metastatic process [61]. Unfortunately, calcitriol and its analogs increased the percent of $Ly6G\text{-}6C^{+}SCC^{high}$ cells (with a parallel decrease in the percentage of activated $CD54^{+}$ (I-CAM) and $CD184^{+}$ (CXCR4) granulocytes, Figure S6) in the blood, which correlated with an increased metastatic potential [22]. The $Ly6G\text{-}6C^{+}SCC^{high}$ granulocytes may serve as the primary source of TGF-$\beta$ [62], which was found to be increased in plasma of 4T1 tumor-bearing mice administered with calcitriol and its analogs in our previous study [22]. The increased percentage of mature granulocytes and the production of cytokines such as TGF-$\beta$ from these cells consequent to the treatment with calcitriol and its analogs may be responsible for lung metastatic niche formation and the enhancement of metastasis of 4T1 cells [61,63].

Another cytokine that was found to be elevated as a result of calcitriol treatment in 4T1 tumor tissue was OPN [22]. OPN is known to contain vitamin D response element (VDRE) in the promoter region of *Spp1* and calcitriol stimulates various cells to secrete OPN [64–66]. OPN stimulates tumor growth and promotes metastasis by influencing tumor angiogenesis [67–69]. Unexpectedly, we found that the stimulated splenocytes from calcitriol- or PRI-2191-treated mice secreted lower levels of OPN than that from control mice, especially during the early stage of the tumor progression. We did not find any literature data indicating inhibitory effects of calcitriol on OPN secretion. One of the possible explanations may be the decreased number of activated T and B cells upon calcitriol treatment, which has been described in the literature [70,71], and only activated T and B cells can produce OPN [72]. The elevated level of OPN in tumor tissue, observed in our previous studies upon treatment with calcitriol and its analogs [22], might also be the result of fibroblast stimulation, which is the major source of OPN in tumor tissue [73,74]. The almost unchanged plasma level of OPN in treated mice [22] is the result of applying such opposing effects of treatment. However, further studies should be performed to explain the effect of calcitriol on the secretion of OPN by T and B lymphocytes and other cells of the immune system in tumor-bearing organisms.

In the OPN knockout mice (*Spp1*−/−), the decreased frequency of Treg cells was found to be related with increased $CD4^+$-activated T cells ($CD44^+CD69^+$ and $CD62L^{low}CD69^+$ cells) in metastatic niche (lungs), indicating OPN as the cytokine responsible for drawing the immunosuppression in the microenvironment of the metastatic niche [69]. In addition, TGF-β itself has a significant effect on $CD4^+$ T-cells. The TGF-β induces the generation of Foxp3 and Treg cells [75]. TGF-β also inhibits the proliferation and functioning of NK-cells. The NK-cells are modulated in part by $CD4^+CD25^+$ regulatory T cells that are known from TGF-β high level production [75]. Our results have shown that the number of $CD4^+$ T cells, $CD4^+CD25^+$ regulatory T cells, $CD335^+$ NK, and $CD8^+$ cells diminished in blood after treatment with calcitriol and its analogs. However, in the spleen, only NK and $CD4^+$ cells were found to be decreased, whereas $CD8^+$ cells were found to be increased, which was clearly visible on day 28. Interestingly, in regional lymph nodes, the number of $CD8^+$ cells was found to be significantly increased by the treatment with calcitriol (from 21–33 day of experiment) (Figure S7F). Moreover, in regional lymph nodes, the level of *Vdr* and immunosuppressive *Spp1* was found to be increased significantly after treatment with calcitriol and its analogs. In the tumor tissue, we also observed changed expressions of cytokines, considered to be involved in antitumor immune response; IFN-γ enhanced secretion, but decreased IL-1β expression after treatment with calcitriol and its analogs. However, the IL-10, involved in the tumor-induced immunosuppression, was found to be increased after treatment with calcitriol and its analogs. These results, taken together with the results from our previous study, which showed an increased level of OPN and decreased level of TGF-β in tumor tissue after treatment [22], revealed a lot of contradictory processes induced by calcitriol and its analogs, giving the negative final result in our experimental model.

In the lung tissue, in correlation with increased *Spp1* and *Tgfb* levels (in this study), we detected an increased lung metastatic foci formation [22]. Therefore, upregulated level of both *Spp1* and *Tgfb* by calcitriol and its analogs treatment in lung tissue, as well as *Spp1* in lymph nodes, are responsible for metastatic immunosuppressive niche formation and facilitation of metastasis. However, the cellular source of an increased OPN and TGF-β level should be analyzed in further studies. We demonstrated that granulocytes (neutrophils), which are the primary cells infiltrating lung tissue in 4T1 tumor-bearing mice [41], were found to be in the same number in the lung of 4T1 tumor-bearing mice after treatment with calcitriol and its analogs as in control mice. However, we observed a diminished percentage of I-CAM ($CD54^+$) and CXCR4 ($CD184^+$) expressing neutrophils in blood (Figure S6), which suggests their diminished antitumor potential [76]. Although the 4T1 cells were not sensitive to the treatment with calcitriol and its analogs in vitro [22], we cannot exclude these cells as the tumor-derived source of OPN and TGF-β in in vivo studies after indirect stimulation by other cells/cytokines in the body. A detailed characteristic of lung infiltrating cells is necessary to indicate the cells responsible for the increased level of pro-metastatic cytokines in this tissue.

## 4. Materials and Methods

### 4.1. Plasma, Tissues, and Cells

We harvested the cells, plasma, and tissues from 4T1 tumor-bearing mice with orthotopically transplanted mammary gland cancer cells ($1 \times 10^4$ cells per mice) [21]. Because of the ethical reasons, we decided to collect all of the biological material from the same mice and perform analysis without repeating animal experiments. All methods used were performed according to European Union (EU)

Directive 2010/63/EU on the protection of animals used for scientific purposes and was approved by the First Local Committee for Experiments with the Use of Laboratory Animals, Wroclaw, Poland (No. of permission: 40/2014, 16/07/2014).

Calcitriol (1,25(OH)2D3) and its analogs, namely PRI-2191 and PRI-2205, were obtained from the Pharmaceutical Research Institute, Warsaw, Poland. We used six- to eight-week-old female BALB/c mice weighing 20–25 g (obtained from the Center of Experimental Medicine of the Medical University of Bialystok, Poland). Mice were subcutaneously (s.c.) treated with 80% propylene glycol (vehicle control), calcitriol (0.5 μg/kg), PRI-2191 (1.0 μg/kg), and PRI-2205 (10.0 μg/kg) thrice a week from the day 7 after transplantation of tumor cells. On days 0, 7, 14, 21, 28, and 33 after transplantation, blood was harvested under anesthesia prior to performing euthanasia of the mice [21]. Subsequently, the tumor, spleen, regional lymph nodes (axillary and inguinal), lungs, and liver were harvested for further analyses.

### 4.2. Flow Cytometry

Blood Cells, Splenocytes, and Lymph Nodes

To obtain the suspension of mononuclear cells and granulocytes, whole blood samples were centrifuged at 400× *g* for 30 min at RT using a gradient density Ficoll Paque Premium (GE Healthcare, Chicago, IL, USA) and the samples were subsequently frozen. The single-cell suspension of spleen and lymph nodes was prepared by passing through sterile nylon filters on a petri dish and then centrifuged twice at 192× *g* for 7 min at 4 °C. The cells ($1 \times 10^6$) suspended in phosphate buffered saline (PBS) with 2% fetal bovine serum (FBS; GE Healthcare, Chicago, IL, USA) were incubated with CD16/CD32 antibody to block the Fc receptors. After incubation, the splenocytes, mononuclear blood cells, and lymph nodes were stained with the anti-mouse conjugated antibodies as follows: rat CD8a-BV510, rat CD4-PE-Cy7, rat CD19-PE, hamster CD3e-APC, rat CD25-BV421, and rat CD335(NKp46)-FITC (BD Biosciences, Franklin Lakes, NJ, USA). Granulocytes were stained with rat CD45-PerCP-Cy5.5, rat Ly6G-Ly6C-APC, rat CD184-PE, and hamster CD54-FITC (BD Biosciences, Franklin Lakes, NJ, USA). Prior to analysis, the cells were washed with PBS containing 2% FBS (centrifuged at 192× *g* for 7 min at 4 °C). For data analysis, a BD LSR Fortessa cytometer with FACSDiva V8.0.1 software (BD Biosciences, Franklin Lakes, NJ, USA) was used.

### 4.3. Splenocytes Culture

Splenocytes ($2 \times 10^6$ cells/mL) were stimulated with lipopolysaccharide (LPS) from *Escherichia coli* (0.5 μg/mL) and concanavalin A (ConA) from *Canavalia ensiformis* (1 μg/mL) (both from Sigma-Aldrich, Saint Louis, MO, USA) for 48 h. Next, the supernatants were collected for further analyses.

### 4.4. Cytokine Array

Supernatants obtained from stimulated splenocytes culture were analyzed with proteome profiler arrays (Proteome Profiler Mouse Cytokine Array Kit, Panel A; R&D Systems, Minneapolis, MN, USA) according to the enclosed instructions. Pixel densities on X-ray film were collected using a multifunctional scanning device Samsung SLC460 (Samsung, Suwon, Korea) or Image Station 4000 MM PRO (Carestream Health, Rochester, NY, USA), after which image analysis was performed (ImageJ 1.48v). For each spot, the final level of optical density was determined as a factor acquired by subtracting the background optical level and dividing by values obtained from the non-treated mice (named as day 0).

### 4.5. Real-Time PCR Array

TRIzol (TRI Reagent; Sigma-Aldrich, Saint Louis, MO, USA) was used for the total RNA extraction according to the manufacturer's recommendations. Quantity and purity of RNA were determined spectrophotometrically at 260 nm using NanoDrop 2000 (Thermo Fisher Scientific, Waltham, MA,

USA). The quality of RNA was verified optionally by agarose gel electrophoresis. Expression of mRNA was quantified by real-time PCR through a ViiA™ 7 Real-Time PCR System (Thermo Fisher Scientific, Waltham, MA, USA) with SYBR green chemistry (Qiagen, Hilden, Germany).

#### 4.5.1. Lymph Nodes and Splenocytes

Synthesis of first-strand cDNA has been done using the compatible RT First Strand Kit (Qiagen, Hilden, Germany), eventually achieving 0.5 µg of cDNA (six mice pooled per group) for a single reaction. Mouse T Helper Cell Differentiation RT$^2$ Profiler Array was acquired from Qiagen Company (Hilden, Germany). A list of genes is shown in Table S8. The conditions for PCR amplification were as follows: 95 °C for 10 min (1 cycle), 95 °C for 15 s (40 cycles), and 60 °C for 1 min. Relative quantification (RQ) values of the target genes were defined using ΔΔCt values with reference to beta-2 microglobulin (*B2m*); glyceraldehyde-3-phosphate dehydrogenase (*Gapdh*); and heat shock protein 90 alpha, class B member 1 (*Hsp90ab1*) genes for lymph nodes, and to actin beta (*Actb*) and *B2m* for splenocytes. Data were analyzed using Qiagen online software suitable for the purchased kit (Qiagen, Hilden, Germany).

#### 4.5.2. Lung Tissue

RNA and cDNA from lung tissue for real-time PCR arrays were prepared according to the method described in our previous study [22]. Data were analyzed using DataAssist 3.01 software (freeware by Applied Biosystems, Foster, CA, USA) with reference to ribosomal protein L13A (*Rpl13a*). The list of genes is shown in Table S9.

### *4.6. Real-Time PCR on Lung, Lymph Nodes, and Splenocytes*

The isolation of RNA and synthesis of cDNA were performed as described previously [21]. Real-time PCR reaction was performed using specific TaqMan primers coding following genes: *Spp1* (Mm00436767_m1) and *Tgfb1* (Mm01178820_m1) for lung cDNA, and *Spp1* (Mm00436767_m1) and *Vdr* (Mm00437297_m1) for cDNA of spleen and lymph nodes (Thermo Fisher Scientific). Briefly, 25 or 40 ng of cDNA (lung or lymph nodes/spleen specimens, respectively) was used for a single reaction. Each sample was performed in triplicates. Data were analyzed using the comparative ΔΔCt method by DataAssist 3.01 software in comparison to endogenous controls: ribosomal protein L13A (*Rpl13a*, Mm01612987_g1) for lung, and beta-2 microglobulin (*B2m*, Mm00437762_m1) for lymph nodes and spleen samples.

### *4.7. ELISA*

Quantitative determination of IFN-γ, TGF-β, IL-1β, IL-2, IL-4, IL-5, IL-6, IL-10, IL-17A, KC, OPN, and G-CSF cytokines using ELISA was performed according to the kit manufacturer's instructions (R&D Systems, Minneapolis, MN, USA; eBioscience, Vienna, Austria, respectively). The expression of proteins was analyzed in plasma, tumor tissue, lung homogenates, and the spleen culture supernatants. Tumor and lung tissue specimens were stored at −80 °C. Preparation of tissue samples for ELISA and the method of protein determination have been described previously [22].

### *4.8. Histopathological Examination of Lung Tissue: Granulocytes Count*

To evaluate infiltration of granulocytes into the lungs, tissue sections were isolated on day 28 and fixed in 4% buffered formalin. Lung tissue specimens were paraffin embedded, cut into 4-µm slices, and stained with hematoxylin and eosin. Granulocyte accumulation was examined based on cell morphology and scored according to the following scheme: 0—none, 1—slight, 2—average, and 3—intense, by an experienced pathologist. Microphotographs were subjected to a computer-assisted image analysis as previously described [77]. BX53 optical microscope (Olympus, Tokyo, Japan) with the CellA software (Olympus Soft Imaging Solution GmbH, Münster, Germany) was used.

### 4.9. Statistical Evaluation

STATISTICA version 10 (StatSoft Inc., Palo Alto, CA, USA) or GraphPad Prism 7.01 (GraphPad Software Inc., La Jolla, CA, USA) were used for statistical analysis. The assumptions of analysis of variance (ANOVA) were tested via Shapiro–Wilk's normality test and Bartlett's test. In the figure legends, specific tests used for data analysis are indicated. $p$ value < 0.05 was considered as significant.

## 5. Conclusions

Thus, we conclude that the treatment of 4T1 tumor-bearing mice with calcitriol and its analogs drive the establishment of tumor-conducive milieu through the impact on immune system. This phenomenon is manifested by increased Th2 and Treg signature in lymph nodes and spleen, which may be supported by the action of calcitriol and its analogs toward granulocytes (decreasing the percentage of activated I-CAM$^+$ and CXCR4$^+$ granulocytes) in the blood. Furthermore, a decrease in NK CD335$^+$ cells in the spleen and lymph nodes may contribute to the stimulated metastatic potential of 4T1 mammary gland tumor cells growing in mice treated with calcitriol and its analogs. Moreover, an increase in the expression of *Spp1* and *Tgfb* in the lung triggered by the applied treatment is responsible for the immunosuppressive metastatic niche formation. Thus, the results of this study indicate that in the case of breast cancer, which shows severe inflammatory response in addition to resistance to treatment with vitamin D compounds, further studies are necessary to determine the efficacy of vitamin D compounds in the treatment of this type of cancer.

**Supplementary Materials:** Supplementary materials can be found at http://www.mdpi.com/1422-0067/19/7/2116/s1.

**Author Contributions:** Conception of the study: J.W. Design of the study: J.W. and A.P. Data acquisition and analysis: A.P., A.A., B.F.-P., M.N., E.T., J.T., J.B., P.G., and J.W. Writing the main part of the manuscript: J.W. All the authors reviewed and approved the final manuscript.

**Funding:** This project was funded by the National Science Center granted on the basis of the decision number 2013/11/B/NZ5/00162. Publication was supported by Wroclaw Center of Biotechnology within a Program—The Leading National Research Center (KNOW) for years 2014–2018.

**Acknowledgments:** We acknowledge Andrzej Kutner (Pharmaceutical Research Institute Warsaw, Poland) for providing us with calcitriol and its analogs.

**Conflicts of Interest:** The authors declare no conflict of interest. The funders had no role in the design of the study; in the collection, analyses, or interpretation of data; in the writing of the manuscript; and in the decision to publish the results.

## Abbreviations

| | |
|---|---|
| ACTB | actin, beta |
| ANGPT1 | angiopoietin 1 |
| ASB2 | ankyrin repeat and SOCS box containing 2 |
| B2M | beta-2 microglobulin |
| BCG | bacillus Calmette-Guerin |
| C/EBP | CCAAT/enhancer binding protein |
| CACNA1F | calcium voltage-gated channel subunit alpha1 F |
| CCL11 | chemokine (C-C motif) ligand 7 |
| CCL7 | chemokine (C-C motif) ligand 7 |
| CCR3 | chemokine (C-C motif) receptor 3 |
| ConA | Concanavalin A |
| CSF2 | colony stimulating factor 2 (granulocyte-macrophage) |
| CST7 | cystatin F (leukocystatin) |
| CTGF | connective tissue growth factor |
| CXCL-1 | chemokine (C-X-C motif) ligand 1 |
| CYP27B1 | cytochrome P450 family 27 subfamily B member 1 |

| | |
|---|---|
| FLT1 | vascular endothelial growth factor receptor 1 |
| FOS1 | FOS like 1 |
| FOXP3 | forkhead box P3 |
| GAPDH | glyceraldehyde-3-phosphate dehydrogenase |
| GATA4 | GATA binding protein 4 |
| G-CSF | granulocyte-colony stimulating factor |
| GM-CSF | granulocyte-macrophage colonystimulating factor |
| GPI | glucose phosphate isomerase |
| HOXA3, HOXA10 | homeobox A3, A10 |
| HSP90AB1 | heat shock protein 90 alpha, class B member 1 |
| IFN-$\gamma$ | interferon gamma |
| IL | interleukin |
| IL-12p70 | interleukin 12 subunit p70 |
| Il12rb2 | interleukin 12 receptor, beta 2 |
| Il17re | interleukin 17 receptor E |
| Il1r1 | interleukin 1 receptor type 1 |
| Irf4, Irf8 | interferon regulatory factor 4 and 8 |
| LPS | lipopolysaccharide |
| MCP-5 | monocyte chemotactic protein 5 |
| M-CSF | macrophage colony-stimulating factor |
| METAP2 | methionyl aminopeptidase 2 |
| MIG | macrophage-induced gene |
| MIP | macrophage inflammatory protein |
| NF2 | neurofibromin 2 |
| NK | natural killer cells |
| Nr4a1, Nr4a3 | nuclear receptor subfamily 4 group A members 1 and 3 |
| OPN | osteopontin |
| p27KIP1 | cyclin dependent kinase inhibitor |
| PSGFA | platelet-derived growth factor $\alpha$ |
| PTGRD2 | prostaglandin D2 receptor 2 |
| PTH | parathormone |
| REL | reticuloendotheliosis oncogene, NF-$\kappa$B subunit |
| RORA | RAR-related orphan receptor alpha |
| RORC | RAR-related orphan receptor gamma |
| SDF-1 | stromal cell-derived factor 1 |
| sICAM-1 | soluble intercellular adhesion molecule-1 |
| Spp1 | osteopontin encoded gene |
| TBX21 | T-box 21 |
| TGF-$\beta$ | transforming growth factor $\beta$ |
| TGIF1 | TGFB-induced factor homeobox 1 |
| TH | T helper lymphocytes |
| TIMP-1 | tissue inhibitor of metalloproteinases |
| TNF | tumor necrosis factor |
| Tregs | regulatory T cells |
| TREM-1 | triggering receptor expressed on myeloid cells |
| UTS2 | urotensin 2 |
| VDR | vitamin D receptor |
| VEGFA | vascular endothelial growth factor A |
| ZEB1 | zinc finger E-box binding homeobox 1 |

## References


1. Coleman, R.E.; Gregory, W.; Marshall, H.; Wilson, C.; Holen, I. The metastatic microenvironment of breast cancer: Clinical implications. *Breast* **2013**, *22*, S50–S56. [CrossRef] [PubMed]
2. Blazejczyk, A.; Papiernik, D.; Porshneva, K.; Sadowska, J.; Wietrzyk, J. Endothelium and cancer metastasis: Perspectives for antimetastatic therapy. *Pharmacol. Rep.* **2015**, *67*, 711–718. [CrossRef] [PubMed]
3. Mantovani, A.; Marchesi, F.; Porta, C.; Sica, A.; Allavena, P. Inflammation and cancer: Breast cancer as a prototype. *Breast* **2007**, *16*, 27–33. [CrossRef] [PubMed]
4. Carron, E.C.; Homra, S.; Rosenberg, J.; Coffelt, S.B.; Kittrell, F.; Zhang, Y.; Creighton, C.J.; Fuqua, S.A.; Medina, D.; Machado, H.L. Macrophages promote the progression of premalignant mammary lesions to invasive cancer. *Oncotarget* **2017**, *8*, 50731. [CrossRef] [PubMed]
5. DeNardo, D.G.; Johansson, M.; Coussens, L.M. Immune cells as mediators of solid tumor metastasis. *Cancer Metastasis Rev.* **2008**, *27*, 11–18. [CrossRef] [PubMed]
6. Shrihari, T.G. Dual role of inflammatory mediators in cancer. *Ecancermedicalscience* **2017**, *11*, 721. [CrossRef] [PubMed]
7. Bouillon, R.; Garmyn, M.; Verstuyf, A.; Segaert, S.; Casteels, K.; Mathieu, C. Paracrine role for calcitriol in the immune system and skin creates new therapeutic possibilities for vitamin D analogs. *Eur. J. Endocrinol.* **1995**, *133*, 7–16. [CrossRef] [PubMed]
8. Dimitrov, V.; Salehi-Tabar, R.; An, B.S.; White, J.H. Non-classical mechanisms of transcriptional regulation by the vitamin D receptor: Insights into calcium homeostasis, immune system regulation and cancer chemoprevention. *J. Steroid Biochem. Mol. Biol.* **2014**, *144*, 74–80. [CrossRef] [PubMed]
9. Chang, J.H.; Cha, H.R.; Lee, D.S.; Seo, K.Y.; Kweon, M.N. 1,25-dihydroxyvitamin D3 inhibits the differentiation and migration of TH17 cells to protect against experimental autoimmune encephalomyelitis. *PLoS ONE* **2010**, *5*, e12925. [CrossRef]
10. Han, Y.-P.; Kong, M.; Zheng, S.; Ren, Y.; Zhu, L.; Shi, H.; Duan, Z. Vitamin D in liver diseases: From mechanisms to clinical trials. *J. Gastroenterol. Hepatol.* **2013**, *28* (Suppl. S1), 49–55. [CrossRef] [PubMed]
11. Overbergh, L.; Stoffels, K.; Waer, M.; Verstuyf, A.; Bouillon, R.; Mathieu, C. Immune Regulation of 25-Hydroxyvitamin D-1$\alpha$-Hydroxylase in Human Monocytic THP1 Cells: Mechanisms of Interferon-$\gamma$-Mediated Induction. *J. Clin. Endocrinol. Metab.* **2006**, *91*, 3566–3574. [CrossRef] [PubMed]
12. Adorini, L.; Daniel, K.C.; Penna, G. Vitamin D receptor agonists, cancer and the immune system: An intricate relationship. *Curr. Top. Med. Chem.* **2006**, *6*, 1297–1301. [CrossRef] [PubMed]
13. Hsu, J.-W.; Yin, P.-N.; Wood, R.; Messing, J.; Messing, E.; Lee, Y.-F.; Hsu, J.-W.; Yin, P.-N.; Wood, R.; Messing, J.; et al. 1 alpha, 25-dihydroxylvitamin D3 promotes Bacillus Calmette-Guérin immunotherapy of bladder cancer. *Oncotarget* **2013**, *4*, 2397–2406. [CrossRef] [PubMed]
14. Guo, J.; Ma, Z.; Ma, Q.; Wu, Z.; Fan, P.; Zhou, X.; Chen, L.; Zhou, S.; Goltzman, D.; Miao, D.; et al. 1, 25$(OH)_2D_3$ inhibits hepatocellular carcinoma development through reducing secretion of inflammatory cytokines from immunocytes. *Curr. Med. Chem.* **2013**, *20*, 4131–4141. [CrossRef] [PubMed]
15. Cao, Y.; Du, Y.; Liu, F.; Feng, Y.; Cheng, S.; Guan, S.; Wang, Y.; Li, X.; Li, B.; Jin, F.; et al. Vitamin D aggravates breast cancer by inducing immunosuppression in the tumor bearing mouse. *Immunotherapy* **2018**, *10*, 555–566. [CrossRef] [PubMed]
16. Jacot, W.; Pouderoux, S.; Thezenas, S.; Chapelle, A.; Bleuse, J.P.; Romieu, G.; Lamy, P.J. Increased prevalence of vitamin D insufficiency in patients with breast cancer after neoadjuvant chemotherapy. *Breast Cancer Res. Treat.* **2012**, *134*, 709–717. [CrossRef] [PubMed]
17. Jacobs, E.T.; Kohler, L.N.; Kunihiro, A.G.; Jurutka, P.W. Vitamin D and Colorectal, Breast, and Prostate Cancers: A Review of the Epidemiological Evidence. *J. Cancer* **2016**, *7*, 232–240. [CrossRef] [PubMed]
18. LaPorta, E.; Welsh, J. Modeling vitamin D actions in triple negative/basal-like breast cancer. *J. Steroid Biochem. Mol. Biol.* **2014**, *144 Pt A*, 65–73. [CrossRef] [PubMed]
19. Williams, J.D.; Aggarwal, A.; Swami, S.; Krishnan, A.V.; Ji, L.; Albertelli, M.A.; Feldman, B.J. Tumor autonomous effects of Vitamin D deficiency promote breast cancer metastasis. *Endocrinology* **2016**, *157*, 1341–1347. [CrossRef] [PubMed]
20. Wietrzyk, J.; Pełczyńska, M.; Madej, J.; Dzimira, S.; Kuśnierczyk, H.; Kutner, A.; Szelejewski, W.; Opolski, A. Toxicity and antineoplastic effect of (24R)-1,24-dihydroxyvitamin D 3 (PRI-2191). *Steroids* **2004**, *69*, 629–635. [CrossRef] [PubMed]

21. Wietrzyk, J.; Chodyński, M.; Fitak, H.; Wojdat, E.; Kutner, A.; Opolski, A. Antitumor properties of diastereomeric and geometric analogs of vitamin D3. *Anticancer Drugs* **2007**, *18*, 447–457. [CrossRef] [PubMed]
22. Anisiewicz, A.; Pawlik, A.; Filip-Psurska, B.; Turlej, E.; Dzimira, S.; Milczarek, M.; Gdesz, K.Z.K.; Papiernik, D.; Jarosz, J.; Kłopotowska, D.; et al. Unfavorable effect of calcitriol and its low-calcemic analogs on metastasis of 4T1 mouse mammary gland cancer. *Int. J. Oncol.* **2018**, *52*, 103–126. [CrossRef] [PubMed]
23. Leyssens, C.; Verlinden, L.; Verstuyf, A. The future of vitamin D analogs. *Front. Physiol.* **2014**, *5*, 122. [CrossRef] [PubMed]
24. Mommers, J.M.; ter Meulen, A.C.; van Erp, P.E.J.; van de Kerkhof, P.C.M. Influence of Tacalcitol on Cell Cycle Kinetics of Human Keratinocytes following Standardized Injury. *Skin Pharmacol. Physiol.* **1999**, *12*, 174–181. [CrossRef] [PubMed]
25. Trynda, J.; Turlej, E.; Milczarek, M.; Pietraszek, A.; Chodyński, M.; Kutner, A.; Wietrzyk, J. Antiproliferative activity and in vivo toxicity of double-point modified analogs of 1,25-dihydroxyergocalciferol. *Int. J. Mol. Sci.* **2015**, *16*, 24873–24894. [CrossRef] [PubMed]
26. Milczarek, M.; Filip-Psurska, B.; Świętnicki, W.; Kutner, A.; Wietrzyk, J. Vitamin D analogs combined with 5-fluorouracil in human HT-29 colon cancer treatment. *Oncol. Rep.* **2014**, *32*, 491–504. [CrossRef] [PubMed]
27. Filip, B.; Milczarek, M.; Wietrzyk, J.; Chodyński, M.; Kutner, A. Antitumor properties of (5E,7E) analogs of vitamin D3. *J. Steroid Biochem. Mol. Biol.* **2010**, *121*, 399–402. [CrossRef] [PubMed]
28. Hisatake, J.; Kubota, T.; Hisatake, Y.; Uskokovic, M.; Tomoyasu, S.; Koeffler, H.P. 5,6-trans-16-ene-vitamin D3: A new class of potent inhibitors of proliferation of prostate, breast, and myeloid leukemic cells. *Cancer Res.* **1999**, *59*, 4023–4029. [PubMed]
29. Milczarek, M.; Psurski, M.; Kutner, A.; Wietrzyk, J. Vitamin D analogs enhance the anticancer activity of 5-fluorouracil in an in vivo mouse colon cancer model. *BMC Cancer* **2013**, *13*, 294. [CrossRef] [PubMed]
30. Krishnan, A.V.; Swami, S.; Feldman, D. Equivalent anticancer activities of dietary vitamin D and calcitriol in an animal model of breast cancer: Importance of mammary CYP27B1 for treatment and prevention. *J. Steroid Biochem. Mol. Biol.* **2013**, *136*, 289–295. [CrossRef] [PubMed]
31. Swami, S.; Krishnan, A.; Wang, J.; Jensen, K.; Horst, R.; Albertelli, M.; Feldman, D. Dietary vitamin $D_3$ and 1,25-dihydroxyvitamin D3 (calcitriol) exhibit equivalent anticancer activity in mouse xenograft models of breast and prostate cancer. *Endocrinology* **2012**, *153*, 2576–2587. [CrossRef] [PubMed]
32. Milczarek, M.; Chodyński, M.; Filip-Psurska, B.; Martowicz, A.; Krupa, M.; Krajewski, K.; Kutner, A.; Wietrzyk, J. Synthesis and Biological Activity of Diastereomeric and Geometric Analogs of Calcipotriol, PRI-2202 and PRI-2205, Against Human HL-60 Leukemia and MCF-7 Breast Cancer Cells. *Cancers* **2013**, *5*, 1355–1378. [CrossRef] [PubMed]
33. Jeong, Y.; Swami, S.; Krishnan, A.V.; Williams, J.D.; Martin, S.; Horst, R.L.; Albertelli, M.A.; Feldman, B.J.; Feldman, D.; Diehn, M. Inhibition of Mouse Breast Tumor-Initiating Cells by Calcitriol and Dietary Vitamin D. *Mol. Cancer Ther.* **2015**, *14*. [CrossRef] [PubMed]
34. Ooi, L.L.; Zhou, H.; Kalak, R.; Zheng, Y.; Conigrave, A.D.; Seibel, M.J.; Dunstan, C.R. Vitamin D deficiency promotes human breast cancer growth in a murine model of bone metastasis. *Cancer Res* **2010**, *70*, 1835–1844. [CrossRef] [PubMed]
35. Rossdeutscher, L.; Li, J.; Luco, A.-L.; Fadhil, I.; Ochietti, B.; Camirand, A.; Huang, D.C.; Reinhardt, T.A.; Muller, W.; Kremer, R. Chemoprevention activity of 25-hydroxyvitamin D in the MMTV-PyMT mouse model of breast cancer. *Cancer Prev. Res.* **2015**, *8*, 120–128. [CrossRef] [PubMed]
36. Maj, E.; Filip-Psurska, B.; Milczarek, M.; Psurski, M.; Kutner, A.; Wietrzyk, J. Vitamin D derivatives potentiate the anticancer and anti-angiogenic activity of tyrosine kinase inhibitors in combination with cytostatic drugs in an A549 non-small cell lung cancer model. *Int. J. Oncol.* **2018**, *52*, 337–366. [CrossRef] [PubMed]
37. Maj, E.; Filip-Psurska, B.; Świtalska, M.; Kutner, A.; Wietrzyk, J. Vitamin D Analogs Potentiate the Antitumor Effect of Imatinib Mesylate in a Human A549 Lung Tumor Model. *Int. J. Mol. Sci.* **2015**, *16*, 27191–27207. [CrossRef] [PubMed]
38. Zhang, Y.; Guo, Q.; Zhang, Z.; Bai, N.; Liu, Z.; Xiong, M.; Wei, Y.; Xiang, R.; Tan, X. VDR Status Arbitrates the Prometastatic Effects of Tumor-Associated Macrophages. *Mol. Cancer Res.* **2014**, *12*, 1181–1191. [CrossRef] [PubMed]
39. Wietrzyk, J.; Nevozhay, D.; Filip, B.; Milczarek, M.; Kutner, A. The antitumor effect of lowered doses of cytostatics combined with new analogs of vitamin D in mice. *Anticancer Res.* **2007**, *27*, 3387–3398. [PubMed]

40. DuPre', S.A.; Hunter, K.W. Murine mammary carcinoma 4T1 induces a leukemoid reaction with splenomegaly: Association with tumor-derived growth factors. *Exp. Mol. Pathol.* **2007**, *82*, 12–24. [CrossRef] [PubMed]
41. DuPré, S.A.; Redelman, D.; Hunter, K.W. The mouse mammary carcinoma 4T1: Characterization of the cellular landscape of primary tumours and metastatic tumour foci. *Int. J. Exp. Pathol.* **2007**, *88*, 351–360. [CrossRef] [PubMed]
42. Varn, F.S.; Mullins, D.W.; Arias-Pulido, H.; Fiering, S.; Cheng, C. Adaptive immunity programmes in breast cancer. *Immunology* **2017**, *150*, 25–34. [CrossRef] [PubMed]
43. Raphael, I.; Nalawade, S.; Eagar, T.N.; Forsthuber, T.G. T cell subsets and their signature cytokines in autoimmune and inflammatory diseases. *Cytokine* **2015**, *74*, 5–17. [CrossRef] [PubMed]
44. Iizuka-Koga, M.; Nakatsukasa, H.; Ito, M.; Akanuma, T.; Lu, Q.; Yoshimura, A. Induction and maintenance of regulatory T cells by transcription factors and epigenetic modifications. *J. Autoimmun.* **2017**, *83*, 113–121. [CrossRef] [PubMed]
45. Grinberg-Bleyer, Y.; Oh, H.; Desrichard, A.; Bhatt, D.M.; Caron, R.; Chan, T.A.; Schmid, R.M.; Klein, U.; Hayden, M.S.; Ghosh, S. NF-κB c-Rel Is Crucial for the Regulatory T Cell Immune Checkpoint in Cancer. *Cell* **2017**, *170*, 1096–1108. [CrossRef] [PubMed]
46. Oh, H.; Grinberg-Bleyer, Y.; Liao, W.; Maloney, D.; Wang, P.; Wu, Z.; Wang, J.; Bhatt, D.M.; Heise, N.; Schmid, R.M.; et al. An NF-κB Transcription-Factor-Dependent Lineage-Specific Transcriptional Program Promotes Regulatory T Cell Identity and Function. *Immunity* **2017**, *47*, 450–465. [CrossRef] [PubMed]
47. Luu, M.; Jenike, E.; Vachharajani, N.; Visekruna, A. Transcription factor c-Rel is indispensable for generation of thymic but not of peripheral Foxp3+ regulatory T cells. *Oncotarget* **2017**, *8*, 52678–52689. [CrossRef] [PubMed]
48. Sekiya, T.; Kashiwagi, I.; Yoshida, R.; Fukaya, T.; Morita, R.; Kimura, A.; Ichinose, H.; Metzger, D.; Chambon, P.; Yoshimura, A. Nr4a receptors are essential for thymic regulatory T cell development and immune homeostasis. *Nat. Immunol.* **2013**, *14*, 230–237. [CrossRef] [PubMed]
49. Dominguez-Villar, M.; Baecher-Allan, C.M.; Hafler, D.A. Identification of T helper type 1–like, Foxp3+ regulatory T cells in human autoimmune disease. *Nat. Med.* **2011**, *17*, 673–675. [CrossRef] [PubMed]
50. Ivanova, E.A.; Orekhov, A.N. T Helper Lymphocyte Subsets and Plasticity in Autoimmunity and Cancer: An Overview. *BioMed Res. Int.* **2015**, *2015*, 1–9. [CrossRef] [PubMed]
51. Liu, Y.; Cao, X. Immunosuppressive cells in tumor immune escape and metastasis. *J. Mol. Med.* **2016**, *94*, 509–522. [CrossRef] [PubMed]
52. Jiang, X.; Shapiro, D.J. The immune system and inflammation in breast cancer. *Mol. Cell. Endocrinol.* **2014**, *382*, 673–682. [CrossRef] [PubMed]
53. Zhu, J.; Yamane, H.; Cote-Sierra, J.; Guo, L.; Paul, W.E. GATA-3 promotes Th2 responses through three different mechanisms: Induction of Th2 cytokine production, selective growth of Th2 cells and inhibition of Th1 cell-specific factors. *Cell Res.* **2006**, *16*, 3–10. [CrossRef] [PubMed]
54. Boonstra, A.; Barrat, F.J.; Crain, C.; Victoria, L.; Savelkoul, H.F.J.; Garra, A.O. 1 α,25-Dihydroxyvitamin D3 Has a Direct Effect on Naive CD4 + T Cells to Enhance the Development of Th2 Cells. *J. Immunol.* **2001**, *167*, 4974–4980. [CrossRef] [PubMed]
55. Borges, M.C.; Martini, L.A.; Rogero, M.M. Current perspectives on vitamin D, immune system, and chronic diseases. *Nutrition* **2011**, *27*, 399–404. [CrossRef] [PubMed]
56. Heine, G.; Niesner, U.; Chang, H.D.; Steinmeyer, A.; Zugel, U.; Zuberbier, T.; Radbruch, A.; Worm, M. 1,25-dihydroxyvitamin D(3) promotes IL-10 production in human B cells. *Eur. J. Immunol.* **2008**, *38*, 2210–2218. [CrossRef] [PubMed]
57. Fujita, H.; Asahina, A.; Komine, M.; Tamaki, K. The direct action of 1α,25(OH)2-vitamin D3 on purified mouse Langerhans cells. *Cell. Immunol.* **2007**, *245*, 70–79. [CrossRef] [PubMed]
58. Matilainen, J.M.; Husso, T.; Toropainen, S.; Seuter, S.; Turunen, M.P.; Gynther, P.; Ylä-Herttuala, S.; Carlberg, C.; Väisänen, S. Primary effect of 1α,25(OH)2D3 on IL-10 expression in monocytes is short-term down-regulation. *Biochim. Biophys. Acta Mol. Cell Res.* **2010**, *1803*, 1276–1286. [CrossRef] [PubMed]
59. Bien, E.; Balcerska, A.; Adamkiewicz-Drozynska, E.; Rapala, M.; Krawczyk, M.; Stepinski, J. Pre-treatment serum levels of interleukin-10, interleukin-12 and their ratio predict response to therapy and probability of event-free and overall survival in childhood soft tissue sarcomas, Hodgkin's lymphomas and acute lymphoblastic leukemias. *Clin. Biochem.* **2009**, *42*, 1144–1157. [CrossRef] [PubMed]

60. Rutella, S.; Zavala, F.; Danese, S.; Kared, H.; Leone, G. Granulocyte Colony-Stimulating Factor: A Novel Mediator of T Cell Tolerance. *J. Immunol.* **2005**, *175*, 7085–7091. [CrossRef] [PubMed]
61. Kowanetz, M.; Wu, X.; Lee, J.; Tan, M.; Hagenbeek, T.; Qu, X.; Yu, L.; Ross, J.; Korsisaari, N.; Cao, T.; et al. Granulocyte-colony stimulating factor promotes lung metastasis through mobilization of Ly6G+Ly6C+ granulocytes. *Proc. Natl. Acad. Sci. USA* **2010**, *107*, 21248–21255. [CrossRef] [PubMed]
62. Yang, L.; Huang, J.; Ren, X.; Gorska, A.E.; Chytil, A.; Aakre, M.; Carbone, D.P.; Matrisian, L.M.; Richmond, A.; Lin, P.C.; et al. Abrogation of TGFβ signaling in mammary carcinomas recruits Gr-1+CD11b+ myeloid cells that promote metastasis. *Cancer Cell* **2008**, *13*, 23–35. [CrossRef] [PubMed]
63. Padua, D.; Zhang, X.H.-F.; Wang, Q.; Nadal, C.; Gerald, W.L.; Gomis, R.R.; Massagué, J. TGFβ Primes Breast Tumors for Lung Metastasis Seeding through Angiopoietin-like 4. *Cell* **2008**, *133*, 66–77. [CrossRef] [PubMed]
64. Chang, P.L.; Prince, C.W. 1α,25-Dihydroxyvitamin D3 Stimulates Synthesis and Secretion of Nonphosphorylated Osteopontin (Secreted Phosphoprotein 1) in Mouse JB6 Epidermal Cells. *Cancer Res.* **1991**, *51*, 2144–2150. [PubMed]
65. Chang, P.L.; Ridall, A.L.; Prince, C.W. Calcitriol regulation of osteopontin expression in mouse epidermal cells. *Endocrinology* **1994**, *135*, 863–869. [CrossRef] [PubMed]
66. Lau, W.L.; Leaf, E.M.; Hu, M.C.; Takeno, M.M.; Kuro-o, M.; Moe, O.W.; Giachelli, C.M. Vitamin D receptor agonists increase klotho and osteopontin while decreasing aortic calcification in mice with chronic kidney disease fed a high phosphate diet. *Kidney Int.* **2012**, *82*, 1261–1270. [CrossRef] [PubMed]
67. Kolb, A.; Kleeff, J.; Guweidhi, A.; Esposito, I.; Giese, N.A.; Adwan, H.; Giese, T.; Büchler, M.W.; Berger, M.R.; Friess, H. Osteopontin influences the invasiveness of pancreatic cancer cells and is increased in neoplastic and inflammatory conditions. *Cancer Biol. Ther.* **2005**, *4*, 740–746. [CrossRef] [PubMed]
68. Li, Y.; Xie, Y.; Cui, D.; Ma, Y.; Sui, L.; Zhu, C.; Kong, H.; Kong, Y. Osteopontin Promotes Invasion, Migration and Epithelial-Mesenchymal Transition of Human Endometrial Carcinoma Cell HEC-1A Through AKT and ERK1/2 Signaling. *Cell. Physiol. Biochem. Int. J. Exp. Cell. Physiol. Biochem. Pharmacol.* **2015**, *37*, 1503–1512. [CrossRef] [PubMed]
69. Sangaletti, S.; Tripodo, C.; Sandri, S.; Torselli, I.; Vitali, C.; Ratti, C.; Botti, L.; Burocchi, A.; Porcasi, R.; Tomirotti, A.; et al. Osteopontin Shapes Immunosuppression in the Metastatic Niche. *Cancer Res.* **2014**, *74*, 4706–4719. [CrossRef] [PubMed]
70. Drozdenko, G.; Scheel, T.; Heine, G.; Baumgrass, R.; Worm, M. Impaired T cell activation and cytokine production by calcitriol-primed human B cells. *Clin. Exp. Immunol.* **2014**, *178*, 364–372. [CrossRef] [PubMed]
71. He, X.-J.; Ding, Y.; Xiang, W.; Dang, X.-Q. Roles of 1,25(OH)2D3 and Vitamin D Receptor in the Pathogenesis of Rheumatoid Arthritis and Systemic Lupus Erythematosus by Regulating the Activation of CD4+ T Cells and the PKCδ/ERK Signaling Pathway. *Cell. Physiol. Biochem.* **2016**, *40*, 743–756. [CrossRef] [PubMed]
72. Guo, B.; Tumang, J.R.; Rothstein, T.L. B cell receptor crosstalk: B cells express osteopontin through the combined action of the alternate and classical BCR signaling pathways. *Mol. Immunol.* **2009**, *46*, 587–591. [CrossRef] [PubMed]
73. Shevde, L.A.; Samant, R.S. Role of osteopontin in the pathophysiology of cancer. *Matrix Biol.* **2014**, *37*, 131–141. [CrossRef] [PubMed]
74. Liu, J.; Xu, K.; Chase, M.; Ji, Y.; Logan, J.K.; Buchsbaum, R.J. Tiam1-regulated osteopontin in senescent fibroblasts contributes to the migration and invasion of associated epithelial cells. *J. Cell Sci.* **2012**, *125*, 376–386. [CrossRef] [PubMed]
75. Yang, L.; Pang, Y.; Moses, H.L. TGF-β and immune cells: An important regulatory axis in the tumor microenvironment and progression. *Trends Immunol.* **2010**, *31*, 220–227. [CrossRef] [PubMed]
76. Eruslanov, E.B.; Bhojnagarwala, P.S.; Quatromoni, J.G.; Stephen, T.L.; Ranganathan, A.; Deshpande, C.; Akimova, T.; Vachani, A.; Litzky, L.; Hancock, W.W.; et al. Tumor-associated neutrophils stimulate T cell responses in early-stage human lung cancer. *J. Clin. Investig.* **2014**, *124*, 5466–5480. [CrossRef] [PubMed]
77. Nowak, M.; Madej, J.A.; Pula, B.; Dziegiel, P.; Ciaputa, R. Expression of matrix metalloproteinase 2 (MMP-2), E-cadherin and Ki-67 in metastatic and non-metastatic canine mammary carcinomas. *Ir. Vet. J.* **2015**, *69*, 9. [CrossRef] [PubMed]

International Journal of
*Molecular Sciences*

MDPI

*Article*

# Calcitriol and Calcidiol Can Sensitize Melanoma Cells to Low–LET Proton Beam Irradiation

**Ewa Podgorska [1,†], Agnieszka Drzal [1,†], Zenon Matuszak [2], Jan Swakon [3], Andrzej Slominski [4], Martyna Elas [1] and Krystyna Urbanska [1,*]**

1 Faculty of Biochemistry, Biophysics and Biotechnology, Jagiellonian University, Gronostajowa 7 Street, Krakow 30-387, Poland; ewa.podgorska@doctoral.uj.edu.pl (E.P.); agnieszka.drzal@doctoral.uj.edu.pl (A.D.); martyna.elas@uj.edu.pl (M.E.)

2 Department of Medical Physics and Biophysics, Faculty of Physics and Applied Computer Science, AGH University of Science and Technology, al. A. Mickiewicza 30, Krakow 30-059, Poland; zenon.matuszak@fis.agh.edu.pl

3 Polish Academy of Sciences, Institute of Nuclear Physics, Krakow 31-342, Poland; jan.swakon@ifj.edu.pl

4 Department of Dermatology, Comprehensive Cancer Center Cancer Chemoprevention Program, University of Alabama at Birmingham, Birmingham, AL 35294, USA; aslominski@uabmc.edu

* Correspondence: krystyna.urbanska@uj.edu.pl; Tel.: +48-12-664-6153

† These authors contributed equally to this work.

Received: 15 June 2018; Accepted: 27 July 2018; Published: 31 July 2018

**Abstract:** Proton beam irradiation promises therapeutic utility in the management of uveal melanoma. Calcitriol (1,25(OH)$_2$D$_3$)—the biologically active metabolite of vitamin D$_3$—and its precursor, calcidiol (25(OH)D$_3$), exert pleiotropic effects on melanoma cells. The aim of the study was to evaluate the effect of both calcitriol and calcidiol on melanoma cell proliferation and their response to proton beam irradiation. Three melanoma cell lines (human SKMEL-188 and hamster BHM Ma and BHM Ab), pre-treated with 1,25(OH)$_2$D$_3$ or 25(OH)D$_3$ at graded concentrations (0, 10, 100 nM), were irradiated with 0–5 Gy and then cultured in vitro. Growth curves were determined by counting the cell number every 24 h up to 120 h, which was used to calculate surviving fractions. The obtained survival curves were analysed using two standard models: linear-quadratic and multi-target single hit. Calcitriol inhibited human melanoma proliferation at 10 nM, while only calcidiol inhibited proliferation of hamster lines at 10 and 100 nM doses. Treatment with either 1,25(OH)$_2$D$_3$ or 25(OH)D$_3$ radio sensitized melanoma cells to low doses of proton beam radiation. The strength of the effect increased with the concentration of vitamin D$_3$. Our data suggest that vitamin D$_3$ may be an adjuvant that modifies proton beam efficiency during melanoma therapy.

**Keywords:** vitamin D$_3$; proton beam radiotherapy; melanoma; in vitro

## 1. Introduction

Because of their high incidence, mortality rates and resistance to the therapy, melanomas are still one the most challenging cancer types for researchers and clinicians [1,2]. Over the last decade, our understanding of the molecular principles regulating melanoma behaviour has improved significantly, leading to new therapies [3,4]. However, even with the new therapeutic approaches the problem of limited efficacy and selective responsiveness of patients still remains [5]. Therefore, new creative approaches and their combinations are required.

Accumulating evidence from a variety of epidemiological and experimental studies confirms in vitro and in vivo anticancer activity of vitamin D$_3$ [6–9]. Those studies indicate that biologically active vitamin D$_3$ derivatives may lower the incidence, and inhibit the progression, of various tumours, including melanoma [10–12] and sensitize them to radiotherapy [13–18]. Moreover, there are reports of an inverse relationship between patient survival and melanoma thickness and 25(OH)$D_3$ serum levels, polymorphisms in the genes encoding the vitamin D receptor (VDR), the vitamin D binding protein, expression of the VDR and CYP27B1 expression [19–25] and complex relations with CYP24A1 [26]. Additionally, it was proposed that VDR plays a role in the development and progression of melanocytic tumours [27].

So far, no studies have been published on a combination of vitamin $D_3$ and proton beam irradiation, a therapy with superior dose distribution compared to photon radiation, which is commonly used in the treatment of uveal melanoma. However, reports on the differences in the level of production of free radicals, cell cycle, cell migration inhibition and apoptotic signalling between photon and proton beam radiotherapy [28–32] suggest possible differences in the effect of their combination with vitamin $D_3$.

This work aims to examine the influence of calcitriol (1,25$(OH)_2D_3$) and calcidiol (25(OH)$D_3$) on the proliferation and response to proton beam radiotherapy of three melanoma cell lines: human SKMEL-188 and hamster BMH Ma and BHM Ab. In the light of existing evidence, we hypothesize that vitamin $D_3$ can sensitize melanoma cells to radiation, thus enhancing its effectiveness. Two different models, linear-quadratic and single hit multi-target, were fitted to analyse the surviving curves of the tested melanoma cell lines exposed to combined treatment and to get the most comprehensive picture of proton radiation dose response.

## 2. Results

### 2.1. Impact of Two Metabolites of Vitamin $D_3$ on Melanoma Cell Proliferation

1,25$(OH)_2D_3$ and 25(OH)$D_3$ affected the growth rate of melanoma cells in culture (Figure 1). In the case of calcitriol, only human SKMEL-188 melanoma cells showed a significant decrease in growth rate, after treatment with the concentration of 10 nM, with BHM Ma and BHM Ab hamster lines showing no major changes in the rate of proliferation. Calcidiol, on the other hand, slightly but significantly stimulated proliferation of SKMEL-188 cells at the concentration of 100 nM, with both BHM Ma and BHM Ab hamster cells showing an inhibition of proliferation at 10 and 100 nM concentrations.

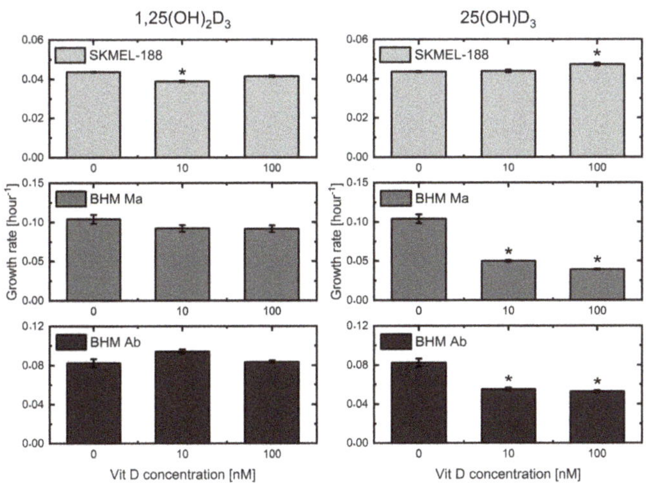


**Figure 1.** The impact of calcitriol (1,25$(OH)_2D_3$) and calcidiol (25(OH)$D_3$) on the proliferation of human SKMel-188 and hamster BHM Ma and BHM Ab melanoma cells. Growth rate was calculated from exponential fitting to cell number, counted every 24 h for 4 days in culture. * denotes statistical significance $p < 0.05$ against control (0 nM of vitamin $D_3$ derivative).

There were also significant differences in basal growth rates between the tested melanoma cell lines. Both hamster lines showed similar proliferation rates (approximately 0.09 $h^{-1}$), while the human SKMEL-188 cells divided almost two times slower. Such differences may be one of the factors influencing response to the vitamin $D_3$ metabolites.

### 2.2. Proton Beam Radiosensitivity of Melanoma Cells

Melanoma cell lines are characterized by different radiosensitivity to proton beam therapy (Figure 2), with the highest level of cell killing seen in BHM amelanotic Ab cells. BHM melanotic Ma and SKMEL-188 cells show similar survival curves. Similar differences in response to radiation between BHM Ma and BHM Ab cells were reported for X ray irradiation, where pigmented cells (BHM Ma) were 2.4 times more radio resistant than the unpigmented (BHM Ab) ones [33]. In the case of proton radiation this characteristic is less pronounced but still marked with the ratio of mean lethal dose of 1.56.

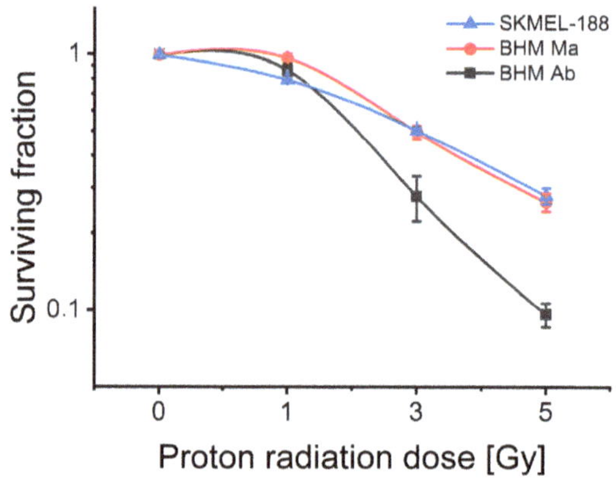


**Figure 2.** Proton beam radiosensitivity of SKMel-188, BHM Ma and BHM Ab melanoma cells. Surviving fraction was determined from the model described by Buch et al. [34], based on the experimental data from cell number, counted every 24 h for 4 days in culture after irradiation.

### 2.3. Vitamin $D_3$ Derivatives Influence Proton Beam Radiosensitivity of Melanoma Cells

Changes in both surviving fraction and survival curve shape can be distinguished after pre-treatment with both vitamin $D_3$ metabolites (Figure 3). The most significant effects are seen for both vitamin $D_3$ derivatives in hamster melanoma lines, with a weaker effect in the human line. Furthermore, higher doses of 25(OH)$D_3$ cause in BHM Ma cells a flattening of the survival curve in the region of higher doses (3–5 Gy). Those changes are consistent with the data from fitted survival models. It has been shown that $\alpha$ and $\beta$ parameters from the linear-quadratic model (LQ model) determine the effectiveness at low and high radiation doses, respectively [35]. Vitamin $D_3$ dose dependent decrease in surviving fraction for proton irradiation dose of 1 Gy, visible on the survival curve, is represented by an increase in the values of parameter $\alpha$ (Figure 4, upper panel). For calcitriol, no significant change in $\alpha$ was detected at 10 nM dose for SKMEL-188 and BHM Ma cells. On survival curves, that concentration was slightly protective for cancer cells, with surviving fraction higher than the control one but the effect was not statistically significant. The most effective dose was 100 nM. Calcidiol has a clear dose dependent effect on the radiosensitivity of the studied melanoma cell lines, with the highest concentration being the most effective one. An increase in parameter $\beta$, indicating higher effectiveness of high doses of radiation, occurs only for calcitriol pre-treatment: 10 nM dose in BHM Ma and BHM Ab cells (Figure 4, lower panel). The same dose leads to a noticeable decrease in surviving fraction on survival curves. The tested doses were within the range of the normal serum level of calcitriol, which is between 50–120 nM.

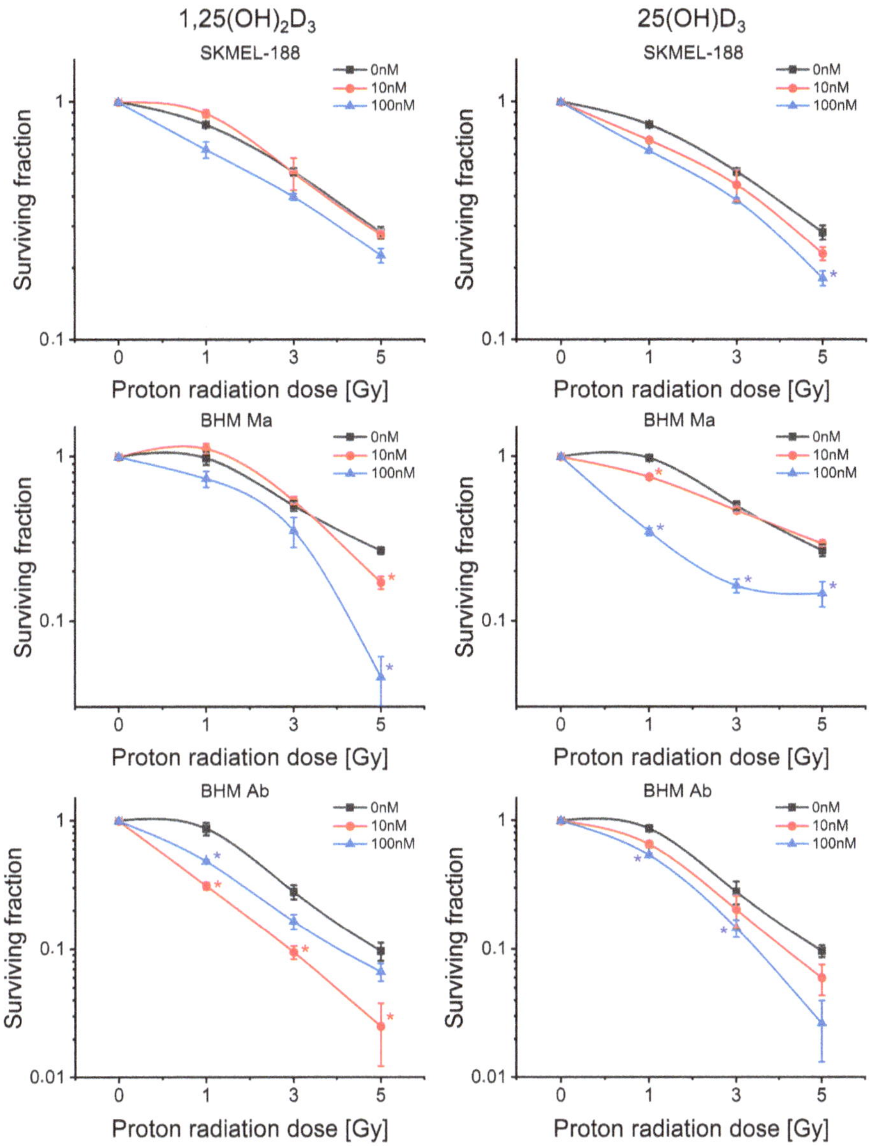


**Figure 3.** The effect of calcitriol (1,25$(OH)_2D_3$) and calcidiol (25$(OH)D_3$) on the cellular response to proton beam irradiation of human SKMel-188 and hamster BHM Ma and BHM Ab melanoma cells. Surviving fraction was determined from the model described by Buch et al. [34], based on the experimental data from cell number, counted every 24 h for 4 days in culture after irradiation. * denotes statistical significance $p < 0.05$ against irradiated control.

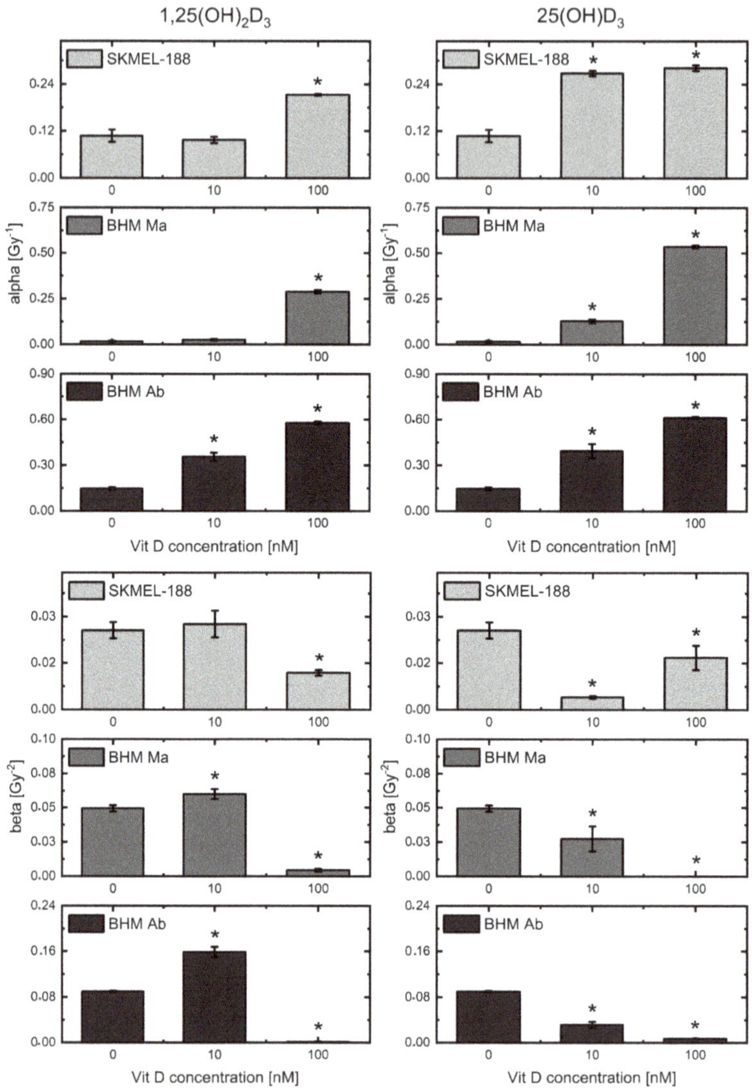


**Figure 4.** Coefficients α (**upper** panel) and β (**lower** panel) from linear-quadratic model calculated for cells pre-treated with calcitriol (1,25(OH)$_2$D$_3$) and calcidiol (25(OH)D$_3$) and irradiated with proton beam. Surviving fraction was determined from the model described by Buch et al. [34], based on the experimental data from SKMel-188, BHM Ma and BHM Ab cell number, counted every 24 h for 4 days in culture after irradiation. * denotes statistical significance $p < 0.05$ against control (0 nM of vitamin D$_3$ derivative).

Results from the second model applied to survival data, single hit multi-target, are in agreement with data from the LQ model. Changes in the calculated parameters $n$ and $D_0$ are shown in Figure 5. Parameter $n$, indicating the required number of hits for cell death and being sensitive to the effectiveness of low radiation doses, shows an opposite trend to parameter α from the LQ model. An increase in α goes together with a decrease in $n$, which is consistent with vitamin D$_3$ metabolites radiosensitizing cells to low doses of proton beam irradiation. The second parameter, mean lethal dose $D_0$, describes the average effect of radiation, without distinction between low and high doses [36]. Only several groups were characterized by a decrease in that parameter in comparison to the control, representing the process of radiosensitization by the tested vitamin $D_3$ analogues. Among them, there were groups treated with 100 nM of calcitriol, the most effective treatment in enhancing the efficiency of low radiation doses. This indicates that the highest influence of vitamin $D_3$ analogues on the averaged effects relates to the response to low doses of proton radiation.

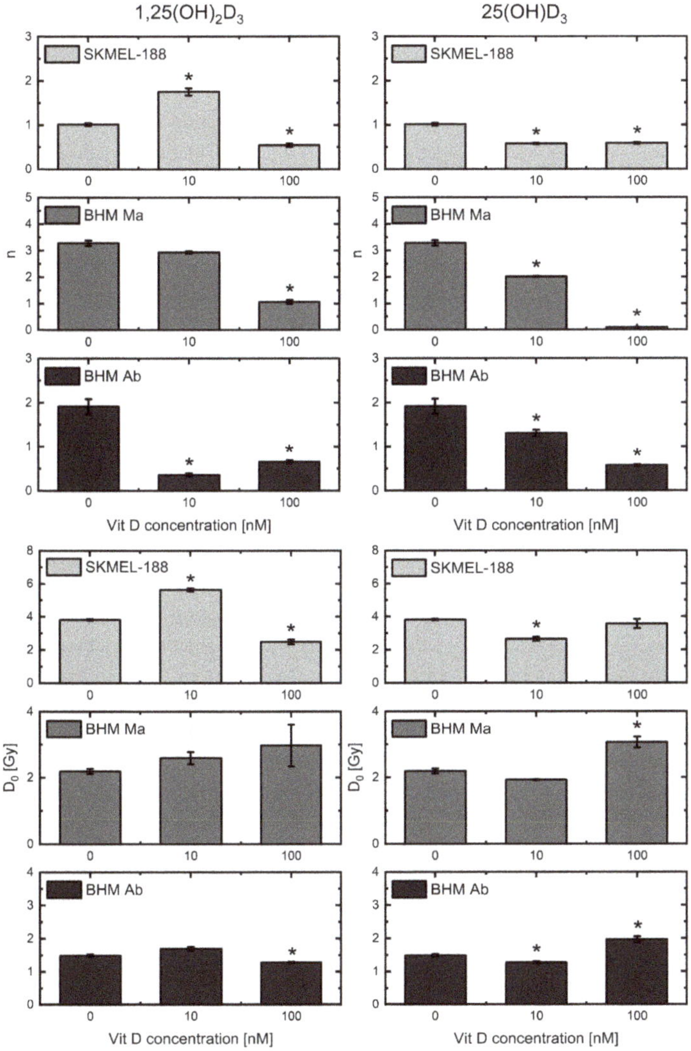


**Figure 5.** Coefficients $n$ (**upper** panel) and $D_0$ (**lower** panel) from multi-target single hit model calculated for cells pre-treated with calcitriol (1,25$(OH)_2D_3$) and calcidiol (25$(OH)D_3$) and irradiated with proton beam. Surviving fraction was determined from the model described by Buch et al. [34], based on the experimental data from SKMel-188, BHM Ma and BHM Ab cell number, counted every 24 h for 4 days in culture after irradiation. * denotes statistical significance $p < 0.05$ against control (0 nM of vitamin $D_3$ derivative).

## 3. Discussion

Current pre-clinical and clinical reports show that vitamin $D_3$ has overall anticancer properties [6–8]. The finding is also supported by a strong correlation between higher serum 25(OH)$D_3$ level and lower incidence of breast [37,38], colon [39], lung [40], prostate [41] and melanoma [10] and by the antiproliferative and pro-differentiative action of vitamin $D_3$ and its analogues towards multiple cancer cell lines [42–47]. Interestingly, not all cancer cell lines do respond to vitamin $D_3$ [20,48–51], which also included melanoma lines [41,50] and such responsiveness can depend on the culture conditions [11].

In the present study, we demonstrate differential effects of the biologically active form of vitamin $D_3$, calcitriol and its precursor calcidiol, on the proliferation rate of three melanoma cell lines: human SKMEL-188 and hamster BHM Ma and BHM Ab (Figure 1). Only SKMEL-188 cells responded to lower doses of 1,25(OH)$_2D_3$ with inhibition of growth rate, which was consistent with other reports [22,52]. However, inhibition of proliferation by calcidiol, with no effect in the case of calcitriol in both BHM Ma and BHM Ab cells requires further explanation. It has been proposed that one of the factors influencing the responsiveness of melanoma cell lines to vitamin $D_3$ is melanin pigmentation [53,54], which would explain the decrease in VDR expression with concomitant decrease in anti-proliferative response to 1,25(OH)$_2D_3$ in human melanoma. It is possible that a similar mechanism could underlie the resistance of BHM Ma, which produces melanin pigment also when cultured in a medium with low tyrosine content [55]. However, since BHM Ab cells were pre-selected for amelanotic phenotype by culturing in DMEM, other factors, including a defect in VDR or a high activity of CYP24A1 that inactivates 1,25(OH)$_2D_3$ could be responsible for the differential effects. It must also be noted that there are alternative receptors for vitamin $D_3$ derivatives that could affect responsiveness to active forms of vitamin $D_3$ (reviewed in [9,10]). Therefore, proper understanding of these differential effects would require additional studies, involving the identification and cloning of hamster nuclear receptors for vitamin $D_3$.

Vitamin $D_3$ enhances the of anticancer drugs such as doxorubicin, cisplatin, gemcitabine or cyclophosphamide [56–58] and sensitizes cancer cells to ionizing radiation [12–18]. In this paper, we focus on combining calcitriol or calcidiol with proton beam therapy. That mode of cancer therapy is commonly used for the treatment of tumours located near or on critical organs, because of its dosimetric benefits. Thanks to the occurrence of a Bragg peak, there is no exit dose normally deposited by photon therapy within the healthy tissue surrounding the tumour [59]. Accumulating evidence suggests that even with relative biological effectiveness of 1.1, there are significant differences in biological effects on tumour cells between low-energy protons and photon radiation [60]. As regards the combination of proton beam radiotherapy and vitamin $D_3$ treatment, the most crucial aspects appear to be the increased production of reactive oxygen species (ROS) and the differences in cell cycle inhibition and apoptotic signalling, all implicated in the response to vitamin $D_3$.

Overall, our results show that vitamin $D_3$ is a potential radiosensitizing agent in proton beam irradiation of melanoma cells. Radiosensitization is a complex concept that has many different interpretations and is used to describe a variety of interactions at biochemical and biological level. Alterations in radiosensitization are demonstrated in the SF curve; a downward or leftward shift of the curve implies a radiosensitizing interaction, while an upward or rightward shift implies a radioprotective influence on the treated cells.

Generally, radiosensitization is conventionally defined as an increased amount of radiation-induced cell death resulting from exposure to a second agent, after correction for the cytotoxicity of the agent (in our case vitamin $D_3$, nontoxic in the tested concentrations). Considering the changes in shape and the above-mentioned shift in SF suggests that both calcitriol and calcidiol are radiosensitizing agents, within the range of the concentrations used. Radiosensitization by vitamin $D_3$ is particularly visible for low radiation doses, for which we a decrease or elimination of the shoulder on the SF curve is observed (Figure 3). The effect of radiosensitization is also visible at high radiation doses of about 5 Gy. The surviving fraction values of the cells are lower than the analogous values for SF curve obtained for the exposure without vitamin $D_3$. Some of the observed survival-dose dependencies, in particular for

BHM Ma cells pre-treated with 100 nM of calcidiol, do not resemble the commonly found SF curves with the shoulder in the low dose area and with a linear course for zero for high doses. In this case, we observe a gradual disappearance of the effect the radiation dose has on the number of killed cells. The SF curve gradually flattens (upward-bending) and the number of surviving cells does not decrease with increasing dose. The effect is the reason for the unusual values of some coefficients characterizing survival curves of such non-classical forms of SF, for example, close to zero $\beta$ values (see Figure 4). The phenomenon (upward-bending of SF) has already been observed during the He-3 irradiation of the V79 cells but the authors interpreted it as an artefact [61]. More recent theoretical studies linked that effect mainly with the repair processes in irradiated cells [62,63]. Nevertheless, the theory is formal and does not explain the upward-bending effect of SF curves with the help of probable molecular mechanisms involved in the repair of irradiated cells. In our work, the upward-bending effect was only observed in the presence of vitamin $D_3$ derivatives. Therefore, it might be due to the radioprotective properties of vitamin $D_3$. Probable targets of such interaction include ATM and mTor, which are directly related to the process of DNA repair and cell proliferation [64–66]. Thus, the radiosensitization effect of high vitamin $D_3$ doses would be classified as radiosensitization by targeting the response to DNA damage.

In conclusion, our data suggest that active forms of vitamin D may improve the effectiveness of proton therapy. They also support further in vivo studies on use of vitamin D as adjuvant during radiotherapy of melanoma.

## 4. Materials and Methods

### 4.1. Cell Lines

Human SKMEL-188 melanoma cell line, a gift from Dr Chakraborty, Yale University, was established from a human metastatic melanoma and then maintained in our laboratories as a continuous cell line [67,68].

BHM Ma, Bomirski Hamster Melanoma—pigmented subline, is a stable transplantable tumour cell line that was derived by Dr Andrzej Bomirski in Gdansk, Poland, from a spontaneous hamster melanoma in 1959 [69].

BHM Ab, Bomirski Hamster Melanoma—non-pigmented subline, arose in 1963 by spontaneous alteration of a black tumour (BHM Ma). The amelanotic subline of Bomirski Hamster Melanoma results in non-pigmented tumours in hamsters. The tumours are malignant, dedifferentiated, fast-growing and metastasizing [69]. The cells isolated from those tumours by means of a non-enzymatic method undergo rapid pigmentation in primary cultures in media containing high concentrations of L-tyrosine [70,71]. The level of pigmentation depends on the culture medium; in DMEM with 10% of foetal bovine serum (FBS) the cells are pigmented, while in media low in L-tyrosine, such as Ham's F10, they are amelanotic [72]. However, culturing for prolonged period in DMEM leads to the selection for an amelanotic phenotype that is stable in cell culture [73]. This type of preselected cells was used in the experiments.

### 4.2. Cell Culture

For each cell line, a different culture medium was used: for BHM Ma—RPMI, for BHM Ab—DMEM and for SKMEL-188 Ham's F10 medium supplemented with glucose, L-glutamine and pyridoxine hydrochloride. The cells were cultured in 75 $cm^2$ flasks in culture medium supplemented with 10% of FBS (Gibco—Thermofisher Scientific, Waltham, MA, USA) and antibiotics (Sigma, St. Luis, MO, USA) and grown until 70% confluent. The calcitriol or calcidiol (Sigma, St. Louis, MO, USA) was dissolved in 100% ethanol to obtain 100 $\mu$M stocks. After that stocks were diluted in media to obtain 10 and 100 nM. Concentration of ethanol in media never exceeded 1%. The cells were incubated with vitamin D compounds for 24 h before irradiation. To study the influence of calcitriol or calcidiol on cell proliferation, cell growth curves were determined by manual counting the cells growing with vitamin $D_3$ metabolites at different concentrations every 24 h up to 120 h. Counts were performed with the use of haemocytometer, by triplicate by one analyst under a 40× objective according to the standard methodology. Influence on cell response to proton radiation was tested by cultivating cells for 24 h prior to irradiation in a culture medium with 10% FBS, antibiotics and vitamin $D_3$ metabolites as indicated.

### 4.3. Irradiation Procedure

The irradiation was performed at the Institute of Nuclear Physics of the Polish Academy of Sciences (IFJ PAN) in Krakow uses he Proteus C-235 cyclotron produced proton beam. The 230 MeV proton beam, after degradation to 70 MeV, was delivered to the treatment room with a small field horizontal beam line. The passive scattering technique and rotating energy modulator were used to forming the irradiation field. A 40 mm diameter, fully modulated proton beam with energy of 61 MeV, spread-out Bragg Peak (SOBP) with 31.5 mm range and 31.5 mm modulation (measured in water), was used for cell irradiation. Dose of 1, 3, or 5 Gy at dose rates of 1 Gy/min, 2 Gy/min and 6.6 Gy/min respectively, were delivered to the samples. The dose averaged $LET_d$ calculated at the depth of 15.8 mm in the SOBP that is, at the centre of the cell container position, was 2.8 keV/μm. The dosimetry was performed according to the recommendations of IAEA TRS-398 protocol [74]. A semiflex ionisation chamber with 0.125 $cm^3$ active volume and a PTW reference class UNIDOS Webline electrometer calibrated at the IFJ PAN with $^{60}Co$ radiation source were used. A dedicated PMMA phantom with a holder for the Eppendorf container was used during irradiation. Cells were irradiated in Eppendorf tubes positioned in the phantom, orthogonally to the direction of the proton beam. Cell suspension in phosphate buffer saline (PBS) at 1 × 106 cells/mL was transported on ice between the facilities, including the untreated (non-irradiated) control. Cells were transferred to the culture medium and placed at 1 × 105 cell/mL in 24-well plates. Every 24 h, for 5 days, cells from 6 wells were removed and counted. The experiment was repeated 3 times for each cell line tested.

### 4.4. Data Analysis

Data from cell counting was used to determine cell growth rates, doubling times and surviving fractions. Since the cells in most of the experimental groups exhibited logistic growth, the first step was to establish the duration of exponential growth phase (96 h). Then, growth rates ($gr$):

$$N_t = N_0 e^{gr \cdot t} \tag{1}$$

and doubling times ($t_{doubling\ time}$):

$$t_{doubling\ time} = \frac{\ln(2)}{gr} \tag{2}$$

where calculated by exponential function fitting, using the nonlinear least squares method. Additionally, time delay for irradiated groups was calculated as previously described [34]. Those values were used to calculate surviving fractions:

$$SF = 2^{-\frac{t_{delay}}{t_{doubling\ time}}} \tag{3}$$

Establishing the relationship between the physical conditions of irradiation and its biological effects is a starting point for any radiobiological experiment. Such relations are known as survival curves and presented as the dependence of cell survival probability (Surviving Fractions—SF) on the absorbed radiation dose (in Gy). Many theoretical models of ionizing radiation–induced cell killing have been proposed and described in literature but because the fundamental mechanisms leading to lethal cell damage are not well understood, the models have semi-empirical character and many limitations [36]. Our experimental survival curves were analysed within framework of two classical models describing the response of cells populations to ionizing radiation (IR). The models have a few basic assumptions in common: cell inactivation is treated as a multistep process, cells are killed by energy absorption deposited in a sensitive volume of the cell and radiation-induced lethal events have Poisson distribution. The first one is a "molecular model," widely known as the linear-quadratic model (LQ):

$$SF(D) = e^{-(\alpha D + \beta D^2)} \quad (4)$$

describing the logarithmic plot of SF in linear and quadratic dose–dependent terms, where $D$ is the absorbed dose. The experimentally determined parameters $\alpha$ and $\beta$ are interpreted as rates of cell annihilation by a single-hit and double-hit mechanism, respectively. Molecular interpretation of LQ model is based on the following assumptions: DNA is a critical target, radiation produces the breakage in DNA strands, the broken bonds in DNA strand can be repaired and the critical damage leading to cell death is a double strand break (DSB), resulting in non-repairable lesions. The linear part ($\alpha$) represents a lethal lesion produced by a direct induction of DSB (one track hit) and the quadratic term ($\beta$) is a result of two single strand breaks (SSB) which could be repaired (a sub-lethal lesion).

The second model used for analysis of our survival curves is a multitarget–single hit (MTSH) model:

$$SF(D) = 1 - (1 - e^{-\frac{D}{D_0}})^n \quad (5)$$

$D_0$ = 1/k is the dose for 1/e survival in the linear portion of the plot, where k—inactivation constant for each target (or mean lethal dose), $n$—target multiplicity, number of targets per cell that must be inactivated for cell death.

Radiation survival curves are presented as log-linear plots of surviving fraction and dose, respectively; surviving fraction is plotted along the vertical axis (normalized to unity when $D = 0$), radiation dose is plotted along the horizontal axis.

Both models were fitted to the dose dependence of survival fraction, using the nonlinear least squares method. All analyses were carried out with in-house written Matlab (2014b, MathWorks, Natick, MA, USA) scripts. Optimization of the fitting procedures (both for growth and survival curves) was conducted until mean R-squared from all curves was higher than 0.9. Besides parameters values, their standard errors were calculated and used for the calculation of weighted mean and its error within experimental groups.

### 4.5. Statistical Analysis

Results were presented as weighted mean and weighted mean error. All analyses were performed with STATISTICA 13 software (Stat-Soft Inc., Tulsa, OK, USA), with the use of weights computed during the fitting procedure. For each parameter, Shapiro-Wilk normality test and Levene's test were used to determine a normal distribution and equality of variances, respectively. Depending on the data, a one-way ANOVA, followed by post-hoc Tukey's HSD test or Kruskal–Wallis H test, followed by Dunn's test was performed. $p$ values smaller than 0.05 were considered statistically significant.

## 5. Conclusions

We conclude that the effect of vitamin $D_3$ metabolites on the proliferation and response to proton radiation in the studied melanoma cell lines is not straightforward. As regards cell proliferation, the impact of melanin content and other factors such as VDR, CYP27A1, CYP27B1 and CYP24A1 gene expressions as well as VDR localization and interaction with other receptors should be considered in the differential responses of cells to calcitriol and calcidiol. The studied vitamin $D_3$ derivatives have proven to be potent radiosensitizers in proton therapy. Radiosensitization by vitamin $D_3$ was particularly visible for low radiation doses. Interestingly, for higher doses the observed upward-bending of SF curves in the presence of vitamin $D_3$ implies involvement of repair mechanisms and interaction between VDR and radiation-induced signalling pathways. Our results indicate the need for a further investigation of the underlying molecular mechanisms and suggest that vitamin $D_3$ can be a promising agent, capable of modifying proton beam therapy efficacy and thus offering a new option in cancer therapy.

**Author Contributions:** E.P. and A.D. performed the experiments and analysed the results, Z.M. tested different models for data fitting, J.S. performed proton beam dosimetry and cell irradiation, A.S. contributed towards discussion of the results and revised the manuscript, M.E. edited and revised the text, K.U. was responsible for the concept of the paper.

**Funding:** This study was supported partially by grant: 2012/07/B/NZ4/01657 from the Polish National Science Centre to Krystyna Urbanska. Faculty of Biochemistry, Biophysics and Biotechnology of the Jagiellonian University is a partner of the Leading National Research Center (KNOW) supported by the Ministry of Science and Higher Education. Partial support from NIH grants R21AR066505, 1R01AR071189-01A1 and R01AR073004 to Andrzej Slominski is also acknowledged. Partial support to Martyna Elas was given by Horizon 2020 grant UMCure2020 no. 667787.

**Conflicts of Interest:** The authors declare no conflict of interest.

## Abbreviations

| | |
|---|---|
| ATM | Ataxia-telangiectasia mutated kinase |
| BHM | Bomirski hamster melanoma |
| CSDA | Continues slowing down approximation |
| DSB | Double strand break |
| IR | Ionizing radiation |
| LQ | Linear-quadratic |
| mTor | Mammalian target of rapamycin |
| MTSH | Multitarget—single hit |
| SF | Surviving fraction |
| SOBP | Spread-out bragg peak |
| SSB | Single strand break |
| VDR | Vitamin D receptor |

## References

1. Matthews, N.H.; Li, W.-Q.; Qureshi, A.A.; Weinstock, M.A.; Cho, E. Epidemiology of Melanoma. In *Cutaneous Melanoma: Etiology and Therapy*; Codon Publications: Brisbane, Australia, 2017.
2. Slominski, A.T.; Carlson, J.A. Melanoma resistance: A bright future for academicians and a challenge for patient advocates. *Mayo Clin. Proc.* **2014**, *89*, 429–433. [CrossRef] [PubMed]
3. Rajkumar, S.; Watson, I.R. Molecular characterisation of cutaneous melanoma: Creating a framework for targeted and immune therapies. *Br. J. Cancer* **2016**, *115*, 145–155. [CrossRef] [PubMed]
4. Su, M.Y.; Fisher, D.E. Immunotherapy in the Precision Medicine Era: Melanoma and Beyond. *PLoS Med.* **2016**, *13*, e1002196. [CrossRef] [PubMed]
5. Levesque, M.P.; Cheng, P.F.; Raaijmakers, M.I.G.; Saltari, A.; Dummer, R. Metastatic melanoma moves on: Translational science in the era of personalized medicine. *Cancer Metastasis Rev.* **2017**, *36*, 7–21. [CrossRef] [PubMed]
6. Ordonez-Moran, P.; Larriba, M.J.; Pendas-Franco, N.; Aguilera, O.; Gonzalez-Sancho, J.M.; Munoz, A. Vitamin D and cancer: An update of in vitro and in vivo data. *Front. Biosci.* **2005**, *10*, 2723–2749. [CrossRef] [PubMed]
7. Feldman, D.; Krishnan, A.V.; Swami, S.; Giovannucci, E.; Feldman, B.J. The role of vitamin D in reducing cancer risk and progression. *Nat. Rev. Cancer* **2014**, *14*, 342–357. [CrossRef] [PubMed]
8. Grant, W.B. Roles of Solar UVB and Vitamin D in Reducing Cancer Risk and Increasing Survival. *Anticancer Res.* **2016**, *36*, 1357–1370. [PubMed]
9. Slominski, A.T.; Brozyna, A.A.; Zmijewski, M.A.; Józwicki, W.; Jetten, A.M.; Mason, R.S.; Tuckey, R.C.; Elmets, C.A. Vitamin D signalling and melanoma: Role of Vitamin D and its receptors in melanoma progression and management. *Lab. Investig.* **2017**, *97*, 706–724. [CrossRef] [PubMed]
10. Slominski, A.T.; Brożyna, A.A.; Skobowiat, C.; Zmijewski, M.A.; Kim, T.K.; Janjetovic, Z.; Oak, A.S.; Jozwicki, W.; Jetten, A.M.; Mason, R.S.; et al. On the role of classical and novel forms of vitamin D in melanoma progression and management. *J. Steroid Biochem. Mol. Biol.* **2018**, *177*, 159–170. [CrossRef] [PubMed]

11. Wasiewicz, T.; Szyszka, P.; Cichorek, M.; Janjetovic, Z.; Tuckey, R.C.; Slominski, A.T.; Zmijewski, M.A. Antitumor effects of vitamin D analogues on hamster and mouse melanoma cell lines in relation to melanin pigmentation. *Int. J. Mol. Sci.* **2015**, *16*, 6645–6667. [CrossRef] [PubMed]
12. Polar, M.K.; Gennings, C.; Park, M.; Gupta, M.S.; Gewirtz, D.A. Effect of the vitamin $D_3$ analog ILX 23-7553 on apoptosis and sensitivity to fractionated radiation in breast tumour cells and normal human fibroblasts. *Cancer Chemother. Pharmacol.* **2003**, *51*, 415–421. [CrossRef] [PubMed]
13. Sundaram, S.; Sea, A.; Feldman, S.; Strawbridge, R.; Hoopes, P.J.; Demidenko, E.; Binderup, L.; Gewirtz, D.A. The combination of a potent vitamin $D_3$ analog, EB 1089, with ionizing radiation reduces tumour growth and induces apoptosis of MCF-7 breast tumour xenografts in nude mice. *Clin. Cancer Res.* **2003**, *9*, 2350–2356. [PubMed]
14. Wang, Z.; Liu, J.; Si, S.; Xu, B.; Wang, Y.; Ren, Y.; Dai, Z.; Cui, D.; Su, X.; Song, S. Vitamin D improves the sensitization of lung cancer to radiotherapy. *Int. J. Clin. Exp. Med.* **2017**, *10*, 4921–4928.
15. Sundaram, S.; Gewirtz, D.A. The vitamin $D_3$ analog EB 1089 enhances the response of human breast tumor cells to radiation. *Radiat. Res.* **1999**, *152*, 479–486. [CrossRef] [PubMed]
16. Chaudhary, M.; Sundaram, S.; Gennings, C.; Carter, H.; Gewirtz, D.A. The vitamin $D_3$ analog, ILX-23-7553, enhances the response to Adriamycin and irradiation in MCF-7 breast tumor cells. *Cancer Chemother. Pharmacol.* **2001**, *47*, 429–436. [CrossRef]
17. Weitsman, G.E.; Koren, R.; Zuck, E.; Rotem, C.; Liberman, U.A.; Ravid, A. Vitamin D sensitizes breast cancer cells to the action of $H_2O_2$: Mitochondria as a convergence point in the death pathway. *Free Radic. Biol. Med.* **2005**, *39*, 266–278. [CrossRef] [PubMed]
18. Dunlap, N.; Schwartz, G.G.; Eads, D.; Cramer, S.D.; Sherk, A.B.; John, V.; Koumenis, C. 1$\alpha$,25-Dihydroxyvitamin $D_3$ (calcitriol) and its analogue, 19-nor-1$\alpha$,25$(OH)_2D_2$, potentiate the effects of ionising radiation on human prostate cancer cells. *Br. J. Cancer* **2003**, *89*, 746–753. [CrossRef] [PubMed]
19. Slominski, A.T.; Brozyna, A.; Jozwicki, W.; Tuckey, R.C. Vitamin D as an adjuvant in melanoma therapy. *Melanoma Manag.* **2015**, *2*, 1–4. [CrossRef] [PubMed]
20. Wu, W.; Beilhartz, G.; Roy, Y.; Richard, C.L.; Curtin, M.; Brown, L.; Cadieux, D.; Coppolino, M.; Farach-Carson, M.C.; Nemere, I.; et al. Nuclear translocation of the 1,25$D_3$-MARRS (membrane associated rapid response to steroids) receptor protein and NFκB in differentiating NB4 leukemia cells. *Exp. Cell Res.* **2010**, *316*, 1101–1108. [CrossRef] [PubMed]
21. Tan, C.; De Noronha, R.G.; Roecker, A.J.; Pyrzynska, B.; Khwaja, F.; Zhang, Z.; Zhang, H.; Teng, Q.; Nicholson, A.C.; Giannakakou, P.; et al. Identification of a novel small-molecule inhibitor of the hypoxia-inducible factor 1 pathway. *Cancer Res.* **2005**, *65*, 605–612. [CrossRef] [PubMed]
22. Skobowiat, C.; Oak, A.S.W.; Kim, T.-K.; Yang, C.H.; Pfeffer, L.M.; Tuckey, R.C.; Slominski, A.T. Noncalcemic 20-hydroxyvitamin $D_3$ inhibits human melanoma growth in in vitro and in vivo models. *Oncotarget* **2016**, *8*, 9823–9834. [CrossRef] [PubMed]
23. Brozyna, A.A.; Jozwicki, W.; Janjetovic, Z.; Slominski, A.T. Expression of vitamin D receptor decreases during progression of pigmented skin lesions. *Hum. Pathol.* **2011**, *42*, 618–631. [CrossRef] [PubMed]
24. Brozyna, A.A.; Jóźwicki, W.; Janjetovic, Z.; Slominski, A.T. Expression of the vitamin D-activating enzyme 1$\alpha$-hydroxylase (CYP27B1) decreases during melanoma progression. *Hum. Pathol.* **2013**, *44*, 374–387. [CrossRef] [PubMed]
25. Brozyna, A.A.; Jóźwicki, W.; Slominski, A.T. Decreased VDR expression in cutaneous melanomas as marker of tumor progression: New data and analyses. *Anticancer Res.* **2014**, *34*, 2735–2744. [PubMed]
26. Brożyna, A.A.; Jochymski, C.; Janjetovic, Z.; Jóźwicki, W.; Tuckey, R.C.; Slominski, A.T. CYP24A1 expression inversely correlates with melanoma progression: Clinic-pathological studies. *Int. J. Mol. Sci.* **2014**, *15*, 19000–19017. [CrossRef] [PubMed]
27. Chagani, S.; Kyryachenko, S.; Yamamoto, Y.; Kato, S.; Ganguli-Indra, G.; Indra, A.K. In vivo Role of Vitamin D Receptor Signaling in UVB-Induced DNA Damage and Melanocyte Homeostasis. *J. Investig. Dermatol.* **2016**, *136*, 2108–2111. [CrossRef] [PubMed]
28. Girdhani, S.; Sachs, R.; Hlatky, L. Biological Effects of Proton Radiation: What We Know and Don't Know. *Radiat. Res.* **2013**, *179*, 257–272. [CrossRef] [PubMed]
29. Gerelchuluun, A.; Hong, Z.; Sun, L.; Suzuki, K.; Terunuma, T.; Yasuoka, K.; Sakae, T.; Moritake, T.; Tsuboi, K. Induction of in situ DNA double-strand breaks and apoptosis by 200 MeV protons and 10 MV X-rays in human tumour cell lines. *Int. J. Radiat. Biol.* **2011**, *87*, 57–70. [CrossRef] [PubMed]

30. Jasińska-Konior, K.; Pochylczuk, K.; Czajka, E.; Michalik, M.; Romanowska-Dixon, B.B.; Swakoń, J.; Urbańska, K.; Elas, M.; Jasinska-Konior, K.; Pochylczuk, K.; et al. Proton beam irradiation inhibits the migration of melanoma cells. *PLoS ONE* **2017**, *12*, e0186002. [CrossRef] [PubMed]
31. Grosse, N.; Fontana, A.O.; Hug, E.B.; Lomax, A.; Coray, A.; Augsburger, M.; Paganetti, H.; Sartori, A.A.; Pruschy, M. Deficiency in homologous recombination renders mammalian cells more sensitive to proton versus photon irradiation. *Int. J. Radiat. Oncol. Biol. Phys.* **2014**, *88*, 175–181. [CrossRef] [PubMed]
32. Antoccia, A.; Sgura, A.; Berardinelli, F.; Cavinato, M.; Cherubini, R.; Gerardi, S.; Tanzarella, C. Cell Cycle Perturbations and Genotoxic Effects in Human Primary Fibroblasts Induced by Low-energy Protons and X/$\gamma$-rays. *J. Radiat. Res.* **2009**, *50*, 457–468. [CrossRef] [PubMed]
33. Śniegocka, M.; Podgórska, E.; Plonka, P.M.; Elas, M.; Romanowska-Dixon, B.; Szczygiel, M.; Żmijewski, M.A.; Cichorek, M.; Markiewicz, A.; Brożyna, A.A.; et al. Transplantable melanomas in hamsters and gerbils as models for human melanoma. Sensitization in melanoma radiotherapy—From animal models to clinical trials. *Int. J. Mol. Sci.* **2018**, *19*, 1048. [CrossRef] [PubMed]
34. Buch, K.; Peters, T.; Nawroth, T.; Sänger, M.; Schmidberger, H.; Langguth, P. Determination of cell survival after irradiation via clonogenic assay versus multiple MTT Assay—A comparative study. *Radiat. Oncol.* **2012**, *7*, 1. [CrossRef] [PubMed]
35. Franken, N.A.P.; Oei, A.L.; Kok, H.P.; Rodermond, H.M.; Sminia, P.; Crezee, J.; Stalpers, L.J.A.; Barendsen, G.W. Cell survival and radiosensitisation: Modulation of the linear and quadratic parameters of the LQ model (Review). *Int. J. Oncol.* **2013**, *42*, 1501–1515. [CrossRef] [PubMed]
36. Bodgi, L.; Canet, A.; Pujo-Menjouet, L.; Lesne, A.; Victor, J.-M.; Foray, N. Mathematical models of radiation action on living cells: From the target theory to the modern approaches. A historical and critical review. *J. Theor. Biol.* **2016**, *394*, 93–101. [CrossRef] [PubMed]
37. Mehta, R.G.; Hussain, E.A.; Mehta, R.R.; Das Gupta, T.K. Chemoprevention of mammary carcinogenesis by 1$\alpha$-hydroxyvitamin D5, a synthetic analog of Vitamin D. *Mutat. Res.* **2003**, *523–524*, 253–264. [CrossRef]
38. Welsh, J.; Wietzke, J.A.; Zinser, G.M.; Byrne, B.; Smith, K.; Narvaez, C.J. Vitamin D-3 Receptor as a Target for Breast Cancer Prevention. *J. Nutr.* **2003**, *133*, 2425S–2433S. [CrossRef] [PubMed]
39. Lamprecht, S.A.; Lipkin, M. Chemoprevention of colon cancer by calcium, vitamin D and folate: Molecular mechanisms. *Nat. Rev. Cancer* **2003**, *3*, 601–614. [CrossRef] [PubMed]
40. Nakagawa, K. 1,25-Dihydroxyvitamin $D_3$ is a preventive factor in the metastasis of lung cancer. *Carcinogenesis* **2004**, *26*, 429–440. [CrossRef] [PubMed]
41. Krishnan, A.V.; Medicine, S.; Feldman, D. Vitamin D Analogs in Cancer Prevention and Therapy. **2003**, *164*. [CrossRef]
42. Szyszka, P.; Zmijewski, M.A.; Slominski, A.T. New vitamin D analogs as potential therapeutics in melanoma. *Expert Rev. Anticancer Ther.* **2012**, *12*, 585–599. [CrossRef] [PubMed]
43. Moreno, J.; Krishnan, A.V.; Feldman, D. Molecular mechanisms mediating the anti-proliferative effects of Vitamin D in prostate cancer. *J. Steroid Biochem. Mol. Biol.* **2005**, *97*, 31–36. [CrossRef] [PubMed]
44. Banerjee, P.; Chatterjee, M. Antiproliferative role of vitamin D and its analogs—A brief overview. *Mol. Cell. Biochem.* **2003**, *253*, 247–254. [CrossRef] [PubMed]
45. Thill, M.; Woeste, A.; Reichert, K.; Fischer, D.; Rody, A.; Friedrich, M.; Köster, F. Vitamin D inhibits ovarian cancer cell line proliferation in combination with celecoxib and suppresses cyclooxygenase-2 expression. *Anticancer Res.* **2015**, *35*, 1197–1203. [PubMed]
46. Chen, J.; Wang, J.; Kim, T.K.; Tieu, E.W.; Tang, E.K.Y.; Lin, Z.; Kovacic, D.; Miller, D.D.; Postlethwaite, A.; Tuckey, R.C.; et al. Novel vitamin D analogs as potential therapeutics: Metabolism, toxicity profiling and antiproliferative activity. *Anticancer Res.* **2014**, *34*, 2153–2164. [PubMed]
47. Kozielewicz, P.; Grafton, G.; Kutner, A.; Curnow, S.J.; Gordon, J.; Barnes, N.M. Novel vitamin D analogues; cytotoxic and anti-proliferative activity against a diffuse large B-cell lymphoma cell line and B-cells from healthy donors. *J. Steroid Biochem. Mol. Biol.* **2016**, *164*, 98–105. [CrossRef] [PubMed]
48. Humeniuk-Polaczek, R.; Marcinkowska, E. Impaired nuclear localization of vitamin D receptor in leukemia cells resistant to calcitriol-induced differentiation. *J. Steroid Biochem. Mol. Biol.* **2004**, *88*, 361–366. [CrossRef] [PubMed]
49. Agadir, A.; Lazzaro, G.; Zheng, Y.; Zhang, X.-K.; Mehta, R. Resistance of HBL100 human breast epithelial cells to vitamin D action. *Carcinogenesis* **1999**, *20*, 577–582. [CrossRef] [PubMed]

50. Seifert, M.; Rech, M.; Meineke, V.; Tilgen, W.; Reichrath, J. Differential biological effects of 1,25-dihydroxyVitamin $D_3$ on melanoma cell lines in vitro. *J. Steroid Biochem. Mol. Biol.* **2004**, *89–90*, 375–379. [CrossRef] [PubMed]
51. Reichrath, J.; Rech, M.; Moeini, M.; Meese, E.; Tilgen, W.; Seifert, M. In vitro comparison of the vitamin D endocrine system in 1,25(OH)$_2$D$_3$-responsive and -resistant melanoma cells. *Cancer Biol. Ther.* **2007**, *6*, 48–55. [CrossRef] [PubMed]
52. Slominski, A.T.; Janjetovic, Z.; Kim, T.K.; Wright, A.C.; Grese, L.N.; Riney, S.J.; Nguyen, M.N.; Tuckey, R.C. Novel vitamin D hydroxyderivatives inhibit melanoma growth and show differential effects on normal melanocytes. *Anticancer Res.* **2012**, *32*, 3733–3742. [PubMed]
53. Sun, J.; Yin, M.; Zhu, S.; Liu, L.; Zhu, Y.; Wang, Z.; Xu, R.X.; Chang, S. Ultrasound-mediated destruction of oxygen and paclitaxel loaded lipid microbubbles for combination therapy in hypoxic ovarian cancer cells. *Ultrason. Sonochem.* **2016**, *28*, 319–326. [CrossRef] [PubMed]
54. Janjetovic, Z.; Brozyna, A.A.; Tuckey, R.C.; Kim, T.K.; Nguyen, M.N.; Jozwicki, W.; Pfeffer, S.R.; Pfeffer, L.M.; Slominski, A.T. High basal NF-κB activity in nonpigmented melanoma cells is associated with an enhanced sensitivity to vitamin $D_3$ derivatives. *Br. J. Cancer* **2011**, *105*, 1874–1884. [CrossRef] [PubMed]
55. Sarna, M.; Zadlo, A.; Czuba-Pelech, B.; Urbanska, K. Nanomechanical phenotype of melanoma cells depends solely on the amount of endogenous pigment in the cells. *Int. J. Mol. Sci.* **2018**, *19*, 607. [CrossRef] [PubMed]
56. Wietrzyk, J.; Nevozhay, D.; Filip, B.; Milczarek, M.; Kutner, A. The antitumor effect of lowered doses of cytostatics combined with new analogs of vitamin D in mice. *Anticancer Res.* **2007**, *27*, 3387–3398. [PubMed]
57. Pelczynska, M.; Switalska, M.; Maciejewska, M.; Jaroszewicz, I.; Kutner, A.; Opolski, A. Antiproliferative activity of vitamin D compounds in combination with cytostatics. *Anticancer Res.* **2006**, *26*, 2701–2705. [PubMed]
58. Ma, Y.; Yu, W.D.; Trump, D.L.; Johnson, C.S. 1,25D$_3$ enhances antitumor activity of gemcitabine and cisplatin in human bladder cancer models. *Cancer* **2010**, *116*, 3294–3303. [CrossRef] [PubMed]
59. Newhauser, W.D.; Zhang, R. The physics of proton therapy. *Phys. Med. Biol.* **2015**, *60*, R155–R209. [CrossRef] [PubMed]
60. Fokas, E.; Kraft, G.; An, H.; Engenhart-Cabillic, R. Ion beam radiobiology and cancer: Time to update ourselves. *Biochim. Biophys. Acta Rev. Cancer* **2009**, *1796*, 216–229. [CrossRef] [PubMed]
61. Folkard, M.; Prise, K.M.; Vojnovic, B.; Newman, H.C.; Roper, M.J.; Michael, B.D. Inactivation of V79 cells by low-energy protons, deuterons and helium-3 ions. *Int. J. Radiat. Biol.* **1996**, *69*, 729–738. [CrossRef] [PubMed]
62. Kundrät, P. Detailed analysis of the cell-inactivation mechanism by accelerated protons and light ions. *Phys. Med. Biol.* **2006**, *51*, 1185–1199. [CrossRef] [PubMed]
63. Kundrát, P.; Lokajíček, M.; Hromčíková, H. Probabilistic two-stage model of cell inactivation by ionizing particles. *Phys. Med. Biol.* **2005**, *50*, 1433–1447. [CrossRef] [PubMed]
64. Trémezaygues, L.; Seifert, M.; Vogt, T.; Tilgen, W.; Reichrath, J. 1,25-Dihydroxyvitamin $D_3$ modulates effects of ionizing radiation (IR) on human keratinocytes: In vitro analysis of cell viability/proliferation, DNA-damage and -repair. *J. Steroid Biochem. Mol. Biol.* **2010**, *121*, 324–327. [CrossRef] [PubMed]
65. Murray, P.J.; Cornelissen, B.; Vallis, K.A.; Chapman, S.J. DNA double-strand break repair: A theoretical framework and its application. *J. R. Soc. Interface* **2016**, *13*, 20150679. [CrossRef] [PubMed]
66. Murray, D.; Mirzayans, R.; Parliament, M. New insights into patient radiosensitivity in relation to the biology of the ATM protein. *Transl. Cancer Res.* **2016**, *5*, S809–S814. [CrossRef]
67. Janczyk, A.; Wolnicka-Glubisz, A.; Chmura, A.; Elas, M.; Matuszak, Z.; Stochel, G.; Urbanska, K. NO-dependent phototoxicity of Roussin's black salt against cancer cells. *Nitric Oxide Biol. Chem.* **2004**, *10*, 42–50. [CrossRef] [PubMed]
68. Brindell, M.; Kuliś, E.; Elmroth, S.K.C.; Urbańska, K.; Stochel, G. Light-induced anticancer activity of [RuCl$_2$(DMSO)$_4$] complexes. *J. Med. Chem.* **2005**, *48*, 7298–7304. [CrossRef] [PubMed]
69. Bomirski, A.; Słominski, A.; Bigda, J. The natural history of a family of transplantable melanomas in hamsters. *Cancer Metastasis Rev.* **1988**, *7*, 95–118. [CrossRef] [PubMed]
70. Słominski, A. Rapid melanization of Bomirski amelanotic melanoma cells in cell culture. *Biosci. Rep.* **1983**, *3*, 189–194. [CrossRef] [PubMed]
71. Słomiński, A. Some properties of Bomirski Ab amelanotic melanoma cells, which underwent spontaneous melanization in primary cell culture—Growth kinetics, cell morphology, melanin content and tumorigenicity. *J. Cancer Res. Clin. Oncol.* **1985**, *109*, 29–37. [CrossRef] [PubMed]

72. Słominski, A.; Moellmann, G.; Kuklinska, E.; Bomirski, A.; Pawelek, J. Positive regulation of melanin pigmentation by two key substrates of the melanogenic pathway, L-tyrosine and L-dopa. *J. Cell Sci.* **1988**, *89*, 287–296. [PubMed]
73. Słomiński, A.; Bomirski, A. Phenotypic changes of Ab hamster melanoma during long-term culture. *Anticancer Res.* **1985**, *5*, 403–409. [PubMed]
74. Boberek, M.; Swakoń, J.; Stolarczyk, L.; Olko Paweł and Waligórski, M. A monitoring system for the 60 MeV radiotherapy proton beam at IFJ PAN using a scintillating screen and a CCD camera. *Rom. Rep. Phys.* **2014**, *66*, 5–15.

International Journal of
*Molecular Sciences*

MDPI

*Review*

# Vitamin D and Uterine Fibroids—Review of the Literature and Novel Concepts

**Michał Ciebiera [1,*], Marta Włodarczyk [2,3], Magdalena Ciebiera [4], Kornelia Zaręba [1], Krzysztof Łukaszuk [5,6] and Grzegorz Jakiel [1]**

1 Department of Obstetrics and Gynecology, Centre of Postgraduate Medical Education, 00-416 Warsaw, Poland; kornelia3@poczta.onet.pl (K.Z.); grzegorz.jakiel1@o2.pl (G.J.)
2 Department of Biochemistry and Pharmacogenomics, Faculty of Pharmacy with Division of Laboratory Medicine, Medical University of Warsaw, 02-097 Warsaw, Poland; mdwlodarczyk@gmail.com
3 Center for Preclinical Research, Medical University of Warsaw, 02-097 Warsaw, Poland
4 Students' Scientific Association at the Department of Obstetrics and Gynecology, Medical University of Warsaw, 02-015 Warsaw, Poland; mciebiera93@gmail.com
5 Department of Obstetrics and Gynecological Nursing, Faculty of Health Sciences, Medical University of Gdansk, 80-210 Gdansk, Poland; luka@gumed.edu.pl
6 INVICTA Fertility and Reproductive Center, 80-172 Gdansk, Poland
* Correspondence: michal.ciebiera@gmail.com; Tel.: +48-607-155-177

Received: 13 June 2018; Accepted: 12 July 2018; Published: 14 July 2018

**Abstract:** This article provides a detailed review of current knowledge on the role of vitamin D and its receptor in the biology and management of uterine fibroids (UFs). Authors present ideas for future steps in this area. A literature search was conducted in PubMed using the following key words: "uterine fibroid" and "vitamin D". The results of the available studies, published in English from January 2002 up to April 2018, have been discussed. Vitamin D is a group of steroid compounds with a powerful impact on many parts of the human body. This vitamin is believed to regulate cell proliferation and differentiation, inhibit angiogenesis, and stimulate apoptosis. Nowadays, hypovitaminosis D is believed to be a major risk factor in the development of UFs. In many studies vitamin D appears to be a powerful factor against UFs, resulting in inhibition of tumor cell division and a significant reduction in its size, however, the exact role of this compound and its receptor in the pathophysiology of UFs is not fully understood. According to available studies, vitamin D and its analogs seem to be promising, effective, and low-cost compounds in the management of UFs and their clinical symptoms, and the anti-tumor activities of vitamin D play an important role in UF biology. The synergy between vitamin D and selected anti-UF drugs is a very interesting issue which requires further research. Further studies about the biological effect of vitamin D on UF biology are essential. Vitamin D preparations (alone or as a co-drugs) could become new tools in the fight with UFs, with the additional beneficial pleiotropic effect.

**Keywords:** uterine fibroid; leiomyoma; vitamin D; vitamin D receptor; vitamin D analogs

## 1. Introduction

Uterine fibroids (UFs) are monoclonal, benign tumors which arise from the smooth muscle cells of the uterus and are predominantly located in the pelvis. They constitute one of the most common pathologies of the female genital tract. UFs occur in 5–70% of all women [1–4], and develop in selected populations, with age and ethnicity as the main risk factors [3,4]. UFs are more prevalent among reproductive aged women and are not observed in pre-pubescent girls, indicating that tumor origin depends on hormonal changes [1,2,5]. Extensive research has identified several factors connected with higher UF occurrence, but data are inconsistent and conflicting [3,6]. In a recently published systematic review, Stewart et al., underlined factors which increase the incidence of UF, and these are: black race, elevated body mass index (BMI), age, premenopausal status, hypertension, positive family history, time elapsed since last labor, consumption of food additives, and soybean milk [3].

Apart from the abovementioned risk factors, recent studies suggested that hypovitaminosis D plays a role in UF development [6–9]. The Study of Environment Lifestyle and Fibroids (SELF) was performed to describe the actual contribution of hypovitaminosis D and other factors to the development of fibroids [10]. Most of the research on this field focuses on Afro-American women, who are at an increased risk for UF occurrence [9,11]. Among Afro-American women, vitamin D deficiency because of higher melanin concentrations results in decreased serum vitamin D levels [7] and reduced expression of the vitamin D receptor (VDR) in the adjacent myometrium [12], compared to white women [13].

UF tumors vary greatly in size, location, and symptoms [2,14]. Most tumors are largely asymptomatic, but they may also cause a wide range of severe and chronic symptoms [3,4,15] in approximately one-quarter to one-third of the affected women [3,14]. The most common symptoms include abnormal and excessive uterine bleeding, secondary iron deficiency anemia, abdominal and pelvic pain, gastric disorders like bloating and constipation, voiding symptoms, infertility and obstetric pathologies (including miscarriage and premature labor) [1,2,14,16,17].

Despite their unquestionable effect on the quality of patient life (QoL), UF-related QoL is very often marginalized [18,19]. Also, the financial burden on the healthcare budget is considerable, including the costs of preoperative diagnosis, surgical treatment, hospitalization time, work absenteeism, medicines, salaries of the medical workers, and the costs of control visits [4,20,21]. The annual direct and indirect costs of UFs in the United States have been estimated at approximately $4.1–9.4 billion and $1.6–17.2 billion, respectively [4,20,21].

Clinically symptomatic UFs are most often treated with surgery [2,22]. Various types of surgical methods are available, both open and endoscopic (hysterectomies, myomectomies, hysteroscopic resections) [22,23]. UFs are the leading reason for the hysterectomy [21,24]. The optimal treatment should reduce blood loss and tumor burden, while preserving fertility [25]. Women who wish to retain their uterus can be offered with less invasive methods [22,23,26]. Nevertheless, many of them will require a re-intervention in the future [27]. Due to the benign nature of the tumors, the first-line treatment should result in lowest morbidity and risk of adverse effects [23,28]. Multiple evidence has suggested progesterone to be the major initiator of UF development and stimulator of their further growth [29,30]. Thus, it is not surprising that ulipristal acetate (UPA), a selective progesterone receptor modulator (SPRM), has become one of the most popular pharmacological treatments of UFs [23,28,31]. Due to its effectiveness, UPA is administered as first-line therapy to prepare UFs for surgery. In some cases, if the effect is satisfying, UPA can be use as the only treatment [31]. However, UPA is not inexpensive, nor is it a substance which can be widely used in prevention for a long time [32]. Also, the European Medicines Agency (EMA) has recently issued warning about the risk of liver failure after UPA use [33], but research is still ongoing and there is not enough information at present about the matter.

The origin of UFs is multifactorial and that is why there are no specific methods of prevention at present [34,35]. Numerous attempts have been made to create inexpensive, safe, and effective methods of prophylaxis but they are still in the early stages [34,35]. In light of this, vitamin D, which plays one of the major roles in UF biology, might be the answer [13].

Vitamin D is a name for a group of steroid compounds, soluble in fats, which exert powerful effects on the human body, and whose receptors are found in various organs [36,37], including the myometrium and UF tumor tissue [38] (Figure 1).

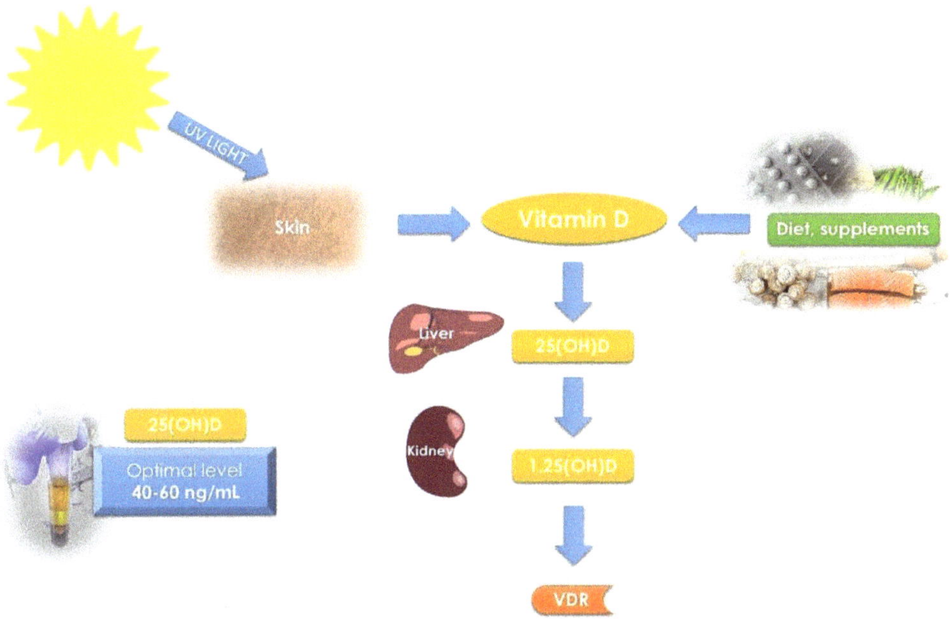


**Figure 1.** Vitamin D metabolism and schematic pathways. Diet, supplements and sunlight are the major source of vitamin D in humans. Vitamin D is synthetized in skin from 7-dehydrocholesterol. In further steps liver converts it to 25(OH)D and then kidney to 1,25(OH)D. Optimal vitamin D serum levels were described as 25(OH)D of 40–60 ng/mL [39].

Vitamin D takes part in cell cycle regulation and cell differentiation, and it also has anti-angiogenic activities [40]. Vitamin D deficiency is an important risk factor in the development process of UFs [6–9,41]. There are several ideas about the use of vitamin D in UF prevention or as a long-term treatment [13,34,42], but ongoing clinical trials in this area remain scarce.

Despite the accumulating data on UFs, information about the involvement of vitamin D in their pathophysiology is limited. Thus, we present an up-to-date review about the role of vitamin D in UF-associated problems, as well as our ideas for future steps.

## 2. Materials and Methods

This article presents an up-to-date review of the publications regarding the current role of vitamin D and its receptor in the pathophysiology and management of uterine fibroids. A literature search was conducted in PubMed using the following key words: "uterine fibroid" and "vitamin D". During our search, we combined the key words into a pair, and found 45 publications. The aim of the review was to evaluate the current state of knowledge about the role of vitamin D in uterine fibroid biology and management. The results of the available studies, published in English from January 2002 up to April 2018, have been discussed. Additional important articles and reviews were considered, when relevant.

## 3. Discussion

### 3.1. Uterine Fibroid Biology—Overview

Genetic studies have proven UFs to be monoclonal hormone-dependent tumors [30,43]. Tumor development begins with the creation of a pathologically changed and transformed primary myometrial cell. Subsequently, all secondary cells divide, making the tumor grow further [44].

Modified cells need proper stimulation in order to divide and produce the extracellular matrix (ECM) [45,46]. The mechanisms controlling the growth of UFs are complex and still not well-recognized [47]. Abnormal and excessive ECM production is a major factor in UF growth [46,47].

The main hormones which simulate UFs development and growth are estrogen [48] and progesterone [29,30,49]. Estrogen [48] and progesterone [49] induce UF formation and growth, affecting them directly and indirectly through various growth factors [50,51]. Numerous experts consider progesterone to be the main steroid initiating uterine muscle differentiation, and its subsequent abnormal growth [30,43,49,52–54]. The effect of progesterone on UF growth has been confirmed by the wide use of its antagonists (SPRMs) in the treatment of UFs [31,55,56]. Estrogens, which play a smaller role in the UF pathophysiology, prepare the tumor to be stimulated by progesterone by upregulating its receptors [48,57].

UFs are greatly affected by genetic abnormalities [58–60]. Since the discovery made by Makinen et al., in 2011 [58], UF genetics has made great advances. Specific mutations within the MED12 gene which encode the mediator complex subunit 12 (MED12) are detected in almost 80% of the UF samples [58,59]. So far, no mutations have been found in MED12 in the healthy myometrium of the women studied [60].

### 3.2. Vitamin D and Its Receptor—Overview

Vitamin D is a group of steroid compounds which have a powerful impact on many parts of the human body, including the musculoskeletal, nervous and immune systems, as well as the genital tract [36,37,61]. The main activity of vitamin D concerns the control of calcium-phosphate balance as well as the correct structure and function of the skeleton [36,62]. Although it is traditionally included in vitamins, vitamin D also fulfills the requirements to be classified as a hormone [63,64]. Vitamin D can occur in several forms—vitamin D1, or calciferol (most often found in fish oils), vitamin D2—ergocalciferol (found in plants) and vitamin D3—cholecalciferol (produced in the skin) [65,66]. Vitamin D is converted to 25-hydroxyvitamin D [25(OH)D] by the 25 $\alpha$-hydroxylase enzyme in the liver, and after that it is hydroxylated in the kidneys to 1,25-dihydroxyvitamin D [1,25(OH)D] [64]. The most active form of this vitamin—1,25(OH)D—presents its activity in almost every tissue in human body [13].

Vitamin D is carried by a specific transportation protein—Vitamin D-binding protein (VDBP)—which belongs to the albumin gene family [67]. This protein transports various forms of vitamin D, including ergocalciferol, cholecalciferol, calcifediol and calcitriol, between the skin, liver and kidneys, and then on to various target tissues [63]. According to Yao et al., similar levels of VDBP were observed in the population of Euro-American and Afro-American women [68]. There are some studies which have already demonstrated the usefulness of VDBP in clinical diagnosis. It might be used as a biomarker for selected diseases, for example, breast cancer [69,70]. Also, in their study from 2012, Lin et al., indicated that VDBP can be used as a potential marker for UFs [70].

Vitamin D is believed to regulate cell proliferation and differentiation, inhibit angiogenesis, and stimulate apoptosis [36,37,61]. Vitamin D works by a specific type of receptor—VDR. It is a mediator of the pleiotropic effect of this vitamin [71]. Vitamin D mediates its metabolic functions through steroid transcriptional mechanisms [64,71]. This vitamin can modulate the expression of various genes in a tissue-specific manner, and then can lead to the inhibition of cell proliferation, differentiation, and apoptosis. These processes can take part in the inhibition of neoplastic transformation as well as tumor growth, such as in UF [13,72].

Endogenous vitamin D production is limited by factors such as geographical location, environmental and individual characteristics (e.g., latitude, season, weather conditions, clothing), as well as the use of sunscreens and other cosmetics [73,74]. Abnormal supplementation and insufficient exposure to solar radiation due to spending the majority of time indoors are believed to be the main reasons for vitamin D deficiency in white female population [73]. People with dark skin, especially black, must spend 5 to 10 times more time outside to produce the same portion of vitamin D as compared to people with fair complexion [73,75], which is the reason why, for example, Afro-Americans are more likely to have low levels of vitamin D [76].

Vitamin D levels defined as "deficient" are the subject of much heated debate among the experts [37,39,77]. According to the Endocrine Society Practice Guidelines on vitamin D status, "deficiency" is defined as 25(OH)D level of <20 ng/mL, insufficiency as 21–29 ng/mL, and sufficiency as at least 30 ng/mL (for the best overall musculoskeletal effect) [78,79]. The actual guidelines suggest a preferred range from 40 to 60 ng/mL when focusing on the pleiotropic effect of vitamin D [39,61,80] (Figure 1).

### 3.3. Vitamin D in Uterine Fibroid Biology

Vitamin D is believed to reduce the risk of chronic illnesses and neoplasms [37,61]. According to the review by Grant, the available scientific evidence supports the notion of vitamin D supplementation as a cancer prevention method [81].

Decreased serum vitamin D levels have been already confirmed in several gynecological and obstetrical pathologies, such as infertility or polycystic ovary syndrome [82–86]. Vitamin D is also known to affect cycle regularity through its effect on hormones such as insulin or androgens. Various studies have confirmed that lower serum 25(OH)D levels were associated with irregular menstrual cycles [87]. Vitamin D may also influence the ovarian reserve and is inversely related to FSH level, as was demonstrated by Jukic et al. [88].

Recent studies have identified abnormal concentrations of vitamin D as important players in the etiology of UFs [6–9,38]. Nowadays, vitamin D deficiency is believed to be also a major risk factor in the development of UFs. Mean 25(OH)D serum levels are significantly lower in UF-positive women as compared to UF-negative controls [6–8]. These findings were also confirmed in Turkish [89] and African-American populations, who are more likely to present both with vitamin D deficiency and presence of UFs [90]. Cultural and environmental differences might play a role in the UF development as well [91], Oskovi Kaplan et al., suggested that traditional clothing style (covering the body), low education or being a housewife are also risk factors for vitamin D deficiency which, at some point in life, might result in UF [89]. Recently, theories about the vital role of vitamin D in the pathogenesis of UFs, and research into the effects of vitamin D on UFs, have gained new momentum. Vitamin D has become one of the key elements of modern theory of UF pathogenesis [6,13]. Epidemiological studies continue to emphasize the role of vitamin D deficiency in the development of UFs. One of the most recently published studies on these correlations was published in 2015. Mitro et al., in their study on 3600 women who took part in the National Health and Nutrition Examination Survey (NHANES) between 2001 and 2006, found no association between low vitamin D levels and the appearance of UFs within the entire population [92]. Interestingly, taking into account only the white population, the decreased serum concentration of vitamin D was a risk factor for UFs, but no such correlation was observed in black women [92]. Thus, larger studies are still necessary to better understand the biology of UF.

The results of the first study conducted to better understand the effect of vitamin D on the growth of UF were published by Blauer et al. [38]. The research was carried out in 2009 and showed the relationship between 1,25(OH)D levels and the growth of UF cells (samples were obtained from women who underwent hysterectomy) [38]. Inhibition of their growth was correlated with vitamin D concentration and increased with increasing vitamin D concentration [38]. In another study, performed later by Sharan et al., 1,25(OH)D caused in vitro inhibition of proliferation of immortal UF cells [93] (Figure 2).

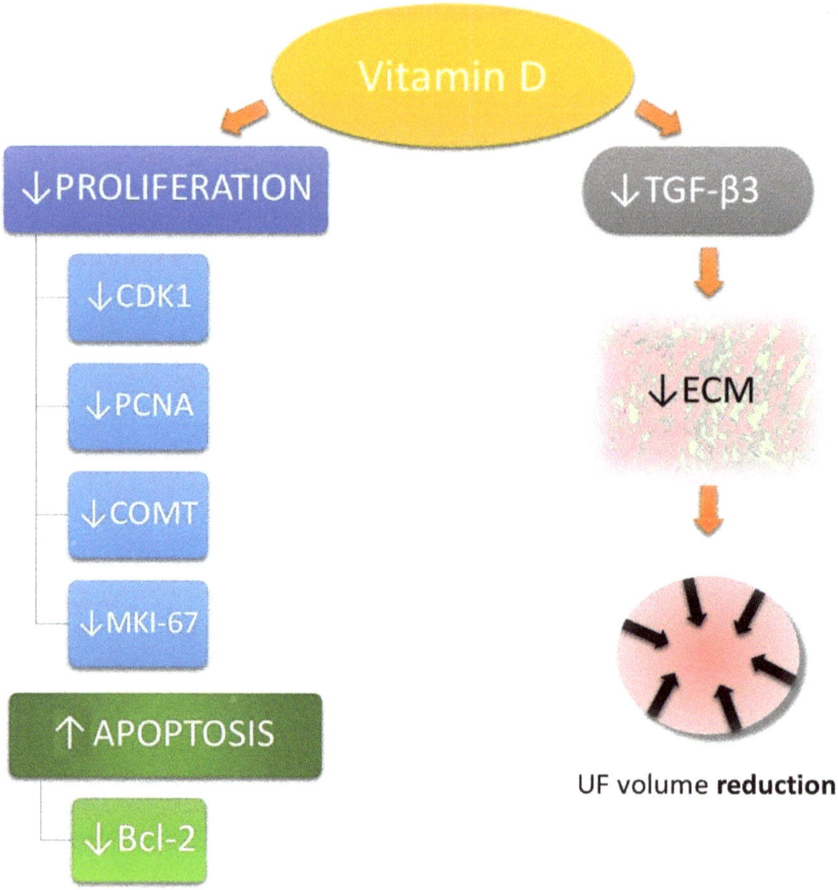


**Figure 2.** Vitamin D influence on uterine fibroid (UF) pathophysiological pathways. Cyclin-dependent kinase 1 (CDK1), proliferating cell nuclear antigen (PCNA), catechol-*O*-methyltransferase (COMT), Bcl-2 protein, proliferation marker protein Ki-67 (MKI-67), extracellular matrix (ECM), transforming growth factor beta 3 (TGF-β3), uterine fibroid (UF).

Their findings were unambiguous: proliferating cell nuclear antigen (PCNA)—known as a molecular marker for proliferation [94], cyclin-dependent kinase 1 (CDK1)—a protein kinase complex known as M-phase promoting factor [95], Bcl-2—considered an important anti-apoptotic protein [96], and catechol-*O*-methyltransferase (COMT)—which is involved in estrogen metabolism [97], were all highly affected by vitamin D compounds [93]. In the same year, Halder et al., published a study which showed the effect of vitamin D3 on the transforming growth factor beta (TGF-β) pathway. In their study, TGF-β3 was inhibited by increased concentrations of vitamin D [98]. Their results further confirmed our belief that research about the role of vitamin D in the UF biology is the right direction because TGF-β is considered to be one of the most relevant factors in the pathogenesis of fibrosis-associated diseases [46], while TGF-β3 is one of the most important TGF-β isoforms in UF biology [99–101]. Its increased serum levels constitute a risk factor for UF incidence [6]. TGF-β3 slows the degradation of ECM [99,102], and plays a vital role in its overproduction by stimulating the expression of selected ECM molecules, such as proteoglycans and proteins [46,103] (Figure 2). In subsequent studies, which have been carried out on animal models (Eker rats), therapeutic doses of vitamin D were found to significantly reduce the size of UFs by suppressing genes responsible for cell growth and cell division, antiapoptotic genes and genes encoding estrogen and progesterone receptors [104,105]. In animal models, vitamin D presented great effect on molecular genetics by suppressing cell growth and proliferation-related genes (e.g., Pcna, Cdk1, Cdk2, Cdk4), antiapoptotic genes (Bcl2 and Bcl2-like1), and estrogen and progesterone receptors [104]. In the same study, immunohistochemical staining revealed decreased expression of additional markers of proliferation (PCNA and MKI-67) [104].

According to an interesting study by Al-Hendy et al., 1,25(OH)D functions as a potent antiestrogenic and antiprogesteronic agent [106]. These authors observed an inverse correlation between the up-regulated estrogen and progesterone receptors and VDR expression in UFs. In the same study, treatment with active vitamin D significantly decreased the levels of estrogen and progesterone receptors [106]. Steroid hormones and their receptors are crucial in UF biology. For example, due to their influence on Wnt/β-catenin and TGF-β pathways [101]. In many tumors, including UFs, cytokines and growth factors play the key role in inflammation and regulation of cell division [40,50,107]. Recently, the elevated expression of activin A and its effect on inflammation and fibrosis have been thoroughly documented as well (e.g., in UFs) [108]. These factors may be also responsible for UF-associated symptoms such as infertility or pain [40], for example, TGF-β (especially the TGF-β3 isoform) which is one of the most important factors in the development and growth and UFs and the related problems [50,98,101]. What is already known is that UFs regulate and stimulate the accumulation of ECM, with TGF-β as the mediator [57,101,109]. Overexpressed TGF-β induces ECM overproduction by stimulating the expression of type I collagen, proteoglycans, and other ECM compounds, what in turn results in abnormal ECM accumulation [101,103,110]. VDR activation by its ligand results in reduced inflammation and fibrosis [34,111]. Owing to the studies by Halder et al., it was possible to prove that a surprisingly beneficial effect in UF growth reduction can be obtained under the influence of vitamin D [98,104]. In their papers, authors concluded that 1,25(OH)D reduces TGF-β3-related gene expression and 1,25(OH)D treatment reducing growth of uterine leiomyoma tumor and inhibition proliferation of leiomyoma cells [98,104]. In 2016, Al-Hendy et al., proved that vitamin D administration reduced the levels of Wnt4 and β-catenin in UF cell cultures [112]. These authors suggested that vitamin D might function as an inhibitor of Wnt4/β-catenin and mechanistic target of rapamycin (mTOR—kinase which regulates cell growth, cell proliferation and cell motility) signaling pathways [112]. In the same study, the authors made an interesting observation about the gene encoding flap structure-specific endonuclease 1 (FEN1) [112,113]. FEN1 is an enzyme involved in DNA damage repair, which was found to be overexpressed in the majority of cancers [114]. When FEN1 is overexpressed, the highly inaccurate DNA repair pathway may be favored, causing a great risk of potential mutation and increased risk of cancer [115]. Al-Hendy et al., tested the effect of vitamin D on the FEN1 protein expression in human UF and found that vitamin to effectively suppress FEN1 in a concentration-dependent manner. They concluded that this was yet another clue indicating the potential use of vitamin D in UF treatment [112].

There are other ways to explain the beneficial effects of vitamin D on UFs. Matrix metalloproteinases (MMPs) are calcium-dependent zinc-containing endopeptidases which play a role in continuous ECM rebuild [116]. MMPs are capable of degrading all kinds of ECM proteins [116]. MMPs enzymes are regulated by tissue inhibitors of metalloproteinases (TIMPs) [117]. In 2013, Halder et al., demonstrated that vitamin D increased TIMP expression in the uterine myometrium. That study revealed that vitamin D plays an important regulatory role in the expression and activities of MMP-2 and MMP-9 [118].

Potential antitumor properties of vitamin D have been covered by other studies, such as on mesenchymal multipotent cells [119]. Many of the pathways in the mesenchymal multipotent cells were found to be similar to those described in UFs. Artaza et al., observed increased expression and nuclear translocation of VDR, decreased expression of TGF-β, collagen I, III and increased expression of bone morphogenic protein 7 (BMP-7) and MMP-8 [119].

### 3.4. Vitamin D Receptor and Uterine Fibroids—Gene Polymorphisms

Data on single nucleotide polymorphisms (SNPs) in UFs are more scarce than other genetic findings.

In 2014, a group of researchers began their work on the study of SNP gene polymorphisms and its correlation with UFs occurrence [120]. Preliminary studies have shown that SNPs related to the metabolism of vitamin D and skin color are associated with the presence of UFs in black women [120]. Among studied SNPs rs12800438 near *DHCR7* and rs6058017 in *ASIP* gene are implicated in vitamin D synthesis in the skin [120]. The relationship between UF and rs739837 and rs886441 polymorphisms in the nuclear hormone receptor for vitamin D has been described [120]. The study by Shahbazi et al. supports the hypothesis that UFs are associated with the VDR rs2228570 polymorphism—correlation between VDR TT genotype and UF occurrence risk [121]. More recently (2018), Yilmaz et al. demonstrated that the presence of the rs2228570 CC genotype may be a risk-reducing factor and the T allele may be a potential risk factor for the development of UFs, which is consistent with the findings of Shahbazi [122]. Both studies had limitations: small sample size and closed populations and their results need to be confirmed on larger populations [120].

### 3.5. Vitamin D—Potential Uterine Fibroid Prophylaxis or Treatment Method

#### 3.5.1. Vitamin D—Optimal Levels and Supplementation against Uterine Fibroids

The least studied factors which affect the risk for UF occurrence are related to lifestyle, diet, nutrition, or place of residence. Especially nutrition and diet can be the gateway to effective prevention of UFs [34,123,124]. As the new guidelines from 2018 defined optimal concentration of vitamin D at 40–60 ng/mL (Figure 1) [39], vitamin D supplementation and sunlight exposure can be the two main clues for UF prevention [125]. In cases of deficiency, vitamin D can be raised to the correct level by taking a supplemental dose—7000 international units (IU)/day or 50,000 IU/week, depending on patient choice [39]. Chronic administration of high doses of vitamin D may lead to its toxic effects, manifested by severe hypercalcemia and functional hypoparathyroidism, resulting in fractures and osteoarticular pain [126]. It seems that the undesirable effects of vitamin D can be bypassed by short-term high-dose therapies instead of chronic administration [37].

In 2016, in a study performed by Ciavattini et al., 53 women received vitamin D supplementation [127], which restored correct vitamin D serum concentrations in women with small burden UFs (<50 mm in diameter and less than 4 tumors). In these women, treatment with vitamin D reduced disease progression. To the best of our knowledge, this is the first study showing beneficial results of vitamin D use in UF management in humans [127].

In our opinion, vitamin D seems to offer a promising, effective, and low-cost prevention or treatment of UFs and their clinical symptoms. Should further findings be positive, vitamin D supplements/drugs could become a new weapon in the battle against UFs, with the additional beneficial pleiotropic effect. Furthermore, skeletal and extra-skeletal advantages support the use of vitamin D as a prophylactic agent in high-risk or UF-positive women [34].

#### 3.5.2. The Use of Paricalcitol in Uterine Fibroid Management

Potential adverse effects of chronic or high-dose vitamin D treatment might be bypassed by using vitamin D analogs [34,105]. The experimental trials performed in animal models provided evidence that VDR agonists have a therapeutic potential in chronic inflammatory diseases and cancer [128]. Interestingly, VDR agonists show agonistic, partial agonistic, or antagonistic activity, depending upon the structure of their side chains [129]. Paricalcitol is a selective vitamin D analog, a VDR activator used mostly in the treatment of secondary hyperparathyroidism [111,130]. These analogs are already present on the market for different indications. However, subsequent studies suggest that they may also have a beneficial anti-proliferative effect on UFs [105,131].

Most of the activities which may be useful in the treatment of UF can be explained on the basis of the observations obtained in nephrological models [132,133]. In kidneys, paricalcitol presented an immunomodulatory effect which can cause limited ECM thickening and may slow down angiogenesis [111,134]. In the same model, paricalcitol interfered with TGF-β1 activation of the TGF-β receptor 1 [135]. The observation is intriguing due to the presented anti-inflammatory and anti-fibrotic properties of this vitamin D analog [135]. According to the data obtained, paricalcitol has an influence on Wnt/β-catenin signaling as well as on NF-κB, which results in decreased expression of ECM [132,133]. The same observations about Wnt/β-catenin signaling are present in UFs, where increased secretion of Wnt ligands under the influence of steroid hormones leads to excessive production of different TGF-β and ECM isoforms, as well as enhanced proliferation of UF stem cells [30,57,101]. In our opinion, we can transfer data about fibrosis to UFs models to a certain extent, because these tumors consist largely of ECM with embedded cells, and excessive ECM production is considered to be one of the key mechanisms of UF formation [101].

### 3.6. Future Concepts in the Area of Uterine Fibroids and Vitamin D

Recent attempts to create a cheap, safe and effective drug targeted at the prevention and treatment of UFs remain in the very early stages, and it is not known whether they will succeed. Vitamin D is a natural supplement which may prevent UF development and growth, and undoubtedly deserves further investigation [136,137].

Vitamin D seems to be a promising, safe and low-cost agent in the prevention or treatment of UFs. Further reports are necessary to prove the efficacy of vitamin D supplementation in women [138]. In cases of further positive observations and effects in randomized trials, vitamin D preparations could become a new generation of anti-UF drugs [42,139] (Figure 3).

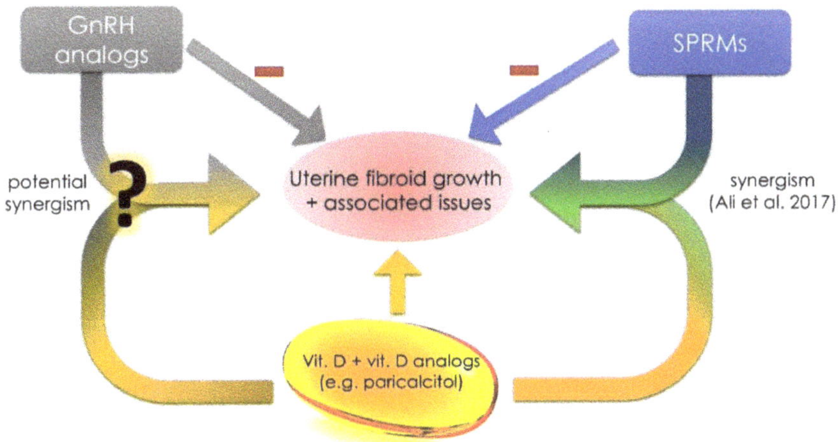


**Figure 3.** Vitamin D, GnRH analogs and SPRMs. According to Ali et al. there is a potential synergism between vitamin D and ulipristal acetate [140]. New data about between vitamin D, vitamin D analogs and other mostly used drugs in UF therapy is necessary to find other synergisms. Gonadotropin releasing hormone (GnRH), selective progesterone receptor modulator (SPRM).

There is evidence supporting the beneficial action of vitamin D supplementation in women with small UFs [127], but further extensive studies are needed to fully understand the exact role of vitamin D in UF biology. Lack of randomized controlled trials on vitamin D use in the prophylaxis or treatment of UFs remains a significant problem. In our opinion, the main reason for that is the lack of unified cut-off thresholds for vitamin D deficiency—they differ in different countries [39].

The consensus in this area can bring tangible benefits to women with UFs. Currently, high-risk patients, those with positive history of UFs, Afro-Americans and those with elevated BMI, should be screened and offered supplementation, if necessary [6]. According to Ali et al., women who would benefit from this management include also these with early menarche, nulliparous, and aged <40 [141].

An additional aspect that should be implemented when constructing subsequent clinical trials and determining recommendations for vitamin D supplementation is an individually differentiated response to vitamin supplementation [142]. According to available data, up to 25% of humanity can be considered as vitamin D low responders (slow response after standard supplementation doses) [142]. In the matter of UFs, the fact of a different response to supplementation doses may be of great importance, both to the effectiveness of treatment and also for economic reasons [20]. For example, similar doses of vitamin D in a high responders group may result in high raises of 25(OH)D serum levels and subsequent retention of tumor growth, whereas low responders will only gain a very low 25(OH)D serum level raise and small pleiotropic effect [142]. It seems, therefore, that the next step in constructing high-quality studies on the influence of vitamin D as a medicine in UFs therapy should be the use of the vitamin D response index when creating patient groups.

Since the correction of vitamin D concentrations has a positive effect on the inhibition of UF growth, we should also consider other therapies (except supplementation) which will increase serum vitamin D levels. Harmon et al. reported that the use of estrogen-containing contraceptives was associated with a 20% increase in serum 25(OH)D concentrations [143]. Our studies have also recently shown that the use of combined oral contraception (COC) with drospirenone results in higher serum vitamin D levels [144]. The exact mechanisms causing the increase are still unknown, so research should be continued. The use of selected form of oral contraception in selected groups could reduce UF-related symptoms as well as help to maintain the correct serum concentrations of vitamin D. It should be emphasized that COC use should not be expected to reduce the tumor volume [138]. In the short-term management, COC can be used to reduce menstrual bleeding associated with UFs [138]. It should be emphasized that according to available data COC use slightly increase the overall risk of breast [145] cancer [146] occurrence. On the other hand, this risk is counterbalanced by the lower risk of endometrial, ovarian, and colorectal cancer in the future in women who used the COC [147].

In light of the above, paricalcitol has a great potential to become an effective drug or co-drug for the conservative treatment of UFs [34]. Paricalcitol effectively reduces the proliferation of human leiomyoma cell cultures and fibroid tumor volumes, and induces apoptosis [105]. Further extensive clinical research is necessary to gain more information about the use of paricalcitol in UF therapy (Figure 3). In the meantime, other VDR analogs should be studied for their potential role in the management of UFs [141].

Due to rare side effects and relatively high safety of vitamin D, we could also consider combination therapies—drugs with additional simultaneous vitamin D supplementation. According to a very recent study by Ali et al., UPA and vitamin D share synergistic anti-fibroid activities [140]. In this study, the combined therapy of UPA and vitamin D resulted in a significant inhibition of UF cell growth (lowest proliferation rate from all studied groups) [140]. This research is a milestone and can bring entirely new perspectives on how to treat UF. We are of the opinion that such treatment would be beneficial in selected populations. Perhaps it could be more effective than the traditional approach in patients with the most severe symptoms, such as in obese African-Americans with vitamin D deficiency. It could also be treated as a type of add-back therapy during gonadotropin releasing hormone (GnRH) analog treatment [148], such as to prevent bone loss [149] or negative effect on mood or cognition [150] caused by estrogen deficiency. Similar studies performed on other substances such as GnRH analogs, for example, leuprolide, goserelin, elagolix or relugolix, might constitute the next step. If the safety of such therapies will be confirmed, the studies should be transferred to the next stages of clinical trials in humans.

Early prevention, appropriate prophylaxis, as well as treatment of UFs at an early stage in high-risk women, are priority actions. Perhaps the solution for the future will be to identify high-risk groups before the appearance of UFs, and then to implement preventive measures. The ideal methods of prevention and early-stage therapy should be inexpensive and relatively free of risk [13,42]. Highly individualized and personalized multi-drug therapies with the use of vitamin D might also be considered.

High-dose vitamin D and vitamin D analogs alone or as co-drugs can sooner or later become optimal, effective, safe drugs for conservative treatment of UFs. First, however, they must undergo advanced clinical trials, where they can confirm their effectiveness.

## 4. Conclusions

Vitamin D plays an essential role in UF biology. Vitamin D and its analogs seem to be promising, effective, and low-cost compounds in the management of UFs and their clinical symptoms. In cases of further positive observations and randomized control trials, vitamin D preparations could become new tools in the fight against UFs, with the additional beneficial pleiotropic effect. Further studies about the biological effect of vitamin D on UF biology are essential. The synergy between vitamin D and selected anti-UF drugs is a very interesting issue which requires further research.

**Author Contributions:** M.C. (Michał Ciebiera), M.C. (Magdalena Ciebiera), K.Ł. and G.J. designed the review. M.C. (Michał Ciebiera), M.W., M.C. (Magdalena Ciebiera), K.Z., K.Ł., G.J. analyzed the data and wrote the paper. M.C. (Michał Ciebiera), K.Ł. and G.J. supervised the work and accepted the final version of the manuscript.

**Funding:** This study was funded by Centre of Postgraduate Medical Education. Grant number 501-1-21-27-18.

**Conflicts of Interest:** The authors declare no conflict of interest.

## Abbreviations

| | |
|---|---|
| 1,25(OH)D | 1,25-dihydroxyvitamin D |
| 25(OH)D | 25-hydroxyvitamin D |
| BMI | body mass index |
| CDK1 | cyclin-dependent kinase 1 |
| COC | combined oral contraception |
| COMT | catechol-*O*-methyltransferase |
| ECM | extracellular matrix |
| EMA | European Medicines Agency |
| FEN-1 | flap structure-specific endonuclease 1 |
| GnRH | Gonadotropin releasing hormone |
| MMP | matrix metalloproteinase |
| mTOR | mechanistic target of rapamycin |
| NHANES | National Health and Nutrition Examination Survey |
| PCNA | proliferating cell nuclear antigen |
| QoL | quality of life |
| SELF | Study of environment lifestyle and fibroids |
| SNP | single nucleotide polymorphism |
| SPRM | selective progesterone receptor modulator |
| TGF-β | transforming growth factor beta |
| TIMP | tissue inhibitor of metalloproteinase |
| UF | uterine fibroid |
| UPA | ulipristal acetate |
| VDBP | vitamin D binding protein |
| VDR | vitamin D receptor |

## References

1. Stewart, E.A. Uterine fibroids. *Lancet* **2001**, *357*, 293–298. [CrossRef]
2. Stewart, E.A.; Laughlin-Tommaso, S.K.; Catherino, W.H.; Lalitkumar, S.; Gupta, D.; Vollenhoven, B. Uterine fibroids. *Nat. Rev. Dis. Primers* **2016**, *2*, 16043. [CrossRef] [PubMed]

3. Stewart, E.A.; Cookson, C.L.; Gandolfo, R.A.; Schulze-Rath, R. Epidemiology of uterine fibroids: A systematic review. *BJOG Int. J. Obstet. Gynaecol.* **2017**, *124*, 1501–1512. [CrossRef] [PubMed]
4. Al-Hendy, A.; Myers, E.R.; Stewart, E. Uterine fibroids: Burden and unmet medical need. *Semin. Reprod. Med.* **2017**, *35*, 473–480. [CrossRef] [PubMed]
5. Baird, D.D.; Dunson, D.B.; Hill, M.C.; Cousins, D.; Schectman, J.M. High cumulative incidence of uterine leiomyoma in black and white women: Ultrasound evidence. *Am. J. Obstet. Gynecol.* **2003**, *188*, 100–107. [CrossRef] [PubMed]
6. Ciebiera, M.; Wlodarczyk, M.; Slabuszewska-Jozwiak, A.; Nowicka, G.; Jakiel, G. Influence of vitamin D and transforming growth factor beta3 serum concentrations, obesity, and family history on the risk for uterine fibroids. *Fertil. Steril.* **2016**, *106*, 1787–1792. [CrossRef] [PubMed]
7. Sabry, M.; Halder, S.K.; Allah, A.S.; Roshdy, E.; Rajaratnam, V.; Al-Hendy, A. Serum vitamin D3 level inversely correlates with uterine fibroid volume in different ethnic groups: A cross-sectional observational study. *Int. J. Women's Health* **2013**, *5*, 93–100. [CrossRef]
8. Paffoni, A.; Somigliana, E.; Vigano, P.; Benaglia, L.; Cardellicchio, L.; Pagliardini, L.; Papaleo, E.; Candiani, M.; Fedele, L. Vitamin D status in women with uterine leiomyomas. *J. Clin. Endocrinol. Metab.* **2013**, *98*, E1374–1378. [CrossRef] [PubMed]
9. Baird, D.D.; Hill, M.C.; Schectman, J.M.; Hollis, B.W. Vitamin D and the risk of uterine fibroids. *Epidemiology* **2013**, *24*, 447–453. [CrossRef] [PubMed]
10. Wise, L.A. Study of environment lifestyle and fibroids (self): Advancing the field of fibroid epidemiology. *J. Women's Health* **2015**, *24*, 862–864. [CrossRef] [PubMed]
11. Baird, D.D.; Harmon, Q.E.; Upson, K.; Moore, K.R.; Barker-Cummings, C.; Baker, S.; Cooper, T.; Wegienka, G. A prospective, ultrasound-based study to evaluate risk factors for uterine fibroid incidence and growth: Methods and results of recruitment. *J. Women's Health* **2015**, *24*, 907–915. [CrossRef] [PubMed]
12. Halder, S.K.; Osteen, K.G.; Al-Hendy, A. 1,25-dihydroxyvitamin D3 reduces extracellular matrix-associated protein expression in human uterine fibroid cells. *Biol. Reprod.* **2013**, *89*, 150. [CrossRef] [PubMed]
13. Brakta, S.; Diamond, J.S.; Al-Hendy, A.; Diamond, M.P.; Halder, S.K. Role of vitamin D in uterine fibroid biology. *Fertil. Steril.* **2015**, *104*, 698–706. [CrossRef] [PubMed]
14. Parker, W.H. Etiology, symptomatology, and diagnosis of uterine myomas. *Fertil. Steril.* **2007**, *87*, 725–736. [CrossRef] [PubMed]
15. Kjerulff, K.H.; Langenberg, P.; Seidman, J.D.; Stolley, P.D.; Guzinski, G.M. Uterine leiomyomas. Racial differences in severity, symptoms and age at diagnosis. *J. Reprod. Med.* **1996**, *41*, 483–490. [PubMed]
16. Buttram, V.C., Jr.; Reiter, R.C. Uterine leiomyomata: Etiology, symptomatology, and management. *Fertil. Steril.* **1981**, *36*, 433–445. [PubMed]
17. Khan, A.T.; Shehmar, M.; Gupta, J.K. Uterine fibroids: Current perspectives. *Int. J. Women's Health* **2014**, *6*, 95–114. [CrossRef] [PubMed]
18. Soave, I.; Marci, R. Uterine leiomyomata: The snowball effect. *Curr. Med. Res. Opin.* **2017**, *33*, 1909–1911. [CrossRef] [PubMed]
19. Soliman, A.M.; Margolis, M.K.; Castelli-Haley, J.; Fuldeore, M.J.; Owens, C.D.; Coyne, K.S. Impact of uterine fibroid symptoms on health-related quality of life of us women: Evidence from a cross-sectional survey. *Curr. Med. Res. Opin.* **2017**, *33*, 1971–1978. [CrossRef] [PubMed]
20. Soliman, A.M.; Yang, H.; Du, E.X.; Kelkar, S.S.; Winkel, C. The direct and indirect costs of uterine fibroid tumors: A systematic review of the literature between 2000 and 2013. *Am. J. Obstet. Gynecol.* **2015**, *213*, 141–160. [CrossRef] [PubMed]
21. Cardozo, E.R.; Clark, A.D.; Banks, N.K.; Henne, M.B.; Stegmann, B.J.; Segars, J.H. The estimated annual cost of uterine leiomyomata in the United States. *Am. J. Obstet. Gynecol.* **2012**, *206*, 211.e1–211.e9. [CrossRef] [PubMed]
22. Wu, H.Y.; Wang, K.C. Minimally invasive approaches to the surgical management of fibroids. *Semin. Reprod. Med.* **2017**, *35*, 533–548. [CrossRef] [PubMed]
23. Donnez, J.; Dolmans, M.M. Uterine fibroid management: From the present to the future. *Hum. Reprod. Update* **2016**, *22*, 665–686. [CrossRef] [PubMed]
24. Farquhar, C.M.; Steiner, C.A. Hysterectomy rates in the United States 1990–1997. *Obstet. Gynecol.* **2002**, *99*, 229–234. [PubMed]

25. Taylor, D.K.; Holthouser, K.; Segars, J.H.; Leppert, P.C. Recent scientific advances in leiomyoma (uterine fibroids) research facilitates better understanding and management. *F1000Research* **2015**, *4*, 183. [CrossRef] [PubMed]
26. Lewis, E.I.; Gargiulo, A.R. The role of hysteroscopic and robot-assisted laparoscopic myomectomy in the setting of infertility. *Clin. Obstet. Gynecol.* **2016**, *59*, 53–65. [CrossRef] [PubMed]
27. Borah, B.J.; Yao, X.; Laughlin-Tommaso, S.K.; Heien, H.C.; Stewart, E.A. Comparative effectiveness of uterine leiomyoma procedures using a large insurance claims database. *Obstet. Gynecol.* **2017**, *130*, 1047–1056. [CrossRef] [PubMed]
28. Faustino, F.; Martinho, M.; Reis, J.; Aguas, F. Update on medical treatment of uterine fibroids. *Eur. J. Obstet. Gynecol. Reprod. Biol.* **2017**, *216*, 61–68. [CrossRef] [PubMed]
29. Kim, J.J.; Sefton, E.C. The role of progesterone signaling in the pathogenesis of uterine leiomyoma. *Mol. Cell. Endocrinol.* **2012**, *358*, 223–231. [CrossRef] [PubMed]
30. Bulun, S.E. Uterine fibroids. *N. Engl. J. Med.* **2013**, *369*, 1344–1355. [CrossRef] [PubMed]
31. Donnez, J.; Donnez, O.; Courtoy, G.E.; Dolmans, M.M. The place of selective progesterone receptor modulators in Myoma therapy. *Minerva Ginecol.* **2016**, *68*, 313–320. [PubMed]
32. Kalampokas, T.; Kamath, M.; Boutas, I.; Kalampokas, E. Ulipristal acetate for uterine fibroids: A systematic review and meta-analysis. *Gynecol. Endocrinol.* **2016**, *32*, 91–96. [CrossRef] [PubMed]
33. Agency, E.M. *Women Taking Esmya for Uterine Fibroids to Have Regular Liver Tests While EMA Review Is Ongoing*; European Medicines Agency: London, UK, 2018.
34. Ciebiera, M.; Lukaszuk, K.; Meczekalski, B.; Ciebiera, M.; Wojtyla, C.; Slabuszewska-Jozwiak, A.; Jakiel, G. Alternative oral agents in prophylaxis and therapy of uterine fibroids-an up-to-date review. *Int. J. Mol. Sci.* **2017**, *18*. [CrossRef] [PubMed]
35. Fritton, K.; Borahay, M.A. New and emerging therapies for uterine fibroids. *Semin. Reprod. Med.* **2017**, *35*, 549–559. [CrossRef] [PubMed]
36. Holick, M.F. Vitamin D: A millenium perspective. *J. Cell. Biochem.* **2003**, *88*, 296–307. [CrossRef] [PubMed]
37. Holick, M.F. Vitamin D deficiency. *N. Engl. J. Med.* **2007**, *357*, 266–281. [CrossRef] [PubMed]
38. Blauer, M.; Rovio, P.H.; Ylikomi, T.; Heinonen, P.K. Vitamin D inhibits myometrial and leiomyoma cell proliferation in vitro. *Fertil. Steril.* **2009**, *91*, 1919–1925. [CrossRef] [PubMed]
39. Pludowski, P.; Holick, M.F.; Grant, W.B.; Konstantynowicz, J.; Mascarenhas, M.R.; Haq, A.; Povoroznyuk, V.; Balatska, N.; Barbosa, A.P.; Karonova, T.; et al. Vitamin D supplementation guidelines. *J. Steroid Biochem. Mol. Biol.* **2018**, *175*, 125–135. [CrossRef] [PubMed]
40. Protic, O.; Toti, P.; Islam, M.S.; Occhini, R.; Giannubilo, S.R.; Catherino, W.H.; Cinti, S.; Petraglia, F.; Ciavattini, A.; Castellucci, M.; et al. Possible involvement of inflammatory/reparative processes in the development of uterine fibroids. *Cell Tissue Res.* **2016**, *364*, 415–427. [CrossRef] [PubMed]
41. Halder, S.; Al-Hendy, A. Hypovitaminosis D and high serum transforming growth factor beta-3: Important biomarkers for uterine fibroids risk. *Fertil. Steril.* **2016**, *106*, 1648–1649. [CrossRef] [PubMed]
42. Wu, J.L.; Segars, J.H. Is vitamin D the answer for prevention of uterine fibroids? *Fertil. Steril.* **2015**, *104*, 559–560. [CrossRef] [PubMed]
43. Elkafas, H.; Qiwei, Y.; Al-Hendy, A. Origin of uterine fibroids: Conversion of myometrial stem cells to tumor-initiating cells. *Semin. Reprod. Med.* **2017**, *35*, 481–486. [CrossRef] [PubMed]
44. Holdsworth-Carson, S.J.; Zaitseva, M.; Vollenhoven, B.J.; Rogers, P.A. Clonality of smooth muscle and fibroblast cell populations isolated from human fibroid and myometrial tissues. *Mol. Hum. Reprod.* **2014**, *20*, 250–259. [CrossRef] [PubMed]
45. Islam, M.S.; Ciavattini, A.; Petraglia, F.; Castellucci, M.; Ciarmela, P. Extracellular matrix in uterine leiomyoma pathogenesis: A potential target for future therapeutics. *Hum. Reprod. Update* **2018**, *24*, 59–85. [CrossRef] [PubMed]
46. Sozen, I.; Arici, A. Interactions of cytokines, growth factors, and the extracellular matrix in the cellular biology of uterine leiomyomata. *Fertil. Steril.* **2002**, *78*, 1–12. [CrossRef]
47. Rafique, S.; Segars, J.H.; Leppert, P.C. Mechanical signaling and extracellular matrix in uterine fibroids. *Semin. Reprod. Med.* **2017**, *35*, 487–493. [CrossRef] [PubMed]
48. Borahay, M.A.; Asoglu, M.R.; Mas, A.; Adam, S.; Kilic, G.S.; Al-Hendy, A. Estrogen receptors and signaling in fibroids: Role in pathobiology and therapeutic implications. *Reprod. Sci.* **2017**, *24*, 1235–1244. [CrossRef] [PubMed]

49. Ishikawa, H.; Ishi, K.; Serna, V.A.; Kakazu, R.; Bulun, S.E.; Kurita, T. Progesterone is essential for maintenance and growth of uterine leiomyoma. *Endocrinology* **2010**, *151*, 2433–2442. [CrossRef] [PubMed]
50. Ciarmela, P.; Islam, M.S.; Reis, F.M.; Gray, P.C.; Bloise, E.; Petraglia, F.; Vale, W.; Castellucci, M. Growth factors and myometrium: Biological effects in uterine fibroid and possible clinical implications. *Hum. Reprod. Update* **2011**, *17*, 772–790. [CrossRef] [PubMed]
51. Islam, M.S.; Protic, O.; Stortoni, P.; Grechi, G.; Lamanna, P.; Petraglia, F.; Castellucci, M.; Ciarmela, P. Complex networks of multiple factors in the pathogenesis of uterine leiomyoma. *Fertil. Steril.* **2013**, *100*, 178–193. [CrossRef] [PubMed]
52. Baird, D.D.; Garrett, T.A.; Laughlin, S.K.; Davis, B.; Semelka, R.C.; Peddada, S.D. Short-term change in growth of uterine leiomyoma: Tumor growth spurts. *Fertil. Steril.* **2011**, *95*, 242–246. [CrossRef] [PubMed]
53. Chill, H.H.; Safrai, M.; Reuveni Salzman, A.; Shushan, A. The rising phoenix-progesterone as the main target of the medical therapy for leiomyoma. *Biomed. Res. Int.* **2017**, *2017*. [CrossRef] [PubMed]
54. Moravek, M.B.; Bulun, S.E. Endocrinology of uterine fibroids: Steroid hormones, stem cells, and genetic contribution. *Curr. Opin. Obstet. Gynecol.* **2015**, *27*, 276–283. [CrossRef] [PubMed]
55. Donnez, J.; Tatarchuk, T.F.; Bouchard, P.; Puscasiu, L.; Zakharenko, N.F.; Ivanova, T.; Ugocsai, G.; Mara, M.; Jilla, M.P.; Bestel, E.; et al. Ulipristal acetate versus placebo for fibroid treatment before surgery. *N. Engl. J. Med.* **2012**, *366*, 409–420. [CrossRef] [PubMed]
56. Chwalisz, K.; Taylor, H. Current and emerging medical treatments for uterine fibroids. *Semin. Reprod. Med.* **2017**, *35*, 510–522. [CrossRef] [PubMed]
57. Borahay, M.A.; Al-Hendy, A.; Kilic, G.S.; Boehning, D. Signaling pathways in leiomyoma: Understanding pathobiology and implications for therapy. *Mol. Med.* **2015**, *21*, 242–256. [CrossRef] [PubMed]
58. Makinen, N.; Mehine, M.; Tolvanen, J.; Kaasinen, E.; Li, Y.; Lehtonen, H.J.; Gentile, M.; Yan, J.; Enge, M.; Taipale, M.; et al. Med12, the mediator complex subunit 12 gene, is mutated at high frequency in uterine leiomyomas. *Science* **2011**, *334*, 252–255. [CrossRef] [PubMed]
59. Heinonen, H.R.; Sarvilinna, N.S.; Sjoberg, J.; Kampjarvi, K.; Pitkanen, E.; Vahteristo, P.; Makinen, N.; Aaltonen, L.A. Med12 mutation frequency in unselected sporadic uterine leiomyomas. *Fertil. Steril.* **2014**, *102*, 1137–1142. [CrossRef] [PubMed]
60. Halder, S.K.; Laknaur, A.; Miller, J.; Layman, L.C.; Diamond, M.; Al-Hendy, A. Novel med12 gene somatic mutations in women from the southern united states with symptomatic uterine fibroids. *Mol. Genet. Genom.* **2015**, *290*, 505–511. [CrossRef] [PubMed]
61. Holick, M.F. The vitamin D deficiency pandemic: Approaches for diagnosis, treatment and prevention. *Rev. Endocr. Metab. Disord.* **2017**, *18*, 153–165. [CrossRef] [PubMed]
62. Chung, M.; Balk, E.M.; Brendel, M.; Ip, S.; Lau, J.; Lee, J.; Lichtenstein, A.; Patel, K.; Raman, G.; Tatsioni, A.; et al. Vitamin D and calcium: A systematic review of health outcomes. *Evid. Rep. Technol. Assess. (Full Rep.)* **2009**, *183*, 1–420.
63. Norman, A.W. From vitamin D to hormone d: Fundamentals of the vitamin D endocrine system essential for good health. *Am. J. Clin. Nutr.* **2008**, *88*, 491S–499S. [CrossRef] [PubMed]
64. Pike, J.W.; Christakos, S. Biology and mechanisms of action of the vitamin D hormone. *Endocrinol. Metab. Clin. N. Am.* **2017**, *46*, 815–843. [CrossRef] [PubMed]
65. Holick, M.F. High prevalence of vitamin D inadequacy and implications for health. *Mayo Clin. Proc.* **2006**, *81*, 353–373. [CrossRef] [PubMed]
66. Webb, A.R. Who, what, where and when-influences on cutaneous vitamin D synthesis. *Prog. Biophys. Mol. Biol.* **2006**, *92*, 17–25. [CrossRef] [PubMed]
67. Verboven, C.; Rabijns, A.; De Maeyer, M.; Van Baelen, H.; Bouillon, R.; De Ranter, C. A structural basis for the unique binding features of the human vitamin d-binding protein. *Nat. Struct. Biol.* **2002**, *9*, 131–136. [CrossRef] [PubMed]
68. Yao, S.; Hong, C.C.; Bandera, E.V.; Zhu, Q.; Liu, S.; Cheng, T.D.; Zirpoli, G.; Haddad, S.A.; Lunetta, K.L.; Ruiz-Narvaez, E.A.; et al. Demographic, lifestyle, and genetic determinants of circulating concentrations of 25-hydroxyvitamin D and vitamin d-binding protein in African American and European American women. *Am. J. Clin. Nutr.* **2017**, *105*, 1362–1371. [CrossRef] [PubMed]
69. Kim, B.K.; Lee, J.W.; Park, P.J.; Shin, Y.S.; Lee, W.Y.; Lee, K.A.; Ye, S.; Hyun, H.; Kang, K.N.; Yeo, D.; et al. The multiplex bead array approach to identifying serum biomarkers associated with breast cancer. *Breast Cancer Res.* **2009**, *11*, R22. [CrossRef] [PubMed]

70. Lin, C.P.; Chen, Y.W.; Liu, W.H.; Chou, H.C.; Chang, Y.P.; Lin, S.T.; Li, J.M.; Jian, S.F.; Lee, Y.R.; Chan, H.L. Proteomic identification of plasma biomarkers in uterine leiomyoma. *Mol. Biosyst.* **2012**, *8*, 1136–1145. [CrossRef] [PubMed]
71. Kato, S. The function of vitamin D receptor in vitamin D action. *J. Biochem.* **2000**, *127*, 717–722. [CrossRef] [PubMed]
72. Ylikomi, T.; Laaksi, I.; Lou, Y.R.; Martikainen, P.; Miettinen, S.; Pennanen, P.; Purmonen, S.; Syvala, H.; Vienonen, A.; Tuohimaa, P. Antiproliferative action of vitamin D. *Vitam. Horm.* **2002**, *64*, 357–406. [PubMed]
73. Nair, R.; Maseeh, A. Vitamin D: The "sunshine" vitamin. *J. Pharmacol. Pharmacother.* **2012**, *3*, 118–126. [CrossRef] [PubMed]
74. Spiro, A.; Buttriss, J.L. Vitamin d: An overview of vitamin D status and intake in Europe. *Nutr. Bull.* **2014**, *39*, 322–350. [CrossRef] [PubMed]
75. Nesby-O'Dell, S.; Scanlon, K.S.; Cogswell, M.E.; Gillespie, C.; Hollis, B.W.; Looker, A.C.; Allen, C.; Doughertly, C.; Gunter, E.W.; Bowman, B.A. Hypovitaminosis D prevalence and determinants among african american and white women of reproductive age: Third national health and nutrition examination survey, 1988–1994. *Am. J. Clin. Nutr.* **2002**, *76*, 187–192. [CrossRef] [PubMed]
76. Zadshir, A.; Tareen, N.; Pan, D.; Norris, K.; Martins, D. The prevalence of hypovitaminosis D among us adults: Data from the Nhanes III. *Ethn. Dis.* **2005**, *15*, S5-97–S5-101. [PubMed]
77. Rosen, C.J. Clinical practice. Vitamin D insufficiency. *N. Engl. J. Med.* **2011**, *364*, 248–254. [CrossRef] [PubMed]
78. Holick, M.F.; Binkley, N.C.; Bischoff-Ferrari, H.A.; Gordon, C.M.; Hanley, D.A.; Heaney, R.P.; Murad, M.H.; Weaver, C.M.; Endocrine, S. Evaluation, treatment, and prevention of vitamin D deficiency: An endocrine society clinical practice guideline. *J. Clin. Endocrinol. Metab.* **2011**, *96*, 1911–1930. [CrossRef] [PubMed]
79. Holick, M.F.; Binkley, N.C.; Bischoff-Ferrari, H.A.; Gordon, C.M.; Hanley, D.A.; Heaney, R.P.; Murad, M.H.; Weaver, C.M. Guidelines for preventing and treating vitamin D deficiency and insufficiency revisited. *J. Clin. Endocrinol. Metab.* **2012**, *97*, 1153–1158. [CrossRef] [PubMed]
80. Pilz, S.; Trummer, C.; Pandis, M.; Schwetz, V.; Aberer, F.; Grubler, M.; Verheyen, N.; Tomaschitz, A.; Marz, W. Vitamin D: Current guidelines and future outlook. *Anticancer Res.* **2018**, *38*, 1145–1151. [CrossRef] [PubMed]
81. Grant, W.B. A review of the evidence supporting the vitamin D-cancer prevention hypothesis in 2017. *Anticancer Res.* **2018**, *38*, 1121–1136. [CrossRef] [PubMed]
82. Grundmann, M.; von Versen-Hoynck, F. Vitamin D—Roles in women's reproductive health? *Reprod. Biol. Endocrinol.* **2011**, *9*, 146. [CrossRef] [PubMed]
83. Lerchbaum, E.; Rabe, T. Vitamin D and female fertility. *Curr. Opin. Obstet. Gynecol.* **2014**, *26*, 145–150. [CrossRef] [PubMed]
84. Dabrowski, F.A.; Grzechocinska, B.; Wielgos, M. The role of vitamin D in reproductive health—A trojan horse or the golden fleece? *Nutrients* **2015**, *7*, 4139–4153. [CrossRef] [PubMed]
85. Skowronska, P.; Pastuszek, E.; Kuczynski, W.; Jaszczol, M.; Kuc, P.; Jakiel, G.; Woclawek-Potocka, I.; Lukaszuk, K. The role of vitamin D in reproductive dysfunction in women—A systematic review. *Ann. Agric. Environ. Med.* **2016**, *23*, 671–676. [CrossRef] [PubMed]
86. Whynott, R.M.; Vaught, K.C.C.; Segars, J.H. The effect of uterine fibroids on infertility: A systematic review. *Semin. Reprod. Med.* **2017**, *35*, 523–532. [CrossRef] [PubMed]
87. Jukic, A.M.; Steiner, A.Z.; Baird, D.D. Lower plasma 25-hydroxyvitamin D is associated with irregular menstrual cycles in a cross-sectional study. *Reprod. Biol. Endocrinol.* **2015**, *13*, 20. [CrossRef] [PubMed]
88. Jukic, A.M.; Steiner, A.Z.; Baird, D.D. Association between serum 25-hydroxyvitamin D and ovarian reserve in premenopausal women. *Menopause* **2015**, *22*, 312–316. [CrossRef] [PubMed]
89. Oskovi Kaplan, Z.A.; Tasci, Y.; Topcu, H.O.; Erkaya, S. 25-hydroxy vitamin D levels in premenopausal turkish women with uterine leiomyoma. *Gynecol. Endocrinol.* **2018**, *34*, 261–264. [CrossRef] [PubMed]
90. Zhao, G.; Ford, E.S.; Tsai, J.; Li, C.; Croft, J.B. Factors associated with vitamin D deficiency and inadequacy among women of childbearing age in the united states. *ISRN Obstet. Gynecol.* **2012**, *2012*. [CrossRef] [PubMed]
91. Catherino, W.H.; Eltoukhi, H.M.; Al-Hendy, A. Racial and ethnic differences in the pathogenesis and clinical manifestations of uterine leiomyoma. *Semin. Reprod. Med.* **2013**, *31*, 370–379. [CrossRef] [PubMed]
92. Mitro, S.D.; Zota, A.R. Vitamin D and uterine leiomyoma among a sample of us women: Findings from nhanes, 2001–2006. *Reprod. Toxicol.* **2015**, *57*, 81–86. [CrossRef] [PubMed]

93. Sharan, C.; Halder, S.K.; Thota, C.; Jaleel, T.; Nair, S.; Al-Hendy, A. Vitamin D inhibits proliferation of human uterine leiomyoma cells via catechol-o-methyltransferase. *Fertil. Steril.* **2011**, *95*, 247–253. [CrossRef] [PubMed]
94. Wang, S.C. Pcna: A silent housekeeper or a potential therapeutic target? *Trends Pharmacol. Sci.* **2014**, *35*, 178–186. [CrossRef] [PubMed]
95. Enserink, J.M.; Kolodner, R.D. An overview of cdk1-controlled targets and processes. *Cell Div.* **2010**, *5*, 11. [CrossRef] [PubMed]
96. Hardwick, J.M.; Soane, L. Multiple functions of bcl-2 family proteins. *Cold Spring Harb. Perspect. Biol.* **2013**, *5*. [CrossRef] [PubMed]
97. De Oliveira, E.; de Aquino Castro, R.; Gomes, M.T.; da Silva, I.D.; Baracat, E.C.; de Lima, G.R.; Sartori, M.G.; Girao, M.J. The catechol-*O*-methyltransferase (comt) gene polymorphism and prevalence of uterine fibroids. *Maturitas* **2008**, *60*, 235–238. [CrossRef] [PubMed]
98. Halder, S.K.; Goodwin, J.S.; Al-Hendy, A. 1,25-dihydroxyvitamin D3 reduces TGF-beta3-induced fibrosis-related gene expression in human uterine leiomyoma cells. *J. Clin. Endocrinol. Metab.* **2011**, *96*, E754–762. [CrossRef] [PubMed]
99. Joseph, D.S.; Malik, M.; Nurudeen, S.; Catherino, W.H. Myometrial cells undergo fibrotic transformation under the influence of transforming growth factor beta-3. *Fertil. Steril.* **2010**, *93*, 1500–1508. [CrossRef] [PubMed]
100. Ciebiera, M.; Wlodarczyk, M.; Wrzosek, M.; Slabuszewska-Jozwiak, A.; Nowicka, G.; Jakiel, G. Ulipristal acetate decreases transforming growth factor beta3 serum and tumor tissue concentrations in patients with uterine fibroids. *Fertil. Steril.* **2018**, *109*, 501–507. [CrossRef] [PubMed]
101. Ciebiera, M.; Wlodarczyk, M.; Wrzosek, M.; Meczekalski, B.; Nowicka, G.; Lukaszuk, K.; Ciebiera, M.; Slabuszewska-Jozwiak, A.; Jakiel, G. Role of transforming growth factor beta in uterine fibroid biology. *Int. J. Mol. Sci.* **2017**, *18*. [CrossRef] [PubMed]
102. Malik, M.; Norian, J.; McCarthy-Keith, D.; Britten, J.; Catherino, W.H. Why leiomyomas are called fibroids: The central role of extracellular matrix in symptomatic women. *Semin. Reprod. Med.* **2010**, *28*, 169–179. [CrossRef] [PubMed]
103. Norian, J.M.; Malik, M.; Parker, C.Y.; Joseph, D.; Leppert, P.C.; Segars, J.H.; Catherino, W.H. Transforming growth factor beta3 regulates the versican variants in the extracellular matrix-rich uterine leiomyomas. *Reprod. Sci.* **2009**, *16*, 1153–1164. [CrossRef] [PubMed]
104. Halder, S.K.; Sharan, C.; Al-Hendy, A. 1,25-dihydroxyvitamin D3 treatment shrinks uterine leiomyoma tumors in the Eker rat model. *Biol. Reprod.* **2012**, *86*, 116. [CrossRef] [PubMed]
105. Halder, S.K.; Sharan, C.; Al-Hendy, O.; Al-Hendy, A. Paricalcitol, a vitamin D receptor activator, inhibits tumor formation in a murine model of uterine fibroids. *Reprod. Sci.* **2014**, *21*, 1108–1119. [CrossRef] [PubMed]
106. Al-Hendy, A.; Diamond, M.P.; El-Sohemy, A.; Halder, S.K. 1,25-dihydroxyvitamin D3 regulates expression of sex steroid receptors in human uterine fibroid cells. *J. Clin. Endocrinol. Metab.* **2015**, *100*, E572–582. [CrossRef] [PubMed]
107. Islam, M.S.; Greco, S.; Janjusevic, M.; Ciavattini, A.; Giannubilo, S.R.; D'Adderio, A.; Biagini, A.; Fiorini, R.; Castellucci, M.; Ciarmela, P. Growth factors and pathogenesis. *Best Pract. Res. Clin. Obstet. Gynaecol.* **2016**, *34*, 25–36. [CrossRef] [PubMed]
108. Protic, O.; Islam, M.S.; Greco, S.; Giannubilo, S.R.; Lamanna, P.; Petraglia, F.; Ciavattini, A.; Castellucci, M.; Hinz, B.; Ciarmela, P. Activin A in inflammation, tissue repair, and fibrosis: Possible role as inflammatory and fibrotic mediator of uterine fibroid development and growth. *Semin. Reprod. Med.* **2017**, *35*, 499–509. [CrossRef] [PubMed]
109. Al-Hendy, A.; Laknaur, A.; Diamond, M.P.; Ismail, N.; Boyer, T.G.; Halder, S.K. Silencing med12 gene reduces proliferation of human leiomyoma cells mediated via WNT/beta-catenin signaling pathway. *Endocrinology* **2017**, *158*, 592–603. [CrossRef] [PubMed]
110. Leppert, P.C.; Baginski, T.; Prupas, C.; Catherino, W.H.; Pletcher, S.; Segars, J.H. Comparative ultrastructure of collagen fibrils in uterine leiomyomas and normal myometrium. *Fertil. Steril.* **2004**, *82*, 1182–1187. [CrossRef] [PubMed]
111. Oblak, M.; Mlinsek, G.; Kandus, A.; Buturovic-Ponikvar, J.; Arnol, M. Effects of paricalcitol on biomarkers of inflammation and fibrosis in kidney transplant recipients: Results of a randomized controlled trial. *Clin. Nephrol.* **2017**, *88*, 119–125. [CrossRef] [PubMed]

112. Al-Hendy, A.; Diamond, M.P.; Boyer, T.G.; Halder, S.K. Vitamin D3 inhibits WNT/beta-catenin and MTOR signaling pathways in human uterine fibroid cells. *J. Clin. Endocrinol. Metab.* **2016**, *101*, 1542–1551. [CrossRef] [PubMed]
113. Gene, N. Fen1 Flap Structure-Specific Endonuclease 1 [Homo Sapiens (Human)]. 8.04.2018. Available online: https://www.ncbi.nlm.nih.gov/gene?Db=gene&Cmd=DetailsSearch&Term=2237 (accessed on 10 June 2018).
114. Singh, P.; Yang, M.; Dai, H.; Yu, D.; Huang, Q.; Tan, W.; Kernstine, K.H.; Lin, D.; Shen, B. Overexpression and hypomethylation of flap endonuclease 1 gene in breast and other cancers. *Mol. Cancer Res.* **2008**, *6*, 1710–1717. [CrossRef] [PubMed]
115. Liang, L.; Deng, L.; Chen, Y.; Li, G.C.; Shao, C.; Tischfield, J.A. Modulation of DNA end joining by nuclear proteins. *J. Biol. Chem.* **2005**, *280*, 31442–31449. [CrossRef] [PubMed]
116. Dou, Q.; Tarnuzzer, R.W.; Williams, R.S.; Schultz, G.S.; Chegini, N. Differential expression of matrix metalloproteinases and their tissue inhibitors in leiomyomata: A mechanism for gonadotrophin releasing hormone agonist-induced tumour regression. *Mol. Hum. Reprod.* **1997**, *3*, 1005–1014. [CrossRef] [PubMed]
117. Brew, K.; Dinakarpandian, D.; Nagase, H. Tissue inhibitors of metalloproteinases: Evolution, structure and function. *Biochim. Biophys. Acta* **2000**, *1477*, 267–283. [CrossRef]
118. Halder, S.K.; Osteen, K.G.; Al-Hendy, A. Vitamin D3 inhibits expression and activities of matrix metalloproteinase-2 and -9 in human uterine fibroid cells. *Hum. Reprod.* **2013**, *28*, 2407–2416. [CrossRef] [PubMed]
119. Artaza, J.N.; Norris, K.C. Vitamin D reduces the expression of collagen and key profibrotic factors by inducing an antifibrotic phenotype in mesenchymal multipotent cells. *J. Endocrinol.* **2009**, *200*, 207–221. [CrossRef] [PubMed]
120. Wise, L.A.; Ruiz-Narvaez, E.A.; Haddad, S.A.; Rosenberg, L.; Palmer, J.R. Polymorphisms in vitamin D-related genes and risk of uterine leiomyomata. *Fertil. Steril.* **2014**, *102*, 503–510. [CrossRef] [PubMed]
121. Shahbazi, S. Exploring the link between VDR rs2228570 and uterine leiomyoma in Iranian women. *Egypt. J. Med. Hum. Genet.* **2016**, *17*, 115–118. [CrossRef]
122. Gulec Yilmaz, S.; Gul, T.; Attar, R.; Yildirim, G.; Isbir, T. Association between fok1 polymorphism of vitamin D receptor gene with uterine leiomyoma in Turkish populations. *J. Turk. Ger. Gynecol. Assoc.* **2018**. [CrossRef]
123. Parazzini, F.; Di Martino, M.; Candiani, M.; Vigano, P. Dietary components and uterine leiomyomas: A review of published data. *Nutr. Cancer* **2015**, *67*, 569–579. [CrossRef] [PubMed]
124. Islam, M.S.; Akhtar, M.M.; Segars, J.H.; Castellucci, M.; Ciarmela, P. Molecular targets of dietary phytochemicals for possible prevention and therapy of uterine fibroids: Focus on fibrosis. *Crit. Rev. Food Sci. Nutr.* **2017**, *57*, 3583–3600. [CrossRef] [PubMed]
125. Pludowski, P.; Karczmarewicz, E.; Bayer, M.; Carter, G.; Chlebna-Sokol, D.; Czech-Kowalska, J.; Debski, R.; Decsi, T.; Dobrzanska, A.; Franek, E.; et al. Practical guidelines for the supplementation of vitamin D and the treatment of deficits in central Europe—Recommended vitamin D intakes in the general population and groups at risk of vitamin D deficiency. *Endokrynol. Pol.* **2013**, *64*, 319–327. [CrossRef] [PubMed]
126. Tebben, P.J.; Singh, R.J.; Kumar, R. Vitamin d-mediated hypercalcemia: Mechanisms, diagnosis, and treatment. *Endocr. Rev.* **2016**, *37*, 521–547. [CrossRef] [PubMed]
127. Ciavattini, A.; Delli Carpini, G.; Serri, M.; Vignini, A.; Sabbatinelli, J.; Tozzi, A.; Aggiusti, A.; Clemente, N. Hypovitaminosis D and "small burden" uterine fibroids: Opportunity for a vitamin D supplementation. *Medicine* **2016**, *95*, e5698. [CrossRef] [PubMed]
128. Vojinovic, J. Vitamin D receptor agonists' anti-inflammatory properties. *Ann. N. Y. Acad. Sci.* **2014**, *1317*, 47–56. [CrossRef] [PubMed]
129. Yamamoto, K.; Anami, Y.; Itoh, T. Development of vitamin D analogs modulating the pocket structure of vitamin D receptor. *Curr. Top. Med. Chem.* **2014**, *14*, 2378–2387. [CrossRef] [PubMed]
130. Trillini, M.; Cortinovis, M.; Ruggenenti, P.; Reyes Loaeza, J.; Courville, K.; Ferrer-Siles, C.; Prandini, S.; Gaspari, F.; Cannata, A.; Villa, A.; et al. Paricalcitol for secondary hyperparathyroidism in renal transplantation. *J. Am. Soc. Nephrol.* **2015**, *26*, 1205–1214. [CrossRef] [PubMed]
131. Zhang, Y.; Kong, J.; Deb, D.K.; Chang, A.; Li, Y.C. Vitamin D receptor attenuates renal fibrosis by suppressing the renin-angiotensin system. *J. Am. Soc. Nephrol.* **2010**, *21*, 966–973. [CrossRef] [PubMed]
132. Sanchez-Nino, M.D.; Bozic, M.; Cordoba-Lanus, E.; Valcheva, P.; Gracia, O.; Ibarz, M.; Fernandez, E.; Navarro-Gonzalez, J.F.; Ortiz, A.; Valdivielso, J.M. Beyond proteinuria: VDR activation reduces renal

inflammation in experimental diabetic nephropathy. *Am. J. Physiol. Ren. Physiol.* **2012**, *302*, F647–657. [CrossRef] [PubMed]
133. Tan, X.; Li, Y.; Liu, Y. Paricalcitol attenuates renal interstitial fibrosis in obstructive nephropathy. *J. Am. Soc. Nephrol.* **2006**, *17*, 3382–3393. [CrossRef] [PubMed]
134. Stavenuiter, A.W.; Farhat, K.; Vila Cuenca, M.; Schilte, M.N.; Keuning, E.D.; Paauw, N.J.; ter Wee, P.M.; Beelen, R.H.; Vervloet, M.G. Protective effects of paricalcitol on peritoneal remodeling during peritoneal dialysis. *Biomed. Res. Int.* **2015**, *2015*. [CrossRef] [PubMed]
135. Poveda, J.; Sanz, A.B.; Fernandez-Fernandez, B.; Carrasco, S.; Ruiz-Ortega, M.; Cannata-Ortiz, P.; Ortiz, A.; Sanchez-Nino, M.D. Mxra5 is a tgf-beta1-regulated human protein with anti-inflammatory and anti-fibrotic properties. *J. Cell. Mol. Med.* **2017**, *21*, 154–164. [CrossRef] [PubMed]
136. Bartels, C.B.; Cayton, K.C.; Chuong, F.S.; Holthouser, K.; Arian, S.E.; Abraham, T.; Segars, J.H. An evidence-based approach to the medical management of fibroids: A systematic review. *Clin. Obstet. Gynecol.* **2016**, *59*, 30–52. [CrossRef] [PubMed]
137. Buggio, L.; Roncella, E.; Somigliana, E.; Vercellini, P. Vitamin D and benign gynaecological diseases: A critical analysis of the current evidence. *Gynecol. Endocrinol.* **2016**, *32*, 259–263. [CrossRef] [PubMed]
138. Kashani, B.N.; Centini, G.; Morelli, S.S.; Weiss, G.; Petraglia, F. Role of medical management for uterine leiomyomas. *Best Pract. Res. Clin. Obstet. Gynaecol.* **2016**, *34*, 85–103. [CrossRef] [PubMed]
139. Al-Hendy, A.; Badr, M. Can vitamin D reduce the risk of uterine fibroids? *Women's Health* **2014**, *10*, 353–358. [CrossRef] [PubMed]
140. Ali, M.L.A.; Shaheen, S.; Sabri, M.; Al-Hendy, A. Vitamin D synergizes the antiproliferative, apoptotic, antifibrotic and anti-inflammatory effects of ulipristal acetate against human uterine fibroids. *Fertil. Steril.* **2017**, *108*, e66. [CrossRef]
141. Ali, M.; Chaudhry, Z.T.; Al-Hendy, A. Successes and failures of uterine leiomyoma drug discovery. *Expert Opin. Drug Discov.* **2018**, *13*, 169–177. [CrossRef] [PubMed]
142. Carlberg, C.; Haq, A. The concept of the personal vitamin D response index. *J. Steroid Biochem. Mol. Biol.* **2018**, *175*, 12–17. [CrossRef] [PubMed]
143. Harmon, Q.E.; Umbach, D.M.; Baird, D.D. Use of estrogen-containing contraception is associated with increased concentrations of 25-hydroxy vitamin D. *J. Clin. Endocrinol. Metab.* **2016**, *101*, 3370–3377. [CrossRef] [PubMed]
144. Ciebiera, M.W.M.; Słabuszewska-Jóźwiak, A.; Ciebiera, M.; Nowicka, G.; Jakiel, G. Vitamin D serum levels in women using contraception containing drospirenone—A preliminary study. *Arch. Med. Sci.* **2017**. [CrossRef]
145. Morch, L.S.; Skovlund, C.W.; Hannaford, P.C.; Iversen, L.; Fielding, S.; Lidegaard, O. Contemporary hormonal contraception and the risk of breast cancer. *N. Engl. J. Med.* **2017**, *377*, 2228–2239. [CrossRef] [PubMed]
146. Smith, J.; Green, J.; Berrington de Gonzalez, A.; Appleby, P.; Peto, J.; Plummer, M.; Franceschi, S.; Beral, V. Cervical cancer and use of hormonal contraceptives: A systematic review. *Lancet* **2003**, *5*, 1159–1167. [CrossRef]
147. Iversen, L.; Sivasubramaniam, S.; Lee, A.J.; Fielding, S.; Hannaford, P.C. Lifetime cancer risk and combined oral contraceptives: The royal college of general practitioners' oral contraception study. *Am. J. Obstet. Gynecol.* **2017**, *216*, 580.e1–580.e9. [CrossRef] [PubMed]
148. McLaren, J.S.; Morris, E.; Rymer, J. Gonadotrophin receptor hormone analogues in combination with add-back therapy: An update. *Menopause Int.* **2012**, *18*, 68–72. [CrossRef] [PubMed]
149. Somekawa, Y.; Chigughi, M.; Harada, M.; Ishibashi, T. Use of vitamin K2 (menatetrenone) and 1,25-dihydroxyvitamin D3 in the prevention of bone loss induced by leuprolide. *J. Clin. Endocrinol. Metab.* **1999**, *84*, 2700–2704. [CrossRef] [PubMed]
150. Lerchbaum, E. Vitamin D and menopause—A narrative review. *Maturitas* **2014**, *79*, 3–7. [CrossRef] [PubMed]

International Journal of
*Molecular Sciences*

MDPI

*Article*

# Antiproliferative Activity of Non-Calcemic Vitamin D Analogs on Human Melanoma Lines in Relation to VDR and PDIA3 Receptors

**Tomasz Wasiewicz [1], Anna Piotrowska [1], Justyna Wierzbicka [1], Andrzej T. Slominski [2,3,4] and Michal A. Zmijewski [1,*]**

1 Department of Histology, Medical University of Gdansk, 80-210 Gdansk, Poland; tomwasxp@poczta.onet.pl (T.W.); annapiotrowska@gumed.edu.pl (A.P.); justyna.wierzbicka@gumed.edu.pl (J.W.)
2 Department of Dermatology, University of Alabama, Birmingham, AL 35294, USA; aslominski@uabmc.edu
3 Comprehensive Cancer Center, Cancer Chemoprevention Program, University of Alabama, Birmingham, AL 35294, USA
4 VA Medical Center, Birmingham, AL 35294, USA
* Correspondence: mzmijewski@gumed.edu.pl; Tel.: +48-58-349-1455; Fax: +48-58-349-1419

Received: 2 August 2018; Accepted: 24 August 2018; Published: 31 August 2018

**Abstract:** Vitamin D is a precursor for secosteroidal hormones, which demonstrate pleiotropic biological activities, including the regulation of growth and the differentiation of normal and malignant cells. Our previous studies have indicated that the inhibition of melanoma proliferation by a short side-chain, low calcemic analog of vitamin D—21(OH)pD is not fully dependent on the expression of vitamin D receptor (VDR). We have examined the effects of classic vitamin D metabolites, 1,25(OH)$_2$D$_3$ and 25(OH)D$_3$, and two low calcemic vitamin D analogs, (21(OH)pD and calcipotriol), on proliferation, mRNA expression and vitamin D receptor (VDR) translocation in three human melanoma cell lines: WM98, A375 and SK-MEL-188b (subline b of SK-MEL-188, which lost responsiveness to 1,25(OH)$_2$D$_3$ and became VDR$^{-/-}$CYP27B1$^{-/-}$). All tested compounds efficiently inhibited the proliferation of WM98 and A375 melanoma cells except SK-MEL-188b, in which only the short side-chain vitamin D analog—21(OH)pD was effective. Overall, 21(OH)pD was the most potent compound in all three melanoma cell lines in the study. The lack of responsiveness of SK-MEL-188b to 1,25(OH)$_2$D$_3$, 25(OH)D$_3$ and calcipotriol is explained by a lack of characteristic transcripts for the *VDR*, its splicing variants as well as for vitamin D-activating enzyme CYP27B1. On the other hand, the expression of *VDR* and its splicing variants and other vitamin D related genes (*RXR*, *PDIA3*, *CYP3A4*, *CYP2R1*, *CYP27B1*, *CYP24A1* and *CYP11A1*) was detected in WM98 and A375 melanomas with the transcript levels being modulated by vitamin D analogs. The expression of *VDR* isoforms in WM98 cells was stimulated strongly by calcipotriol. The antiproliferative activities of 21(OH)pD appear not to require VDR translocation to the nucleus, which explains the high efficacy of this noncalcemic pregnacalciferol analog in SK-MEL-188b melanoma, that is, VDR$^{-/-}$. Therefore, we propose that 21(OH)pD is a good candidate for melanoma therapy, although the mechanism of its action remains to be defined.

**Keywords:** vitamin D; 1,25(OH)$_2$D$_3$; calcipotriol; 21(OH)pD; vitamin D analogs; melanoma; human melanoma cell lines; VDR translocation; anti-melanoma activity

## 1. Introduction

Vitamin D is a naturally occurring hormone precursor, which after activation serves as a pleiotropic regulator of homeostasis on cellular and whole organism levels [1,2]. It is produced in the skin which is subjected to the ultraviolet B (UVB) fraction of sunlight or could be acquired from nutritional sources and supplementation. UVB exposure results in the conversion of epidermal 7-dehydrocholesterol (7DHC) into pre-vitamin $D_3$, which undergoes subsequent isomerization to vitamin $D_3$ [2–5]. Vitamin $D_3$ has to be hydroxylated in order to implement its biological activities. The first step takes place in the liver, where vitamin $D_3$ is converted to 25(OH)$D_3$ (25-hydroxyvitamin $D_3$, calcifediol) by CYP2R1, CYP27A1 or CYP3A4 (CYP3A4 belongs to cytochrome P450 superfamily and is involved in drug metabolism, but also has a capacity for vitamin D hydroxylation), and 25(OH)$D_3$ is subsequently hydroxylated by 1$\alpha$-hydroxylase (CYP27B1) in the kidney. In addition, 1,25$(OH)_2D_3$ can be generated locally in many cell types (keratinocytes, dendritic cells, melanocytes, lymphocytes and cancer cells) expressing appropriate enzymatic machinery [6,7]. The level of 1,25$(OH)_2D_3$ and its metabolites in circulation are tightly regulated through a negative feedback loop by 24-hydroxylase (CYP24A1), which metabolizes calcitriol into water-soluble and inactive calcitroic acid [8]. The catalytically inactive splicing variant of *CYP24A1* (*CYP24SV*) may serve as dominant negative regulator of vitamin $D_3$ catabolism and possibly contribute to the extracellular accumulation of 1,25(OH)2$D_3$ [9]. Most recently, it was discovered that CYP11A1, the rate limiting enzyme of steroidogenesis, also metabolizes 7DHC, vitamin $D_3$ and lumisterol to their corresponding hydroxyderivatives with full and short side-chains [10–15].

According to the genomic pathway, active forms of vitamin $D_3$ exert their biological activities through interactions with the VDR, which is expressed by a majority of cells in the body [16,17]. Activated VDR forms a heterodimer with retinoid X receptor (RXR), and this complex is translocated to the nucleus where it binds to the VDREs (vitamin D response elements) and activates the expression of hundreds of human genes [17,18]. The activity of this complex is also regulated by the recruitment of co-activators or co-repressors to modulate the expression of selected genes, including those participating in the inhibition of cell cycle progression and the stimulation of differentiation and apoptosis [19–22].

In addition to the well-characterized vitamin $D_3$ genomic actions, its analogs may also exert a rapid, non-genomic response (RR: rapid response) [23–25]. There is growing evidence that the co-localization of VDR [26,27] with caveolae of plasma membrane, as well as the binding of vitamin D to PDIA3 (Protein disulfide-isomerase A3), is responsible for the alternative (non-genomic) action of vitamin $D_3$ and its analogs [23,28].

According to epidemiological studies, the incidence of human malignant melanoma has been increasing steadily since the 1970s [29–32]. Early detection of melanoma results in a good prognosis for patients; however, the survival rate and therapeutic procedures for advanced or metastatic melanomas are very limited. Ultraviolet radiation, in addition to being fully carcinogenic and a main cause of melanoma [32–34], is also the most important factor promoting the formation of vitamin $D_3$. The anti-melanoma activity of vitamin D derivatives have been reported previously [35–40]. In addition, the presence of specific polymorphisms of *VDR* [41–43] or a decreased expression of *VDR*, *CYP27B1*, *CYP24A1* and defects in vitamin D signaling are linked to more advanced stages of melanoma or poorer prognosis [44–48]. Furthermore, proper supplementation with vitamin D is believed to be an important factor in cancer prevention [47,49–51].

Active forms of vitamin $D_3$ show antiproliferative properties against several types of cancer, including colorectal cancer [52], breast cancer [53], prostate cancer [54] and melanoma [39,40,55–59]. Furthermore, the use of 1,25$(OH)_2D_3$ as an anticancer drug in very high concentrations (above 50,000 units/day) is currently under clinical investigation; however, it has to be acknowledged that it could potentially cause hypercalcemia [40,50,60]. Since vitamin $D_3$ analogs with modified or shortened side-chains were shown to have low or no effect on calcium levels [59,61–63], they are potentially better alternatives for calcitriol. So far, more than 3000 vitamin $D_3$ analogs have been synthesized, and their biological activity is still intensively investigated as a single agent or in combination with other cytostatics [40,64–66]. In this study, we have investigated the response of three human melanoma cell lines against four vitamin $D_3$ analogs. We have analyzed the antiproliferative potential of analogs

(Figure 1) and found that 21(OH)pD inhibits melanoma growth through a mechanism independent from VDR or PDIA3.

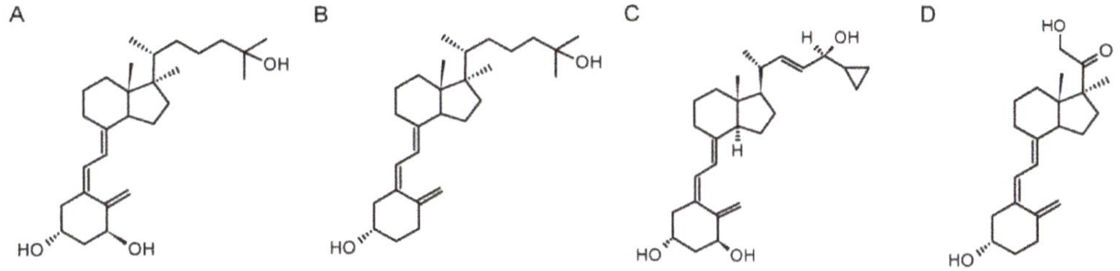


**Figure 1.** Chemical structure of vitamin $D_3$ analogs: 1,25$(OH)_2D_3$ (**A**), 25(OH)$D_3$ (**B**), calcipotriol (**C**) and 21(OH)pD (**D**).

## 2. Results

### 2.1. New Vitamin $D_3$ Analogs Effectively Inhibit A375 Cell Proliferation

Our previous studies on human melanomas SK-MEL-188 [10,39,57,59,62,67–69], A375 [55], and WM98 [39,68] demonstrated the sensitivity of melanoma cells to vitamin $D_3$ analogs with short and full side-chains. The efficacy of compounds was attenuated by melanogenesis, which was associated with the downregulation of VDR expression [56,68]. However, it was also shown that the antiproliferative activity of short side-chain analogs of vitamin $D_3$, such as 21(OH)pD, were not fully dependent on VDR expression in rodent melanomas [56], consistent with the poor docking score on the ligand binding domain of the VDR and the poor translocation of VDR to the nucleus [69].

To better understand the effect of VDR expression on the differential action of vitamin $D_3$ analogs, we used three human melanoma lines, A375 and WM98 and a subline b of SKMEL-188 which lost responsiveness to vitamin D during in vitro passaging [55], and detected the VDR expression and sensitivity to selected vitamin D analogs (1,25$(OH)_2D_3$, 25(OH)$D_3$, 21(OH)pD and calcipotriol). As expected, these compounds [40,65] effectively inhibited the proliferation of A375 and WM98 melanomas expressing VDR receptor. $IC_{50}$ values ranged from pM to μM, and the effects on cell proliferation were strongly dependent on the melanoma cell line and the nature of the compound. Interestingly, non-pigmented SK-MEL-188b was found to be resistant to vitamin $D_3$ analogs with a full-length side-chain (1,25$(OH)_2D_3$, 25(OH)$D_3$ and calcipotriol) and sensitive to short side-chained 21(OH)pD (Figure 2). Overall, 21(OH)pD was found to be the most potent inhibitor of growth in tested melanoma cell lines.

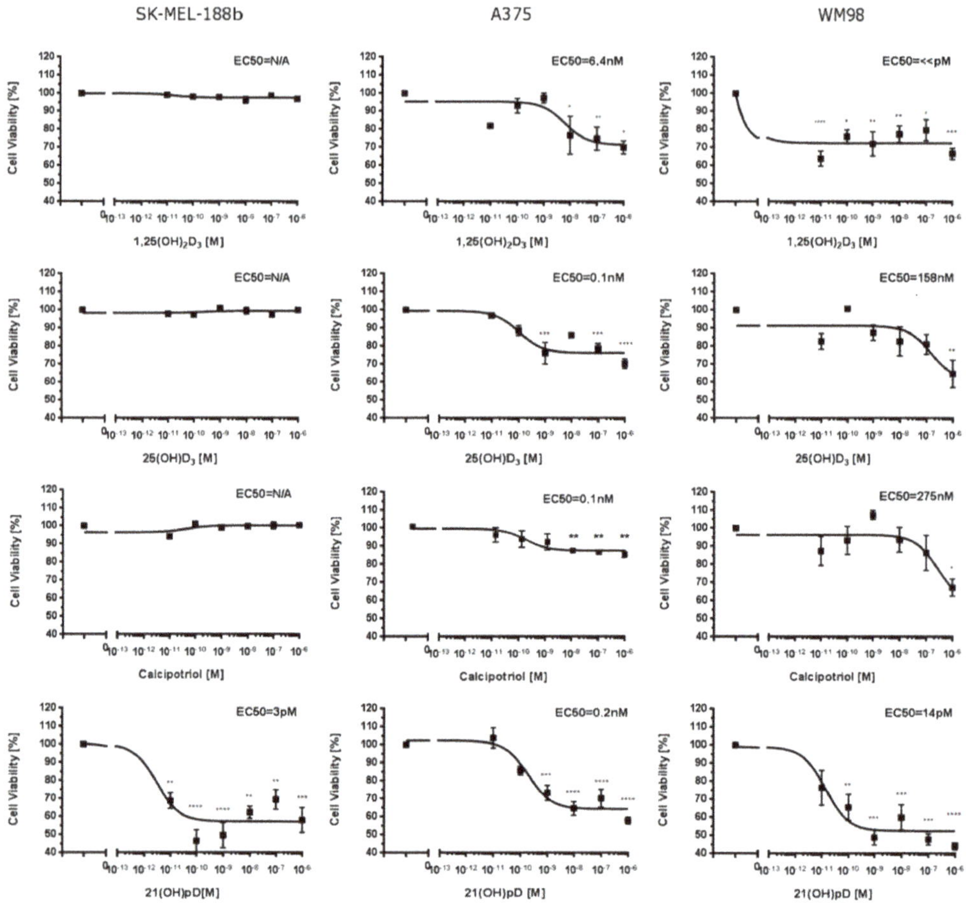


**Figure 2.** Effects of 1,25(OH)$_2$D$_3$, 25(OH)D$_3$, calcipotriol and 21(OH)pD on growth of human SK-MEL-188b, A375 and WM98 melanoma cells. Cells were seeded into 96-well plates and incubated in a medium supplemented with serial dilution of vitamin D analogs from 0.01 nM to 1000 nM concentration (as described in Material and Methods). The statistical significance of results has been analyzed using one-way ANOVA (GraphPad Software, San Diego, CA, USA) and data are presented as means ± SEM for at least three independent measurements. The cutoff point of significance was defined as $p < 0.05$ (* $p < 0.05$, ** $p < 0.01$, *** $p < 0.001$, **** $p < 0.0001$).

### 2.2. The Effect of Vitamin D$_3$ Compounds on the Expression of Genes Related to Vitamin D$_3$ Action or Metabolism

First, we investigated the expression of vitamin D$_3$-related genes in untreated melanoma cells. As shown in Figure 3, PCR fragments characteristic for *VDR*, *CYP27A1* and *CYP27B1* mRNAs were below the level of detectability in SK-MEL-188b melanoma, which did not respond to vitamin D. Only the mRNA of *PDIA3* was detected in this line. On the other hand, in A375 and WM98, cell lines corresponding to transcripts for all tested genes were detected.

Next, we tested the effects of vitamin D$_3$ analogues on the relative mRNA levels of vitamin D$_3$-related genes in responsive melanoma cell lines (A375 and WM98). The treatment of WM98 melanoma with full length side-chain vitamin D$_3$ analogs (1,25(OH)$_2$D$_3$, 25(OH)D$_3$ and calcipotriol) is presented in Figure 4. Only 1,25(OH)$_2$D$_3$ stimulated the expression of *VDR* Co-receptor *RXR* and of *CYP27A1* (Figure 4B). All three analogs, however, stimulated the expression of *CYP3A4*, *CYP2R1*, *CYP24A1* and its splicing variant *CYP24A1sv*. Interestingly, the strongest effect (approximately a 6-fold induction) on *CYP2R1* mRNA level was observed in cells treated with calcipotriol (Figure 4C). Furthermore, the level of *CYP27B1* mRNA was not affected by tested secosteroids (Figure 4G), while the expression of *PDIA3* was stimulated only by 25(OH)$D_3$ (Figure 4E).

The treatment of A375 melanoma with four vitamin $D_3$ analogs (1,25$(OH)_2D_3$, calcipotriol and 21(OH)pD) decreased the expression of *VDR, RXR, PDIA3, CYP2R1* genes (Figure 5). Interestingly, the strongest decrease in the mRNA levels of *CYP3A4* and *CYP2R1* was observed in A375 melanoma cells treated with 21(OH)pD. The induction of *CYP24A1* gene was hundreds of times folds higher in cells treated with 1,25$(OH)_2D_3$ and calcipotriol when compared to 21(OH)pD, showing minimal effect. Finally, the mRNA level of *CYP27B1* was not affected by vitamin D analogues and *CYP11A1* was elevated in A375 melanoma cells only after treatment with calcipotriol.

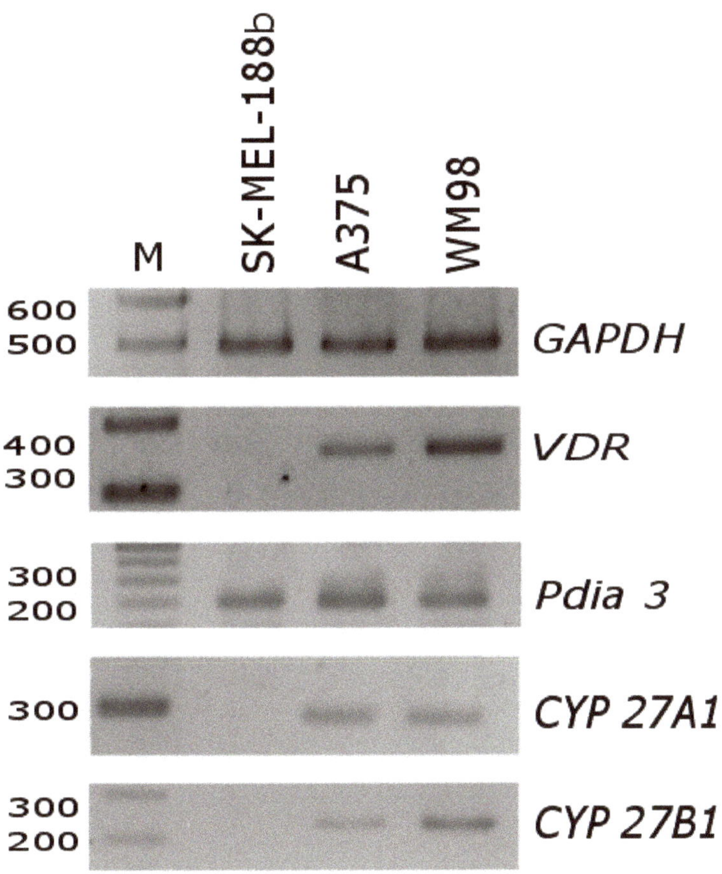


**Figure 3.** Differences in mRNA level of *VDR*, *PDIA3*, *CYP27A1* and *CYP27B1* genes in SK-MEL-188b cells in comparison with A375 and WM98 cells. Glyceraldehyde 3-phosphate dehydrogenase (GAPDH) was used as a reporter gene to normalize all samples. See Table S1 for primer sequences and the predicted length of PCR fragments.

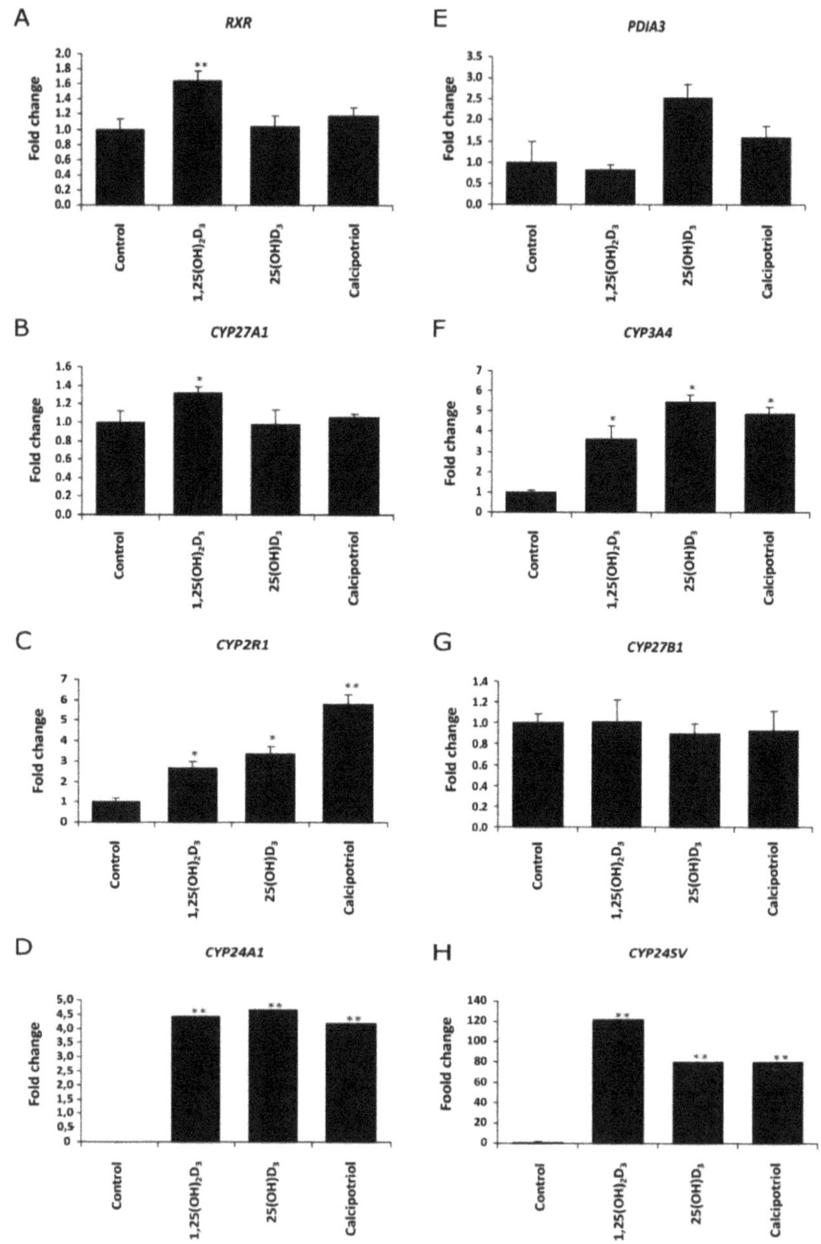


**Figure 4.** Vitamin D analogs treatment modulate the expression of (**A**) retinoid X receptor (*RXR*), (**E**) disulfide isomerase (*PDIA3*), 25-hydoxylases ((**B**) *CYP27A1*, (**C**) *CYP2R1* or (**F**) *CYP3A4*, (**G**) 1α-hydroxylase (*CYP27B1*)) and 24-hydroxylases ((**D**) *CYP24A1* or (**H**) *CYP24SV*) in WM98 melanoma cell line. Cells were stimulated with 1 μM $1,25(OH)_2D_3$, $25(OH)D_3$ or calcipotriol for 24 h. Quantitative PCR analyses were performed as described in Materials and Methods. Statistical significance was estimated using *t*-test and data are presented as means ± SD ($n = 3$). The cutoff point for significance is defined as $p < 0.05$ (* $p < 0.05$, ** $p < 0.01$).

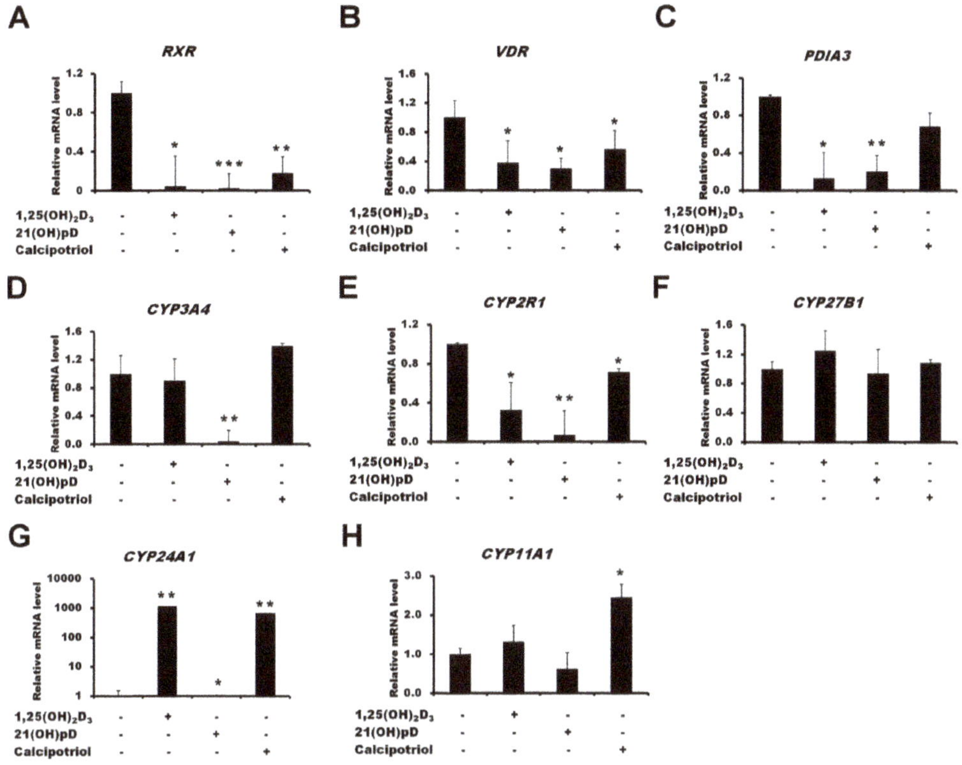


**Figure 5.** Effects of vitamin D compounds (1,25(OH)$_2$D$_3$, 21(OH)pD or calcipotriol) on *RXR* (**A**), *VDR* (**B**), *PDIA3* (**C**), *CYP3A4* (**D**), *CYP2R1* (**E**), *CYP27B1* (**F**), *CYP24A1* (**G**) and *CYP11A1* (**H**) genes expression in A375 melanoma cells. A375 melanoma cells were incubated with 100 nM of 1,25(OH)$_2$D$_3$, 21(OH)pD or calcipotriol for 24 h. mRNA levels were measured by qPCR. Data are shown as means ± S.D of three independent experiments carried out in duplicate. The cutoff point for significance is defined as $p < 0.05$ (* $p < 0.05$, ** $p < 0.01$, *** $p < 0.001$).

It is well established that the antiproliferative effects of vitamin D$_3$ analogs strongly depend on the expression of vitamin D receptor VDR [40,47,65]. As shown in Figure 3, the melanoma cell lines used in the study differ in their basal level of *VDR* expression, and the transcript is absent in SK-MEL-188b, which is not responsive to vitamin D. Theoretically, an alternative splicing of *VDR* pre-mRNA may results in the production of at least five alternative transcripts (Figure 6A); thus, the unique set of primers was designed in order to detect specific variants of *VDR* mRNA (Table S1). Detection of specific variants of *VDR* was confirmed by sequencing of PCR fragments previously separated on agarose gel (Figure 6B). WM98 and A375 melanoma cell lines express all four PCR fragments (Pr 1, Pr 2, Pr 4 and Pr 5) corresponding to *VDR* splicing variant c (Figure 6A). In the WM98 melanoma, the cell line presence of an mRNA characteristic for isoforms a (Pr 3) and b (Pr 4, upper band) was also detected. Interestingly, in SK-MEL-188b melanoma cells, only one fragment of mRNA corresponding to the alternative promoter region of *VDR* (isoform c) was detected. Finally, the effects of two vitamin D$_3$ analogs 25(OH)D$_3$ and calcipotriol on the alternative splicing of *VDR* pre-mRNA was tested in WM98 cell line (Figure 6C). It was show that treatment with calcipotriol, but not 25(OH)D$_3$, resulted in at least 7-fold elevation of mRNA levels corresponding to all fragments of the *VDR* tested (fragments detected by a set of primers: Pr 1 to Pr 5).

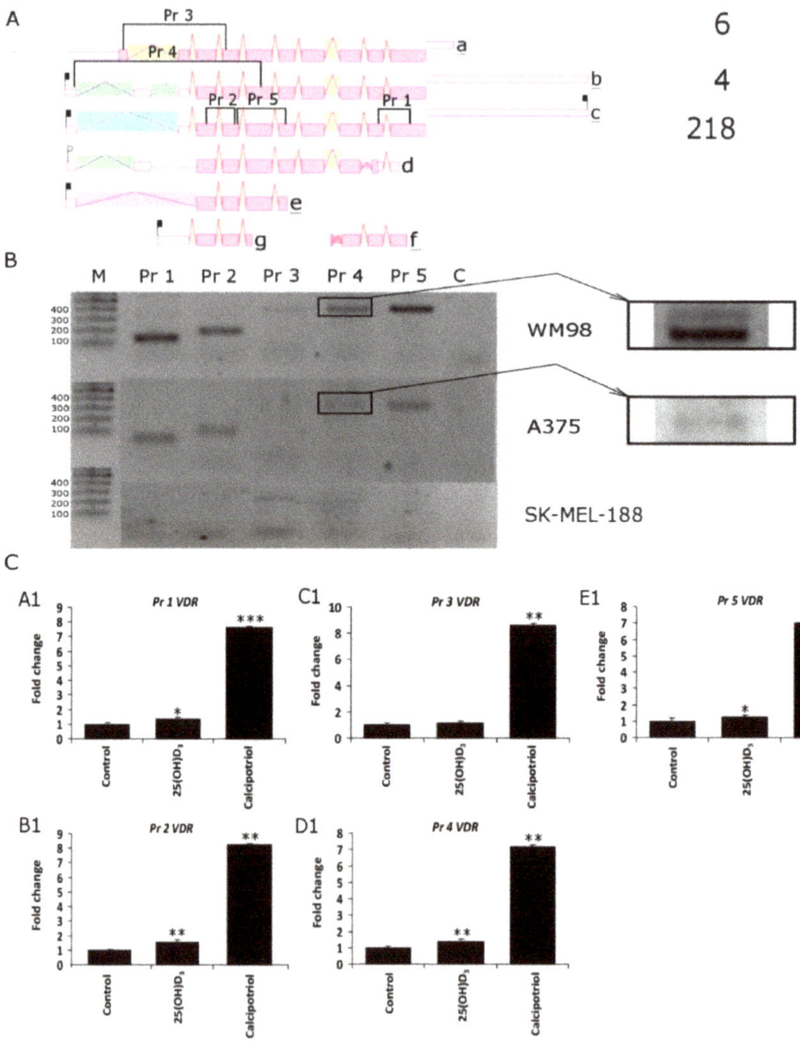


**Figure 6.** (**A**) Five different PCR primers sets were designed (Pr 1–Pr 5) in order to detect the expression of *VDR* splicing variants in WM98, A375 and SK-MEL-188b cell lines. The most common isoforms are a b and c and the rarely identified isoforms are d, e, f and g. The positions of sets of primers designed to differentiate *VDR* isoforms is also shown. Product 1 (Pr 1) of 132 b.p. is characteristic for isoforms a, b, c and f; Product 2 (Pr 2) of 180 b.p. is universal for all isoforms accept f; Product 3 (Pr 3) of 386 b.p. is unique for isoform a; Product 4 (Pr 4), depending on length of the PCR fragment, indicates isoform "b" 532 b.p. or "c" 410 b.p.; Product 5 (Pr 5) of 384 b.p. is characteristic for isoforms a, b, c, d and e, but not g and f (diagram of VDR splicing variants taken from AceView https://www.ncbi.nlm.nih.gov/IEB/Research/Acembly/). (**B**) Semiquntitative PCR was used to differentiate the *VDR* isoforms (see Material and Methods for details); molecular weight marker (M) as described in. (**C**) Effects of vitamin D analogs on the expression of *VDR* isoforms were analyzed in WM98 melanoma cells. Melanoma cells were treated with $25(OH)D_3$ or calcipotriol at 1 µM concentration for 24 h. Statistical significance was estimated using *t*-test and data are presented as means ± SD ($n = 3$). The cutoff point for significance is defined as $p < 0.05$ (* $p < 0.05$, ** $p < 0.01$, *** $p < 0.001$).

### 2.3. *Translocation of Vitamin D Receptor by Secosteroids*

The ligand-induced translocation of VDR receptor into the nucleus was studied using A375 cells stably transduced with pLenti-CMV-VDR-EGFP lentiviral construct [70]. 1,25(OH)$_2$D$_3$ or calcipotriol induced a dose-dependent and efficient translocation of VDR-GFP fusion protein into the nucleus (Figure 7). A lack of VDR-GFP translocation to the nucleus was observed in cells treated with 21(OH)pD.

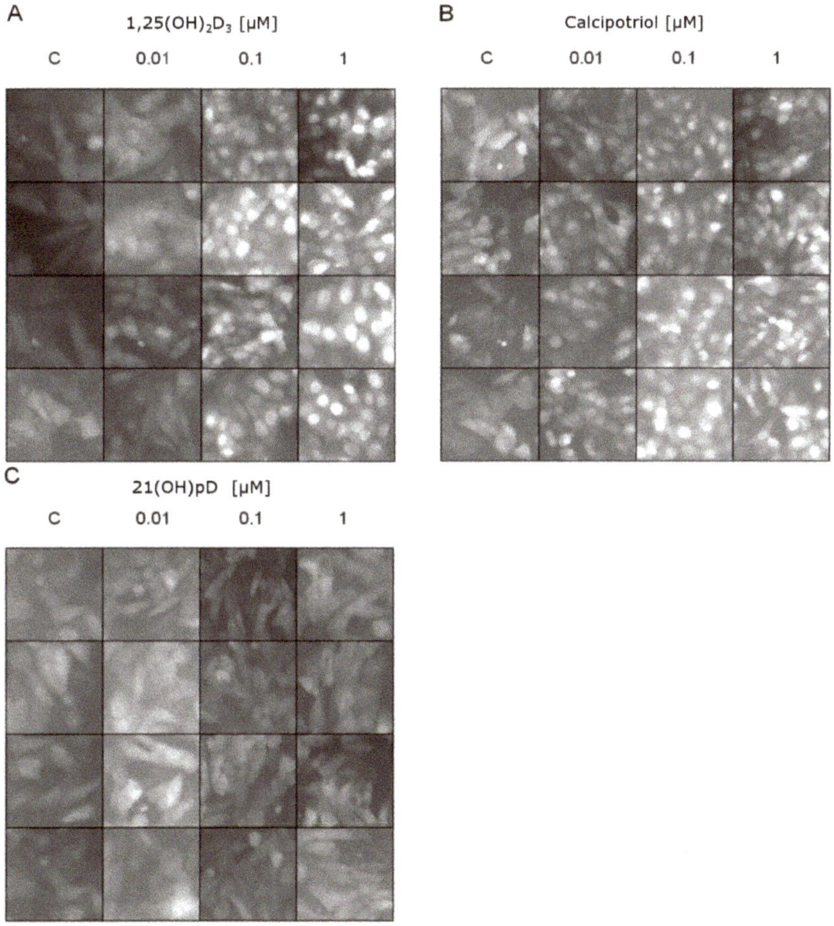


**Figure 7.** Effects of vitamin D analogs: (**A**) 1,25(OH)$_2$D$_3$, (**B**) calcipotriol or (**C**) 21(OH)pD$_3$ at 0 (C—control), 0.01, 0.1 or 1 μM concentrations on VDR translocation to the nucleus (green fluorescence shown in a gray scale). Melanoma A375 cells were transduced using pLenti-CMV-VDR-GFP-pgkpuro construct: see Materials and Methods. Vitamin D analogs except 21(OH)pD induce VDR translocation to the nucleus after 1 h of incubation in a concentration-dependent manner. Each panel shows four random micrographs for each concentration taken by fluorescent microscope under 100× magnification.

## 3. Discussion

There is growing evidence that vitamin D$_3$ analogs are excellent candidates for melanoma therapy [40,47,65,71,72]. This is supported by the strong correlation between an adequate level of 25(OH)D$_3$ (>30 ng/mL in the serum) and a decreased incidence or severity of several cancers, including melanoma [62,73–77]. However, the use of an active form of vitamin $D_3$, $1,25(OH)_2D_3$ is limited due to its potential hypercalcemic effects; thus, the development and biological evaluation of new low calcemic vitamin $D_3$ derivatives is highly desirable [40,55,65,72].

Here, we present results showing the different sensitivities of three melanoma cell lines (SK-MEL-188b, A375 and WM98) to vitamin $D_3$ analogs. Recently, we described vitamin $D_3$-resistant clone of SK-MEL1-88 (SK-MEL-188b) [55]. Contrary to parental SK-MEL-188, this clone was insensitive to the treatment with $1,25(OH)_2D_3$, or with several analogs of vitamin $D_2$, except of PRI-1731 with the inverse orientation of A ring in comparison to parental $1,25(OH)_2D_3$ by the introduction of (5Z,7Z) modification [55,78,79]. This clone does not express the coding region of *VDR* and of *CYP27B1*. Therefore, we call it SK-MEL-188*VDR*$^{-/-}$*CYP27B1*$^{-/-}$. Interestingly, a short side-chain analog of vitamin $D_3$ with a pregnacalciferol-type configuration, such as 21(OH)pD, was shown to inhibit the growth of the SK-MEL-188b line. It has to be underlined that this analog, also showed high potency against two other melanoma cell lines in the study (A375 and WM98) and minimal stimulation of CYP24A1. This indicates that the antiproliferative properties of 21-OHpD are independent on its action on the VDR, which is consistent with our previous findings on pregnacalciferol derivatives acting as poor activators of VDR per VDR translocation studies and molecular modeling [69]. One of the possible mechanisms of 21(OH)pD action could include an activation of non-genomic pathways [25,80], including those linked to the PDIA3 protein [81,82]. However, interaction with other nuclear steroid receptors cannot be excluded. In addition, other receptors such as retinoic orphan acid receptors (ROR)$\alpha$ and $\gamma$ can be considered since vitamin D hydroxyderivatives can act as reverse agonists on these receptors [62,83] and related pregnalumisterol derivatives can also act on ROR$\alpha$ and ROR$\gamma$ [15]. These considerations require future in-depth investigations to define specific receptors for pD compounds (secosteroids with shortened side-chain).

Two other melanoma lines were responsive to vitamin D (WM98 and A375) and showed relatively high expression levels of *VDR* transcript. Thus, the studies presented above on three melanoma lines are consistent with reports showing that the expression of *VDR* is the key factor responsible for the antitumor activities of $1,25(OH)_2D_3$ [18,47,84,85]. Of note is that its decreased expression correlates with advanced melanoma staging, progression and decreased overall patient survival and disease-free survival time [45,46]. Accordingly, *VDR* polymorphism [41,42,72] and decreased levels of $25(OH)D_3$ in the serum [33,86] positively correlate with melanoma prevalence and poor prognosis.

Here, we show that sensitivity to tested vitamin $D_3$ hydroxyderivatives with full side-chains depends on *VDR* expression and its alternative splicing. Interestingly, sensitive melanoma lines (A375, WM98) expressed all three major *VDR* splicing variants, suggesting that they may play an important but not necessarily identical function in vitamin $D_3$ signaling. Further studies are required to elucidate the significance of the expression of *VDR* splicing variants. Furthermore, the expression of vitamin $D_3$-related genes was altered by vitamin $D_3$ analogs, and the effects on the expression of *VDR, RXR, PDIA3, CYP2R1* or *CYP24A1* were stronger in WM98 melanoma cells in comparison to the A375 line. This phenomenon could explain higher sensitivity of WM98 to $1,25(OH)_2D_3$ in comparison to A375 line.

In summary, low calcemic vitamin $D_3$ analogs such as $25(OH)D_3$, 21(OH)pD or calcipotriol showed similar antiproliferative activity to $1,25(OH)_2D_3$ in melanoma cell lines expressing VDR spliced variants. Furthermore, the short side-chain analog 21(OH)pD was found to be superior among vitamin D analogs and was the only one, which inhibited the growth of the melanoma subline negative for VDR, indicating a mechanism of action that is VDR-independent.

## 4. Materials and Methods

### 4.1. Cell Lines and Vitamin D Analogs

In our study, we determined the inhibitory effects of vitamin $D_3$ analogs against immortalized human melanoma cell lines SK-MEL-188b, WM98 and A375. SK-MEL-188b cells were cultured in F10 medium supplemented with 10% fetal bovine serum (FBS, Sigma, Poznan, Poland) and 1× antibiotic-antimycotic solution (Anti-Anti, Sigma). A375 and WM98 cell lines were cultured in Dulbecco's Modified Eagle Medium (DMEM) supplemented with 10% FBS and 1% anti-anti. To eliminate the influence of sterols present in fetal bovine serum in the experiments, we used 5% charcoal-stripped FBS (GE Healthcare Life Sciences, Warsaw, Poland). Vitamin $D_3$ analogs $1,25(OH)_2D_3$, $25(OH)D_3$ and calcipotriol were acquired from the Pharmaceutical Research Institute (Warsaw, Poland), while the new short side analog 21(OH)pD was synthesized in collaboration with ProChimia Surfaces (Gdynia, Poland) as described previously [49,50]. The chemical structures of these secosteroids are presented in Figure 1.

### 4.2. SRB Assay

Melanoma cells were seeded on 96-well plates at a density of 8000 per well in an appropriate medium supplemented with 5% charcoal-stripped FBS and 1× anti-anti solution. After 24 h, the medium was replaced with the fresh one containing vehicle and serial dilutions of vitamin $D_3$ analogs at concentrations from 10 pM to 1 μM. Following incubation for 48 h, 100 μL of 20% TCA (trichloroacetic acid) was added and cells were incubated for 1 h in 4 °C. The medium was discarded and cells were washed 5 times with sterile water. Afterwards, cells were dried and 100 μL of SRB (0.4 g sulforhodamin B in 100 mL 1% acetic acid) was added into the plate wells for 15 min incubation at room temperature. Next, cells were washed 5 times with 1% acetic acid, dried and resolved with 150 μL of 10 mM Tris (pH = 10.5) for 10 min in room temperature. The absorbance was measured spectroscopically at 570 nm with 96-well plate reader (BioTek, Winooski, VT, USA).

### 4.3. Classical PCR and Real-Time PCR Analysis

SK-MEL-188b, A375 and WM98 melanoma cells were treated with $1,25(OH)_2D_3$, $25(OH)D_3$ and calcipotriol (only WM98 line) for 24 h (Figure 5 and Figure S1) or collected without treatment (Figures 3 and 4). RNA was isolated using a Total RNA Kit (A&A Biotechnology, Gdynia, Poland). Reverse transcription (500 ng RNA/reaction) was carried out with a RevertAid™ First Strand cDNA Synthesis Kit (Fermentas, Vilnius, Lithuania). Classic PCR and real-time PCR were performed using 5-fold diluted cDNA and 2× PCR Master Mix (A&A Biotechnology) or real time 2× PCR Master Mix SYBR Set A, B (A&A Biotechnology). The primers to amplify fragments of *ACTB*, *VDR*, *RXR*, *PDIA3*, *CYP27A1*, *CYP2R1*, *CYP3A4*, *CYP27B1*, *CYP24A1*, *CYP24SV* genes were designed with Primer Quest software (Integrated Device Technology, San Jose, CA, USA) (Table S1). The data was collected on a MJ Mini BioRad cycler (BioRad, Hercules, CA, USA) or Termocycler StepOne Real-Time PCR Systems (Life Technologies, Carlsbad, CA, USA). PCR products were visualized using the Mupid-One electrophoresis system (BioRad) and ethidium bromide staining.

### 4.4. VDR Translocation

The melanoma A375 cell line was used to study the VDR receptor translocation to nucleus after vitamin $D_3$ analog treatment. Cells were transduced by pLenti-CMV-VDR-EGFP-pgkpuro plasmid, where VDR and GFP were expressed as a fusion protein in cytoplasmic compartment. Cells were cultured on 12 wells plate ($0.5 \times 10^5$ cells/well) in DMEM supplemented with 5% charcoal-stripped FBS and 24 h later medium was replaced with DMEM containing $1,25(OH)_2D_3$, calcipotriol and 21(OH)pD at 0.01 μM, 0.1 μM or 1 μM concentrations. After 1 h of incubation VDR translocation from cytoplasm to nucleus was determined with fluorescent Eclipse TE300 microscope (Nikon, Tokyo, Japan).

### 4.5. Statistical Analysis

SRB viability data was presented using one-way Anova test to compare experimental groups: cells were treated with vitamin $D_3$ analogs with a concentration from 0.01 nM to 1000 nM with control samples (GraphPad Software). The antiproliferative potency of vitamin $D_3$ analogs were compared by the calculation of EC50 for every compound (half maximal effective concentration).

Real-time PCR data was analyzed with the comparative ΔΔ-Ct method normalized to the reference gene *ACTB*. As a control, a probe with sterile water was used instead of cDNA. Data are presented as mean ± SD ($n$ = 4–6). Student's $t$-test (for two groups) or one-way ANOVA with appropriate post-hoc test (for more than two groups) were used to analyze data using Excel (Microsoft) or Prism 7.00 (GraphPad Software), respectively. Statistically significant differences are denoted with asterisks: $p < 0.05$ (* $p < 0.05$; ** $p < 0.01$; *** $p < 0.0001$).

**Supplementary Materials:** Supplementary materials can be found at http://www.mdpi.com/1422-0067/19/9/2583/s1.

**Author Contributions:** T.W. and A.P. and J.W. performed the main experiments, took part in data analysis and preparation of the manuscript. M.A.Z. and A.T.S. designed experiments; M.A.Z., T.W., J.W. and A.T.S. analyzed data and prepared final version of the manuscript.

**Acknowledgments:** The study was supported by a grant of the Polish Ministry of Science and Higher Education, Project No. N405 623238 and National Science Center OPUS Program under contract 2017/25/B/NZ3/00431 to MAZ; and NIH grants 1 R01AR073004-01A1. 1 RO1 AR071189-01A1 and R21AR066505 to ATS. Authors would like to thank Adam Figarski, MD, for help in experiment with VDR translocation.

**Conflicts of Interest:** The authors declare no conflict of interest.

## Abbreviations

| | |
|---|---|
| $1\alpha,25(OH)_2D_3$ | $1\alpha$,25-dihydroxyvitamin $D_3$ (calcitriol) |
| 21(OH)pD | 21-hydroxypregnacalciferol |
| $25(OH)D_3$ | 25-hydroxyvitamin $D_3$ (calcifediol) |
| 7-DHC | 7-dehydrocholesterol (provitamin $D_3$, cholesta-5,7-dien-3β-ol) |
| 7-DHP | 7-dehydropregnenolone |
| MARRS receptor | Membrane-Associated Rapid Response to Steroid binding protein (other names: ERp57, GRp58, Pdia3) |
| PDIA3 | Protein disulfide-isomerase A3 |
| ROS | reactive oxygen species |
| UVA/B | ultraviolet radiation A and B |
| VDR | vitamin D receptor |
| VDRE | vitamin D response elements |

## References

1. Holick, M.F. Vitamin D: Evolutionary, physiological and health perspectives. *Curr. Drug Targets* **2011**, *12*, 4–18. [CrossRef] [PubMed]
2. Bikle, D.D. Vitamin D and the skin: Physiology and pathophysiology. *Rev. Endocr. Metab. Dis.* **2012**, *13*, 3–19. [CrossRef] [PubMed]
3. Wierzbicka, J.; Piotrowska, A.; Żmijewski, M.A. The renaissance of vitamin D. *Acta Biochim. Pol.* **2014**, *61*, 679–686. [PubMed]
4. Holick, M.F. Sunlight, UV-radiation, vitamin D and skin cancer: How much sunlight do we need? *Adv. Exp. Med. Biol.* **2014**, *810*, 1–16. [PubMed]
5. Wacker, M.; Holick, M.F. Sunlight and vitamin D: A global perspective for health. *Dermatoendocrinol* **2013**, *5*, 51–108. [CrossRef] [PubMed]
6. Tangpricha, V.; Flanagan, J.N.; Whitlatch, L.W.; Tseng, C.C.; Chen, T.C.; Holt, P.R.; Lipkin, M.S.; Holick, M.F. 25-hydroxyvitamin D-1α-hydroxylase in normal and malignant colon tissue. *Lancet* **2001**, *357*, 1673–1674. [CrossRef]
7. Radermacher, J.; Diesel, B.; Seifert, M.; Tilgen, W.; Reichrath, J.; Fischer, U.; Meese, E. Expression analysis of CYP27B1 in tumor biopsies and cell cultures. *Anticancer Res.* **2006**, *26*, 2683–2686. [PubMed]
8. Yu, O.B.; Arnold, L.A. Calcitroic acid-a review. *ACS Chem. Biol.* **2016**, *11*, 2665–2672. [CrossRef] [PubMed]
9. Ren, S.; Nguyen, L.; Wu, S.; Encinas, C.; Adams, J.S.; Hewison, M. Alternative splicing of vitamin D-24-hydroxylase: A novel mechanism for the regulation of extrarenal 1,25-dihydroxyvitamin D synthesis. *J. Biol. Chem.* **2005**, *280*, 20604–20611. [CrossRef] [PubMed]

10. Slominski, A.T.; Zmijewski, M.A.; Semak, I.; Sweatman, T.; Janjetovic, Z.; Li, W.; Zjawiony, J.K.; Tuckey, R.C. Sequential metabolism of 7-dehydrocholesterol to steroidal 5,7-dienes in adrenal glands and its biological implication in the skin. *PLoS ONE* **2009**, *4*, e4309. [CrossRef] [PubMed]
11. Slominski, A.T.; Li, W.; Kim, T.K.; Semak, I.; Wang, J.; Zjawiony, J.K.; Tuckey, R.C. Novel activities of cyp11a1 and their potential physiological significance. *J. Steroid Biochem.* **2015**, *151*, 25–37. [CrossRef] [PubMed]
12. Slominski, A.T.; Kim, T.K.; Li, W.; Postlethwaite, A.; Tieu, E.W.; Tang, E.K.; Tuckey, R.C. Detection of novel cyp11a1-derived secosteroids in the human epidermis and serum and pig adrenal gland. *Sci. Rep.* **2015**, *5*, 14875. [CrossRef] [PubMed]
13. Slominski, A.T.; Kim, T.K.; Shehabi, H.Z.; Semak, I.; Tang, E.K.; Nguyen, M.N.; Benson, H.A.; Korik, E.; Janjetovic, Z.; Chen, J.; et al. In vivo evidence for a novel pathway of vitamin $D_3$ metabolism initiated by P450scc and modified by CYP27B1. *FASEB J.* **2012**, *26*, 3901–3915. [CrossRef] [PubMed]
14. Slominski, A.T.; Kim, T.K.; Chen, J.; Nguyen, M.N.; Li, W.; Yates, C.R.; Sweatman, T.; Janjetovic, Z.; Tuckey, R.C. Cytochrome P450scc-dependent metabolism of 7-dehydrocholesterol in placenta and epidermal keratinocytes. *Int. J. Biochem. Cell Biol.* **2012**, *44*, 2003–2018. [CrossRef] [PubMed]
15. Slominski, A.T.; Kim, T.K.; Hobrath, J.V.; Janjetovic, Z.; Oak, A.S.W.; Postlethwaite, A.; Lin, Z.; Li, W.; Takeda, Y.; Jetten, A.M.; et al. Characterization of a new pathway that activates lumisterol in vivo to biologically active hydroxylumisterols. *Sci. Rep.* **2017**, *7*, 11434. [CrossRef] [PubMed]
16. Cavalier, E.; Delanaye, P.; Chapelle, J.P.; Souberbielle, J.C. Vitamin D: Current status and perspectives. *Clin. Chem. Lab. Med.* **2009**, *47*, 120–127. [CrossRef] [PubMed]
17. Campbell, M.J. Vitamin D and the RNA transcriptome: More than mRNA regulation. *Front. Physiol.* **2014**, *5*, 181. [CrossRef] [PubMed]
18. Long, M.D.; Sucheston-Campbell, L.E.; Campbell, M.J. Vitamin D receptor and RXR in the post-genomic era. *J. Cell. Physiol.* **2015**, *230*, 758–766. [CrossRef] [PubMed]
19. Liaw, C.W.; Grigoriadis, D.E.; Lovenberg, T.W.; De Souza, E.B.; Maki, R.A. Localization of ligand-binding domains of human corticotropin-releasing factor receptor: A chimeric receptor approach. *Mol. Endocrinol.* **1997**, *11*, 980–985. [CrossRef] [PubMed]
20. Kerry, D.M.; Dwivedi, P.P.; Hahn, C.N.; Morris, H.A.; Omdahl, J.L.; May, B.K. Transcriptional synergism between vitamin D-responsive elements in the rat 25-hydroxyvitamin $D_3$ 24-hydroxylase (*CYP24*) promoter. *J. Biol. Chem.* **1996**, *271*, 29715–29721. [CrossRef] [PubMed]
21. Wang, T.T.; Tavera-Mendoza, L.E.; Laperriere, D.; Libby, E.; MacLeod, N.B.; Nagai, Y.; Bourdeau, V.; Konstorum, A.; Lallemant, B.; Zhang, R.; et al. Large-scale in silico and microarray-based identification of direct 1,25-dihydroxyvitamin $D_3$ target genes. *Mol. Endocrinol.* **2005**, *19*, 2685–2695. [CrossRef] [PubMed]
22. Reichrath, J. Will analogs of 1,25-dihydroxyvitamin $D_3$ (calcitriol) open a new era in cancer therapy? *Onkologie* **2001**, *24*, 128–133. [PubMed]
23. Nemere, I.; Garbi, N.; Hämmerling, G.J.; Khanal, R.C. Intestinal cell calcium uptake and the targeted knockout of the 1,25$D_3$-MARRS (membrane-associated, rapid response steroid-binding) receptor/PDIA3/Erp57. *J. Biol. Chem.* **2010**, *285*, 31859–31866. [CrossRef] [PubMed]
24. Cui, X.; Gooch, H.; Petty, A.; McGrath, J.J.; Eyles, D. Vitamin D and the brain: Genomic and non-genomic actions. *Mol. Cell. Endocrinol.* **2017**, *453*, 131–143. [CrossRef] [PubMed]
25. Hii, C.S.; Ferrante, A. The non-genomic actions of vitamin D. *Nutrients* **2016**, *8*, 135. [CrossRef] [PubMed]
26. Huhtakangas, J.A.; Olivera, C.J.; Bishop, J.E.; Zanello, L.P.; Norman, A.W. The vitamin D receptor is present in caveolae-enriched plasma membranes and binds 1$\alpha$,25(OH)2-vitamin $D_3$ in vivo and in vitro. *Mol. Endocrinol.* **2004**, *18*, 2660–2671. [CrossRef] [PubMed]
27. Zhao, G.; Simpson, R.U. Membrane localization, caveolin-3 association and rapid actions of vitamin D receptor in cardiac myocytes. *Steroids* **2010**, *75*, 555–559. [CrossRef] [PubMed]
28. Nemere, I.; Safford, S.E.; Rohe, B.; DeSouza, M.M.; Farach-Carson, M.C. Identification and characterization of 1,25$D_3$-membrane-associated rapid response, steroid (1,25$D_3$-MARRS) binding protein. *J. Steroid Biochem.* **2004**, *89–90*, 281–285. [CrossRef] [PubMed]
29. Strouse, J.J.; Fears, T.R.; Tucker, M.A.; Wayne, A.S. Pediatric melanoma: Risk factor and survival analysis of the surveillance, epidemiology and end results database. *J. Clin. Oncol.* **2005**, *23*, 4735–4741. [CrossRef] [PubMed]

30. Barroso, E.; Fernandez, L.P.; Milne, R.L.; Pita, G.; Sendagorta, E.; Floristan, U.; Feito, M.; Aviles, J.A.; Martin-Gonzalez, M.; Arias, J.I.; et al. Genetic analysis of the vitamin D receptor gene in two epithelial cancers: Melanoma and breast cancer case-control studies. *BMC Cancer* **2008**, *8*, 385. [CrossRef] [PubMed]
31. Yan, S.F.; Ramasamy, R.; Naka, Y.; Schmidt, A.M. Glycation, inflammation, and rage: A scaffold for the macrovascular complications of diabetes and beyond. *Circ. Res.* **2003**, *93*, 1159–1169. [CrossRef] [PubMed]
32. Arnold, M.; de Vries, E.; Whiteman, D.C.; Jemal, A.; Bray, F.; Parkin, D.M.; Soerjomataram, I. Global burden of cutaneous melanoma attributable to ultraviolet radiation in 2012. *Int. J. Cancer* **2018**, *143*, 1305–1314. [CrossRef] [PubMed]
33. Reichrath, J.; Reichrath, S. Sunlight, vitamin D and malignant melanoma: An update. *Adv. Exp. Med. Biol.* **2014**, *810*, 390–405. [PubMed]
34. Lo, J.A.; Fisher, D.E. The melanoma revolution: From UV carcinogenesis to a new era in therapeutics. *Science* **2014**, *346*, 945–949. [CrossRef] [PubMed]
35. Mansur, C.P.; Gordon, P.R.; Ray, S.; Holick, M.F.; Gilchrest, B.A. Vitamin D, its precursors, and metabolites do not affect melanization of cultured human melanocytes. *J. Investing. Dermatol.* **1988**, *91*, 16–21. [CrossRef]
36. Reichrath, J. Skin cancer prevention and UV-protection: How to avoid vitamin D-deficiency? *Brit. J. Dermatol.* **2009**, *161*, 54–60. [CrossRef] [PubMed]
37. Wagner, C.L.; McNeil, R.B.; Johnson, D.D.; Hulsey, T.C.; Ebeling, M.; Robinson, C.; Hamilton, S.A.; Hollis, B.W. Health characteristics and outcomes of two randomized vitamin D supplementation trials during pregnancy: A combined analysis. *J. Steroid Biochem.* **2013**, *136*, 313–320. [CrossRef] [PubMed]
38. Pelczynska, M.; Switalska, M.; Maciejewska, M.; Jaroszewicz, I.; Kutner, A.; Opolski, A. Antiproliferative activity of vitamin D compounds in combination with cytostatics. *Anticancer Res.* **2006**, *26*, 2701–2705. [PubMed]
39. Slominski, A.T.; Janjetovic, Z.; Kim, T.K.; Wright, A.C.; Grese, L.N.; Riney, S.J.; Nguyen, M.N.; Tuckey, R.C. Novel vitamin D hydroxyderivatives inhibit melanoma growth and show differential effects on normal melanocytes. *Anticancer Res.* **2012**, *32*, 3733–3742. [PubMed]
40. Slominski, A.T.; Brożyna, A.A.; Skobowiat, C.; Zmijewski, M.A.; Kim, T.K.; Janjetovic, Z.; Oak, A.S.; Jozwicki, W.; Jetten, A.M.; Mason, R.S.; et al. On the role of classical and novel forms of vitamin D in melanoma progression and management. *J. Steroid Biochem.* **2018**, *177*, 159–170. [CrossRef] [PubMed]
41. Cauci, S.; Maione, V.; Buligan, C.; Linussio, M.; Serraino, D.; Stinco, G. Bsmi (rs1544410) and foki (rs2228570) vitamin D receptor polymorphisms, smoking, and body mass index as risk factors of cutaneous malignant melanoma in northeast italy. *Cancer Biol. Med.* **2017**, *14*, 302–318. [CrossRef] [PubMed]
42. La Marra, F.; Stinco, G.; Buligan, C.; Chiriacò, G.; Serraino, D.; Di Loreto, C.; Cauci, S. Immunohistochemical evaluation of vitamin D receptor (VDR) expression in cutaneous melanoma tissues and four *VDR* gene polymorphisms. *Cancer Biol. Med.* **2017**, *14*, 162–175. [PubMed]
43. Orlow, I.; Reiner, A.S.; Thomas, N.E.; Roy, P.; Kanetsky, P.A.; Luo, L.; Paine, S.; Armstrong, B.K.; Kricker, A.; Marrett, L.D.; et al. Vitamin D receptor polymorphisms and survival in patients with cutaneous melanoma: A population-based study. *Carcinogenesis* **2016**, *37*, 30–38. [CrossRef] [PubMed]
44. Brożyna, A.A.; Jochymski, C.; Janjetovic, Z.; Jóźwicki, W.; Tuckey, R.C.; Slominski, A.T. CYP24A1 expression inversely correlates with melanoma progression: Clinic-pathological studies. *Int. J. Mol. Sci.* **2014**, *15*, 19000–19017. [CrossRef] [PubMed]
45. Brożyna, A.A.; Jozwicki, W.; Janjetovic, Z.; Slominski, A.T. Expression of vitamin D receptor decreases during progression of pigmented skin lesions. *Hum. Pathol.* **2011**, *42*, 618–631. [CrossRef] [PubMed]
46. Brożyna, A.A.; Jóźwicki, W.; Slominski, A.T. Decreased VDR expression in cutaneous melanomas as marker of tumor progression: New data and analyses. *Anticancer Res.* **2014**, *34*, 2735–2743. [PubMed]
47. Slominski, A.T.; Brożyna, A.A.; Zmijewski, M.A.; Jóźwicki, W.; Jetten, A.M.; Mason, R.S.; Tuckey, R.C.; Elmets, C.A. Vitamin D signaling and melanoma: Role of vitamin D and its receptors in melanoma progression and management. *Lab. Invest.* **2017**, *97*, 706–724. [CrossRef] [PubMed]
48. Brożyna, A.A.; Jóźwicki, W.; Janjetovic, Z.; Slominski, A.T. Expression of the vitamin D-activating enzyme 1α-hydroxylase (CYP27B1) decreases during melanoma progression. *Hum. Pathol.* **2013**, *44*, 374–387. [CrossRef] [PubMed]
49. Timerman, D.; McEnery-Stonelake, M.; Joyce, C.J.; Nambudiri, V.E.; Hodi, F.S.; Claus, E.B.; Ibrahim, N.; Lin, J.Y. Vitamin D deficiency is associated with a worse prognosis in metastatic melanoma. *Oncotarget* **2017**, *8*, 6873–6882. [CrossRef] [PubMed]

50. Grant, W.B. Roles of solar UVB and vitamin D in reducing cancer risk and increasing survival. *Anticancer Res.* **2016**, *36*, 1357–1370. [PubMed]
51. Płudowski, P.; Karczmarewicz, E.; Bayer, M.; Carter, G.; Chlebna-Sokół, D.; Czech-Kowalska, J.; Dębski, R.; Decsi, T.; Dobrzańska, A.; Franek, E.; et al. Practical guidelines for the supplementation of vitamin D and the treatment of deficits in central europe–recommended vitamin D intakes in the general population and groups at risk of vitamin D deficiency. *Endokrynol. Pol.* **2013**, *64*, 319–327. [CrossRef] [PubMed]
52. Wierzbicka, J.M.; Binek, A.; Ahrends, T.; Nowacka, J.D.; Szydłowska, A.; Turczyk, Ł.; Wąsiewicz, T.; Wierzbicki, P.M.; Sądej, R.; Tuckey, R.C.; et al. Differential antitumor effects of vitamin d analogues on colorectal carcinoma in culture. *Int. J. Oncol.* **2015**, *47*, 1084–1096. [CrossRef] [PubMed]
53. Nag, K.; Sultana, N.; Kato, A.; Hirose, S. Headless splice variant acting as dominant negative calcitonin receptor. *Biochem. Biophys. Res. Commun.* **2007**, *362*, 1037–1043. [CrossRef] [PubMed]
54. Giardino, W.J.; Mark, G.P.; Stenzel-Poore, M.P.; Ryabinin, A.E. Dissociation of corticotropin-releasing factor receptor subtype involvement in sensitivity to locomotor effects of methamphetamine and cocaine. *Psychopharmacology* **2012**, *219*, 1055–1063. [CrossRef] [PubMed]
55. Piotrowska, A.; Wierzbicka, J.; Nadkarni, S.; Brown, G.; Kutner, A.; Żmijewski, M.A. Antiproliferative activity of double point modified analogs of 1,25-dihydroxyvitamin $D_2$ against human malignant melanoma cell lines. *Int. J. Mol. Sci.* **2016**, *17*, 76. [CrossRef] [PubMed]
56. Wasiewicz, T.; Szyszka, P.; Cichorek, M.; Janjetovic, Z.; Tuckey, R.C.; Slominski, A.T.; Zmijewski, M.A. Antitumor effects of vitamin D analogs on hamster and mouse melanoma cell lines in relation to melanin pigmentation. *Int. J. Mol. Sci.* **2015**, *16*, 6645–6667. [CrossRef] [PubMed]
57. Zmijewski, M.A.; Li, W.; Chen, J.; Kim, T.K.; Zjawiony, J.K.; Sweatman, T.W.; Miller, D.D.; Slominski, A.T. Synthesis and photochemical transformation of 3α21-dihydroxypregna-5,7-dien-20-one to novel secosteroids that show anti-melanoma activity. *Steroids* **2011**, *76*, 193–203. [CrossRef] [PubMed]
58. Skobowiat, C.; Oak, A.S.; Kim, T.K.; Yang, C.H.; Pfeffer, L.M.; Tuckey, R.C.; Slominski, A.T. Noncalcemic 20-hydroxyvitamin $D_3$ inhibits human melanoma growth in in vitro and in vivo models. *Oncotarget* **2017**, *8*, 9823–9834. [CrossRef] [PubMed]
59. Slominski, A.; Kim, T.K.; Zmijewski, M.A.; Janjetovic, Z.; Li, W.; Chen, J.; Kusniatsova, E.I.; Semak, I.; Postlethwaite, A.; Miller, D.D.; et al. Novel vitamin D photoproducts and their precursors in the skin. *Dermatoendocrinol* **2013**, *5*, 7–19. [CrossRef] [PubMed]
60. Zmijewski, M.A.; Fisches, T.W.; Slominski, A.T. Melatonin regulates expression and activity of androgen related genes in human skin cells. *J. Invest. Dermatol.* **2008**, *128*, 893.
61. Plum, L.A.; Prahl, J.M.; Ma, X.; Sicinski, R.R.; Gowlugari, S.; Clagett-Dame, M.; DeLuca, H.F. Biologically active noncalcemic analogs of 1α,25-dihydroxyvitamin D with an abbreviated side chain containing no hydroxyl. *Proc. Natl. Acad. Sci. USA* **2004**, *101*, 6900–6904. [CrossRef] [PubMed]
62. Slominski, A.T.; Kim, T.K.; Hobrath, J.V.; Oak, A.S.W.; Tang, E.K.Y.; Tieu, E.W.; Li, W.; Tuckey, R.C.; Jetten, A.M. Endogenously produced nonclassical vitamin D hydroxy-metabolites act as "Biased" Agonists on VDR and inverse agonists on roralpha and rorgamma. *J. Steroid Biochem.* **2017**, *173*, 42–56. [CrossRef] [PubMed]
63. Wang, J.; Slominski, A.; Tuckey, R.C.; Janjetovic, Z.; Kulkarni, A.; Chen, J.; Postlethwaite, A.E.; Miller, D.; Li, W. 20-hydroxyvitamin $D_3$ inhibits proliferation of cancer cells with high efficacy while being non-toxic. *Anticancer Res.* **2012**, *32*, 739–746. [PubMed]
64. Duffy, M.J.; Murray, A.; Synnott, N.C.; O'Donovan, N.; Crown, J. Vitamin D analogues: Potential use in cancer treatment. *Crit. Rev. Oncol. Hematol.* **2017**, *112*, 190–197. [CrossRef] [PubMed]
65. Szyszka, P.; Zmijewski, M.A.; Slominski, A.T. New vitamin D analogs as potential therapeutics in melanoma. *Expert Rev. Anticancer Ther.* **2012**, *12*, 585–599. [CrossRef] [PubMed]
66. Reichrath, J.; Zouboulis, C.C.; Vogt, T.; Holick, M.F. Targeting the vitamin D endocrine system (VDES) for the management of inflammatory and malignant skin diseases: An historical view and outlook. *Rev. Endocr. Metab. Disord.* **2016**, *17*, 405–417. [CrossRef] [PubMed]
67. Zmijewski, M.A.; Li, W.; Zjawiony, J.K.; Sweatman, T.W.; Chen, J.; Miller, D.D.; Slominski, A.T. Photo-conversion of two epimers (20R and 20S) of pregna-5,7-diene-3β, 17α, 20-triol and their bioactivity in melanoma cells. *Steroids* **2009**, *74*, 218–228. [CrossRef] [PubMed]
68. Janjetovic, Z.; Brozyna, A.A.; Tuckey, R.C.; Kim, T.K.; Nguyen, M.N.; Jozwicki, W.; Pfeffer, S.R.; Pfeffer, L.M.; Slominski, A.T. High basal NF-kB activity in nonpigmented melanoma cells is associated with an enhanced sensitivity to vitamin $D_3$ derivatives. *Brit. J. Cancer.* **2011**, *105*, 1874–1884. [CrossRef] [PubMed]

69. Kim, T.K.; Wang, J.; Janjetovic, Z.; Chen, J.; Tuckey, R.C.; Nguyen, M.N.; Tang, E.K.; Miller, D.; Li, W.; Slominski, A.T. Correlation between secosteroid-induced vitamin D receptor activity in melanoma cells and computer-modeled receptor binding strength. *Mol. Cell. Endocrinol.* **2012**, *361*, 143–152. [CrossRef] [PubMed]
70. Slominski, A.T.; Kim, T.K.; Janjetovic, Z.; Tuckey, R.C.; Bieniek, R.; Yue, J.; Li, W.; Chen, J.; Nguyen, M.N.; Tang, E.K.; et al. 20-hydroxyvitamin $D_2$ is a noncalcemic analog of vitamin D with potent antiproliferative and prodifferentiation activities in normal and malignant cells. *Am. J. Physiol. Cell Physiol.* **2011**, *300*, C526–C541. [CrossRef] [PubMed]
71. Carlberg, C.; Seuter, S. A genomic perspective on vitamin D signaling. *Anticancer Res.* **2009**, *29*, 3485–3493. [PubMed]
72. Piotrowska, A.; Wierzbicka, J.; Żmijewski, M.A. Vitamin D in the skin physiology and pathology. *Acta. Biochim. Pol.* **2016**, *63*, 17–29. [CrossRef] [PubMed]
73. Hansen, C.M.; Madsen, M.W.; Arensbak, B.; Skak-Nielsen, T.; Latini, S.; Binderup, L. Down-regulation of laminin-binding integrins by 1$\alpha$,25-dihydroxyvitamin $D_3$ in human melanoma cells in vitro. *Cell Adhes. Commun.* **1998**, *5*, 109–120. [CrossRef] [PubMed]
74. Evans, S.R.; Houghton, A.M.; Schumaker, L.; Brenner, R.V.; Buras, R.R.; Davoodi, F.; Nauta, R.J.; Shabahang, M. Vitamin D receptor and growth inhibition by 1,25-dihydroxyvitamin $D_3$ in human malignant melanoma cell lines. *J. Surg. Res.* **1996**, *61*, 127–133. [CrossRef] [PubMed]
75. Essa, S.; Denzer, N.; Mahlknecht, U.; Klein, R.; Collnot, E.M.; Schäfer, U.; Tilgen, W.; Reichrath, J. VDR microrna expression and epigenetic silencing of vitamin D signaling in melanoma cells. *J. Steroid Biochem.* **2010**, *121*, 110–113. [CrossRef] [PubMed]
76. Krishnan, A.V.; Peehl, D.M.; Feldman, D. The role of vitamin D in prostate cancer. *Recent Results Cancer Res.* **2003**, *164*, 205–221. [PubMed]
77. Seifert, M.; Rech, M.; Meineke, V.; Tilgen, W.; Reichrath, J. Differential biological effects of 1,25-dihydroxyvitamin $D_3$ on melanoma cell lines in vitro. *J. Steroid Biochem.* **2004**, *89–90*, 375–379. [CrossRef] [PubMed]
78. Corcoran, A.; Nadkarni, S.; Yasuda, K.; Sakaki, T.; Brown, G.; Kutner, A.; Marcinkowska, E. Biological evaluation of double point modified analogues of 1,25-dihydroxyvitamin $D_2$ as potential anti-leukemic agents. *Int. J. Mol. Sci.* **2016**, *17*, 91. [CrossRef] [PubMed]
79. Nadkarni, S.; Chodyński, M.; Krajewski, K.; Cmoch, P.; Marcinkowska, E.; Brown, G.; Kutner, A. Convergent synthesis of double point modified analogs of 1$\alpha$,25-dihydroxyvitamin $D_2$ for biological evaluation. *J. Steroid Biochem.* **2016**, *164*, 45–49. [CrossRef] [PubMed]
80. Trochoutsou, A.I.; Kloukina, V.; Samitas, K.; Xanthou, G. Vitamin-D in the immune system: Genomic and non-genomic actions. *Mini Rev. Med. Chem.* **2015**, *15*, 953–963. [CrossRef] [PubMed]
81. Doroudi, M.; Plaisance, M.C.; Boyan, B.D.; Schwartz, Z. Membrane actions of 1$\alpha$,25$(OH)_2D_3$ are mediated by $Ca^{2+}$/calmodulin-dependent protein kinase II in bone and cartilage cells. *J. Steroid Biochem.* **2015**, *145*, 65–74. [CrossRef] [PubMed]
82. Chen, J.; Doroudi, M.; Cheung, J.; Grozier, A.L.; Schwartz, Z.; Boyan, B.D. Plasma membrane PDIA3 and VDR interact to elicit rapid responses to 1$\alpha$,25$OH_2D_3$. *Cell. Signal.* **2013**, *25*, 2362–2373. [CrossRef] [PubMed]
83. Slominski, A.T.; Kim, T.K.; Takeda, Y.; Janjetovic, Z.; Brozyna, A.A.; Skobowiat, C.; Wang, J.; Postlethwaite, A.; Li, W.; Tuckey, R.C.; et al. ROR$\alpha$ and ROR $\gamma$ are expressed in human skin and serve as receptors for endogenously produced noncalcemic 20-hydroxy-and 20,23-dihydroxyvitamin D. *FASEB J.* **2014**, *28*, 2775–2789. [CrossRef] [PubMed]
84. Bikle, D.D.; Jiang, Y.; Nguyen, T.; Oda, Y.; Tu, C.L. Disruption of vitamin D and calcium signaling in keratinocytes predisposes to skin cancer. *Front Physiol.* **2016**, *7*, 296. [CrossRef] [PubMed]
85. Slominski, A.T.; Zmijewski, M.A.; Plonka, P.M.; Szaflarski, J.P.; Paus, R. How UV light touches the brain and endocrine system through skin, and why. *Endocrinology* **2018**, *159*, 1992–2007. [CrossRef] [PubMed]
86. Wyatt, C.; Lucas, R.M.; Hurst, C.; Kimlin, M.G. Vitamin D deficiency at melanoma diagnosis is associated with higher breslow thickness. *PLoS ONE* **2015**, *10*, e0126394. [CrossRef] [PubMed]

International Journal of
*Molecular Sciences*

MDPI

*Article*

# Differential and Overlapping Effects of 20,23(OH)$_2$D3 and 1,25(OH)$_2$D3 on Gene Expression in Human Epidermal Keratinocytes: Identification of AhR as an Alternative Receptor for 20,23(OH)$_2$D3

**Andrzej T. Slominski [1,2,3,*], Tae-Kang Kim [1], Zorica Janjetovic [1], Anna A. Brożyna [4,5], Michal A. Żmijewski [6], Hui Xu [1], Thomas R. Sutter [7], Robert C. Tuckey [8], Anton M. Jetten [9] and David K. Crossman [10]**

1. Department of Dermatology, University of Alabama at Birmingham, Birmingham, AL 35294, USA; tkim@uabmc.edu (T.-K.K.); zjanjetovic@uabmc.edu (Z.J.); huixu@uabmc.edu (H.X.)
2. Comprehensive Cancer Center, Cancer Chemoprevention Program, University of Alabama at Birmingham, Birmingham, AL 35294, USA
3. Veteran Administration Medical Center, Birmingham, AL 35294, USA
4. Department of Medical Biology, Faculty of Biology and Environment Protection, Nicolaus Copernicus University, 87-100 Toruń, Poland; brozyna.anna@gmail.com
5. Department of Tumor Pathology and Pathomorphology, Oncology Centre-Prof. Franciszek Łukaszczyk Memorial Hospital, 85-796 Bydgoszcz, Poland
6. Department of Histology, Medical University of Gdańsk, 80-211 Gdańsk, Poland; mzmijewski@gumed.edu.pl
7. Feinstone Center for Genomic Research, University of Memphis, Memphis, TN 38152 USA; tsutter@memphis.edu
8. School of Molecular Sciences, The University of Western Australia, Perth, WA 6009, Australia; robert.tuckey@uwa.edu.au
9. Immunity, Inflammation, and Disease Laboratory/Cell Biology Group, National Institute of Environmental Health Sciences, NIH, Research Triangle Park, NC 27709, USA; jetten@niehs.nih.gov
10. Howell and Elizabeth Heflin Center for Human Genetics, Genomic Core Facility, University of Alabama at Birmingham, Birmingham, AL 35294, USA; dcrossman@uabmc.edu

* Correspondence: aslominski@uabmc.edu; Tel.: +1-205-934-5245; Fax: +1-205-996-0302

Received: 30 August 2018; Accepted: 3 October 2018; Published: 8 October 2018

**Abstract:** A novel pathway of vitamin D activation by CYP11A has previously been elucidated. To define the mechanism of action of its major dihydroxy-products, we tested the divergence and overlap between the gene expression profiles of human epidermal keratinocytes treated with either CYP11A1-derived 20,23(OH)$_2$D3 or classical 1,25(OH)$_2$D3. Both secosteroids have significant chemical similarity with the only differences being the positions of the hydroxyl groups. mRNA was isolated and examined by microarray analysis using Illumina's HumanWG-6 chip/arrays and subsequent bioinformatics analyses. Marked differences in the up- and downregulated genes were observed between 1,25(OH)$_2$D3- and 20,23(OH)$_2$D3-treated cells. Hierarchical clustering identified both distinct, opposite and common (overlapping) gene expression patterns. *CYP24A1* was a common gene strongly activated by both compounds, a finding confirmed by qPCR. Ingenuity pathway analysis identified VDR/RXR signaling as the top canonical pathway induced by 1,25(OH)$_2$D3. In contrast, the top canonical pathway induced by 20,23(OH)$_2$D3 was AhR, with VDR/RXR being the second nuclear receptor signaling pathway identified. QPCR analyses validated the former finding by revealing that 20,23(OH)$_2$D3 stimulated *CYP1A1* and *CYP1B1* gene expression, effects located downstream of AhR. Similar stimulation was observed with 20(OH)D3, the precursor to 20,23(OH)$_2$D3, as well as with its downstream metabolite, 17,20,23(OH)$_3$D3. Using a Human AhR Reporter Assay System we showed marked activation of AhR activity by 20,23(OH)$_2$D3, with weaker stimulation by 20(OH)D3. Finally, molecular modeling using an AhR LBD model predicted vitamin D3 hydroxyderivatives to be good ligands for this receptor. Thus, our microarray, qPCR, functional studies and molecular modeling indicate that AhR is the major receptor target for 20,23(OH)$_2$D3, opening an exciting area of investigation on the interaction of different vitamin D3-hydroxyderivatives with AhR and the subsequent downstream activation of signal transduction pathways in a cell-type-dependent manner.

**Keywords:** vitamin D; dihydroxyvitamin D; epidermal keratinocytes; nuclear receptor signaling; microarray

---

## 1. Introduction

Vitamin D3 (D3) is formed by ultraviolet B radiation (UVB)-mediated breaking of the B ring of 7-dehydrocholesterol (7DHC) followed by thermal isomerization of the resulting pre-vitamin D3 to D3 [1,2]. The vast majority of circulating D3 is generated in epidermal keratinocytes [3]. D3 is a prohormone that is activated by sequential hydroxylations at C25 (by CYP2R1 or CYP27A1) and C1$\alpha$ (by CYP27B1) to 1,25-dihydroxyvitamin D3 (1,25(OH)$_2$D3), the hormonally active form, referred to as the canonical pathway [4–7]. At the systemic level, C25 hydroxylation takes place in the liver with the resulting 25(OH)D3 being hydroxylated at C1$\alpha$ in kidneys [3–7]. The same pathway operates in peripheral tissues including epidermal keratinocytes, the major site of D3 formation [1].

In addition to the canonical pathway of vitamin D activation described above, novel CYP11A-mediated pathways have been discovered (reviewed in [8]). Specifically, CYP11A1, the first enzyme of steroidogenesis that hydroxylates and then cleaves the side chain of cholesterol to produce pregnenolone (reviewed in [9,10]), can also hydroxylate and cleave the side chain of 7DHC, and hydroxylate the side chain of D3 and D2 without cleavage [11–16]. The two major products of CYP11A1 action on D3, with defined stereochemistry, are 20*S*-hydroxyvitamin D3 (20(OH)D3) and 20*S*,23*S*-dihydroxyvitamin D3 (20,23(OH)$_2$D3) [17]. These pathways operate in cultured epidermal, human and pig keratinocytes, dermal fibroblasts, colon cancer cells, and have also been described ex vivo for placenta and adrenal glands [18–23]. Importantly, the major products of these pathways are detectable in vivo in human serum, epidermis and adrenal glands [24].

The classical, hormonally-active dihydroxy form of vitamin D3, 1,25(OH)$_2$D3, in addition to playing a fundamental role in body calcium and phosphorous homeostasis and in the proper functioning of the skeletomuscular system, has pleiotropic effects on different organs and cell functions (reviewed in [3,6,25–29]). These studies show that 1,25(OH)$_2$D3 has immunomodulatory properties, is involved in the regulation of reproduction, pregnancy, child development, neurodevelopment, regulation of global metabolic and endocrine homeostasis and functions of the cardiovascular system, and has anticancer activities (reviewed in [2,6,30–50]). At the cellular level, it regulates proliferation, differentiation, apoptosis, senescence, metabolism, migration, secretory activities, and protective and reparative mechanisms against oxidative stress and radiation. It is widely accepted that these functions are regulated by different signal transduction pathways initiated by 1,25(OH)$_2$D3 binding to the vitamin D receptor (VDR) at the genomic binding site, and to some degree at a nongenomic binding site, in a cell-type dependent manner (reviewed [6,45–47,51–54]). In the skin, 1,25(OH)$_2$D3 regulates the epidermal barrier and hair cycling and has radioprotective, anti-cancer and anti-inflammatory properties [1,3,52,53,55–59].

The novel secosteroids, produced by the non-canonical activation pathways initiated by CYP11A1, inhibit the proliferation of epidermal keratinocytes, melanocytes and dermal fibroblasts and promote the differentiation of keratinocytes. Furthermore, they inhibit fibrotic activities of fibroblasts and have immunomodulatory properties (reviewed in [19,49,60]). Importantly, 20(OH)D3 and 20,23(OH)$_2$D3 are non-calcemic at pharmacological doses [61–63] which is in contrast to the highly calcemic effects of 1,25(OH)$_2$D3 and 25(OH)D3. 20(OH)D3 and 20,23(OH)$_2$D3 also attenuate the symptoms of skin fibrosis, rheumatoid arthritis and have photoprotective properties [8,19,23,64,65]. The CYP11A1-derived secosteroids have pleiotropic phenotypic effects that are cell-type–dependent [19,23,60–62,65–75]. They can act as biased agonists of the VDR [19,60,76,77] and can act as inverse agonists on retinoic acid orphan receptors (ROR) α and γ [60,78].

To better define the signaling pathways and mechanisms underlying the similarities and differences between phenotypic activities of classical 1,25(OH)$_2$D3 and the major dihydroxy product of CYP11A1 action on vitamin D3, 20,23(OH)$_2$D3, we examined and compared the gene expression profiles of human keratinocytes exposed to these secosteroids. Bioinformatics analysis was performed and differences and similarities in the activities of these structurally similar but distinct dihydroxy-D3 species were compared.

## 2. Results and Discussion

The structures and sequences of the reactions producing 1,25(OH)$_2$D3 and 20,23(OH)$_2$D3 in the epidermis are shown in Figure 1.

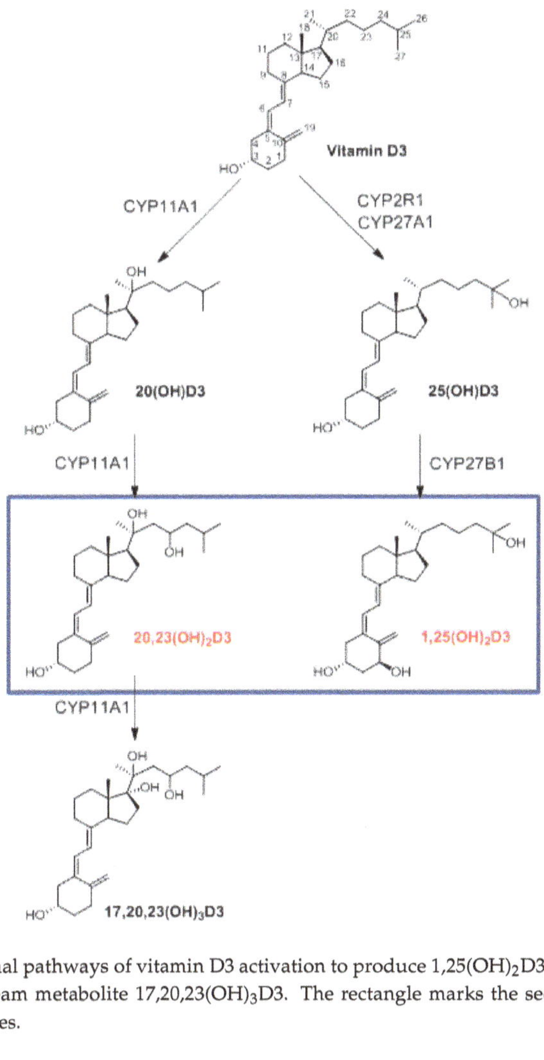


**Figure 1.** Epidermal pathways of vitamin D3 activation to produce 1,25(OH)$_2$D3 and 20,23(OH)$_2$D3 and the downstream metabolite 17,20,23(OH)$_3$D3. The rectangle marks the secosteroids used for microarray analyses.

The schematic outline of the experimental design is presented in Figure 2. Briefly, to test the divergence and overlap between the gene expression patterns, human neonatal epidermal keratinocytes combined from four African-American [79] donors were treated with 1,25(OH)$_2$D3 or 20,23(OH)$_2$D3 for 6 or 24 h. Microarray assays were performed using Illumina's HumanWG-6_V2 (Platform GPL13376) chip/array as described in Materials and Methods and the raw data has been deposited at the NCBI GEO (GSE117351).

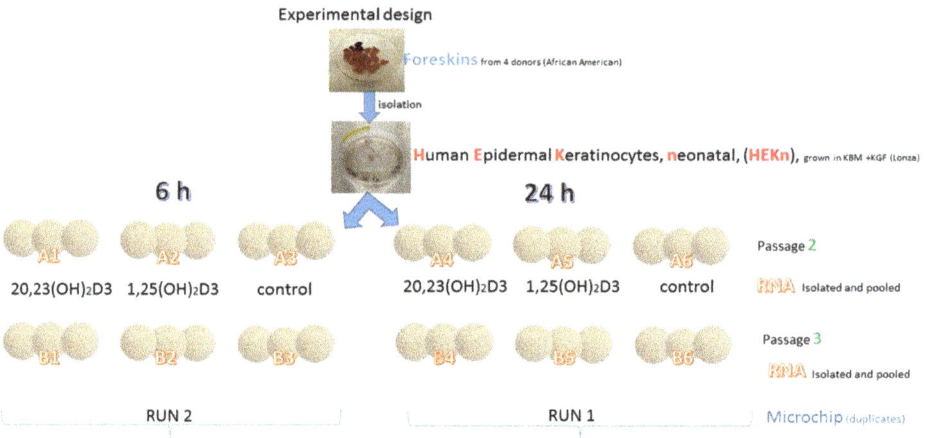


**Figure 2.** Outline of the experimental design.

The relative changes in gene expression (average of two independent experiments that used triplicate cell cultures), were normalized vs. vehicle control (0.1% ethanol). Table 1 shows marked differences in the number of genes up- or downregulated by either 1,25(OH)$_2$D3 or 20,23(OH)$_2$D3 using 1.5-, 2- and 4-fold cut-off values (FC). Average signal values for filtered gene clusters with FC $\geq \pm 1.5$ are shown in Supplemental excel file #1. Briefly, treatment with 1,25(OH)$_2$D3 for 6 h leads to changes in the expression of 148 vs. 37 genes for 20,23(OH)$_2$D3 when using 1.5-FC, and 38 vs. 21 and 3 vs. 0 when using 2- and 4-FC, respectively. After 24 h, this trend changed to 410 and 4079 genes regulated, respectively, by 1,25(OH)$_2$D3 and 20,23(OH)$_2$D3 with 1.5-FC value, and 119 and 1611 for 2-FC value and 12 and 199 genes for 4-FC value, respectively (Table 1).

**Table 1.** Number of genes up or downregulated in keratinocytes by 1,25(OH)$_2$D3 or 20,23(OH)$_2$D3 in comparison to vehicle control using 1.5-, 2- and 4-fold cut-off values.

| Time | Genes | 1,25(OH)$_2$D3 | | | 20,23(OH)$_2$D3 | | |
|---|---|---|---|---|---|---|---|
| | | >1.5-Fold | >2-Fold | >4-Fold | >1.5-Fold | >2-Fold | >4-Fold |
| 6 h | Upregulated | 116 | 35 | 3 | 33 | 21 | 0 |
| | Downregulated | 32 | 3 | 0 | 4 | 0 | 0 |
| 24 h | Upregulated | 266 | 98 | 12 | 2013 | 763 | 98 |
| | Downregulated | 144 | 21 | 0 | 2066 | 848 | 101 |

Hierarchical clustering identified patterns of genes responding to either 1,25(OH)$_2$D3 or 20,23(OH)$_2$D3, or to both. Selected gene clusters, representing the altered expression after 6 h of incubation as well as Venn diagrams are shown in Figure 3A. The heat maps corresponding to relative gene expression levels displayed both distinct or opposite, or common (overlapping) gene expression. For 2-FC there was only 1 common gene (CYP24A1) stimulated by both 1,25(OH)$_2$D3 (82 fold) and 20,23(OH)$_2$D3 (3.4 fold). This differential stimulation of CYP24A1 was further confirmed by qPCR

(Figure 3B) and is consistent with the literature on 1,25(OH)$_2$D3 [1,3,6,27,51] and 20,23(OH)$_2$D3 [69,80]. For 1.5-FC there were two common genes, CYP24A1 and the gene with a target id ILMN_131812 (identified as small ILF3/NF90-associated RNA A1 (SNAR-A1)), for which expression was stimulated.

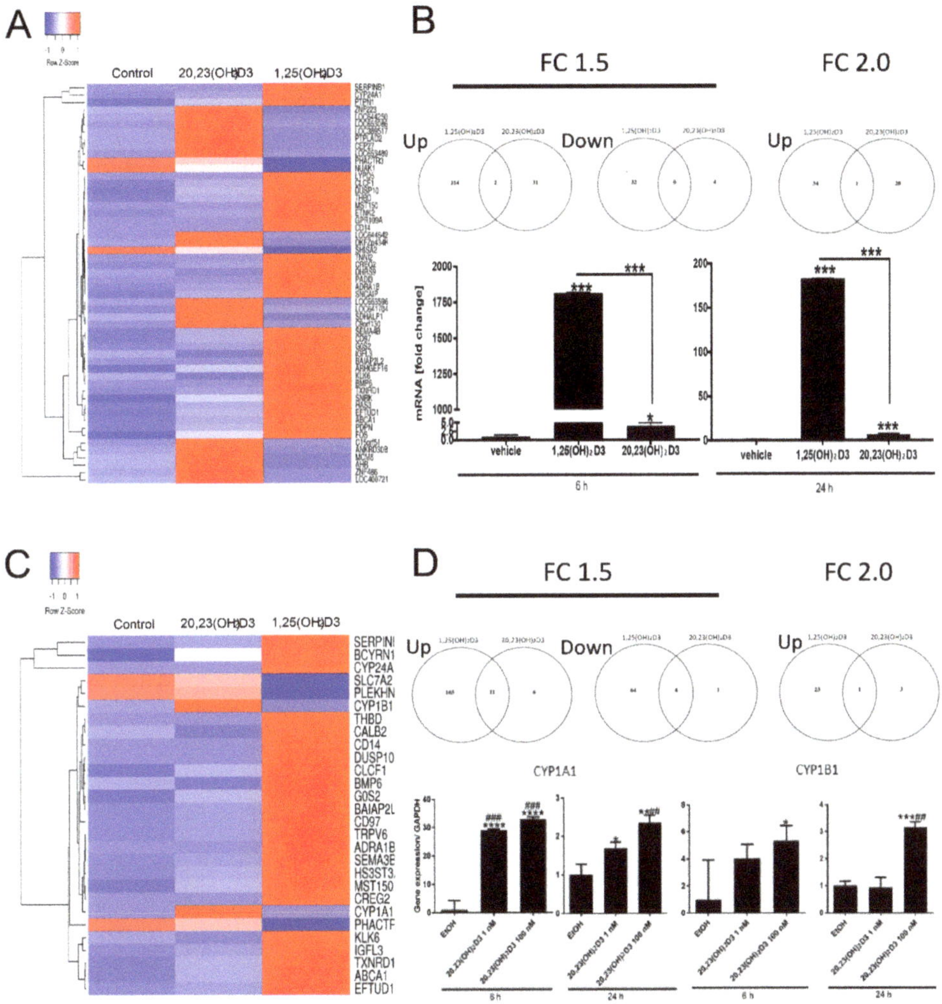


**Figure 3.** Changes in gene expression in human epidermal keratinocytes treated with 1,25(OH)$_2$D3 or 20,23(OH)$_2$D3 for 6 h. (**A**) Heat map of the gene expression pattern for experiment #1 with corresponding Venn diagrams for FC $\geq$ 2 and 1.5. (**B**) Effect of $10^{-7}$ M of 1,25(OH)$_2$D3 or 20,23(OH)$_2$D3 on *CYP24A1* expression in keratinocytes after 6 and 24 h treatment. (**C**) Heat map of the gene expression pattern for experiment #2 with corresponding Venn diagrams for FC-2 and 1.5. (**D**) Effect of $10^{-9}$ and $10^{-7}$ M of 20,23(OH)$_2$D3 on CYP1A1 and CYP1B1 expression in keratinocytes after 6 or 24 h treatment. Data represent means ± SD ($n$ = 3) where * $p < 0.05$, ** $p < 0.01$, *** $p < 0.001$ and **** $p < 0.0001$ at student $t$-test; ## $p < 0.01$ and ### $p < 0.001$ at one-way ANOVA test.

Ingenuity pathway analysis using FC $\geq \pm 1.5$ was performed. The top canonical pathway induced by 1,25(OH)$_2$D3 was VDR/RXR signaling (Table 2) (Supplemental Figure S1A), which was expected [3,6,51,81,82]. This was followed by the roles of osteoblasts, osteoclasts and chondrocytes in rheumatoid arthritis; the role of macrophages, fibroblasts and endothelial cells in rheumatoid arthritis; and Toll-like receptor signaling (Table 2), which is consistent with previously reported functions of 1,25(OH)$_2$D3 [3,6,30,31,45,51,54]. Interestingly, the next top nuclear receptor signaling pathway activated by 1,25(OH)$_2$D3 was linked to the glucocorticoid receptor (GR) followed by the aryl hydrocarbon receptor [74], PPAR, PPAR$\alpha$/RXR$\alpha$, LXR/RXR, and RAR (Table 2). The inclusion of these additional pathways could be secondary to the use of the same dimeric partner, RXR, and communication between receptors, or alternatively by activation by signaling pathways downstream of VDR. For example, it is already known that 1,25(OH)$_2$D3 can selectively activate local elements of hypothalamo-pituitary adrenal axis in keratinocytes [71]. The significance of additional nuclear receptor signaling is out of the scope of this paper and is a goal of our future research.

**Table 2.** Canonical pathways activated by 1,25(OH)$_2$D3 in human epidermal keratinocytes after 6 h of treatment. Nuclear receptors are marked in bold.

| Ingenuity Canonical Pathways | *p*-Value | Overlap (%) | Downregulated | No Change | Upregulated | No Overlap with Dataset |
|---|---|---|---|---|---|---|
| **VDR/RXR Activation** | $1.0 \times 10^{-10}$ | 12.8 | 34/78 (44%) | 0/78 (0%) | 43/78 (55%) | 1/78 (1%) |
| Role of Osteoblasts, Osteoclasts and Chondrocytes in Rheumatoid Arthritis | $2.2 \times 10^{-7}$ | 5.02 | 100/219 (46%) | 0/219 (0%) | 114/219 (52%) | 5/219 (2%) |
| Role of Macrophages, Fibroblasts and Endothelial Cells in Rheumatoid Arthritis | $4.4 \times 10^{-6}$ | 3.72 | 144/296 (49%) | 0/296 (0%) | 138/296 (47%) | 14/296 (5%) |
| Toll-like Receptor Signaling | $9.3 \times 10^{-6}$ | 8.11 | 39/74 (53%) | 0/74 (0%) | 33/74 (45%) | 2/74 (3%) |
| Hepatic Cholestasis | $1.2 \times 10^{-5}$ | 4.94 | 83/162 (51%) | 0/162 (0%) | 75/162 (46%) | 4/162 (2%) |
| **Glucocorticoid Receptor Signaling** | $1.5 \times 10^{-5}$ | 3.64 | 152/275 (55%) | 0/275 (0%) | 117/275 (43%) | 6/275 (2%) |
| Role of Cytokines in Mediating Communication between Immune Cells | $3.1 \times 10^{-5}$ | 9.09 | 32/55 (58%) | 0/55 (0%) | 20/55 (36%) | 3/55 (5%) |
| IL-10 Signaling | $8.5 \times 10^{-5}$ | 7.35 | 33/68 (49%) | 0/68 (0%) | 34/68 (50%) | 1/68 (1%) |
| IL-6 Signaling | 0.00012 | 5.17 | 57/116 (49%) | 0/116 (0%) | 59/116 (51%) | 0/116 (0%) |
| p38 MAPK Signaling | 0.000126 | 5.13 | 62/117 (53%) | 0/117 (0%) | 52/117 (44%) | 3/117 (3%) |
| MIF Regulation of Innate Immunity | 0.000151 | 9.76 | 19/41 (46%) | 0/41 (0%) | 21/41 (51%) | 1/41 (2%) |
| Molecular Mechanisms of Cancer | 0.000155 | 2.74 | 201/365 (55%) | 0/365 (0%) | 157/365 (43%) | 7/365 (2%) |
| iNOS Signaling | 0.0002 | 9.09 | 23/44 (52%) | 0/44 (0%) | 20/44 (45%) | 1/44 (2%) |
| **Aryl Hydrocarbon Receptor Signaling** | 0.000331 | 4.29 | 73/140 (52%) | 0/140 (0%) | 61/140 (44%) | 6/140 (4%) |
| **PPAR Signaling** | 0.000398 | 5.32 | 52/94 (55%) | 0/94 (0%) | 38/94 (40%) | 4/94 (4%) |
| LPS/IL-1 Mediated Inhibition of RXR Function | 0.000631 | 3.2 | 119/219 (54%) | 0/219 (0%) | 88/219 (40%) | 12/219 (5%) |
| TNFR2 Signaling | 0.000891 | 10.3 | 14/29 (48%) | 0/29 (0%) | 14/29 (48%) | 1/29 (3%) |
| HMGB1 Signaling | 0.001202 | 4.17 | 54/120 (45%) | 0/120 (0%) | 63/120 (53%) | 3/120 (3%) |
| MIF-mediated Glucocorticoid Regulation | 0.001318 | 9.09 | 16/33 (48%) | 0/33 (0%) | 16/33 (48%) | 1/33 (3%) |
| ILK Signaling | 0.001479 | 3.23 | 102/186 (55%) | 0/186 (0%) | 77/186 (41%) | 7/186 (4%) |
| IL-17A Signaling in Fibroblasts | 0.001549 | 8.57 | 20/35 (57%) | 0/35 (0%) | 15/35 (43%) | 0/35 (0%) |
| Role of JAK2 in Hormone-like Cytokine Signaling | 0.001549 | 8.57 | 18/35 (51%) | 0/35 (0%) | 14/35 (40%) | 3/35 (9%) |
| PI3K Signaling in B Lymphocytes | 0.001585 | 3.91 | 66/128 (52%) | 0/128 (0%) | 57/128 (45%) | 5/128 (4%) |

**Table 2.** *Cont.*

| Ingenuity Canonical Pathways | p-Value | Overlap (%) | Downregulated | No Change | Upregulated | No Overlap with Dataset |
|---|---|---|---|---|---|---|
| Factors Promoting Cardiogenesis in Vertebrates | 0.003236 | 4.35 | 52/92 (57%) | 0/92 (0%) | 37/92 (40%) | 3/92 (3%) |
| TNFR1 Signaling | 0.004169 | 6.12 | 27/49 (55%) | 0/49 (0%) | 20/49 (41%) | 2/49 (4%) |
| **Antioxidant Action of Vitamin C** | 0.004169 | 4.04 | 51/99 (52%) | 0/99 (0%) | 42/99 (42%) | 6/99 (6%) |
| Acute Phase Response Signaling | 0.005248 | 2.96 | 88/169 (52%) | 0/169 (0%) | 79/169 (47%) | 2/169 (1%) |
| Differential Regulation of Cytokine Production in Macrophages and T Helper Cells by IL-17A and IL-17F | 0.006166 | 11.1 | 7/18 (39%) | 0/18 (0%) | 11/18 (61%) | 0/18 (0%) |
| **PPARα/RXRα Activation** | 0.006607 | 2.79 | 98/179 (55%) | 0/179 (0%) | 67/179 (37%) | 14/179 (8%) |
| Hepatic Fibrosis/Hepatic Stellate Cell Activation | 0.007244 | 2.73 | 96/183 (52%) | 0/183 (0%) | 84/183 (46%) | 3/183 (2%) |
| Type II Diabetes Mellitus Signaling | 0.007586 | 3.42 | 50/117 (43%) | 0/117 (0%) | 64/117 (55%) | 3/117 (3%) |
| Estrogen-Dependent Breast Cancer Signaling | 0.008318 | 4.76 | 30/63 (48%) | 0/63 (0%) | 33/63 (52%) | 0/63 (0%) |
| **LXR/RXR Activation** | 0.008511 | 3.31 | 68/121 (56%) | 0/121 (0%) | 53/121 (44%) | 0/121 (0%) |
| **RAR Activation** | 0.008511 | 2.63 | 86/190 (45%) | 0/190 (0%) | 100/190 (53%) | 4/190 (2%) |

The top canonical nuclear receptor pathway induced by 20,23(OH)$_2$D3 was AhR signaling (Supplemental Figure S2A) with VDR/RXR being next (Supplemental Figure S3A) (Table 3). While the identification of the VDR/RXR as the target for 20,23(OH)$_2$D3 is consistent with previously reported functional data and molecular modeling [60,65,69,83], identification of the AhR as its primary target was unexpected and hence it was further analyzed in detail as described below. Table 4 shows that for 1,25(OH)$_2$D3 the nuclear signaling pathways VDR/RXR, followed by AhR, PPARα/RXRα, RAR and LXR/RXR, were among the top toxicity-related pathways identified. The top signaling pathways for 20,23(OH)$_2$D3 were linked to the activation of AhR and VDR/RXR (Table 5). Tables 6 and 7 show certain functional similarities between top diseases and bifunctions affected by both molecules. For example, cancer, and organismal injury and abnormalities, are the top two diseases affected by both molecules. These phenotypic similarities are consistent with previously reported studies comparing the biological effects of 1,25(OH)$_2$D3 and CYP11A1-derived D3-hydroxyderivatives, including 20,23(OH)$_2$D3, and indicate similarities between the effects on cell proliferation and differentiation, as well as similar anti-inflammatory, photoprotective and anti-cancer actions [23,60–62,64,72,80,84].

**Table 3.** Canonical pathways activated by 20,23(OH)$_2$D3 in human epidermal keratinocytes after 6 h of treatment. Nuclear receptors are marked in bold.

| Ingenuity Canonical Pathways | p-Value | Overlap (%) | Downregulated | No Change | Upregulated | No Overlap with Dataset |
|---|---|---|---|---|---|---|
| 2-ketoglutarate Dehydrogenase Complex | 0.004898 | 25 | 2/4 (50%) | 0/4 (0%) | 2/4 (50%) | 0/4 (0%) |
| **Aryl Hydrocarbon Receptor Signaling** | 0.012589 | 1.43 | 64/140 (46%) | 0/140 (0%) | 70/140 (50%) | 6/140 (4%) |
| Aldosterone Signaling in Epithelial Cells | 0.014791 | 1.32 | 80/152 (53%) | 0/152 (0%) | 69/152 (45%) | 3/152 (2%) |
| TCA Cycle II (Eukaryotic) | 0.027542 | 4.35 | 14/23 (61%) | 0/23 (0%) | 9/23 (39%) | 0/23 (0%) |
| Bupropion Degradation | 0.0302 | 4 | 12/25 (48%) | 0/25 (0%) | 12/25 (48%) | 1/25 (4%) |
| D-myo-inositol (1,4,5)-Trisphosphate Biosynthesis | 0.032359 | 3.7 | 13/27 (48%) | 0/27 (0%) | 13/27 (48%) | 1/27 (4%) |

**Table 3.** *Cont.*

| Ingenuity Canonical Pathways | $p$-Value | Overlap (%) | Downregulated | No Change | Upregulated | No Overlap with Dataset |
|---|---|---|---|---|---|---|
| Acetone Degradation I (to Methylglyoxal) | 0.032359 | 3.7 | 13/27 (48%) | 0/27 (0%) | 13/27 (48%) | 1/27 (4%) |
| Xenobiotic Metabolism Signaling | 0.042658 | 0.738 | 133/271 (49%) | 0/271 (0%) | 122/271 (45%) | 16/271 (6%) |
| Estrogen Biosynthesis | 0.045709 | 2.63 | 17/38 (45%) | 0/38 (0%) | 20/38 (53%) | 1/38 (3%) |
| Nicotine Degradation III | 0.064565 | 1.85 | 23/54 (43%) | 0/54 (0%) | 22/54 (41%) | 9/54 (17%) |
| Melatonin Degradation I | 0.067608 | 1.75 | 25/57 (44%) | 0/57 (0%) | 24/57 (42%) | 8/57 (14%) |
| Superpathway of Melatonin Degradation | 0.072444 | 1.61 | 27/62 (44%) | 0/62 (0%) | 27/62 (44%) | 8/62 (13%) |
| GM-CSF Signaling | 0.072444 | 1.61 | 26/62 (42%) | 0/62 (0%) | 36/62 (58%) | 0/62 (0%) |
| Nicotine Degradation II | 0.074131 | 1.59 | 26/63 (41%) | 0/63 (0%) | 25/63 (40%) | 12/63 (19%) |
| **VDR/RXR Activation** | 0.091201 | 1.28 | 34/78 (44%) | 0/78 (0%) | 43/78 (55%) | 1/78 (1%) |
| Acute Myeloid Leukemia Signaling | 0.091201 | 1.27 | 39/79 (49%) | 0/79 (0%) | 38/79 (48%) | 2/79 (3%) |
| **TR/RXR Activation** | 0.097724 | 1.18 | 39/85 (46%) | 0/85 (0%) | 46/85 (54%) | 0/85 (0%) |
| Regulation of Actin-based Motility by Rho | 0.105196 | 1.1 | 45/91 (49%) | 0/91 (0%) | 39/91 (43%) | 7/91 (8%) |
| Antioxidant Action of Vitamin C | 0.114025 | 1.01 | 53/99 (54%) | 0/99 (0%) | 40/99 (40%) | 6/99 (6%) |
| Rac Signaling | 0.119399 | 0.962 | 47/104 (45%) | 0/104 (0%) | 55/104 (53%) | 2/104 (2%) |
| Type I Diabetes Mellitus Signaling | 0.125893 | 0.909 | 54/110 (49%) | 1/110 (1%) | 46/110 (42%) | 9/110 (8%) |
| RhoA Signaling | 0.138676 | 0.82 | 56/122 (46%) | 0/122 (0%) | 61/122 (50%) | 5/122 (4%) |
| 3-phosphoinositide Biosynthesis | 0.175792 | 0.633 | 72/158 (46%) | 0/158 (0%) | 76/158 (48%) | 10/158 (6%) |
| RhoGDI Signaling | 0.190985 | 0.578 | 75/173 (43%) | 0/173 (0%) | 94/173 (54%) | 4/173 (2%) |
| Superpathway of Inositol Phosphate Compounds | 0.212814 | 0.513 | 87/195 (45%) | 0/195 (0%) | 96/195 (49%) | 12/195 (6%) |
| Actin Cytoskeleton Signaling | 0.233884 | 0.461 | 93/217 (43%) | 0/217 (0%) | 115/217 (53%) | 9/217 (4%) |
| Signaling by Rho Family GTPases | 0.249459 | 0.427 | 106/234 (45%) | 0/234 (0%) | 124/234 (53%) | 4/234 (2%) |
| Protein Ubiquitination Pathway | 0.268534 | 0.392 | 121/255 (47%) | 0/255 (0%) | 126/255 (49%) | 8/255 (3%) |

**Table 4.** Ingenuity toxicity list secondary to keratinocytes treatment with 1,25(OH)$_2$D3 in humans for 6 h. Nuclear receptors are marked in bold.

| Ingenuity Toxicity Lists | $p$-Value | Overlap (%) | Downregulated | No Change | Upregulated | No Overlap with Dataset |
|---|---|---|---|---|---|---|
| **VDR/RXR Activation** | $1.0 \times 10^{-10}$ | 12.8 | 34/78 (44%) | 0/78 (0%) | 43/78 (55%) | 1/78 (1%) |
| Hepatic Cholestasis | $1.4 \times 10^{-5}$ | 4.85 | 85/165 (52%) | 0/165 (0%) | 76/165 (46%) | 4/165 (2%) |
| Renal Necrosis/Cell Death | $2.8 \times 10^{-5}$ | 2.59 | 241/501 (48%) | 0/501 (0%) | 242/501 (48%) | 18/501 (4%) |
| **Aryl Hydrocarbon Receptor Signaling** | $9.6 \times 10^{-5}$ | 4.35 | 79/161 (49%) | 0/161 (0%) | 67/161 (42%) | 15/161 (9%) |
| Liver Necrosis/Cell Death | 0.000525 | 2.87 | 131/279 (47%) | 0/279 (0%) | 136/279 (49%) | 12/279 (4%) |
| Liver Proliferation | 0.000776 | 3.08 | 110/227 (48%) | 0/227 (0%) | 106/227 (47%) | 11/227 (5%) |
| Cardiac Hypertrophy | 0.00138 | 2.24 | 207/401 (52%) | 0/401 (0%) | 171/401 (43%) | 23/401 (6%) |
| LPS/IL-1 Mediated Inhibition of RXR Function | 0.00138 | 2.79 | 124/251 (49%) | 0/251 (0%) | 93/251 (37%) | 34/251 (14%) |
| Hepatic Stellate Cell Activation | 0.001549 | 8.57 | 17/35 (49%) | 0/35 (0%) | 18/35 (51%) | 0/35 (0%) |
| Increases Liver Steatosis | 0.002188 | 4.82 | 43/83 (52%) | 0/83 (0%) | 37/83 (45%) | 3/83 (4%) |

**Table 4.** *Cont.*

| Ingenuity Toxicity Lists | p-Value | Overlap (%) | Downregulated | No Change | Upregulated | No Overlap with Dataset |
|---|---|---|---|---|---|---|
| Mechanism of Gene Regulation by Peroxisome Proliferators via **PPARα** | 0.003631 | 4.21 | 54/95 (57%) | 0/95 (0%) | 40/95 (42%) | 1/95 (1%) |
| Hepatic Fibrosis | 0.004169 | 4.04 | 42/99 (42%) | 0/99 (0%) | 54/99 (55%) | 3/99 (3%) |
| Increases Liver Damage | 0.005495 | 3.74 | 48/107 (45%) | 0/107 (0%) | 56/107 (52%) | 3/107 (3%) |
| **PPARα/RXRα Activation** | 0.007244 | 2.73 | 100/183 (55%) | 0/183 (0%) | 69/183 (38%) | 14/183 (8%) |
| Acute Renal Failure Panel (Rat) | 0.007943 | 4.84 | 31/62 (50%) | 0/62 (0%) | 24/62 (39%) | 7/62 (11%) |
| **RAR Activation** | 0.008511 | 2.63 | 86/190 (45%) | 0/190 (0%) | 100/190 (53%) | 4/190 (2%) |
| Cardiac Necrosis/Cell Death | 0.008913 | 2.23 | 141/269 (52%) | 0/269 (0%) | 114/269 (42%) | 14/269 (5%) |
| Cardiac Fibrosis | 0.008913 | 2.6 | 104/192 (54%) | 0/192 (0%) | 72/192 (38%) | 16/192 (8%) |
| **LXR/RXR Activation** | 0.008913 | 3.25 | 69/123 (56%) | 0/123 (0%) | 54/123 (44%) | 0/123 (0%) |
| Nongenotoxic Hepatocarcinogenicity Biomarker Panel | 0.00912 | 9.09 | 11/22 (50%) | 0/22 (0%) | 10/22 (45%) | 1/22 (5%) |
| Increases Renal Damage | 0.016218 | 3.7 | 38/81 (47%) | 0/81 (0%) | 36/81 (44%) | 7/81 (9%) |
| NRF2-mediated Oxidative Stress Response | 0.019498 | 2.14 | 99/234 (42%) | 0/234 (0%) | 103/234 (44%) | 32/234 (14%) |
| TGF-β Signaling | 0.021878 | 3.33 | 54/90 (60%) | 0/90 (0%) | 35/90 (39%) | 1/90 (1%) |

**Table 5.** Ingenuity toxicity list secondary to keratinocytes treatment with 20,23(OH)$_2$D3 in humans for 6 h. Nuclear receptors are marked in bold.

| Ingenuity Toxicity Lists | p-Value | Overlap (%) | Downregulated | No Change | Upregulated | No Overlap with Dataset |
|---|---|---|---|---|---|---|
| Cytochrome P450 Panel—Substrate is a Vitamin (Human) | 0.007244 | 16.7 | 2/6 (33%) | 0/6 (0%) | 4/6 (67%) | 0/6 (0%) |
| **Aryl Hydrocarbon Receptor Signaling** | 0.016218 | 1.24 | 71/161 (44%) | 0/161 (0%) | 75/161 (47%) | 15/161 (9%) |
| Cytochrome P450 Panel—Substrate is a Sterol (Human) | 0.016982 | 7.14 | 5/14 (36%) | 0/14 (0%) | 9/14 (64%) | 0/14 (0%) |
| Cytochrome P450 Panel—Substrate is a Xenobiotic (Human) | 0.021878 | 5.56 | 7/18 (39%) | 0/18 (0%) | 9/18 (50%) | 2/18 (11%) |
| Nongenotoxic Hepatocarcinogenicity Biomarker Panel | 0.026303 | 4.55 | 12/22 (55%) | 0/22 (0%) | 9/22 (41%) | 1/22 (5%) |
| Cytochrome P450 Panel—Substrate is a Xenobiotic (Mouse) | 0.0302 | 4 | 5/25 (20%) | 0/25 (0%) | 7/25 (28%) | 13/25 (52%) |
| Cytochrome P450 Panel—Substrate is a Xenobiotic (Rat) | 0.030903 | 3.85 | 5/26 (19%) | 0/26 (0%) | 7/26 (27%) | 14/26 (54%) |
| Xenobiotic Metabolism Signaling | 0.063096 | 0.595 | 155/336 (46%) | 0/336 (0%) | 139/336 (41%) | 42/336 (13%) |
| **VDR/RXR Activation** | 0.091201 | 1.28 | 34/78 (44%) | 0/78 (0%) | 43/78 (55%) | 1/78 (1%) |
| **TR/RXR Activation** | 0.097724 | 1.18 | 39/85 (46%) | 0/85 (0%) | 46/85 (54%) | 0/85 (0%) |
| Hepatic Fibrosis | 0.114025 | 1.01 | 49/99 (49%) | 0/99 (0%) | 47/99 (47%) | 3/99 (3%) |
| Increases Liver Hyperplasia/ Hyperproliferation | 0.116145 | 0.99 | 39/101 (39%) | 0/101 (0%) | 52/101 (51%) | 10/101 (10%) |

**Table 5.** *Cont.*

| Ingenuity Toxicity Lists | *p*-Value | Overlap (%) | Downregulated | No Change | Upregulated | No Overlap with Dataset |
|---|---|---|---|---|---|---|
| Renal Necrosis/Cell Death | 0.12388 | 0.399 | 224/501 (45%) | 1/501 (0%) | 258/501 (51%) | 18/501 (4%) |
| Fatty Acid Metabolism | 0.133352 | 0.855 | 57/117 (49%) | 0/117 (0%) | 39/117 (33%) | 21/117 (18%) |
| Cardiac Fibrosis | 0.209894 | 0.521 | 89/192 (46%) | 0/192 (0%) | 87/192 (45%) | 16/192 (8%) |
| Liver Proliferation | 0.24322 | 0.441 | 98/227 (43%) | 0/227 (0%) | 118/227 (52%) | 11/227 (5%) |

**Table 6.** Categories of biological functions with diseases or function annotation activated by 1,25(OH)$_2$D3 in human epidermal keratinocytes after 6 h of treatment.

| Categories of Biological Function | Diseases or Functions Annotation | *p*-Value | Activation z-Score | # of Genes |
|---|---|---|---|---|
| Cancer, Organismal Injury and Abnormalities | growth of tumor | $1.5 \times 10^{-12}$ | 0.033 | 31 |
| Cellular Movement | cell movement | $4.5 \times 10^{-12}$ | 1.845 | 52 |
| Cancer, Cellular Development, Cellular Growth and Proliferation, Organismal Injury and Abnormalities, Tumor Morphology | proliferation of tumor cells | $7.5 \times 10^{-12}$ | −0.27 | 23 |
| Cellular Movement | migration of cells | $1.3 \times 10^{-11}$ | 1.801 | 48 |
| Cellular Growth and Proliferation | proliferation of cells | $4.2 \times 10^{-11}$ | 0.347 | 70 |
| Carbohydrate Metabolism | metabolism of polysaccharide | $4.4 \times 10^{-11}$ | 0.755 | 16 |
| Cell Death and Survival | apoptosis of tumor cell lines | $8.3 \times 10^{-11}$ | 1.089 | 36 |
| Cellular Movement | invasion of cells | $8.7 \times 10^{-11}$ | 0.571 | 30 |
| Inflammatory Response | inflammatory response | $1.4 \times 10^{-10}$ | 0.766 | 28 |
| Cellular Development | differentiation of cells | $1.5 \times 10^{-10}$ | 2.774 | 51 |
| Cell Death and Survival | cell survival | $3.1 \times 10^{-10}$ | 2.346 | 38 |
| Cell Death and Survival | apoptosis | $5.7 \times 10^{-10}$ | 1.307 | 55 |
| Cell Death and Survival | necrosis | $6.3 \times 10^{-10}$ | 1.7 | 54 |
| Cell Death and Survival | cell viability | $6.4 \times 10^{-10}$ | 2.547 | 36 |
| Tissue Morphology | quantity of cells | $7.1 \times 10^{-10}$ | 1.065 | 43 |
| Cardiovascular System Development and Function, Organismal Development | vasculogenesis | $9.4 \times 10^{-10}$ | 0.844 | 26 |
| Carbohydrate Metabolism | synthesis of polysaccharide | $9.6 \times 10^{-10}$ | 0.297 | 13 |
| Cellular Growth and Proliferation, Tissue Development | proliferation of connective tissue cells | $1.2 \times 10^{-9}$ | 1.232 | 23 |
| Cellular Movement | cell movement of tumor cell lines | $1.2 \times 10^{-9}$ | 0.955 | 28 |
| Cell Death and Survival | cell death of tumor cell lines | $1.2 \times 10^{-9}$ | 0.81 | 39 |
| Cellular Development, Cellular Growth and Proliferation | proliferation of tumor cell lines | $1.9 \times 10^{-9}$ | 0.324 | 39 |
| Organismal Survival | morbidity or mortality | $2.3 \times 10^{-9}$ | −1.006 | 51 |
| Embryonic Development, Organismal Development | development of body trunk | $2.7 \times 10^{-9}$ | 0.06 | 32 |
| Cell Death and Survival | cell death | $3.1 \times 10^{-9}$ | 1.556 | 62 |
| Cancer, Organismal Injury and Abnormalities | growth of malignant tumor | $3.3 \times 10^{-9}$ | 0.518 | 19 |
| Cellular Development, Cellular Growth and Proliferation, Connective Tissue Development and Function, Tissue Development | proliferation of fibroblasts | $4.0 \times 10^{-9}$ | 0.604 | 17 |

**Table 6.** *Cont.*

| Categories of Biological Function | Diseases or Functions Annotation | *p*-Value | Activation z-Score | # of Genes |
|---|---|---|---|---|
| Cellular Function and Maintenance, Hematological System Development and Function | function of myeloid cells | $4.5 \times 10^{-9}$ | | 14 |
| Cell Signaling, Small Molecule Biochemistry | synthesis of nitric oxide | $4.7 \times 10^{-9}$ | 0.89 | 15 |
| Dermatological Diseases and Conditions | psoriasis | $5.0 \times 10^{-9}$ | | 22 |

**Table 7.** Categories of biological functions with diseases or function annotation activated by 20,23(OH)$_2$D3 in human epidermal keratinocytes after 6 h of treatment.

| Categories of Biological Function | Diseases or Functions Annotation | *p*-Value | # of Genes |
|---|---|---|---|
| Cancer, Organismal Injury and Abnormalities, Respiratory Disease | carcinoma in lung | $6.5 \times 10^{-5}$ | 8 |
| Vitamin and Mineral Metabolism | metabolism of vitamin | 0.000254 | 3 |
| Developmental Disorder, Skeletal and Muscular Disorders | hypertrophy of smooth muscle | 0.000325 | 2 |
| Cancer, Organismal Injury and Abnormalities, Respiratory Disease | non-small cell lung cancer | 0.000334 | 7 |
| Infectious Diseases | internalization of virus | 0.000355 | 2 |
| Cancer, Organismal Injury and Abnormalities | adenocarcinoma | 0.000603 | 20 |
| Cancer, Organismal Injury and Abnormalities, Reproductive System Disease | prostate cancer | 0.000618 | 7 |
| Cancer, Gastrointestinal Disease, Hepatic System Disease, Organismal Injury and Abnormalities | liver adenoma | 0.000692 | 2 |
| Cancer, Endocrine System Disorders, Organismal Injury and Abnormalities | endocrine gland tumor | 0.000718 | 9 |
| Cell Death and Survival | apoptosis of germ cells | 0.000799 | 3 |
| Lipid Metabolism, Small Molecule Biochemistry, Vitamin and Mineral Metabolism | catabolism of terpenoid | 0.000828 | 2 |
| Cancer, Organismal Injury and Abnormalities | epithelial cancer | 0.000989 | 23 |
| Endocrine System Development and Function, Lipid Metabolism, Small Molecule Biochemistry, Vitamin and Mineral Metabolism | synthesis of estrogen | 0.00114 | 2 |
| Ophthalmic Disease, Organismal Injury and Abnormalities | age-related macular degeneration type 6 | 0.00122 | 1 |
| Cell Cycle | arrest in sub-G1 phase of endometrial cancer cell lines | 0.00122 | 1 |
| Cell Morphology, Connective Tissue Development and Function | blebbing of pulmonary fibroblasts | 0.00122 | 1 |
| Organismal Injury and Abnormalities | calcification of uterus | 0.00122 | 1 |
| Cell Cycle, Cell Death and Survival | chromosome condensation of pulmonary fibroblasts | 0.00122 | 1 |

**Table 7.** *Cont.*

| Categories of Biological Function | Diseases or Functions Annotation | *p*-Value | # of Genes |
|---|---|---|---|
| Cell-To-Cell Signaling and Interaction, Inflammatory Response | cytotoxic reaction of bone marrow cells | 0.00122 | 1 |
| Tissue Morphology | deficiency of mast cells | 0.00122 | 1 |
| Cell Cycle | delay in G1/S phase transition of hepatoma cell lines | 0.00122 | 1 |
| Embryonic Development, Organ Development, Organismal Development, Tissue Development, Visual System Development and Function | development of outflow pathway | 0.00122 | 1 |
| Embryonic Development, Organ Development, Organismal Development, Reproductive System Development and Function, Tissue Development | development of placenta decidua | 0.00122 | 1 |
| Embryonic Development, Organ Development, Organismal Development, Reproductive System Development and Function, Tissue Development | development of placental spongiotrophoblast layer | 0.00122 | 1 |
| Cardiovascular System Development and Function, Tissue Morphology | diameter of portal vein | 0.00122 | 1 |
| Cardiovascular System Development and Function, Tissue Morphology | diameter of umbilical vein | 0.00122 | 1 |
| Hereditary Disorder, Ophthalmic Disease, Organismal Injury and Abnormalities | digenic early-onset glaucoma | 0.00122 | 1 |
| Cancer, Organismal Injury and Abnormalities, Reproductive System Disease | estrogen receptor positive endometrial cancer | 0.00122 | 1 |
| Connective Tissue Disorders, Organismal Injury and Abnormalities | fibrosis of submucosa | 0.00122 | 1 |
| Digestive System Development and Function, Embryonic Development, Organ Development, Organismal Development, Tissue Development | formation of salivary duct | 0.00122 | 1 |
| Cardiovascular System Development and Function, Embryonic Development, Lymphoid Tissue Structure and Development, Organ Development, Organismal Development, Respiratory System Development and Function, Tissue Development | formation of tracheal duct | 0.00122 | 1 |
| Cancer, Gastrointestinal Disease, Organismal Injury and Abnormalities | hyperplasia of pylorus | 0.00122 | 1 |
| Cancer, Cardiovascular Disease, Organismal Injury and Abnormalities | hyperplasia of vasculature | 0.00122 | 1 |
| Developmental Disorder, Gastrointestinal Disease | hypertrophy of gastric epithelium | 0.00122 | 1 |
| Dermatological Diseases and Conditions, Developmental Disorder | hypertrophy of skin | 0.00122 | 1 |

Because of the unexpected differences between 1,25(OH)$_2$D3 and 20,23(OH)$_2$D3, the 6 h incubation experiment was repeated in a similar manner as shown in Figure 2 and microarray analyses were performed using Illumina's HumanWG-6_V2 (Platform GPL13376) chip/array. Average signal values for filtered gene clusters with FC $\geq \pm 1.5$ are shown in Supplemental excel file #2. The heat maps corresponding to relative gene expression and Venn diagrams are shown in Figure 3C. Again, for a 2-fold cut-off value there was only one common gene (*CYP24A1*) whose expression was stimulated by both 1,25(OH)$_2$D3 (80-fold) and 20,23(OH)$_2$D3 (2.9-fold). For FC $\geq \pm 1.5$ there were 11 common genes upregulated and 4 downregulated. Again, ingenuity pathway analysis showed that VDR/RXR was the top canonical pathway induced by 1,25(OH)$_2$D3, followed by the role of osteoblasts, osteoclasts and chondrocytes in rheumatoid arthritis. As before, other nuclear receptor signaling pathways included LXR/RXR, GR and AhR. For 20,23(OH)$_2$D3 the top nuclear receptor signaling pathways were again AhR and VDR/RXR. Of note, this microarray showed that 20,23(OH)$_2$D3 upregulated two genes downstream of AhR signaling, *CYP1A1* and *CYP1B1*, by factors of 2.4 and 2.6, respectively. This stimulation was confirmed by qPCR (Figure 3D). VDR/RXR was identified as the top toxicity pathway for 1,25(OH)$_2$D3 and AhR for 20,23(OH)$_2$D3.

More robust data were obtained with 24 h of treatment for which the average signal values for filtered gene clusters with FC $\geq \pm 1.5$ are shown in Supplemental excel file #3. Because of the large number of genes affected (Table 1), the heat map of differentially expressed genes and Venn diagrams were generated using the 4-FC value which show three overlapping genes (*CYP24A1*, *MMP3* and *SERPINB1*) as well as distinct gene expression patterns (Figure 4). For FC $\geq \pm 1.5$, 93 and 72 common genes were up- and downregulated, respectively. Ingenuity pathway analysis using FC $\geq \pm 2.0$ was consistent with results obtained after 6 h of treatment. Again, the top canonical pathway for 1,25(OH)$_2$D3 was VDR/RXR (Supplemental Figure S1B) followed by MIF-related glucocorticoid regulation and regulation of the innate immunity system (Table 8). AhR signaling was also listed. The top canonical pathways induced by 20,23(OH)$_2$D3 were AhR signaling (Supplemental Figure S2B) and the cholesterol biosynthesis pathway (Table 9). Interestingly, the involvement of a second nuclear receptor complex was emphasized by VDR/RXR activation (Supplemental Figure S3B), with p53 signaling also being listed. The latter is consistent with the photoprotective properties of 20,23(OH)$_2$D3 and activation of p53 by its direct precursor, 20(OH)D3 [23]. The top affected toxicity pathways for 1,25(OH)$_2$D3 included VDR/RXR, xenobiotic metabolism, cardiac fibrosis and cytochrome P450s (Table 10). AhR signaling was also listed. For 20,23(OH)$_2$D3, AhR signaling was again listed as the top toxicity pathway followed by cholesterol synthesis, p53 signaling and again VDR/RXR (Table 11). The top upstream gene regulation pathways for 1,25(OH)$_2$D3 included vitamin D3-VDR-RXR, calcitriol, dexamethasone, progesterone and β-estradiol, while for 20,23(OH)$_2$D3, included TP53 (p53 tumor suppressor), β-estradiol, lipopolysaccharide, TNF (tumor necrosis factor) and TGF β1 (transforming growth factor-β1) (not shown).

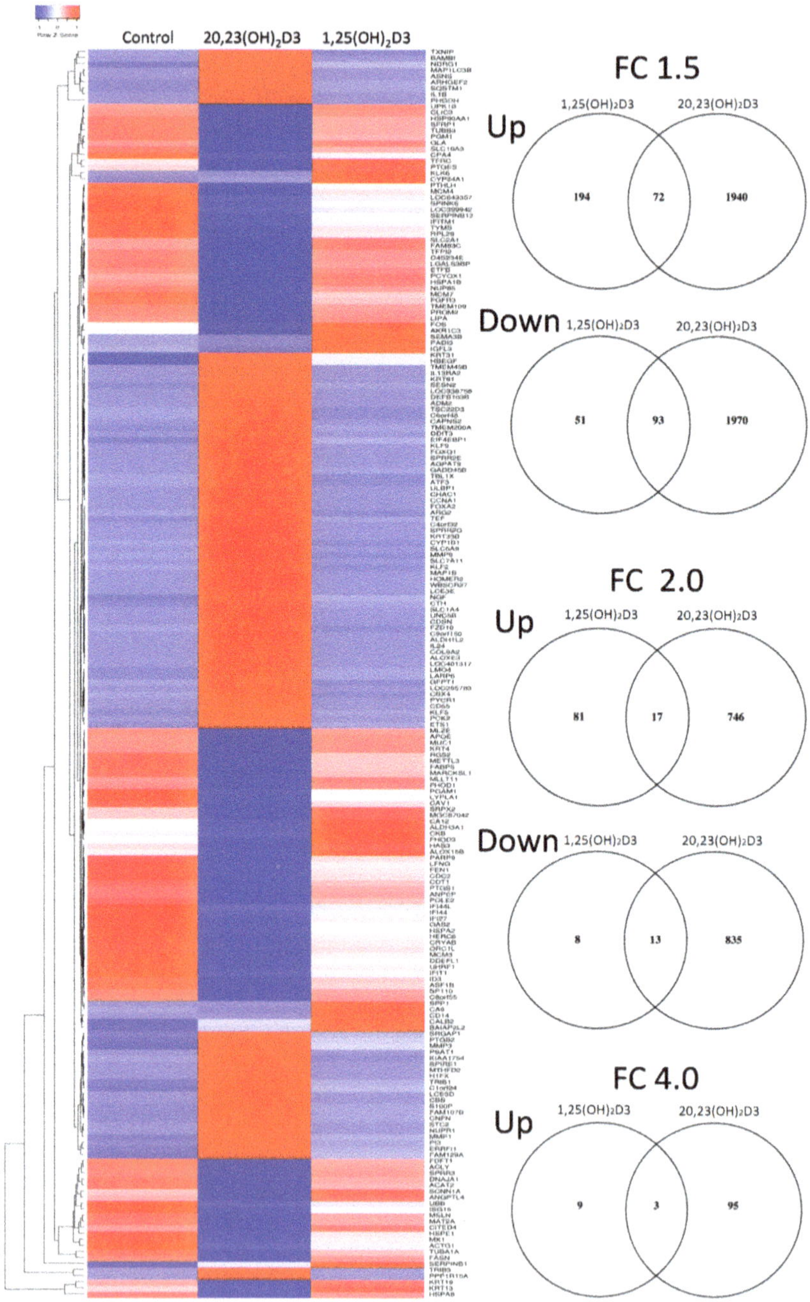


**Figure 4.** Heat map of gene expression pattern in human epidermal keratinocytes treated with $10^{-7}$ M of 1,25(OH)$_2$D3 or 20,23(OH)$_2$D3 for 24 h. On the right are the corresponding Venn diagrams for FC $\geq$ 4, 2 and 1.5.

**Table 8.** Canonical pathways activated by 1,25(OH)$_2$D3 in human epidermal keratinocytes after 24 h of treatment. Nuclear receptors are marked in bold.

| Ingenuity Canonical Pathways | *p*-Value | Overlap (%) | Downregulated | No Change | Upregulated | No Overlap with Dataset |
|---|---|---|---|---|---|---|
| **VDR/RXR Activation** | $7.9 \times 10^{-15}$ | 15.4 | 23/78 (29%) | 0/78 (0%) | 54/78 (69%) | 1/78 (1%) |
| MIF-mediated Glucocorticoid Regulation | $2.6 \times 10^{-5}$ | 12.1 | 12/33 (36%) | 0/33 (0%) | 20/33 (61%) | 1/33 (3%) |
| MIF Regulation of Innate Immunity | $6.3 \times 10^{-5}$ | 9.76 | 19/41 (46%) | 0/41 (0%) | 21/41 (51%) | 1/41 (2%) |
| α-tocopherol Degradation | 0.000162181 | 50 | 0/4 (0%) | 0/4 (0%) | 4/4 (100%) | 0/4 (0%) |
| Antioxidant Action of Vitamin C | 0.000177828 | 5.05 | 43/99 (43%) | 0/99 (0%) | 50/99 (51%) | 6/99 (6%) |
| Retinoate Biosynthesis I | 0.000691831 | 9.09 | 9/33 (27%) | 0/33 (0%) | 20/33 (61%) | 4/33 (12%) |
| Coagulation System | 0.000831764 | 8.57 | 17/35 (49%) | 0/35 (0%) | 18/35 (51%) | 0/35 (0%) |
| Estrogen Biosynthesis | 0.001047129 | 7.89 | 18/38 (47%) | 0/38 (0%) | 19/38 (50%) | 1/38 (3%) |
| iNOS Signaling | 0.00162181 | 6.82 | 17/44 (39%) | 0/44 (0%) | 26/44 (59%) | 1/44 (2%) |
| Role of IL-17A in Arthritis | 0.002884032 | 5.56 | 27/54 (50%) | 0/54 (0%) | 27/54 (50%) | 0/54 (0%) |
| Parkinson's Signaling | 0.003162278 | 12.5 | 9/16 (56%) | 0/16 (0%) | 7/16 (44%) | 0/16 (0%) |
| **LXR/RXR Activation** | 0.003890451 | 3.31 | 61/121 (50%) | 0/121 (0%) | 60/121 (50%) | 0/121 (0%) |
| CD40 Signaling | 0.004897788 | 4.62 | 31/65 (48%) | 0/65 (0%) | 33/65 (51%) | 1/65 (2%) |
| IL-10 Signaling | 0.005495409 | 4.41 | 32/68 (47%) | 0/68 (0%) | 35/68 (51%) | 1/68 (1%) |
| Role of MAPK Signaling in the Pathogenesis of Influenza | 0.005754399 | 4.35 | 27/69 (39%) | 0/69 (0%) | 39/69 (57%) | 3/69 (4%) |
| LPS/IL-1 Mediated Inhibition of RXR Function | 0.006025596 | 2.28 | 102/219 (47%) | 0/219 (0%) | 105/219 (48%) | 12/219 (5%) |
| Role of Osteoblasts, Osteoclasts and Chondrocytes in Rheumatoid Arthritis | 0.006025596 | 2.28 | 111/219 (51%) | 0/219 (0%) | 103/219 (47%) | 5/219 (2%) |
| IL-17 Signaling | 0.006456542 | 4.17 | 33/72 (46%) | 0/72 (0%) | 39/72 (54%) | 0/72 (0%) |
| LPS-stimulated MAPK Signaling | 0.00676083 | 4.11 | 33/73 (45%) | 0/73 (0%) | 40/73 (55%) | 0/73 (0%) |
| Toll-like Receptor Signaling | 0.007079458 | 4.05 | 32/74 (43%) | 0/74 (0%) | 40/74 (54%) | 2/74 (3%) |
| BMP signaling pathway | 0.007585776 | 3.95 | 43/76 (57%) | 0/76 (0%) | 31/76 (41%) | 2/76 (3%) |
| Intrinsic Prothrombin Activation Pathway | 0.01023293 | 6.9 | 12/29 (41%) | 0/29 (0%) | 16/29 (55%) | 1/29 (3%) |
| 4-1BB Signaling in T Lymphocytes | 0.011481536 | 6.45 | 19/31 (61%) | 0/31 (0%) | 12/31 (39%) | 0/31 (0%) |
| Acute Phase Response Signaling | 0.012302688 | 2.37 | 76/169 (45%) | 0/169 (0%) | 91/169 (54%) | 2/169 (1%) |
| Endothelin-1 Signaling | 0.013182567 | 2.33 | 89/172 (52%) | 0/172 (0%) | 77/172 (45%) | 6/172 (3%) |
| Inhibition of Angiogenesis by TSP1 | 0.013803843 | 5.88 | 21/34 (62%) | 0/34 (0%) | 11/34 (32%) | 2/34 (6%) |
| Xenobiotic Metabolism Signaling | 0.014454398 | 1.85 | 127/271 (47%) | 0/271 (0%) | 128/271 (47%) | 16/271 (6%) |
| IL-17A Signaling in Fibroblasts | 0.014454398 | 5.71 | 15/35 (43%) | 0/35 (0%) | 20/35 (57%) | 0/35 (0%) |
| Interferon Signaling | 0.015488166 | 5.56 | 28/36 (78%) | 0/36 (0%) | 8/36 (22%) | 0/36 (0%) |

**Table 8.** *Cont.*

| Ingenuity Canonical Pathways | *p*-Value | Overlap (%) | Downregulated | No Change | Upregulated | No Overlap with Dataset |
|---|---|---|---|---|---|---|
| Thyroid Hormone Biosynthesis | 0.015848932 | 33.3 | 1/3 (33%) | 0/3 (0%) | 2/3 (67%) | 0/3 (0%) |
| April Mediated Signaling | 0.016982437 | 5.26 | 18/38 (47%) | 0/38 (0%) | 20/38 (53%) | 0/38 (0%) |
| Inhibition of Matrix Metalloproteases | 0.017782794 | 5.13 | 17/39 (44%) | 0/39 (0%) | 21/39 (54%) | 1/39 (3%) |
| **RAR Activation** | 0.018197009 | 2.11 | 87/190 (46%) | 0/190 (0%) | 99/190 (52%) | 4/190 (2%) |
| Thrombin Signaling | 0.018620871 | 2.09 | 90/191 (47%) | 0/191 (0%) | 96/191 (50%) | 5/191 (3%) |
| B Cell Activating Factor Signaling | 0.018620871 | 5 | 21/40 (53%) | 0/40 (0%) | 19/40 (48%) | 0/40 (0%) |
| Dermatan Sulfate Biosynthesis (Late Stages) | 0.022387211 | 4.55 | 16/44 (36%) | 0/44 (0%) | 26/44 (59%) | 2/44 (5%) |
| IL-6 Signaling | 0.023442288 | 2.59 | 52/116 (45%) | 0/116 (0%) | 64/116 (55%) | 0/116 (0%) |
| p38 MAPK Signaling | 0.023988329 | 2.56 | 52/117 (44%) | 0/117 (0%) | 62/117 (53%) | 3/117 (3%) |

**Table 9.** Canonical pathways activated by 20,23$(OH)_2$D3 in human epidermal keratinocytes after 24 h of treatment. Nuclear receptors are marked in bold.

| Ingenuity Canonical Pathways | *p*-Value | Overlap (%) | Downregulated | No Change | Upregulated | No Overlap with Dataset |
|---|---|---|---|---|---|---|
| **Aryl Hydrocarbon Receptor Signaling** | $9.5 \times 10^{-10}$ | 22.9 | 80/140 (57%) | 0/140 (0%) | 54/140 (39%) | 6/140 (4%) |
| Superpathway of Cholesterol Biosynthesis | $9.1 \times 10^{-9}$ | 46.4 | 22/28 (79%) | 0/28 (0%) | 4/28 (14%) | 2/28 (7%) |
| Cell Cycle Control of Chromosomal Replication | $6.3 \times 10^{-8}$ | 44.4 | 21/27 (78%) | 0/27 (0%) | 6/27 (22%) | 0/27 (0%) |
| Mismatch Repair in Eukaryotes | $2.3 \times 10^{-7}$ | 56.2 | 15/16 (94%) | 0/16 (0%) | 1/16 (6%) | 0/16 (0%) |
| Unfolded protein response | $3.5 \times 10^{-7}$ | 29.6 | 24/54 (44%) | 0/54 (0%) | 29/54 (54%) | 1/54 (2%) |
| Fatty Acid α-oxidation | $1.5 \times 10^{-6}$ | 47.4 | 10/19 (53%) | 0/19 (0%) | 6/19 (32%) | 3/19 (16%) |
| Ethanol Degradation IV | $6.8 \times 10^{-6}$ | 40.9 | 17/22 (77%) | 0/22 (0%) | 2/22 (9%) | 3/22 (14%) |
| Cholesterol Biosynthesis I | $7.9 \times 10^{-6}$ | 53.8 | 12/13 (92%) | 0/13 (0%) | 1/13 (8%) | 0/13 (0%) |
| p53 Signaling | $8.3 \times 10^{-6}$ | 20.4 | 53/98 (54%) | 0/98 (0%) | 45/98 (46%) | 0/98 (0%) |
| **VDR/RXR Activation** | $1.5 \times 10^{-5}$ | 21.8 | 26/78 (33%) | 0/78 (0%) | 51/78 (65%) | 1/78 (1%) |
| GADD45 Signaling | $1.7 \times 10^{-5}$ | 42.1 | 14/19 (74%) | 0/19 (0%) | 5/19 (26%) | 0/19 (0%) |
| Putrescine Degradation III | $2.8 \times 10^{-5}$ | 40 | 12/20 (60%) | 0/20 (0%) | 5/20 (25%) | 3/20 (15%) |
| Histamine Degradation | $4.5 \times 10^{-5}$ | 43.8 | 10/16 (63%) | 0/16 (0%) | 3/16 (19%) | 3/16 (19%) |
| Dopamine Degradation | $4.7 \times 10^{-5}$ | 33.3 | 16/27 (59%) | 0/27 (0%) | 6/27 (22%) | 5/27 (19%) |
| Tryptophan Degradation X (Mammalian, via Tryptamine) | $6.2 \times 10^{-5}$ | 36.4 | 13/22 (59%) | 0/22 (0%) | 5/22 (23%) | 4/22 (18%) |

**Table 9.** *Cont.*

| Ingenuity Canonical Pathways | $p$-Value | Overlap (%) | Downregulated | No Change | Upregulated | No Overlap with Dataset |
|---|---|---|---|---|---|---|
| Xenobiotic Metabolism Signaling | $9.3 \times 10^{-5}$ | 13.3 | 131/271 (48%) | 0/271 (0%) | 124/271 (46%) | 16/271 (6%) |
| Oxidative Ethanol Degradation III | 0.000109648 | 38.9 | 13/18 (72%) | 0/18 (0%) | 2/18 (11%) | 3/18 (17%) |
| Mevalonate Pathway I | 0.000112202 | 46.2 | 8/13 (62%) | 0/13 (0%) | 3/13 (23%) | 2/13 (15%) |
| Estrogen-mediated S-phase Entry | 0.000125893 | 33.3 | 18/24 (75%) | 0/24 (0%) | 6/24 (25%) | 0/24 (0%) |
| Hereditary Breast Cancer Signaling | 0.000165959 | 16.3 | 77/129 (60%) | 0/129 (0%) | 45/129 (35%) | 7/129 (5%) |
| **Glucocorticoid Receptor Signaling** | 0.000269153 | 12.7 | 122/275 (44%) | 0/275 (0%) | 147/275 (53%) | 6/275 (2%) |
| Adipogenesis pathway | 0.000371535 | 15.7 | 66/127 (52%) | 0/127 (0%) | 55/127 (43%) | 6/127 (5%) |
| Interferon Signaling | 0.000537032 | 25 | 24/36 (67%) | 0/36 (0%) | 12/36 (33%) | 0/36 (0%) |
| Superpathway of Serine and Glycine Biosynthesis I | 0.000630957 | 57.1 | 3/7 (43%) | 0/7 (0%) | 4/7 (57%) | 0/7 (0%) |
| Superpathway of Geranylgeranyldi phosphate Biosynthesis I (via Mevalonate) | 0.000630957 | 35.3 | 11/17 (65%) | 0/17 (0%) | 4/17 (24%) | 2/17 (12%) |
| Semaphorin Signaling in Neurons | 0.000758578 | 20.8 | 27/53 (51%) | 0/53 (0%) | 24/53 (45%) | 2/53 (4%) |
| Glutaryl-CoA Degradation | 0.000776247 | 41.7 | 7/12 (58%) | 0/12 (0%) | 4/12 (33%) | 1/12 (8%) |
| Pancreatic Adenocarcinoma Signaling | 0.000794328 | 16 | 53/106 (50%) | 0/106 (0%) | 53/106 (50%) | 0/106 (0%) |
| Role of CHK Proteins in Cell Cycle Checkpoint Control | 0.001047129 | 20 | 35/55 (64%) | 0/55 (0%) | 20/55 (36%) | 0/55 (0%) |
| Glycolysis I | 0.001096478 | 28 | 20/25 (80%) | 0/25 (0%) | 4/25 (16%) | 1/25 (4%) |
| NRF2-mediated Oxidative Stress Response | 0.001230269 | 13.3 | 89/180 (49%) | 0/180 (0%) | 87/180 (48%) | 4/180 (2%) |
| HIF1α Signaling | 0.001412538 | 15.7 | 52/102 (51%) | 0/102 (0%) | 48/102 (47%) | 2/102 (2%) |
| Aldosterone Signaling in Epithelial Cells | 0.001548817 | 13.8 | 82/152 (54%) | 0/152 (0%) | 67/152 (44%) | 3/152 (2%) |
| Serotonin Degradation | 0.001737801 | 17.9 | 38/67 (57%) | 0/67 (0%) | 16/67 (24%) | 13/67 (19%) |

**Table 10.** Toxicity-related pathways identified by Ingenuity in human keratinocytes treated with $1,25(OH)_2D3$ for 24 h. Nuclear receptors are marked in bold.

| Ingenuity Toxicity Lists | $p$-Value | Overlap (%) | Downregulated | No Change | Upregulated | No Overlap with Dataset |
|---|---|---|---|---|---|---|
| **VDR/RXR Activation** | $7.9 \times 10^{-15}$ | 15.4 | 23/78 (29%) | 0/78 (0%) | 54/78 (69%) | 1/78 (1%) |
| Xenobiotic Metabolism Signaling | 0.00040738 | 2.38 | 144/336 (43%) | 0/336 (0%) | 150/336 (45%) | 42/336 (13%) |
| Cardiac Fibrosis | 0.000537032 | 3.12 | 85/192 (44%) | 0/192 (0%) | 91/192 (47%) | 16/192 (8%) |
| Cytochrome P450 Panel—Substrate is an Eicosanoid (Human) | 0.000562341 | 28.6 | 1/7 (14%) | 0/7 (0%) | 6/7 (86%) | 0/7 (0%) |
| Cytochrome P450 Panel—Substrate is a Fatty Acid (Human) | 0.001202264 | 20 | 3/10 (30%) | 0/10 (0%) | 7/10 (70%) | 0/10 (0%) |

**Table 10.** *Cont.*

| Ingenuity Toxicity Lists | *p*-Value | Overlap (%) | Downregulated | No Change | Upregulated | No Overlap with Dataset |
|---|---|---|---|---|---|---|
| Cardiac Hypertrophy | 0.00128825 | 2 | 193/401 (48%) | 0/401 (0%) | 185/401 (46%) | 23/401 (6%) |
| Liver Proliferation | 0.00128825 | 2.64 | 103/227 (45%) | 0/227 (0%) | 113/227 (50%) | 11/227 (5%) |
| Hepatic Fibrosis | 0.001862087 | 4.04 | 49/99 (49%) | 1/99 (1%) | 46/99 (46%) | 3/99 (3%) |
| **LXR/RXR Activation** | 0.004073803 | 3.25 | 62/123 (50%) | 0/123 (0%) | 61/123 (50%) | 0/123 (0%) |
| Renal Necrosis/Cell Death | 0.005011872 | 1.6 | 238/501 (48%) | 0/501 (0%) | 245/501 (49%) | 18/501 (4%) |
| LPS/IL-1 Mediated Inhibition of RXR Function | 0.010715193 | 1.99 | 107/251 (43%) | 0/251 (0%) | 110/251 (44%) | 34/251 (14%) |
| Positive Acute Phase Response Proteins | 0.010964782 | 6.67 | 11/30 (37%) | 0/30 (0%) | 19/30 (63%) | 0/30 (0%) |
| Liver Necrosis/Cell Death | 0.016218101 | 1.79 | 118/279 (42%) | 0/279 (0%) | 149/279 (53%) | 12/279 (4%) |
| **RAR Activation** | 0.018197009 | 2.11 | 87/190 (46%) | 0/190 (0%) | 99/190 (52%) | 4/190 (2%) |
| Increases Liver Damage | 0.019054607 | 2.8 | 46/107 (43%) | 0/107 (0%) | 58/107 (54%) | 3/107 (3%) |
| Fatty Acid Metabolism | 0.023988329 | 2.56 | 48/117 (41%) | 0/117 (0%) | 48/117 (41%) | 21/117 (18%) |
| Increases Liver Hepatitis | 0.030902954 | 3.85 | 22/52 (42%) | 0/52 (0%) | 29/52 (56%) | 1/52 (2%) |

**Table 11.** Toxicity-related pathways identified by Ingenuity in human keratinocytes treated with 20,23(OH)$_2$D3 for 24 h. Nuclear receptors are marked in bold.

| Ingenuity Toxicity Lists | *p*-Value | Overlap (%) | Downregulated | No Change | Upregulated | No Overlap with Dataset |
|---|---|---|---|---|---|---|
| **Aryl Hydrocarbon Receptor Signaling** | $2.6 \times 10^{-9}$ | 21.1 | 90/161 (56%) | 0/161 (0%) | 56/161 (35%) | 15/161 (9%) |
| Cholesterol Biosynthesis | $1.1 \times 10^{-8}$ | 62.5 | 13/16 (81%) | 0/16 (0%) | 3/16 (19%) | 0/16 (0%) |
| Renal Necrosis/Cell Death | $1.9 \times 10^{-7}$ | 13.2 | 228/501 (46%) | 0/501 (0%) | 255/501 (51%) | 18/501 (4%) |
| Primary Glomerulonephritis Biomarker Panel (Human) | $1.7 \times 10^{-6}$ | 63.6 | 5/11 (45%) | 0/11 (0%) | 6/11 (55%) | 0/11 (0%) |
| p53 Signaling | $2.6 \times 10^{-6}$ | 21.2 | 54/99 (55%) | 0/99 (0%) | 45/99 (45%) | 0/99 (0%) |
| **VDR/RXR Activation** | $1.5 \times 10^{-5}$ | 21.8 | 26/78 (33%) | 0/78 (0%) | 51/78 (65%) | 1/78 (1%) |
| Liver Proliferation | $3.0 \times 10^{-5}$ | 14.5 | 96/227 (42%) | 0/227 (0%) | 120/227 (53%) | 11/227 (5%) |
| Cardiac Hypertrophy | $5.1 \times 10^{-5}$ | 12.2 | 176/401 (44%) | 0/401 (0%) | 202/401 (50%) | 23/401 (6%) |
| Liver Necrosis/Cell Death | $7.8 \times 10^{-5}$ | 13.3 | 117/279 (42%) | 0/279 (0%) | 150/279 (54%) | 12/279 (4%) |
| Oxidative Stress | 0.000380189 | 21.1 | 39/57 (68%) | 0/57 (0%) | 17/57 (30%) | 1/57 (2%) |
| **Mechanism of Gene Regulation by Peroxisome Proliferators via PPARα** | 0.000645654 | 16.8 | 38/95 (40%) | 0/95 (0%) | 56/95 (59%) | 1/95 (1%) |
| Xenobiotic Metabolism Signaling | 0.000794328 | 11.6 | 152/336 (45%) | 0/336 (0%) | 142/336 (42%) | 42/336 (13%) |
| Increases Renal Proliferation | 0.002398833 | 13.9 | 68/137 (50%) | 0/137 (0%) | 62/137 (45%) | 7/137 (5%) |
| Fatty Acid Metabolism | 0.002398833 | 14.5 | 65/117 (56%) | 0/117 (0%) | 31/117 (26%) | 21/117 (18%) |

**Table 11.** *Cont.*

| Ingenuity Toxicity Lists | p-Value | Overlap (%) | Downregulated | No Change | Upregulated | No Overlap with Dataset |
|---|---|---|---|---|---|---|
| Increases Liver Steatosis | 0.003890451 | 15.7 | 32/83 (39%) | 0/83 (0%) | 48/83 (58%) | 3/83 (4%) |
| Decreases Depolarization of Mitochondria and Mitochondrial Membrane | 0.004570882 | 25 | 18/24 (75%) | 0/24 (0%) | 5/24 (21%) | 1/24 (4%) |
| Cell Cycle: G1/S Checkpoint Regulation | 0.004677351 | 16.7 | 35/66 (53%) | 0/66 (0%) | 28/66 (42%) | 3/66 (5%) |

Tables 12 and 13 show some similarities and differences with dermatological diseases and conditions; with cancer, organismal injury and abnormalities being the main diseases affected by 20,23$(OH)_2$D3 and 1,25$(OH)_2$D3. With regard to molecular and cellular functions, cellular growth and proliferation, cell death and survival, cellular movement and cell cycle were the major functions for 20,23$(OH)_2$D3, and cellular movement, cell signaling, small molecule biochemistry, lipid metabolism and cellular development for 1,25$(OH)_2$D3. Among the 25 networks activated by 20,23$(OH)_2$D3, the top five included: (1) connective tissue disorders, neurological diseases, organismal injuries and abnormalities, (2) RNA post-transcriptional modification, carbohydrate metabolism and lipid metabolism, (3) connective tissue, developmental, skeletal and muscular disorders, (4) cellular movement, endocrine system disorders, gastrointestinal diseases and (5) nucleic acid metabolism, small molecules biochemistry and dermatological diseases and conditions. Among the 15 networks activated by 1,25$(OH)_2$D3, the top five included: (1) cancer, organismal functions, organismal injuries and abnormalities, (2) cell-to-cell signaling and interaction, cellular assembly and organization, cellular development, (3) cellular growth and proliferation, tissue development and cancer, (4) molecular transport, carbohydrate and lipid metabolism and (5) protein degradation, protein synthesis, cellular assembly and organization.

**Table 12.** Categories of biological functions with diseases or function annotation activated by 1,25$(OH)_2$D3 in human epidermal keratinocytes after 24 of treatment.

| Categories of Biological Function | Diseases or Functions Annotation | p-Value | Activation z-Score | # of Genes |
|---|---|---|---|---|
| Cancer, Organismal Injury and Abnormalities | benign neoplasia | $1.0 \times 10^{-9}$ | 0.927 | 26 |
| Dermatological Diseases and Conditions | psoriasis | $2.6 \times 10^{-9}$ | | 20 |
| Cardiovascular System Development and Function, Cellular Movement | cell movement of endothelial cells | $8.8 \times 10^{-9}$ | 1.37 | 15 |
| Cancer, Cellular Movement, Organismal Injury and Abnormalities, Tumor Morphology | invasion of tumor cells | $1.2 \times 10^{-8}$ | 1.596 | 11 |
| Cell Signaling, Small Molecule Biochemistry | synthesis of nitric oxide | $1.9 \times 10^{-8}$ | −0.217 | 13 |
| Cardiovascular System Development and Function, Cellular Movement | homing of endothelial cells | $3.2 \times 10^{-8}$ | 1.597 | 7 |
| Cancer, Organismal Injury and Abnormalities, Tumor Morphology | invasion of tumor | $3.4 \times 10^{-8}$ | 1.63 | 12 |
| Lipid Metabolism, Small Molecule Biochemistry | metabolism of eicosanoid | $3.4 \times 10^{-8}$ | 2.747 | 12 |

**Table 12.** *Cont.*

| Categories of Biological Function | Diseases or Functions Annotation | *p*-Value | Activation z-Score | # of Genes |
|---|---|---|---|---|
| Cardiovascular Disease, Hematological Disease | Thrombosis | $5.9 \times 10^{-8}$ | −0.946 | 10 |
| Organismal Injury and Abnormalities | Fibrosis | $1.1 \times 10^{-7}$ | −1.401 | 17 |
| Lipid Metabolism, Small Molecule Biochemistry | metabolism of prostaglandin | $1.2 \times 10^{-7}$ | 2.589 | 10 |
| Immunological Disease | hypersensitive reaction | $1.5 \times 10^{-7}$ | 0.914 | 15 |
| Cardiovascular System Development and Function, Organismal Development | vasculogenesis | $1.6 \times 10^{-7}$ | 1.825 | 20 |
| Inflammatory Response | inflammation of organ | $1.8 \times 10^{-7}$ | −0.022 | 26 |
| Dermatological Diseases and Conditions, Inflammatory Disease, Inflammatory Response | Dermatitis | $1.8 \times 10^{-7}$ | −0.355 | 15 |
| Cardiovascular System Development and Function, Organismal Development | vascularization of hindlimb | $2.3 X 10^{-7}$ | 1.994 | 4 |
| Cellular Movement | homing | $3.4 \times 10^{-7}$ | 1.468 | 17 |
| Cancer, Cellular Development, Cellular Growth and Proliferation, Organismal Injury and Abnormalities, Tumor Morphology | proliferation of tumor cells | $5.4 \times 10^{-7}$ | −0.189 | 15 |
| Cancer, Organismal Injury and Abnormalities | growth of tumor | $6.0 \times 10^{-7}$ | 0.402 | 20 |
| Cellular Movement | invasion of cells | $6.3 \times 10^{-7}$ | 1.731 | 21 |

**Table 13.** Categories of biological functions with diseases or function annotation activated by 20,23$(OH)_2$D3 in human epidermal keratinocytes after 24 h of treatment.

| Categories of Biological Function | Diseases or Functions Annotation | *p*-Value | Activation z-Score | # of Genes |
|---|---|---|---|---|
| Cellular Growth and Proliferation | proliferation of cells | $7.0 \times 10^{-33}$ | −3.052 | 551 |
| Dermatological Diseases and Conditions | psoriasis | $4.6 \times 10^{-29}$ | | 142 |
| Cell Death and Survival | cell death | $1.2 \times 10^{-27}$ | 1.116 | 493 |
| Cell Death and Survival | necrosis | $1.9 \times 10^{-26}$ | 1.15 | 400 |
| Cell Death and Survival | apoptosis | $1.9 \times 10^{-25}$ | 0.619 | 405 |
| Cell Death and Survival | cell death of tumor cell lines | $7.2 \times 10^{-25}$ | 0.971 | 266 |
| Cell Death and Survival | apoptosis of tumor cell lines | $7.6 \times 10^{-23}$ | 0.672 | 219 |
| Cellular Movement | cell movement | $1.3 \times 10^{-22}$ | −0.187 | 334 |
| Cellular Movement | migration of cells | $6.7 \times 10^{-21}$ | −0.554 | 301 |
| Cancer, Organismal Injury and Abnormalities | abdominal neoplasm | $8.7 \times 10^{-20}$ | −1.733 | 1030 |
| Infectious Diseases | Viral Infection | $1.2 \times 10^{-19}$ | 0.737 | 261 |
| Cancer, Organismal Injury and Abnormalities | tumorigenesis of tissue | $6.5 \times 10^{-19}$ | −0.349 | 1047 |
| Cancer, Organismal Injury and Abnormalities | abdominal cancer | $2.3 \times 10^{-18}$ | −1.938 | 1014 |
| Cancer, Organismal Injury and Abnormalities | cancer | $4.2 \times 10^{-18}$ | 1.528 | 1215 |
| Cancer, Organismal Injury and Abnormalities | neoplasia of epithelial tissue | $2.3 \times 10^{-17}$ | −0.365 | 1026 |

**Table 13.** *Cont.*

| Categories of Biological Function | Diseases or Functions Annotation | p-Value | Activation z-Score | # of Genes |
|---|---|---|---|---|
| Cellular Development, Cellular Growth and Proliferation | proliferation of tumor cell lines | $3.0 \times 10^{-17}$ | −2.431 | 245 |
| Cancer, Organismal Injury and Abnormalities | benign neoplasia | $3.5 \times 10^{-17}$ | −0.029 | 165 |
| Cell Death and Survival | cell survival | $4.7 \times 10^{-17}$ | −0.537 | 225 |
| Cancer, Organismal Injury and Abnormalities | advanced stage solid tumor | $7.1 \times 10^{-17}$ | −0.397 | 106 |

Because of the unexpected finding that AhR signaling represented the top regulatory pathway activated by 20,23$(OH)_2$D3, and is validated by qPCR analysis of *CYP1A1* and *CYP1B1* genes expression (Figure 3D), we examined whether 20(OH)D3, which is the precursor to 20,23$(OH)_2$D3, and 17,20,23$(OH)_2$D3 and 1,20$(OH)_2$D3, which are downstream metabolites (see Figure 1), also affected the expression of genes linked to AhR in HaCaT keratinocytes. Figure 5A shows that 20(OH)D3 stimulated the expression of *CYP1A1* and *CYP1B1* in a dose-dependent fashion, with a stimulatory effect also seen for the *AhR* gene. 17,20,23$(OH)_3$D3 (1 μM) could also stimulate *CYPA1*, *CYP1B1* and *AhR* expression, while 1,20$(OH)_2$D3 had only a small effect on *CYPB1* and no effect on *CYP1A1* and *AhR*. Finally, we used a Human AhR Reporter Assay System (INDIGO, Biosciences) to analyze the effect of several D3-hydroxyderivatives on AhR-mediated transactivation. The kit contains AhR Reporter Cells that contain the luciferase reporter gene functionally linked to an AhR-responsive promoter, which provides a sensitive surrogate measure of the changes in AhR-mediated activation of luciferase reporter. Figure 6 shows that there was marked activation of AhR activity by 20,23$(OH)_2$D3 with weaker but significant activation by 20(OH)D3 or 1,25$(OH)_2$D3. Thus, the functional studies support the microarray analysis indicating that hydroxyderivatives of D3 can act on AhR. This finding can be explained by the promiscuous nature of AhR and its activity [85].

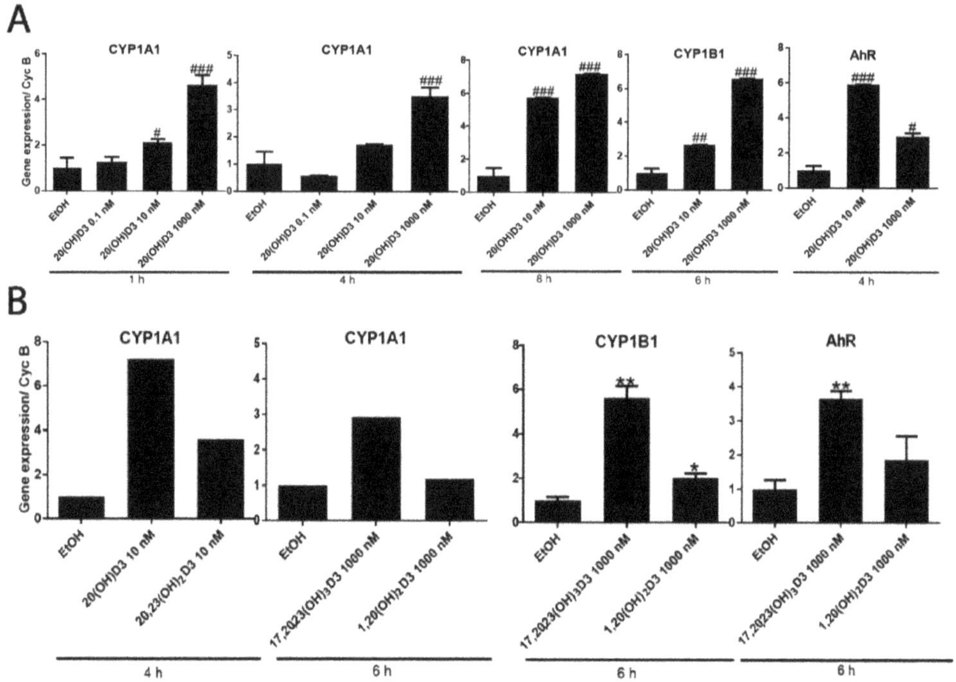


**Figure 5.** Changes in CYP1A1, CYP1B1 and AhR gene expression in HaCaT keratinocytes treated with vitamin D3 hydroxyderivatives as a function of the time of treatment. **A.** Dose-dependent effect of 20(OH)D3 on the gene expression. Data represent means ± SD ($n$ = 3) where $^{\#}$ $p < 0.05$, $^{\#\#}$ $p < 0.01$ and $^{\#\#\#}$ $p < 0.001$ at one-way ANOVA test. **B.** Effect of 1,25(OH)$_2$D3 and 20,23(OH)$_2$D3 on the gene expression as indicated. Data represent means ($n$ = 2) for CYP1A1, or means ± SD ($n$ = 3) for CYP1B1 and AhR where * $p < 0.05$, ** $p < 0.01$, at student $t$-test.

An additional mechanistic insight into the above interactions was provided by modeling using the crystal structure of the ligand-binding domain (LBD) of human AhR. The presently available crystal structure of the human AhR (PDB: 5NJ8) is missing the LBD region. A model of the human AhR LBD with bound 20S,23S(OH)$_2$D3 was developed as described under Methods. Briefly, the final model was based on the homology modelling template of C-terminal Per-ARNT-Sim domain of Hypoxia-Inducible Factor-2α, PDB entry code 3H82. The sequence identity between human AhR and the modelled sequence is 27%; the alignment is shown in Supplemental Figure S4. Short molecular dynamic simulation runs were performed on selected docked poses of 20S,23R/S(OH)$_2$D3 epimers in order to identify binding modes most favorably accommodated in the binding site of the homology model. The selected complex with 20S,23S(OH)$_2$D3 was simulated for 100 ns to allow for local structural adjustments of flexible regions to the presence of the vitamin D3 scaffold. The final conformer obtained is referred to as the 'refined AhR model'. Further ligand-induced effects were explored through a 250 ns simulation production run starting with this model. Over the first 130 ns the ligand-induced conformational changes were in the vicinity of F295 and S320. The latter is in a flexible region with two adjacent glycine residues while F295 is part of a loop structure 'covering' the binding pocket. The conformation adopted by 130 ns in these regions were maintained for the rest of the simulation time, likely stabilized by a hydrogen bonding network that formed, involving ligand hydroxyl groups, T289, S320 side chains and the backbone of F295 as shown in a representative simulation snapshot at 230 ns in Figure 7. Interactions of this network link the more rigid beta-sheet structure of the pocket containing T289 with two loop regions. The flexible 'belt' between G309-H326 includes a short helical segment near S320 that also shifted due to the presence of the ligand. This binding mode also changes the preferred orientation of H291 which by 130 ns simulation time forms a stable hydrogen bond with the backbone carbonyl of K292, an interaction not present in the initial or refined AhR models. Alanine mutation of T281, H285 of mouse AhR corresponding to the human residues T289 and H291 was shown to dramatically decrease Hsp90 binding [86]. Figure 7 illustrates that differences in the structural fold of AhR between the homology model, the refined AhR model and the simulation conformer are mainly within loops and the flexible 'belt' region.

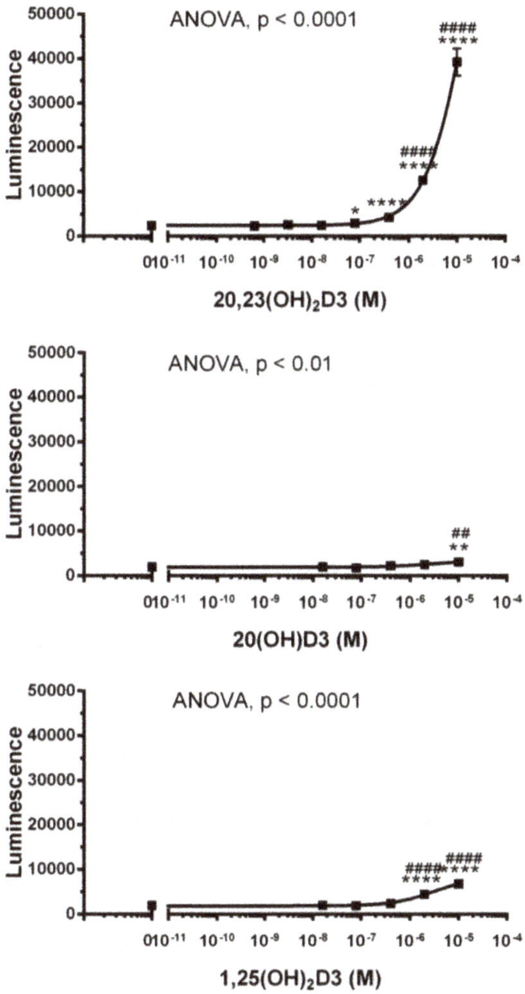


**Figure 6.** Stimulation of AhR activity by 20(OH)D3, 22,23(OH)$_2$D3, and 1,25(OH)$_2$D3. The assays for 22,23(OH)$_2$D3 were performed in quadruplicate, while for 20(OH)D3 and 1,25(OH)$_2$D3 in triplicate. Data represent means ± SD where * $p < 0.05$, ** $p < 0.01$ and **** $p < 0.0001$ at student $t$-test; ## $p < 0.01$ and #### $p < 0.0001$ at one-way ANOVA test.

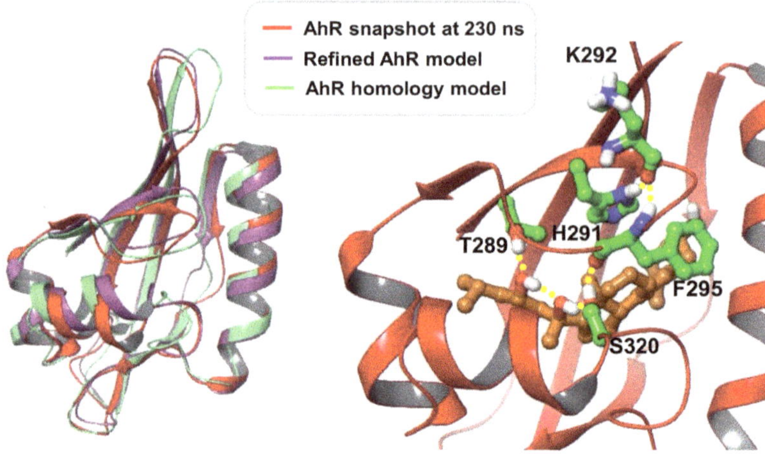


**Figure 7.** Structural fold of AhR models. To the left: Superimposed are the initial homology model, the refined AhR model and a molecular dynamic simulation snapshot at 230 ns. To the right: Close-up view of the simulation snapshot at 230 ns, displaying the ligand and AhR residues involved in an interaction network, as discussed in the text. 20*S*,23*S*$(OH)_2$D3 is shown with carbon atoms colored light brown, AhR residue carbons colored green; all other atoms are colored by atom type (O: red, N: blue, S: yellow). Hydrogen bonding interactions are indicated with yellow spheres.

The proposed binding model of 20S,23S$(OH)_2$D3 is shown in Figure 8A through a representative simulation conformer at 230 ns. Figure 8B shows the fraction of simulation time during which interactions are present with each AhR residue, as averaged over 130—250 ns. The most stable polar interactions are hydrogen bonding between 23-OH and T289 at 90% and between 3-OH and S336 at 85%, with S346 also contributing 26% of the simulation time. These interactions anchor the two end regions of the scaffold in the pocket. 20-OH is hydrogen bonding with S320 for 39% of the stimulation time. Due to intra-molecular hydrogen bonding between the ligand hydroxyls, the 20-OH group is positioned to act as a hydrogen bond donor to S320, which allows S320 to interact with F295. This interaction is likely important for these loop conformational changes and their effect on H291. Ligand–protein contacts versus simulation time are shown in Supplemental Figure S5.

Loop conformational changes induced by 20S,23S$(OH)_2$D3 are likely specific to this ligand. Therefore, for docking other vitamin D3 analogs the refined AhR model was utilized, applying the Induced Fit method. As shown in Table 14, Glide XP docking scores of three analogs are notably lower compared to other compounds within this set, 20(OH)D3, 1,25$(OH)_2$D3 and 1,20$(OH)_2$D3. Docked poses of all analogues are very similar, and also closely overlap with 20S,23S$(OH)_2$D3 in the refined AhR model. Docked poses for all analogs are displayed in Figure 9, along with the binding mode of 20S,23S$(OH)_2$D3 for comparison. Residues from Induced Fit structures contributing to polar interactions with ligands are shown only; all residues in proximity of docked ligands are included in Supplemental Figure S6. Docked Vitamin D3 analogs share similar hydrogen bonding interactions through hydroxyl groups: 1-OH interacts with S365, 3-OH with S336 and possibly S346, 17-OH and 20-OH with S320, 23-OH with T289. Docking results predict that 25-OH interacts with T289. A short, 20 ns molecular dynamic simulation was performed on 20(OH)D3, 1,25$(OH)_2$D3, 17,20,23S$(OH)_3$D3, starting with docked poses. The ligands maintained the binding mode and predicted interactions during simulation except for 1,25$(OH)_2$D3. Therefore, simulation of the latter was extended another 50 ns, during which the pose of 1,25$(OH)_2$D3 shifted, disrupting hydrogen bonding between 3-OH and S336 that was only present for 35% of simulation time. In comparison, in the case of 20(OH)D3 and 17,20,23S$(OH)_3$D3 the same interaction was present 85% and 66% of time, respectively. Due to mobility of the aliphatic chain in the binding site, 25-OH formed contacts with T289 30% and Y310 37% of the simulation time. 1,25$(OH)_2$D3 may have a distinct binding mode and interaction with AhR than the other analogs. Interactions of 17,20,23S$(OH)_3$D3 are analogous to those of 20*S*,23*S*$(OH)_2$D3. However, hydrogen bonding between 17-OH with S320 may interfere with structural changes such as those induced by 20*S*,23*S*$(OH)_2$D3 during the 250 ns simulation production run. While 20(OH)D3 is also predicted to form analogous contacts, the absence of 23(OH) interactions is likely significant.

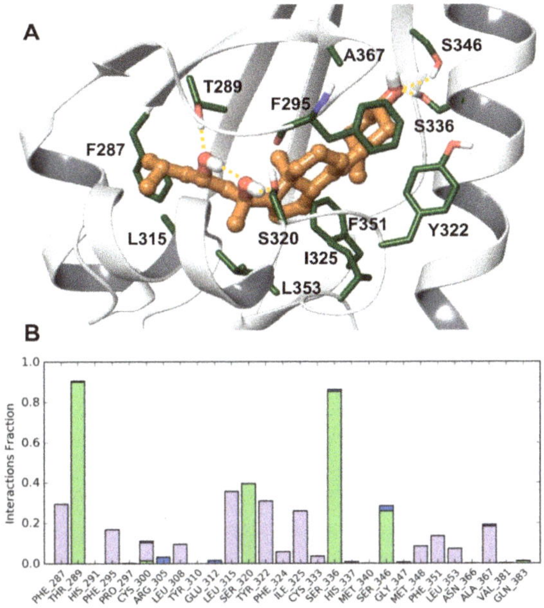


**Figure 8.** (**A**) The proposed binding model 20*S*,23*S*$(OH)_2$D3 at human AhR. (**A**) Representative simulation snapshot at 230 ns. Shown residues contribute to the binding of the ligand over simulation time. (**B**) Fraction of simulation time during which interactions are present with each AhR residue, averaged over 130–250 ns of the simulation production run.

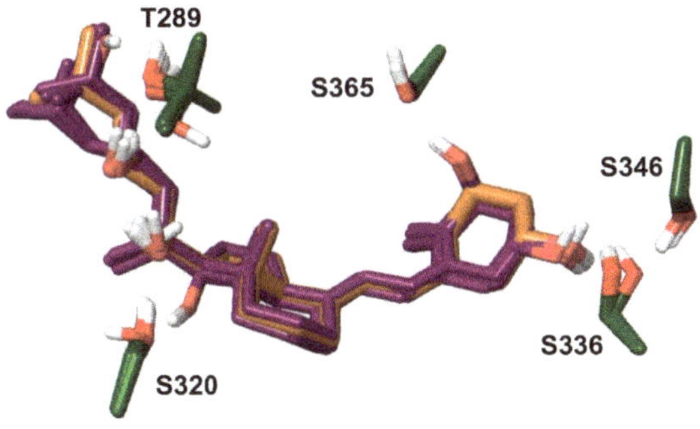


**Figure 9.** Induced Fit docked vitamin D3 analogs displayed simultaneously. The pose of 20*S*,23*S*$(OH)_2$D3 from the refined AhR model is shown for comparison, with carbon atoms colored light brown. Only AhR residues involved in polar interactions are shown.

**Table 14.** Glide XP scores of vitamin D3 analogs docked into the refined human AhR LBD model.

| Compound | Score | Compound | Score |
|---|---|---|---|
| 20*S*OHD3 | −13.3 | 1,20*S*,23*S*(OH)$_3$D3 | −16.1 |
| 1,25(OH)$_2$D3 | −13.1 | 1,20*S*,23*R*(OH)$_3$D3 | −16.4 |
| 20*S*,23*S*(OH)$_2$D3 | −15.1 | 17,20*S*,23*S*(OH)$_3$D3 | −14.9 |
| 20*S*,23*R*(OH)$_2$D3 | −15.7 | 17,20*S*,23*R*(OH)$_3$D3 | −15.4 |
| 1,20*S*(OH)$_2$D3 | −13.0 | | |

*Modelling Conclusions*

Molecular dynamic simulation of the developed AhR-20*S*,23*S*(OH)$_2$D3 model predicts strong hydrogen bonding interactions between this ligand and T289, S336. A hydrogen bond formed with S320 is also well maintained during simulation. A number of AhR residues have favorable non-polar contacts with the ligand (Figure 8). The simulation trajectory predicts that ligand-specific interactions induce a conformational change in the region in the vicinity of S320 and F295, also leading to a distinct position and interaction of H291. The interaction network that forms during simulation due to the ligand links the beta-sheet structure of the pocket with two loops, restraining the conformation of flexible regions in the binding site. The presented model is also consistent with the observed effect of 20*S*,23*S*(OH)$_2$D3 on AhR since, in particular, T289 and H291 are essential residues for Hsp90 binding.

Docking of a set of D3 analogs predicts ligand binding modes close to that of 20*S*,23*S*(OH)$_2$D3, as well as analogous interactions with AhR. Short simulation runs of docked poses of 20(OH)D3 and 17,20,23*S*(OH)$_3$D3 predict stability of the starting ligand poses. While forming interactions analogous to those of 20*S*,23*S*(OH)$_2$D3, these two analogs lack features that contribute to the induced effects of 20*S*,23*S*(OH)$_2$D3 during simulation. The 20(OH)D3 analog lacks hydrogen bonding through 23-OH and in the case of 17,20,23*S*(OH)$_3$D3 the 17-OH group may interfere with the interactions between S320 and the F295 backbone. Stability of the docked pose of 1,25(OH)$_2$D3 was also explored through molecular dynamic simulation. Shifting and fluctuations of 1,25(OH)$_2$D3 over simulation time suggests that this ligand would not adopt a binding mode close to that of 20*S*,23*S*(OH)$_2$D3 in the AhR binding site. Thus, modelling predictions are consistent with the distinct effects of these D3 analogs on AhR.

## 3. Materials and Methods

### 3.1. Materials

Vitamin D3 (D3) and 1,25(OH)$_2$D3 were purchased from Sigma-Aldrich (St. Louis, MO, USA). 20,23(OH)$_2$D3 was produced by hydroxylation of D3 by CYP11A, extracted with dichloromethane and purified as described in References [14,66]. 20*S*-Hydroxyvitamin D3 (20(OH)D3, 1α,20*S*-dihydroxyvitamin D3 (1,20(OH)$_2$D3) and 17,20S,23*S*-trihydroxyvitamin D3 (17,20,23(OH)$_3$D3) were also synthesized using CYP11A1 as described before [66,87]. An extinction coefficient of 18,000 $M^{-1}$ $cm^{-1}$ at 263 nm was used to quantify concentrations of 20,23(OH)$_2$D3 [88] and the secosteroids were divided, dried and stored at −80 °C until use. Secosteroids were dissolved in ethanol prior to experiments to obtain stock solutions of $10^{-4}$ M.

The structures of the secosteroids tested and the routes of enzymatic synthesis that include C25 and C1 hydroxylation for 1,25(H)$_2$D3 [5], and the sequential hydroxylation of the D3 side chain by CYP11A1 producing 20,23(OH)$_2$D3 and 17,20.23(OH)$_3$D3 [14,66], are shown in Figure 1.

### 3.2. Cell Culture

Neonatal foreskins of African American [79] donors were used to isolate neonatal human epidermal keratinocytes (HEK) following standard protocols described previously [69,89]. The use of human tissues were approved both by the IRB at the UTHSC as an exempt protocol #4 and by the IRB at the University of Alabama Birmingham, as they are not subject to FDA regulation and not Human Subject Research. Cells were grown in keratinocyte basal medium (KBM) supplemented with keratinocyte growth factors (KGF) (Lonza, Walkersville, MD, USA) on collagen coated plates [68] and second and third passages were used for the experiments [69]. Human epidermal HaCaT keratinocytes were cultured in Dulbecco's Modified Eagle Medium (DMEM) supplemented with glucose, L-glutamine, pyridoxine hydrochloride (Cell Grow), 5% fetal bovine serum (FBS) (Atlanta Biologicals, Flowery Branch, GA, USA) and 1% penicillin/streptomycin/amphotericin antibiotic solution (Thermo Fisher Scientific, Waltham, MA, USA). Human cells were cultured at 37 °C, with a $CO_2$ concentration of 5%, 100% humidity, and media were changed every second and/or third day.

Prior to treatment with secosteroids, HaCaT cells were serum deprived for 24 h and the medium was changed to DMEM medium containing 5% charcoal-treated FBS (ctFBS) (Atlanta Biologicals, Flowery Branch, GA, USA) to which D3 hydroxymetabolites from the stock solutions were added. For epidermal neonatal keratinocytes, the KBM with KGF was supplemented with 0.5% bovine serum albumin (BSA) prior to the addition of D3 derivatives.

### 3.3. Microarray Assays

Petri dishes (100 mm in diameter) were seeded with human neonatal keratinocytes that were combined from five different black donors at either passage 2 or 3. After reaching 70–80% of confluence, cells were treated with $10^{-7}$ M of either 20,23(OH)$_2$D3 or 1,25(OH)$_2$D3, or with 0.1% ethanol (EtOH) as a solvent control for 6 or 24 h. After, these cells were isolated from three plates per each experimental condition and combined for passage 2 and 3, separately (Figure 2).

The RNA from HEK treated with either 20,23(OH)$_2$D3 or 1,25(OH)$_2$D3, or 0.1% ethanol control, was isolated using the Absolutely RNA Miniprep Kit (Qiagen, Germantown, MD, USA). High purity RNA samples were subjected to microarray analysis at the Molecular Resources Center at the UTHSC. Expression profiling was accomplished using whole-genome gene expression direct hybridization assay using Illumina's HumanWG-6_V2 (Platform GPL13376) chip/array (Illumina, San Diego, CA, USA). Each array contains full-length 50-mer probes representing more than 22,000 well-annotated RefSeq transcripts, including up-to-date genes derived from the National Center for Biotechnology Information Reference Sequence (NCBI RefSeq) database. Initially, 250 ng total RNA was converted to cDNA, followed by an in vitro transcription step to generate labeled cRNA following the manufacturer's recommendations (Applied Biosystems, Foster City, CA, USA).

The labeled probes were then mixed with hybridization reagents and hybridized overnight to the Human BeadChips. Following washing and staining, the BeadChips were imaged using the Illumina BeadArray Reader to measure fluorescence intensity at each probe. The intensity of the signal corresponds to the quantity of the respective mRNA in the original sample.

### 3.4. Bioinformatics Analysis

For generating networks, a data set containing gene identifiers and corresponding expression values was uploaded into the application. Each identifier was mapped to its corresponding object in Ingenuity's Knowledge Base. A FC of ±2 or ±1.5, where indicated, was set to identify molecules whose expression was significantly differentially regulated. These molecules, called Network Eligible molecules, were overlaid onto a global molecular network developed from information contained in Ingenuity's Knowledge Base. Networks of Network Eligible Molecules were then algorithmically generated based on their connectivity. The Functional Analysis identified the biological functions and/or diseases that were most significant to the entire data set. Molecules from the dataset that met the FC cutoff of ±2 or ±1.5 and were associated with biological functions and/or diseases in Ingenuity's Knowledge Base were considered for the analysis. Right-tailed Fisher's exact test was used to calculate a p-value determining the probability that each biological function and/or disease assigned to that data set is due to chance alone.

### 3.5. Real-Time Reverse Transcription Polymerase Chain Reaction (qRT-PCR)

Semiconfluent cultures of human neonatal keratinocytes or HaCaT cells were treated for 6 h, 24 h or as indicated in the figure legends with vitamin D3 hydroxyderivatives or ethanol, and RNA isolated as described above. Reverse transcription was done using the Transcriptor First Strand cDNA Synthesis Kit (Applied Biosystems, Foster City, CA, USA) with 100 ng RNA per reaction. qRT-PCR was performed using cDNA diluted 10-fold in sterile water and a TaqMan PCR Master Mix. Reactions (in triplicate) were performed at 50 °C for 2 min, 95 °C for 10 min and then 50 cycles of 95 °C for 15 s and 60 °C for 1 min. The primers and probes were designed with the universal probe library (Roche). Data were collected on a Roche Light Cycler 480. The amount of amplified product for each gene was compared to that of Cyclophilin B or GAPDH using a comparative $C_T$ ($\Delta\Delta C_T$) method. Supplemental Table S1 lists the primers used for qRT-PCR amplifications.

### 3.6. Interaction of Hydroxyvitamin D Derivatives with AhR

Interaction of 20(OH)D3, 20,23$(OH)_2$D3 and 1,25$(OH)_2$D3 with AhR was evaluated using the Human AhR Reporter Assay System (INDIGO Biosciences, State College, PA, USA) according to the manufacture's protocol. Briefly, AhR reporter cells were recovered on a 96-well plate frame using the cell recovery medium for 5 h, followed by treatment with vitamin D3 hydroxyderivatives in the compound screening medium for 22 h. After removing the media from the wells, luciferase detection reagent was added to the wells and luminescence was measured using a Cytation 5 Cell Imaging Multi-Mode Reader (Winooski, VT, USA).

### 3.7. Statistical Analyses

Data are presented as means ± SD ($n$ = 3–4), and were analyzed with a Student's $t$-test (for two groups) or ANOVA using Prism 4.00 (GraphPad Software, San Diego, CA, USA). Statistically significant differences are denoted with asterisks for $t$-tests or for one way ANOVA with # as indicated in the figure legends.

### 3.8. Data Deposition

The data reported in this paper have been deposited in the Gene Expression Omnibus (GEO) database, https://www.ncbi.nlm.nih.gov/geo (accession no. GEO: GSE117351).

### 3.9. Development of a Human AhR LBD Model Complexed with 20S,23S$(OH)_2$D3

The strategy applied utilized tools implemented in the Schrödinger software package, version 2017-4 (Schrödinger, LLC, New York, NY, USA). Homology modelling of the human AhR ligand-binding domain was based on crystal structures of the C-terminal Per-ARNT-Sim domain of Hypoxia-Inducible Factor-2α (HIF-2α PAS-B), PDB entry codes 3H82 and 4XT2. Based on the two templates, two homology models were built using the energy-based homology model building method in Schrödinger. The sequence identity is 27% between human AhR and template sequences in the modelled LBD region. The sequence alignment is shown in Supplemental Figure S4. Residues were numbered according to the human AhR sequence (Uniprot ID P35869). Co-crystallized ligands were included. Modeled loops that contained gaps in the sequence alignment were refined through default loop refinement options. The models were relaxed through restrained energy minimization in Protein Preparation Wizard (OPLS3 force field).

Initial binding mode hypotheses were generated through docking 20*S*,23*R*/*S*$(OH)_2$D3 into the two obtained AhR models. Out of the top scoring poses at both AhR models, eight were selected for protein-ligand complex refinement (Prime tool in Schrödinger software), followed by a 10 ns molecular dynamic simulation run for each complex using Desmond. Four poses induced distortions in rigid, AhR beta-sheet/helical backbone structures within 10 ns simulation and were not considered further. Out of the remaining poses, the most favorable contacts were formed by two similar poses of the epimers: 20*S*,23*S*/*R*$(OH)_2$D3. Simulation of these poses was extended to 20 ns, which suggested that 20*S*,23*S*$(OH)_2$D3 is more favorably accommodated than its *R* epimer. The 20*S*,23*S*$(OH)_2$D3 complex conformer at 20 ns showed an overall RMSD (root-mean-square deviation) of 1.48 from its homology modelling template (PDB: 3H82). In order to allow flexible regions to adjust to the presence of the bound vitamin D3 scaffold, the model was further simulated for 100 ns or for 230 or 25 ns as indicated. The final conformer was relaxed through restrained energy minimization and is referred to as the 'refined AhR model'. The overall RMSD of this model from its homology modelling template (PDB: 3H82) is 1.58, suggesting stability of the AhR structure over simulation time. The model structure contains only two residues with backbone dihedrals in disallowed regions, both of which are glycines: Gly309 and Gly374.

### *3.10. Docking Method*

The induced Fit docking method was used as implemented in Schrödinger, version 2017-4 (Schrödinger, LLC, New York, NY, USA). This method combines Prime tools and Glide docking, taking into account the flexibility of residues in proximity to the ligand. 20,23$(OH)_2$D3 was Induced Fit docked into the two human AhR homology models using default parameters except optimization of side chains was extended to 6 Å around the ligand and Glide re-docking was done in extra-precision mode. Docking of vitamin D3 analogs into the refined AhR model also utilized Induced Fit with default options, except that van der Waals scaling parameters for docking were set to 1.0 for both protein and ligand (no scaling), and re-docking was done in extra-precision mode.

### *3.11. Molecular Dynamic Simulation Method*

Molecular dynamic simulations were performed using Desmond (Schrödinger, LLC, New York, NY, USA) with the OPLS3 force field. Structures were solved in TIP3P explicit waters with boundary conditions in a 10 Å buffered orthorhombic system. Counter-ions were added. The NPT ensemble was employed with temperature fixed at 300 K and pressure at 1.01 bar. The cutoff radius for Coulombic interactions was set to 10 Å. The trajectory was recorded at 10 ps intervals.

## 4. Conclusions

Gene expression profile analysis demonstrated that 20,23$(OH)_2$D3 and 1,25$(OH)_2$D3 induce distinct and overlapping gene expression patterns in keratinocytes linked to the activation of common (VDR-dependent) and distinct (involving other nuclear receptors) signal transduction pathways. Taking into consideration the strong chemical similarity between 20,23$(OH)_2$D3 and 1,25$(OH)_2$D3 (Figure 1), the marked differences in gene expression panels (Table 1, Figures 3 and 4) were unexpected. This is because our previous studies predominantly showed phenotypic similarities between the effects of both dihydroxy-vitamin D3 species, such as regulation of cell proliferation and differentiation, and anti-inflammatory, photoprotective and anticancer functions, with only a couple of notable differences where 20,23$(OH)_2$D3 displayed no calcemic effects and poor activation of CYP24A1. These predominantly overlapping effects are most likely secondary to the redundancy of downstream phenotypic regulators and intercommunication between distinct transduction pathways at the cell or organ levels. The similarities are likely related to the activation of the VDR. The most significant and unexpected finding is the identification of AhR as the major receptor for 20,23$(OH)_2$D3, which also appears to be activated by other CYP11A1-derived vitamin D3 derivatives, and possibly by 1,25$(OH)_2$D3 to some extent, as predicted by molecular modeling. The future challenge is to precisely define the interaction of different vitamin D3 hydroxyderivatives with the ligand binding domain of AhR and how it is affected by the location of the OH-group on the side chain or at C1$\alpha$, and how the activation of downstream signal transduction pathways occurs.

**Supplementary Materials:** Supplementary materials can be found at http://www.mdpi.com/1422-0067/19/10/3072/s1.

**Author Contributions:** A.T.S. designed experiments, analyzed data and wrote the manuscript. T.-K.K. performed the main experiments, took part in data analysis and preparation of the manuscript. Z.J. performed the main experiments, took part in data analysis and preparation of the manuscript. A.A.B. performed experiments, took part in data analysis and preparation of the manuscript. M.A.Ż. performed the experiments, took part in data analysis and preparation of the manuscript. H.X. took part in data analysis and preparation of the manuscript. T.R.S. performed bioinformatics analysis and took part in preparation of the manuscript. R.C.T. took part in data analysis and preparation of the manuscript. A.M.J. took part in data analysis and preparation of the manuscript. D.K.C. performed bioinformatics analysis and took part in preparation of the manuscript.

**Funding:** The studies were supported by grants R21AR066505, 1R01AR073004-01A1 and 1 RO1 AR071189-01A1 from NIH and VA Merit award 1I01BX004293-01A1 to ATS, grant R01ES017014 to TRS, N402 662840 from the Polish Ministry of Science and Higher Education to MAZ and internal funds from the University of Western Australia, Australia and the University of Memphis Feinstone Center for Genomic Research.

**Acknowledgments:** We thank Quynh Tran for initial normalization of the microarray data.

**Conflicts of Interest:** The authors declare no conflict of interest.

## Abbreviations

| | |
|---|---|
| 1,20(OH)$_2$D3 | 1,20-dihyhydroxyvitamin D3 |
| 17,20,23(OH)$_2$D3 | 17,20,23-trihyhydroxyvitamin D3 |
| 1$\alpha$,25(OH)$_2$D3 | 1$\alpha$,25-dihydroxyvitamin D3 |
| 20(OH)D3 | 20-hydroxyvitamin D3 |
| 20,23(OH)$_2$D3 | 20,23-dihyhydroxyvitamin D3 |
| 25(OH)D3 | 25-hydroxyvitamin D3 |
| 7-DHC | 7-dehydrocholesterol |
| AA | African-American |
| AhR | aryl hydrocarbon receptor |
| GR | glucocorticoid receptor |
| LBD | ligand binding domain |
| LXR | liver X receptor |
| PPAR | peroxisome proliferator-activated receptor |
| RAR | retinoic acid receptor |
| RXR | retinoid X receptor |
| VDR | vitamin D receptor |
| UVB | ultraviolet B radiation |

## References

1. Holick, M.F. Vitamin D: A millenium perspective. *J. Cell. Biochem.* **2003**, *88*, 296–307. [CrossRef] [PubMed]
2. Wacker, M.; Holick, M.F. Sunlight and vitamin D: A global perspective for health. *Dermato-Endocrinology* **2013**, *5*, 51–108. [CrossRef] [PubMed]
3. Holick, M.F. Vitamin D deficiency. *N. Engl. J. Med.* **2007**, *357*, 266–281. [CrossRef] [PubMed]
4. Zhu, J.; DeLuca, H.F. Vitamin d 25-hydroxylase—Four decades of searching, are we there yet? *Arch. Biochem. Biophys.* **2012**, *523*, 30–36. [CrossRef] [PubMed]
5. Jones, G.; Prosser, D.E.; Kaufmann, M. Cytochrome P450-mediated metabolism of vitamin D. *J. Lipid Res.* **2014**, *55*, 13–31. [CrossRef] [PubMed]
6. Christakos, S.; Dhawan, P.; Verstuyf, A.; Verlinden, L.; Carmeliet, G. Vitamin D: Metabolism, molecular mechanism of action, and pleiotropic effects. *Physiol. Rev.* **2016**, *96*, 365–408. [CrossRef] [PubMed]
7. Adams, J.S.; Hewison, M. Extrarenal expression of the 25-hydroxyvitamin D-1-hydroxylase. *Arch. Biochem. Biophys.* **2012**, *523*, 95–102. [CrossRef] [PubMed]
8. Slominski, A.T.; Li, W.; Kim, T.K.; Semak, I.; Wang, J.; Zjawiony, J.K.; Tuckey, R.C. Novel activities of CYP11A1 and their potential physiological significance. *J. Steroid Biochem. Mol. Biol.* **2015**, *151*, 25–37. [CrossRef] [PubMed]
9. Tuckey, R.C. Progesterone synthesis by the human placenta. *Placenta* **2005**, *26*, 273–281. [CrossRef] [PubMed]
10. Miller, W.L.; Auchus, R.J. The molecular biology, biochemistry, and physiology of human steroidogenesis and its disorders. *Endocr. Rev.* **2011**, *32*, 81–151. [CrossRef] [PubMed]

11. Guryev, O.; Carvalho, R.A.; Usanov, S.; Gilep, A.; Estabrook, R.W. A pathway for the metabolism of vitamin D3: Unique hydroxylated metabolites formed during catalysis with cytochrome P450scc (CYP11A1). *Proc. Natl. Acad. Sci. USA* **2003**, *100*, 14754–14759. [CrossRef] [PubMed]
12. Slominski, A.; Zjawiony, J.; Wortsman, J.; Semak, I.; Stewart, J.; Pisarchik, A.; Sweatman, T.; Marcos, J.; Dunbar, C.; R, C.T. A novel pathway for sequential transformation of 7-dehydrocholesterol and expression of the P450scc system in mammalian skin. *Eur. J. Biochem.* **2004**, *271*, 4178–4188. [CrossRef] [PubMed]
13. Slominski, A.; Semak, I.; Zjawiony, J.; Wortsman, J.; Li, W.; Szczesniewski, A.; Tuckey, R.C. The cytochrome P450scc system opens an alternate pathway of vitamin D3 metabolism. *FEBS J.* **2005**, *272*, 4080–4090. [CrossRef] [PubMed]
14. Tuckey, R.C.; Li, W.; Zjawiony, J.K.; Zmijewski, M.A.; Nguyen, M.N.; Sweatman, T.; Miller, D.; Slominski, A. Pathways and products for the metabolism of vitamin D3 by cytochrome P450scc. *FEBS J.* **2008**, *275*, 2585–2596. [CrossRef] [PubMed]
15. Slominski, A.; Semak, I.; Wortsman, J.; Zjawiony, J.; Li, W.; Zbytek, B.; Tuckey, R.C. An alternative pathway of vitamin D metabolism. Cytochrome P450scc (CYP11A1)-mediated conversion to 20-hydroxyvitamin D2 and 17,20-dihydroxyvitamin D2. *FEBS J.* **2006**, *273*, 2891–2901. [CrossRef] [PubMed]
16. Nguyen, M.N.; Slominski, A.; Li, W.; Ng, Y.R.; Tuckey, R.C. Metabolism of vitamin D2 to 17,20,24-trihydroxyvitamin D2 by cytochrome P450scc (CYP11A1). *Drug Metab. Dispos.* **2009**, *37*, 761–767. [CrossRef] [PubMed]
17. Lin, Z.; Marepally, S.R.; Ma, D.; Myers, L.K.; Postlethwaite, A.E.; Tuckey, R.C.; Cheng, C.Y.; Kim, T.K.; Yue, J.; Slominski, A.T.; et al. Chemical synthesis and biological activities of 20s,24s/r-dihydroxyvitamin D3 epimers and their 1alpha-hydroxyl derivatives. *J. Med. Chem.* **2015**, *58*, 7881–7887. [CrossRef] [PubMed]
18. Slominski, A.T.; Kim, T.K.; Shehabi, H.Z.; Semak, I.; Tang, E.K.; Nguyen, M.N.; Benson, H.A.; Korik, E.; Janjetovic, Z.; Chen, J.; et al. In vivo evidence for a novel pathway of vitamin D(3) metabolism initiated by P450scc and modified by CYP27B1. *FASEB J.* **2012**, *26*, 3901–3915. [CrossRef] [PubMed]
19. Slominski, A.T.; Kim, T.K.; Shehabi, H.Z.; Tang, E.K.; Benson, H.A.; Semak, I.; Lin, Z.; Yates, C.R.; Wang, J.; Li, W.; et al. In vivo production of novel vitamin D2 hydroxy-derivatives by human placentas, epidermal keratinocytes, Caco-2 colon cells and the adrenal gland. *Mol. Cell. Endocrinol.* **2014**, *383*, 181–192. [CrossRef] [PubMed]
20. Slominski, A.T.; Kim, T.K.; Li, W.; Tuckey, R.C. Classical and non-classical metabolic transformation of vitamin D in dermal fibroblasts. *Exp. Dermatol.* **2016**, *25*, 231–232. [CrossRef] [PubMed]
21. Slominski, A.T.; Zmijewski, M.A.; Semak, I.; Sweatman, T.; Janjetovic, Z.; Li, W.; Zjawiony, J.K.; Tuckey, R.C. Sequential metabolism of 7-dehydrocholesterol to steroidal 5,7-dienes in adrenal glands and its biological implication in the skin. *PLoS ONE* **2009**, *4*, e4309. [CrossRef] [PubMed]
22. Slominski, A.T.; Kim, T.K.; Chen, J.; Nguyen, M.N.; Li, W.; Yates, C.R.; Sweatman, T.; Janjetovic, Z.; Tuckey, R.C. Cytochrome P450scc-dependent metabolism of 7-dehydrocholesterol in placenta and epidermal keratinocytes. *Int. J. Biochem. Cell Biol.* **2012**, *44*, 2003–2018. [CrossRef] [PubMed]
23. Slominski, A.T.; Janjetovic, Z.; Kim, T.K.; Wasilewski, P.; Rosas, S.; Hanna, S.; Sayre, R.M.; Dowdy, J.C.; Li, W.; Tuckey, R.C. Novel non-calcemic secosteroids that are produced by human epidermal keratinocytes protect against solar radiation. *J. Steroid Biochem. Mol. Biol.* **2015**, *148*, 52–63. [CrossRef] [PubMed]
24. Slominski, A.T.; Kim, T.K.; Li, W.; Postlethwaite, A.; Tieu, E.W.; Tang, E.K.; Tuckey, R.C. Detection of novel CYP11A1-derived secosteroids in the human epidermis and serum and pig adrenal gland. *Sci. Rep.* **2015**, *5*, 14875. [CrossRef] [PubMed]
25. DeLuca, H.F. Overview of general physiologic features and functions of vitamin D. *Am. J. Clin. Nutr.* **2004**, *80*, 1689S–1696S. [CrossRef] [PubMed]
26. Holick, M.F. Vitamin D and bone health. *J. Nutr.* **1996**, *126*, 1159S–1164S. [CrossRef] [PubMed]
27. Bikle, D.D. Vitamin D: An ancient hormone. *Exp. Dermatol.* **2011**, *20*, 7–13. [CrossRef] [PubMed]
28. Pike, J.W.; Christakos, S. Biology and mechanisms of action of the vitamin D hormone. *Endocrinol. Metab. Clin. N. Am.* **2017**, *46*, 815–843. [CrossRef] [PubMed]
29. Van Driel, M.; van Leeuwen, J. Vitamin D endocrinology of bone mineralization. *Mol. Cell. Endocrinol.* **2017**, *453*, 46–51. [CrossRef] [PubMed]
30. Hewison, M. An update on vitamin D and human immunity. *Clin. Endocrinol.* **2012**, *76*, 315–325. [CrossRef] [PubMed]

31. Vanherwegen, A.S.; Gysemans, C.; Mathieu, C. Regulation of immune function by vitamin d and its use in diseases of immunity. *Endocrinol. Metab. Clin. N. Am.* **2017**, *46*, 1061–1094. [CrossRef] [PubMed]
32. Berridge, M.J. Vitamin D deficiency: Infertility and neurodevelopmental diseases (attention deficit hyperactivity disorder, autism, and schizophrenia). *Am. J. Physiol. Cell Physiol.* **2018**, *314*, C135–C151. [CrossRef] [PubMed]
33. Liu, N.Q.; Hewison, M. Vitamin d, the placenta and pregnancy. *Arch. Biochem. Biophys.* **2012**, *523*, 37–47. [CrossRef] [PubMed]
34. Ganguly, A.; Tamblyn, J.A.; Finn-Sell, S.; Chan, S.Y.; Westwood, M.; Gupta, J.; Kilby, M.D.; Gross, S.R.; Hewison, M. Vitamin D, the placenta and early pregnancy: Effects on trophoblast function. *J. Endocrinol.* **2018**, *236*, R93–R103. [CrossRef] [PubMed]
35. Shin, J.S.; Choi, M.Y.; Longtine, M.S.; Nelson, D.M. Vitamin d effects on pregnancy and the placenta. *Placenta* **2010**, *31*, 1027–1034. [CrossRef] [PubMed]
36. Knabl, J.; Vattai, A.; Ye, Y.; Jueckstock, J.; Hutter, S.; Kainer, F.; Mahner, S.; Jeschke, U. Role of placental VDR expression and function in common late pregnancy disorders. *Int. J. Mol. Sci.* **2017**, *18*, 2340. [CrossRef] [PubMed]
37. Lorenzen, M.; Boisen, I.M.; Mortensen, L.J.; Lanske, B.; Juul, A.; Blomberg Jensen, M. Reproductive endocrinology of vitamin D. *Mol. Cell. Endocrinol.* **2017**, *453*, 103–112. [CrossRef] [PubMed]
38. Cui, X.; Gooch, H.; Petty, A.; McGrath, J.J.; Eyles, D. Vitamin D and the brain: Genomic and non-genomic actions. *Mol. Cell. Endocrinol.* **2017**, *453*, 131–143. [CrossRef] [PubMed]
39. Di Somma, C.; Scarano, E.; Barrea, L.; Zhukouskaya, V.V.; Savastano, S.; Mele, C.; Scacchi, M.; Aimaretti, G.; Colao, A.; Marzullo, P. Vitamin D and neurological diseases: An endocrine view. *Int. J. Mol. Sci.* **2017**, *18*, 2482. [CrossRef] [PubMed]
40. Welsh, J. Function of the vitamin D endocrine system in mammary gland and breast cancer. *Mol. Cell. Endocrinol.* **2017**, *453*, 88–95. [CrossRef] [PubMed]
41. Angellotti, E.; Pittas, A.G. The role of vitamin d in the prevention of type 2 diabetes: To D or not to D? *Endocrinology* **2017**, *158*, 2013–2021. [CrossRef] [PubMed]
42. Dimova, R.; Tankova, T.; Chakarova, N. Vitamin d in the spectrum of prediabetes and cardiovascular autonomic dysfunction. *J. Nutr.* **2017**, *147*, 1607–1615. [CrossRef] [PubMed]
43. Berridge, M.J. Vitamin D deficiency and diabetes. *Biochem. J.* **2017**, *474*, 1321–1332. [CrossRef] [PubMed]
44. Kassi, E.; Nasiri-Ansari, N.; Papavassiliou, A.G. Vitamin d affects glucocorticoid action in target cells. *Oncotarget* **2017**, *8*, 7220–7221. [CrossRef] [PubMed]
45. Plum, L.A.; DeLuca, H.F. Vitamin D, disease and therapeutic opportunities. *Nat. Rev. Drug Discov.* **2010**, *9*, 941–955. [CrossRef] [PubMed]
46. Feldman, D.; Krishnan, A.V.; Swami, S.; Giovannucci, E.; Feldman, B.J. The role of vitamin d in reducing cancer risk and progression. *Nat. Rev. Cancer* **2014**, *14*, 342–357. [CrossRef] [PubMed]
47. Fleet, J.C.; DeSmet, M.; Johnson, R.; Li, Y. Vitamin d and cancer: A review of molecular mechanisms. *Biochem. J.* **2012**, *441*, 61–76. [CrossRef] [PubMed]
48. Masuda, S.; Jones, G. Promise of vitamin d analogues in the treatment of hyperproliferative conditions. *Mol. Cancer Ther.* **2006**, *5*, 797–808. [CrossRef] [PubMed]
49. Slominski, A.T.; Brozyna, A.A.; Skobowiat, C.; Zmijewski, M.A.; Kim, T.K.; Janjetovic, Z.; Oak, A.S.; Jozwicki, W.; Jetten, A.M.; Mason, R.S.; et al. On the role of classical and novel forms of vitamin d in melanoma progression and management. *J. Steroid Biochem. Mol. Biol.* **2018**, *177*, 159–170. [CrossRef] [PubMed]
50. Slominski, A.T.; Brozyna, A.A.; Zmijewski, M.A.; Jozwicki, W.; Jetten, A.M.; Mason, R.S.; Tuckey, R.C.; Elmets, C.A. Vitamin D signaling and melanoma: Role of vitamin d and its receptors in melanoma progression and management. *Lab. Investig.* **2017**, *97*, 706–724. [CrossRef] [PubMed]
51. Norman, A.W. Minireview: Vitamin d receptor: New assignments for an already busy receptor. *Endocrinology* **2006**, *147*, 5542–5548. [CrossRef] [PubMed]
52. Mason, R.S.; Reichrath, J. Sunlight vitamin d and skin cancer. *Anticancer Agents Med. Chem.* **2013**, *13*, 83–97. [CrossRef] [PubMed]
53. Bikle, D.D.; Elalieh, H.; Welsh, J.; Oh, D.; Cleaver, J.; Teichert, A. Protective role of vitamin D signaling in skin cancer formation. *J. Steroid Biochem. Mol. Biol.* **2013**, *136*, 271–279. [CrossRef] [PubMed]

54. Neme, A.; Seuter, S.; Carlberg, C. Selective regulation of biological processes by vitamin d based on the spatio-temporal cistrome of its receptor. *Biochim. Biophys. Acta* **2017**, *1860*, 952–961. [CrossRef] [PubMed]
55. Bikle, D.D. Vitamin D receptor, UVR, and skin cancer: A potential protective mechanism. *J. Investig. Dermatol.* **2008**, *128*, 2357–2361. [CrossRef] [PubMed]
56. Elias, P.M. Structure and function of the stratum corneum extracellular matrix. *J. Investig. Dermatol.* **2012**, *132*, 2131–2133. [CrossRef] [PubMed]
57. Dixon, K.M.; Norman, A.W.; Sequeira, V.B.; Mohan, R.; Rybchyn, M.S.; Reeve, V.E.; Halliday, G.M.; Mason, R.S. 1alpha,25(OH)(2)-vitamin D and a nongenomic vitamin d analogue inhibit ultraviolet radiation-induced skin carcinogenesis. *Cancer Prev. Res. (Phila)* **2011**, *4*, 1485–1494. [CrossRef] [PubMed]
58. Makarova, A.; Wang, G.; Dolorito, J.A.; Kc, S.; Libove, E.; Epstein, E.H., Jr. Vitamin D3 produced by skin exposure to UVR inhibits murine basal cell carcinoma carcinogenesis. *J. Investig. Dermatol.* **2017**, *137*, 2613–2619. [CrossRef] [PubMed]
59. Slominski, A.T.; Zmijewski, M.A.; Semak, I.; Zbytek, B.; Pisarchik, A.; Li, W.; Zjawiony, J.; Tuckey, R.C. Cytochromes P450 and skin cancer: Role of local endocrine pathways. *Anticancer Agents Med. Chem.* **2014**, *14*, 77–96. [CrossRef] [PubMed]
60. Slominski, A.T.; Kim, T.K.; Hobrath, J.V.; Oak, A.S.W.; Tang, E.K.Y.; Tieu, E.W.; Li, W.; Tuckey, R.C.; Jetten, A.M. Endogenously produced nonclassical vitamin D hydroxy-metabolites act as "Biased" Agonists on VDR and inverse agonists on RORalpha and RORgamma. *J. Steroid Biochem. Mol. Biol.* **2017**, *173*, 42–56. [CrossRef] [PubMed]
61. Wang, J.; Slominski, A.; Tuckey, R.C.; Janjetovic, Z.; Kulkarni, A.; Chen, J.; Postlethwaite, A.E.; Miller, D.; Li, W. 20-Hydroxyvitamin D(3) inhibits proliferation of cancer cells with high efficacy while being non-toxic. *Anticancer Res.* **2012**, *32*, 739–746. [PubMed]
62. Slominski, A.T.; Janjetovic, Z.; Fuller, B.E.; Zmijewski, M.A.; Tuckey, R.C.; Nguyen, M.N.; Sweatman, T.; Li, W.; Zjawiony, J.; Miller, D.; et al. Products of vitamin D3 or 7-dehydrocholesterol metabolism by cytochrome P450scc show anti-leukemia effects, having low or absent calcemic activity. *PLoS ONE* **2010**, *5*, e9907. [CrossRef] [PubMed]
63. Chen, J.; Wang, J.; Kim, T.K.; Tieu, E.W.; Tang, E.K.; Lin, Z.; Kovacic, D.; Miller, D.D.; Postlethwaite, A.; Tuckey, R.C.; et al. Novel vitamin d analogs as potential therapeutics: Metabolism, toxicity profiling, and antiproliferative activity. *Anticancer Res.* **2014**, *34*, 2153–2163. [PubMed]
64. Tongkao-On, W.; Carter, S.; Reeve, V.E.; Dixon, K.M.; Gordon-Thomson, C.; Halliday, G.M.; Tuckey, R.C.; Mason, R.S. Cyp11a1 in skin: An alternative route to photoprotection by vitamin D compounds. *J. Steroid Biochem. Mol. Biol.* **2015**, *148*, 72–78. [CrossRef] [PubMed]
65. Slominski, A.; Janjetovic, Z.; Tuckey, R.C.; Nguyen, M.N.; Bhattacharya, K.G.; Wang, J.; Li, W.; Jiao, Y.; Gu, W.; Brown, M.; et al. 20s-hydroxyvitamin D3, noncalcemic product of CYP11A1 action on vitamin D3, exhibits potent antifibrogenic activity in vivo. *J. Clin. Endocrinol. Metab.* **2013**, *98*, E298–E303. [CrossRef] [PubMed]
66. Tuckey, R.C.; Li, W.; Shehabi, H.Z.; Janjetovic, Z.; Nguyen, M.N.; Kim, T.K.; Chen, J.; Howell, D.E.; Benson, H.A.; Sweatman, T.; et al. Production of 22-hydroxy metabolites of vitamin D3 by cytochrome P450scc (CYP11A1) and analysis of their biological activities on skin cells. *Drug Metab. Dispos.* **2011**, *39*, 1577–1588. [CrossRef] [PubMed]
67. Zbytek, B.; Janjetovic, Z.; Tuckey, R.C.; Zmijewski, M.A.; Sweatman, T.W.; Jones, E.; Nguyen, M.N.; Slominski, A.T. 20-Hydroxyvitamin D3, a product of vitamin D3 hydroxylation by cytochrome P450scc, stimulates keratinocyte differentiation. *J. Investig. Dermatol.* **2008**, *128*, 2271–2280. [CrossRef] [PubMed]
68. Janjetovic, Z.; Zmijewski, M.A.; Tuckey, R.C.; DeLeon, D.A.; Nguyen, M.N.; Pfeffer, L.M.; Slominski, A.T. 20-Hydroxycholecalciferol, product of vitamin D3 hydroxylation by P450scc, decreases NF-kappaB activity by increasing IkappaB alpha levels in human keratinocytes. *PLoS ONE* **2009**, *4*, e5988. [CrossRef] [PubMed]
69. Janjetovic, Z.; Tuckey, R.C.; Nguyen, M.N.; Thorpe, E.M., Jr.; Slominski, A.T. 20,23-dihydroxyvitamin D3, novel P450scc product, stimulates differentiation and inhibits proliferation and NF-kappaB activity in human keratinocytes. *J. Cell. Physiol.* **2010**, *223*, 36–48. [PubMed]
70. Wasiewicz, T.; Szyszka, P.; Cichorek, M.; Janjetovic, Z.; Tuckey, R.C.; Slominski, A.T.; Zmijewski, M.A. Antitumor effects of vitamin D analogs on hamster and mouse melanoma cell lines in relation to melanin pigmentation. *Int. J. Mol. Sci.* **2015**, *16*, 6645–6667. [CrossRef] [PubMed]

71. Wierzbicka, J.M.; Zmijewski, M.A.; Piotrowska, A.; Nedoszytko, B.; Lange, M.; Tuckey, R.C.; Slominski, A.T. Bioactive forms of vitamin d selectively stimulate the skin analog of the hypothalamus-pituitary-adrenal axis in human epidermal keratinocytes. *Mol. Cell. Endocrinol.* **2016**, *437*, 312–322. [CrossRef] [PubMed]
72. Skobowiat, C.; Oak, A.S.; Kim, T.K.; Yang, C.H.; Pfeffer, L.M.; Tuckey, R.C.; Slominski, A.T. Noncalcemic 20-hydroxyvitamin D3 inhibits human melanoma growth in in vitro and in vivo models. *Oncotarget* **2017**, *8*, 9823–9834. [CrossRef] [PubMed]
73. Janjetovic, Z.; Brozyna, A.A.; Tuckey, R.C.; Kim, T.K.; Nguyen, M.N.; Jozwicki, W.; Pfeffer, S.R.; Pfeffer, L.M.; Slominski, A.T. High basal NF-kappaB activity in nonpigmented melanoma cells is associated with an enhanced sensitivity to vitamin D3 derivatives. *Br. J. Cancer* **2011**, *105*, 1874–1884. [CrossRef] [PubMed]
74. Wierzbicka, J.M.; Binek, A.; Ahrends, T.; Nowacka, J.D.; Szydlowska, A.; Turczyk, L.; Wasiewicz, T.; Wierzbicki, P.M.; Sadej, R.; Tuckey, R.C.; et al. Differential antitumor effects of vitamin d analogues on colorectal carcinoma in culture. *Int. J. Oncol.* **2015**, *47*, 1084–1096. [CrossRef] [PubMed]
75. Piotrowska, A.; Wierzbicka, J.; Slebioda, T.; Wozniak, M.; Tuckey, R.C.; Slominski, A.T.; Zmijewski, M.A. Vitamin D derivatives enhance cytotoxic effects of $H_2O_2$ or cisplatin on human keratinocytes. *Steroids* **2016**, *110*, 49–61. [CrossRef] [PubMed]
76. Lin, Z.; Marepally, S.R.; Goh, E.S.Y.; Cheng, C.Y.S.; Janjetovic, Z.; Kim, T.K.; Miller, D.D.; Postlethwaite, A.E.; Slominski, A.T.; Tuckey, R.C.; et al. Investigation of 20s-hydroxyvitamin D3 analogs and their 1alpha-OH derivatives as potent vitamin d receptor agonists with anti-inflammatory activities. *Sci. Rep.* **2018**, *8*, 1478. [CrossRef] [PubMed]
77. Lin, Z.; Chen, H.; Belorusova, A.Y.; Bollinger, J.C.; Tang, E.K.Y.; Janjetovic, Z.; Kim, T.K.; Wu, Z.; Miller, D.D.; Slominski, A.T.; et al. 1alpha,20S-dihydroxyvitamin D3 interacts with vitamin D receptor: Crystal structure and route of chemical synthesis. *Sci. Rep.* **2017**, *7*, 10193. [CrossRef] [PubMed]
78. Slominski, A.T.; Kim, T.K.; Takeda, Y.; Janjetovic, Z.; Brozyna, A.A.; Skobowiat, C.; Wang, J.; Postlethwaite, A.; Li, W.; Tuckey, R.C.; et al. RORalpha and ROR gamma are expressed in human skin and serve as receptors for endogenously produced noncalcemic 20-hydroxy- and 20,23-dihydroxyvitamin D. *FASEB J.* **2014**, *28*, 2775–2789. [CrossRef] [PubMed]
79. Le Poole, I.C.; van den Berg, F.M.; van den Wijngaard, R.M.; Galloway, D.A.; van Amstel, P.J.; Buffing, A.A.; Smits, H.L.; Westerhof, W.; Das, P.K. Generation of a human melanocyte cell line by introduction of HPV16 E6 and E7 genes. *In Vitro Cell. Dev. Biol. Anim.* **1997**, *33*, 42–49. [CrossRef] [PubMed]
80. Slominski, A.T.; Kim, T.K.; Li, W.; Yi, A.K.; Postlethwaite, A.; Tuckey, R.C. The role of CYP11A1 in the production of vitamin D metabolites and their role in the regulation of epidermal functions. *J. Steroid Biochem. Mol. Biol.* **2014**, *144PA*, 28–39. [CrossRef] [PubMed]
81. Bikle, D.D.; Oda, Y.; Tu, C.L.; Jiang, Y. Novel mechanisms for the vitamin d receptor (VDR) in the skin and in skin cancer. *J. Steroid Biochem. Mol. Biol.* **2015**, *148*, 47–51. [CrossRef] [PubMed]
82. Carlberg, C. What do we learn from the genome-wide perspective on vitamin D3? *Anticancer Res.* **2015**, *35*, 1143–1151. [PubMed]
83. Kim, T.K.; Wang, J.; Janjetovic, Z.; Chen, J.; Tuckey, R.C.; Nguyen, M.N.; Tang, E.K.; Miller, D.; Li, W.; Slominski, A.T. Correlation between secosteroid-induced vitamin D receptor activity in melanoma cells and computer-modeled receptor binding strength. *Mol. Cell. Endocrinol.* **2012**, *361*, 143–152. [CrossRef] [PubMed]
84. Slominski, A.T.; Janjetovic, Z.; Kim, T.K.; Wright, A.C.; Grese, L.N.; Riney, S.J.; Nguyen, M.N.; Tuckey, R.C. Novel vitamin D hydroxyderivatives inhibit melanoma growth and show differential effects on normal melanocytes. *Anticancer Res.* **2012**, *32*, 3733–3742. [PubMed]
85. Soshilov, A.A.; Denison, M.S. Ligand promiscuity of aryl hydrocarbon receptor agonists and antagonists revealed by site-directed mutagenesis. *Mol. Cell. Biol.* **2014**, *34*, 1707–1719. [CrossRef] [PubMed]
86. Soshilov, A.; Denison, M.S. Ligand displaces heat shock protein 90 from overlapping binding sites within the aryl hydrocarbon receptor ligand-binding domain. *J. Biol. Chem.* **2011**, *286*, 35275–35282. [CrossRef] [PubMed]
87. Tuckey, R.C.; Janjetovic, Z.; Li, W.; Nguyen, M.N.; Zmijewski, M.A.; Zjawiony, J.; Slominski, A. Metabolism of 1alpha-hydroxyvitamin D3 by cytochrome P450scc to biologically active 1alpha,20-dihydroxyvitamin D3. *J. Steroid Biochem. Mol. Biol.* **2008**, *112*, 213–219. [CrossRef] [PubMed]

88. Hiwatashi, A.; Nishii, Y.; Ichikawa, Y. Purification of cytochrome P-450d1 alpha (25-hydroxyvitamin D3-1 alpha-hydroxylase) of bovine kidney mitochondria. *Biochem. Biophys. Res. Commun.* **1982**, *105*, 320–327. [CrossRef]
89. Janjetovic, Z.; Nahmias, Z.P.; Hanna, S.; Jarrett, S.G.; Kim, T.K.; Reiter, R.J.; Slominski, A.T. Melatonin and its metabolites ameliorate ultraviolet b-induced damage in human epidermal keratinocytes. *J. Pineal Res.* **2014**, *57*, 90–102. [CrossRef] [PubMed]

International Journal of
*Molecular Sciences*

MDPI

*Article*

# Vitamin D Receptor Is Necessary for Mitochondrial Function and Cell Health

**Chiara Ricca [1,†], Alessia Aillon [1,†], Loredana Bergandi [1] and Francesca Silvagno [1,*,†]**

1 Department of Oncology, University of Torino, Via Santena 5 bis, 10126 Torino, Italy; chiara.ricca48@edu.unito.it (C.R.); alessia.aillon@edu.unito.it (A.A.); loredana.bergandi@gmail.com (L.B.)

2 Department of Chirurgia Generale e Specialistiche, Banca della Cute, AOU Città della Salute e della Scienza Torino, Via Zuretti 29, 10126 Torino, Italy; daniela.alotto@gmail.com (D.A.); ccastagnoli@cittadellasalute.to.it (C.C.)

* Correspondence: francesca.silvagno@unito.it

† These authors contributed equally to this work.

Received: 8 May 2018; Accepted: 3 June 2018; Published: 5 June 2018

**Abstract:** Vitamin D receptor (VDR) mediates many genomic and non-genomic effects of vitamin D. Recently, the mitochondrial effects of vitamin D have been characterized in many cell types. In this article, we investigated the importance of VDR not only in mitochondrial activity and integrity but also in cell health. The silencing of the receptor in different healthy, non-transformed, and cancer cells initially decreased cell growth and modulated the cell cycle. We demonstrated that, in silenced cells, the increased respiratory activity was associated with elevated reactive oxygen species (ROS) production. In the long run, the absence of the receptor caused impairment of mitochondrial integrity and, finally, cell death. Our data reveal that VDR plays a central role in protecting cells from excessive respiration and production of ROS that leads to cell damage. Because we confirmed our observations in different models of both normal and cancer cells, we conclude that VDR is essential for the health of human tissues.

**Keywords:** vitamin D receptor; silencing; mitochondria; reactive oxygen species; respiratory chain; cytochrome C; cell proliferation; cell death

## 1. Introduction

The active form of vitamin D (1,25(OH)$_2$D$_3$) exerts its beneficial effects on the whole organism by regulating calcium homeostasis and by modulating a large set of genes involved in the differentiation and function of virtually every tissue. The transcriptional control is mediated by the vitamin D receptor (VDR), which promotes ligand-dependent induction or repression of gene transcription together with its binding partner retinoid X receptor (RXR) and many recruited activators or repressors. The incredibly high number of target genes accounts for the pleiotropic functions of VDR.

Besides the long-recognized role of 1,25(OH)$_2$D$_3$ in regulating calcium and phosphate metabolism, many biological networks are influenced by VDR, including bone remodeling [1], xenobiotic detoxification [2], cell physiology (reviewed in [3,4]), immunity [5], and metabolism [6–8]. Recently, a novel mitochondrial localization of VDR has been described [9,10], and the characterized mitochondrial function of 1,25(OH)$_2$D$_3$/VDR has been depicted as the hub linking the control of cell metabolism to the transcriptional status of the cell. In fact, the work of our group [11,12] and others [13] has demonstrated that, through VDR activity, 1,25(OH)$_2$D$_3$ reduces mitochondrial respiration and rewires cell metabolism toward the biosynthetic pathways. This metabolic control sustains both the proliferative asset [11] and more specialized functions of the cells [12], depending on the cellular phenotype. On the basis of these recent observations, it is clear that not only the defective activity of VDR affects the expression of many genes, and thus the levels of many proteins, but also mitochondrial metabolism and function must be profoundly altered by VDR failure. The mitochondrial compartment is central in many processes; besides being the powerhouse of the cell, mitochondria are also important reservoirs of metabolic intermediates, are considered calcium and iron stores [14,15], and behave as molecular factories (for example they are the site of iron insertion in organic molecules). Given their central role, a severe mitochondrial damage leads to apoptotic cell death.

The aim of this work was to explore the results of a defective expression of VDR in cell health and function. We silenced the receptor in different cell types and observed a severe reduction in cell proliferation followed by cell death. We investigated the molecular mechanisms governing the increased vulnerability of the silenced cells and demonstrated the involvement of the mitochondrial compartment.

## 2. Results

### 2.1. Two Different Human Cell Lines and Human Primary Cells Silenced for VDR Strongly Reduce Their Proliferation Rate

With the aim of investigating the effects of a severe reduction of VDR activity on cell physiology, we silenced the receptor by lentiviral delivery of shRNA against human VDR. Three different cell types were selected as examples of malignant, non-malignant, and healthy phenotypes: the human breast cancer cell line MCF7, the human proliferating keratinocyte cell line HaCaT, and primary healthy human fibroblasts, respectively. The genetic ablation of VDR expression by this technique was very effective, as previously demonstrated [11], and the suppression of the protein was confirmed in all cell types by western blotting analysis. The abatement of VDR expression is shown in Figure 1. One week after infection, cell proliferation was investigated either by crystal violet staining (MCF7 and HaCaT cells) or by BrdU incorporation (fibroblasts), and the results are presented in Figure 2A,B. In all cell types, the silencing of VDR caused a great reduction of growth. Accordingly, the analysis of their cell cycle showed a remarkably reduced S phase and a decreased G0/G1 phase, and the cells accumulated in the G2/M phase (Figure 2C).

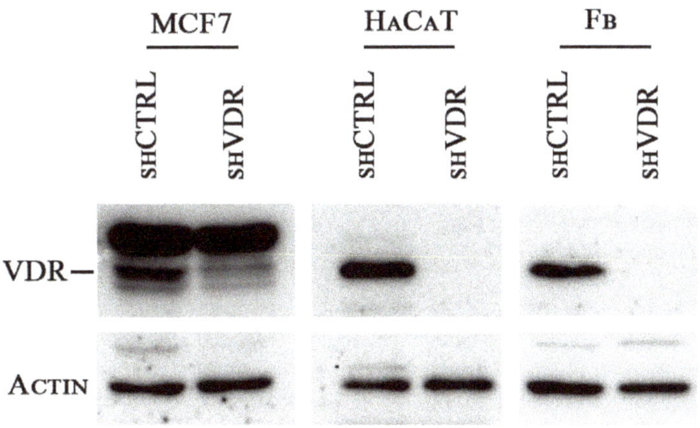


**Figure 1.** shRNA-mediated Vitamin D receptor (VDR) knockdown in the human cells MCF7, HaCaT, and primary fibroblasts (Fb) abrogates VDR expression. The cells were silenced by lentiviral infection with an shRNA against VDR (shVDR) or with a scrambled non-targeting shRNA as control (shCTRL). Seven days after infection, VDR expression was evaluated in the cellular whole extracts by western blot analysis. Actin was detected as an internal control for protein loading.

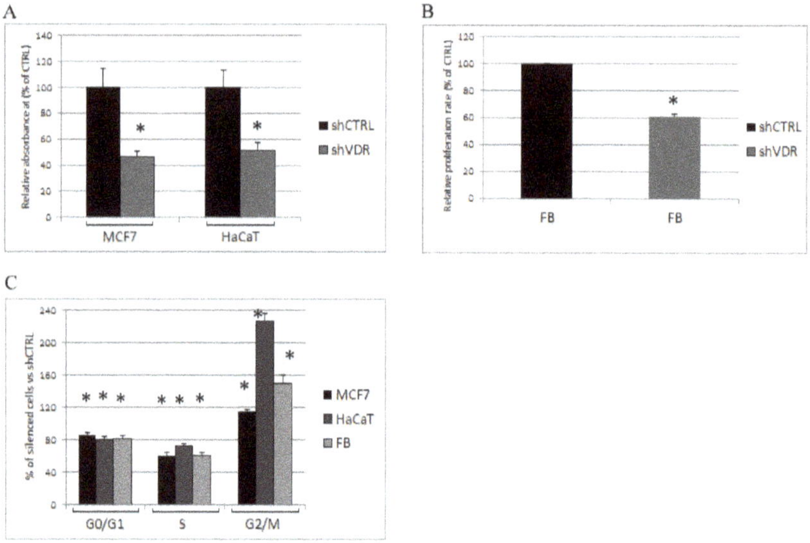


**Figure 2.** Analysis of cell proliferation in silenced cells. One week after infection, the control (shCTRL) and VDR knockdown cells (shVDR) were seeded and assayed for (**A**) proliferation rate, measured by crystal violet staining or (**B**) BrdU incorporation; (**C**) The cell cycle of MCF7, HaCaT, and fibroblasts (Fb) was evaluated by cytofluorimetry, and the distribution of the silenced cells throughout the cell cycle was expressed as percentage of the shCTRL cells in the same phase. The data are expressed as the means ± SD of three independent experiments; * $p < 0.05$ compared to the control.

### *2.2. The Ablation of VDR Enhances Mitochondrial Respiratory Activity and the Production of Reactive Oxygen Species*

We have previously demonstrated that VDR controls mitochondrial respiratory activity [11]. Here, we confirmed that VDR silencing enhanced the respiratory activity of HaCaT cells by measuring the increment of mitochondrial membrane potential (Figure 3A); moreover, by real-time PCR analysis of MCF7 and HaCaT transcripts, we detected the increased expression of several components of the respiratory chain coupled to oxidative phosphorylation. Because both nuclear- and mitochondrial-encoded proteins are required for the formation of active respiratory complexes, we evaluated the transcription of two subunits of cytochrome C oxidase (COX or respiratory complex IV) and two subunits of ATP synthase whose transcripts are both of mitochondrial and nuclear origin: COX2 and MT-ATP6 (a mitochondrial gene encoding the ATP synthase Fo subunit 6) are markers of mitochondrial transcription activity, and COX4 and ATP5B (a nuclear gene encoding ATP synthase subunit beta) are markers of the nuclear contribution to respiratory chain modulation. Their increased expression, shown in Figure 3B, was in agreement with the observed enhanced respiratory membrane potential. One of the consequences of the respiratory burst is the production of reactive oxygen species (ROS); therefore, we measured ROS production in control cells and VDR knockdown cells. We demonstrated the increase of ROS levels in all silenced cells (Figure 3C) and detected the highest increment in primary fibroblasts.

On the basis of our observations, we concluded that, in all cell types analyzed, VDR was an essential negative modulator of mitochondrial respiration, and its ablation increased both the expression and the activity of the respiratory chain, and the consequent ROS production.

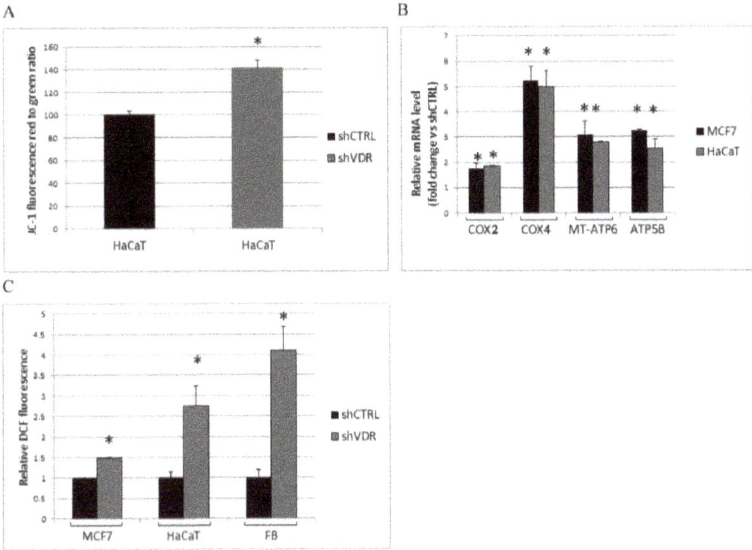


**Figure 3.** The silencing of VDR induces mitochondrial respiration and enhances the production of ROS. The metabolic assays and the extraction of mRNA were carried out one week after silencing the cells with shRNA control (shCTRL) or VDR shRNA (shVDR). (**A**) The mitochondrial respiratory activity was assessed in HaCaT cells by cytofluorimetric evaluation of the mitochondrial dye JC-1, and (**B**) the expression of the respiratory chain complexes was analyzed by real-time PCR. The values plotted on the graph represent the fold change in transcript expression in silenced versus control cells and are displayed as the means ± SD of three independent experiments; (**C**) reactive oxygen species (ROS) production was measured and expressed relative to control cells. The data represent the means ± SD of three independent experiments; * $p < 0.05$ compared to the control.

### 2.3. *The Silencing of VDR Triggers Long-Term Cellular Damage and Cell Death*

Two weeks after delivery of shRNA particles, the silenced cells lost their healthy phenotype and looked damaged when observed under the microscope. We hypothesized a massive apoptotic death caused by the measured increase of ROS; therefore, we decided to quantify the cellular damage and to verify the mitochondrial origin of the death process. First, we assessed the release of the enzyme lactate dehydrogenase (LDH) as a marker of lost cell integrity. The results of our analysis are displayed in Figure 4A. As expected, all the silenced cells accumulated great amounts of LDH in their supernatants in comparison to control cells. This increase in LDH was particularly evident in the medium of the silenced fibroblasts. Next, we evaluated the signs of the lost mitochondrial integrity by western blotting analysis of cytochrome C content in subcellular fractions. Total lysates were prepared along with mitochondrial and cytosolic extracts, and the levels of cytochrome C were quantified with a specific antibody. The results of our analysis are displayed in Figure 4B, and the data were quantified and plotted in Figure 4C. Both in MCF7 and in HaCaT cells, we found a decreased content of cytochrome C in the mitochondrial compartment and, alongside this loss, we detected the increase of the mitochondrial protein in the cytosolic fractions. At the same time, the expression of cytochrome C in the whole lysates was unchanged, demonstrating the release of the mitochondrial protein pool into the cytosolic milieu. The intracellular trafficking of an essential element of the respiratory chain is considered the hallmark of apoptosis driven by a defective mitochondrial function. We also investigated another marker of an ongoing apoptotic death: the cleavage of the nuclear enzyme poly ADP ribose polymerase (PARP). In fact, the proteolytic cleavage of PARP into 89 and 24 kDa fragments by caspases is an early indicator of apoptosis [16]. We analyzed the protein content of nuclear preparations from control and silenced MCF7 and HaCaT cells and were able to detect a decreased amount of the 116 kDa PARP protein in the nuclear extracts of silenced cells, which was the demonstration of the occurring cleavage and loss of the full-length enzyme (Figure 4D). The results of these experiments were quantified and plotted on graph, as shown in Figure 4E.

All together, our data demonstrated that the silencing of VDR led to a severe cell damage that had all the signs of a mitochondrial-mediated apoptotic death.

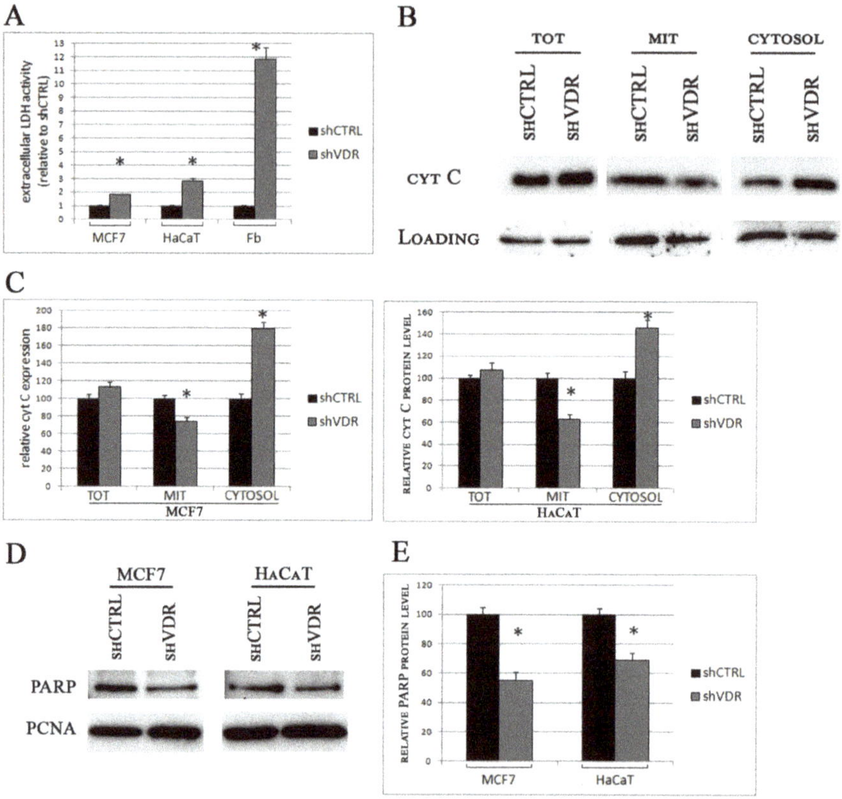


**Figure 4.** VDR silencing leads to a long-term cellular damage, loss of mitochondrial integrity, and apoptotic death. Two weeks after silencing, cell damage was evaluated. (**A**) Toxicity was assayed by lactate dehydrogenase (LDH) release in the extracellular medium; (**B**) The intracellular levels of cytochrome C were evaluated by western blotting analysis of total extracts (TOT), mitochondrial (MIT), and cytosolic fractions. Actin expression was used as a control of equal loading in total and cytosolic extracts, whereas equal mitochondrial loading was verified by VDAC detection; (**C**) Bands from three different experiments were quantified and normalized for loading, and the data were plotted on graph as percentage of control; (**D**) Full-length PARP (116 kDa) was detected in nuclear extracts of control (shCTRL) or silenced cells (shVDR) by western blotting analysis, and PCNA was used as a loading control. The blots are representative of three independent experiments; (**E**) The bands were quantified and normalized for loading, and the data were plotted on graph as percentage of shCTRL. The data represent the means ± SD of three independent experiments; * $p < 0.05$ compared to the control.

## 3. Discussion

Vitamin D is active in every tissue, and the perturbation of its signaling is involved in many diseases [17]. The importance and the pleiotropic effects of the activity of its receptor VDR have been elucidated by the murine knockout models, which have displayed a modified phenotype in many tissues, such as bone, intestine (reviewed in [18]), skin [19], lung [20], muscle [21], endothelium [22], and adipose tissue, and in metabolism [23,24]. Although the knockout models are useful to investigate the general pathways controlled by vitamin D, at the cellular level they have a major flaw: they originate cells that, since their formation and embryonic development, have never relied on VDR for transcriptional control and modulation of metabolism. If we wonder about the importance of VDR activity and whether it is essential for cellular tasks, another approach is to knockdown the receptor in a cell population that expresses VDR to modulate growth, differentiation, and many other functions. When the deranged signaling of $1,25(OH)_2D_3$ is investigated, it must be considered that the insufficiency of vitamin D and the loss of the receptor have some similar but partly distinct consequences. The hypocalcemic phenotype shared by vitamin D deficiency and VDR knockout models can be reversed by a high-calcium diet [25], but even at very low levels of $1,25(OH)_2D_3$, the VDR can still operate in a ligand-independent modality [26,27] or can respond to other molecules [28–30]. Therefore, in order to analyze fully and unambiguously the perturbation of $1,25(OH)_2D_3$ signaling, the best approach is to delete the receptor, and we followed this line of investigation.

In this work, we carried out a genetic silencing of VDR in three cell types different from each other, for the purpose of testing the general importance of VDR in cell physiology. MCF7 were selected as an example of malignant human cancer cell lines, HaCaT cells were a good model of proliferating but not transformed human keratinocytes, and primary human fibroblasts were chosen as a model of healthy human cells.

The first important finding of this work was that the ablation of the receptor resulted in increased respiratory activity that enhanced the production of intracellular ROS. Interestingly, in all cell types, we found that VDR controls both the mitochondrial (COX2 and MT-ATP6) and the nuclear transcription (COX4 and ATP5B) of the proteins involved in respiratory activity and ATP synthesis, in agreement with the necessity of coordinating the nuclear and the mitochondrial transcription of the components of the respiratory process. It is known that the respiratory chain is a major source of ROS; in particular, the complexes I, III, and IV are involved in radical biosynthesis [31,32]. ROS production is beneficial to some extent and is involved in cell cycle progression [33], but an excessive boost can be detrimental and can trigger cell damage. In all three silenced cellular models, the increase in ROS levels was remarkable, especially in primary fibroblasts, and could exceed the antioxidant defenses. Indeed, while the initial effect of silencing was growth arrest and the modulation of the cell cycle, the long-term effect of VDR loss was cell damage, measured as LDH release. The increase of ROS levels and the toxicity were directly proportional, since we observed the smallest rise of ROS and toxicity in silenced MCF7 and the highest effects in silenced primary fibroblasts. This observation is reasonably accounted for by some reported characteristics of cancer cells. In fact, it is known that the transformed cells use ROS signals to drive proliferation and other events required for tumor progression and that the elevated ROS levels are balanced by an increased activity of antioxidant enzymes in cancer cells [34]; accordingly, the protective role of VDR was particularly evident in healthy fibroblasts, although, in our experimental setting, VDR defended even the most transformed MCF7 cells against oxidative stress.

In our previous work, we demonstrated the metabolic importance of VDR and its effect on proliferation: the receptor curbs the respiratory activity and allows the rewiring of metabolic intermediates toward biosynthesis, thus sustaining proliferation [11]. The results of the present research unveil a novel role for VDR in cellular physiology, namely, the protection from the excessive respiratory activity and the limitation of ROS production. In this study, we observed the negative effects of the derangement of the metabolic control exerted by VDR; indeed, in the long run, the excessive production of ROS consequent to VDR ablation had deleterious effects on the mitochondrial function and survival of cells. One of the consequences of the excessive damage caused by ROS is the variation in mitochondrial membrane permeability that results in cytochrome C release and apoptotic death [35]. In line with this, we demonstrated that, after two weeks of absence of VDR, mitochondrial integrity was lost, and the cells showed the signs of an apoptotic fate. The interplay between vitamin D–VDR,

ROS signaling, and the antioxidant system is complex; on the one hand, it has been demonstrated that vitamin D and its analogues can increase the cytotoxicity mediated by ROS [36], while, on the other hand, few reports have proved that vitamin D–VDR is able to inhibit the apoptosis triggered by oxidative stress [37–41]; in addition, in this study, for the first time, we showed the protective role exerted by VDR itself, without additional stressors. Moreover, our study investigated a novel mechanism involved in the antiapoptotic effects of VDR, previously ascribed only to its transcriptional activity [42]. In fact, not only the loss of VDR disrupts the traditional pathways regulated by its transcriptional control [42], but also, as we demonstrated, the silencing of VDR generates an unbalanced metabolism that leads to cytotoxicity. Although we have displayed the essential role of VDR in cell metabolism, health, and survival, our data and conclusions are not necessarily in contrast with the fact that the VDR knockout phenotype in animals is not lethal. Obviously, the tissues lacking VDR since their embryonic development have found compensatory mechanisms to balance the effects of VDR deletion.

In conclusion, in the present study, we discovered a novel important role for VDR in cell health. We demonstrated that the mitochondrial effects of the receptor not only regulate the respiratory activity but also protect from oxidative damage and preserve mitochondrial integrity and cell survival. Our data were obtained in different cell types, cancerous as well as healthy cells, rendering the discovered novel function a general feature of vitamin D–VDR role in many tissues.

Another intriguing consideration about this study is based on the fact that the phenotype obtained by the experimental silencing of the receptor could mimic the pathological situations in which the expression of VDR is downregulated (for example by epigenetic mechanisms [43,44]) or its activity is compromised (for example by polymorphisms [45,46]). It is interesting to highlight that many respiratory chain dysfunctions and deleterious ROS overproduction are recurrent themes in human pathologies, ranging from neurodegenerative diseases to cancer [47,48], and may be of paramount importance in ageing [49]. The results of this study demonstrate the protective role of VDR and raise the possibility that the loss of VDR function could be partly responsible of, or at least could be an adverse event in such diseases.

## 4. Materials and Methods

### 4.1. Cell Culture

The immortalized human epidermal keratinocyte cell line (HaCaT) and the MCF7 human breast cancer cell line were purchased from American Type Culture Collection (ATCC) (Manassas, VA, USA). Dermal primary fibroblasts from healthy donors were obtained from Banca della Cute, Turin, Italy, and were used in early passages. The cells were cultured in Dulbecco's modified Eagle's medium (DMEM) that had been supplemented with 10% fetal bovine serum and 1% antibiotics (penicillin, streptomycin), at 37 °C in a humidified atmosphere containing 5% $CO_2$. All culture reagents were from Sigma-Aldrich (Sigma, St. Louis, MO, USA).

### 4.2. Lentiviral-Mediated shRNA Targeting

PLKO.1 lentiviral shRNA clones targeting the human VDR (TRCN0000276543) and a scrambled nontargeting control were purchased from Sigma (Sigma Mission shRNA) (Sigma, St. Louis, MO, USA) and were previously described and characterized in terms of efficiency [11]. Lentiviral transduction particles were produced in HEK293T cells as previously reported [11]. Briefly, the cotransfection of the shRNA plasmid together with the packaging vectors was carried out by lipofectamine reagent, and the supernatants were used for overnight transduction of the cells. Puromycin selection began 24 h after infection. Within one week from infection or after two weeks, the cells were seeded for experimental assays or harvested for RNA and protein analysis.

### 4.3. Extract Preparation and Western Blotting Analysis

Subcellular fractionation and Western blotting analyses were conducted as previously described [10]. The protein content of the total extracts and mitochondrial fractions was quantified using the DC protein assay (Bio-Rad, Hercules, CA, USA); 50 μg of total lysates and 30 μg of the mitochondrial or nuclear fractions were separated by 10% SDS-PAGE and analyzed using Western blotting. The analysis of cytochrome C was carried out after a 12% SDS-PAGE. The proteins were immunostained with the indicated primary antibodies for 1 h at room temperature, and detection of the proteins of interest was performed using peroxidase-conjugated secondary antibodies (Pierce, Rockford, IL, USA), followed by ECL detection (ECL detection kit, Perkin Elmer Life Science, Foster City, CA USA). Mouse anti-VDR (sc-13133), anti-actin (sc-8432), anti-PCNA (sc-56), and rabbit anti-PARP (sc-7150) antibodies were purchased from Santa Cruz Biotechnology (Santa Cruz, CA, USA). The anti-VDAC (anti-porin 31HL) monoclonal antibody was purchased from Calbiochem (La Jolla, CA, USA). The mouse anti-cytochrome C (65981A) antibody was from BD Biosciences Pharmingen (San Diego, CA, USA).

### 4.4. Real-Time Polymerase Chain Reaction (qRT-PCR)

Total RNA was extracted with TRIzol® (Invitrogen, Thermo Fisher Scientific, Waltham, MA, USA). One μg of total RNA was reversely transcribed into cDNA, in a final volume of 20 μL, using the High-Capacity cDNA Reverse Transcription Kit (Thermo Fisher Scientific, Waltham, MA, USA), according to the manufacturer's instructions. Quantitative PCR was carried out in a final volume of 20 μL, using the SensiFASTTM SYBR® Hi-ROX Kit (Bioline Srl, Trento, Italy) with the following primers:

COX2, fwd 5′-CGACTACGGCGGACTAATCT-3′, rev 5′-TCGATTGTCAACGTCAAGGA-3′; COX4, fwd 5′-CGAGCAATTTCCACCTCTGT-3′, rev 5′-GGTCAGCCGATCCATATAA-3′;

ATP5B, fwd 5′-GTGGGCTATCAGCCTACCCT-3′, rev 5′-CAAGTCATCAGCAGGCACAT-3′; MT-ATP6, fwd 5′-CCAATAGCCCTGGCCGTAC-3′, rev 5′-CGCTTCCAATTAGGTGCATGA-3′; β2M, fwd 5′-AGCAAGGACTGGTCTTTCTATCTC-3′, rev 5′-ATGTCTCGATCCCACTTAACTA-3′. Beta 2-microglobulin β2M was used as an internal control. PCR amplification was one cycle of denaturation at 95 °C for 2 min, 40 cycles of amplification, including denaturation at 95 °C for 5 s and annealing/ extension at 60 °C for 30 s. The $2^{-\Delta\Delta Ct}$ method was used to analyze the data.

### 4.5. Proliferation Assay

Within one week from silencing, the effect of VDR silencing on the growth of the different human cells was determined either by colorimetric measurement of cell numbers by crystal violet staining (MCF7 and HaCaT cells) or by BrdU incorporation (primary fibroblasts). The primary human fibroblasts were poorly stained by the crystal violet method, and the detection by spectrophotometer was inadequate to quantify these cells; therefore, the more sensitive BrdU assay was chosen. The same number of control or silenced cells (2000, 1000, or 500 cells per well) was seeded on 96-multiwell plates, and the cells were either cultured for five days and then stained with crystal violet or assayed after two days for BrdU incorporation. At the end of this period, MCF7 and HaCaT cells were fixed for 15 min with 11% glutaraldehyde, and the plates were washed three times, air dried, and stained for 20 min with a 0.1% crystal violet solution. The plates were then extensively washed and air-dried prior to solubilization of the bound dye with a 10% acetic acid solution. The absorbance was determined at 595 nm. The proliferation of primary fibroblasts was evaluated by the Cell Proliferation ELISA BrdU kit (Roche Applied Science, Penzberg, Germany), used according to the supplied instructions. The data collected from twelve wells were averaged for each experimental condition, and each experiment was repeated three times.

### 4.6. Cell Cycle Analysis

Within one week from silencing, the control and silenced cells were seeded at the same density and, after 48 h, were detached in 1 mL PBS-EDTA 5 mM by scraping, collected, and fixed in 70% cold ethanol. After 3 h at −30 °C, the cells were centrifuged at 2000 rpm for 5 min, washed twice with PBS, resuspended in 200 μL MUSE® cell cycle reagent, and then incubated for 30 min in the dark at room temperature. Cellular DNA content was analyzed by Muse Cell Analyzer (Merck S.p.a., Milan, Italy). To quantify the relative percentage of cells in the G0/G1, S, and G2/M phases of the cell cycle, the Muse™ Cell Analyzer software was used.

### 4.7. Cytofluorimetric Evaluation of the Mitochondrial Membrane Potential

JC-1, a mitochondrial dye that stains the mitochondria in living cells in a membrane potential-dependent fashion, was used as previously reported [11]. Within one week from silencing, HaCaT cells were harvested by trypsinization, washed with PBS, and incubated with JC-1 (2 mg/mL final concentration) at 37 °C for 30 min. After washing, JC-1 accumulation was determined using flow cytometric analysis. The amount of JC-1 retained by 10,000 cells per sample was measured at 530 nm (FL-1 green fluorescence) and 590 nm (FL-2 red fluorescence) using a flow cytometer and analyzed using Cell Quest Alias software. The ratio FL2/FL1 was evaluated to determine the mitochondrial membrane potential.

### 4.8. Measurement of Intracellular ROS Production

After one week from silencing, the cells were harvested and loaded for 15 min with 10 μM 2′,7′-dichlorodihydrofluorescein diacetate (DCFH-DA, Sigma). DCFH-DA is a cell-permeable probe that is cleaved intracellularly by nonspecific esterases to form DCFH, which is further oxidized by ROS to form the fluorescent compound dichlorofluorescein (DCF) [50]. DCF fluorescence was determined at an excitation wavelength of 504 nm and an emission wavelength of 529 nm, using a Packard EL340 microplate reader (Bio-Tek Instruments, Winooski, VT, USA). The fluorescence values were normalized to the protein content and expressed as values relative to the control.

### 4.9. Toxicity Assay (LDH Release)

Two weeks after silencing, cell damage was evaluated by measuring the release of lactate dehydrogenase in the growth medium. The medium was collected, and the cells were harvested by scraping and sonicated on ice with two 10 s bursts. Protein content was quantified by the DC protein assay. Aliquots of growth medium were supplemented with a reaction mixture for the measurement of LDH, as previously described [51]. The enzymatic activity in the extracellular medium was measured spectrophotometrically as absorbance variation at 340 nm (37 °C) and was expressed as μmol NADH oxidized/min/mg cell protein, to normalize the extracellular activity to the cell number. The data were plotted relative to control values.

### 4.10. Bands Quantification and Statistical Analysis

The bBands from protein electrophoresis were quantified by scanning digital densitometry using an ImageJ software analysis (ImageJ version 1.29, Sun Microsystems Inc., Palo Alto, CA, USA). All data were expressed as mean ± S.D of three independent experiments. Statistical analysis of the data was performed using an unpaired, two-tailed Student's $t$-test; $p < 0.05$ was considered to be significant.

**Author Contributions:** F.S. conceived and designed the study; F.S., C.R., A.A., L.B., and D.A. performed the experiments; F.S., L.B., and C.C. analysed the data; F.S. and C.C. supervised the study; F.S., C.R., A.A., and L.B. wrote the manuscript. All authors approved the final version of the manuscript prior to submission.

**Acknowledgments:** This work was supported by Ministero dell'Istruzione Universita' e Ricerca.

**Conflicts of Interest:** The authors declare no conflict of interest. The founding sponsors had no role in the design of the study; in the collection, analyses, or interpretation of data; in the writing of the manuscript, and in the decision to publish the results.

## References


1. Barthel, T.K.; Mathern, D.R.; Whitfield, G.K.; Haussler, C.A.; Hopper, H.A.; Hsieh, J.-C.; Slater, S.A.; Hsieh, G.; Kaczmarska, M.; Jurutka, P.W.; et al. 1,25-Dihydroxyvitamin D3/VDR-mediated induction of FGF23 as well as transcriptional control of other bone anabolic and catabolic genes that orchestrate the regulation of phosphate and calcium mineral metabolism. *J. Steroid Biochem. Mol. Biol.* **2007**, *103*, 381–388. [CrossRef] [PubMed]
2. Thompson, P.D.; Jurutka, P.W.; Whitfield, G.K.; Myskowski, S.M.; Eichhorst, K.R.; Dominguez, C.E.; Haussler, C.A.; Haussler, M.R. Liganded VDR induces CYP3A4 in small intestinal and colon cancer cells via DR3 and ER6 vitamin D responsive elements. *Biochem. Biophys. Res. Commun.* **2002**, *299*, 730–738. [CrossRef]
3. Haussler, M.R.; Jurutka, P.W.; Mizwicki, M.; Norman, A.W. Vitamin D receptor (VDR)-mediated actions of $1\alpha$,25(OH)$_2$vitamin D$_3$: Genomic and non-genomic mechanisms. *Best Pract. Res. Clin. Endocrinol. Metab.* **2011**, *25*, 543–559. [CrossRef] [PubMed]
4. Samuel, S.; Sitrin, M.D. Vitamin D's role in cell proliferation and differentiation. *Nutr. Rev.* **2008**, *66*, S116–S124. [CrossRef] [PubMed]
5. Trochoutsou, A.I.; Kloukina, V.; Samitas, K.; Xanthou, G. Vitamin-D in the Immune System: Genomic and Non-Genomic Actions. *Mini Rev. Med. Chem.* **2015**, *15*, 953–963. [CrossRef] [PubMed]
6. Kriebitzsch, C.; Verlinden, L.; Eelen, G.; van Schoor, N.M.; Swart, K.; Lips, P.; Meyer, M.B.; Pike, J.W.; Boonen, S.; Carlberg, C.; et al. 1,25-Dihydroxyvitamin D3 influences cellular homocysteine levels in murine preosteoblastic MC3T3-E1 cells by direct regulation of cystathionine β-synthase. *J. Bone Miner. Res.* **2011**, *26*, 2991–3000. [CrossRef] [PubMed]
7. Bozic, M.; Guzmán, C.; Benet, M.; Sánchez-Campos, S.; García-Monzón, C.; Gari, E.; Gatius, S.; Valdivielso, J.M.; Jover, R. Hepatocyte vitamin D receptor regulates lipid metabolism and mediates experimental diet-induced steatosis. *J. Hepatol.* **2016**, *65*, 748–757. [CrossRef] [PubMed]
8. Zheng, W.; Tayyari, F.; Gowda, G.A.N.; Raftery, D.; McLamore, E.S.; Shi, J.; Porterfield, D.M.; Donkin, S.S.; Bequette, B.; Teegarden, D. 1,25-dihydroxyvitamin D regulation of glucose metabolism in Harvey-ras transformed MCF10A human breast epithelial cells. *J. Steroid Biochem. Mol. Biol.* **2013**, *138*, 81–89. [CrossRef] [PubMed]
9. Silvagno, F.; De Vivo, E.; Attanasio, A.; Gallo, V.; Mazzucco, G.; Pescarmona, G. Mitochondrial localization of vitamin D receptor in human platelets and differentiated megakaryocytes. *PLoS ONE* **2010**, *5*, e8670. [CrossRef] [PubMed]
10. Silvagno, F.; Consiglio, M.; Foglizzo, V.; Destefanis, M.; Pescarmona, G. Mitochondrial translocation of vitamin D receptor is mediated by the permeability transition pore in human keratinocyte cell line. *PLoS ONE* **2013**, *8*, e54716. [CrossRef] [PubMed]
11. Consiglio, M.; Destefanis, M.; Morena, D.; Foglizzo, V.; Forneris, M.; Pescarmona, G.; Silvagno, F. The vitamin D receptor inhibits the respiratory chain, contributing to the metabolic switch that is essential for cancer cell proliferation. *PLoS ONE* **2014**, *9*, e115816. [CrossRef] [PubMed]
12. Consiglio, M.; Viano, M.; Casarin, S.; Castagnoli, C.; Pescarmona, G.; Silvagno, F. Mitochondrial and lipogenic effects of vitamin D on differentiating and proliferating human keratinocytes. *Exp. Dermatol.* **2015**, *24*, 748–753. [CrossRef] [PubMed]
13. Ricciardi, C.J.; Bae, J.; Esposito, D.; Komarnytsky, S.; Hu, P.; Chen, J.; Zhao, L. 1,25-Dihydroxyvitamin D3/vitamin D receptor suppresses brown adipocyte differentiation and mitochondrial respiration. *Eur. J. Nutr.* **2015**, *54*, 1001–1012. [CrossRef] [PubMed]
14. Carafoli, E. The interplay of mitochondria with calcium: An historical appraisal. *Cell Calcium* **2012**, *52*, 1–8. [CrossRef] [PubMed]
15. Napier, I.; Ponka, P.; Richardson, D.R. Iron trafficking in the mitochondrion: Novel pathways revealed by disease. *Blood* **2005**, *105*, 1867–1874. [CrossRef] [PubMed]
16. Soldani, C.; Lazzè, M.C.; Bottone, M.G.; Tognon, G.; Biggiogera, M.; Pellicciari, C.E.; Scovassi, A.I. Poly(ADP-ribose) polymerase cleavage during apoptosis: When and where? *Exp. Cell Res.* **2001**, *269*, 193–201. [CrossRef] [PubMed]
17. Basit, S. Vitamin D in health and disease: A literature review. *Br. J. Biomed. Sci.* **2013**, *70*, 161–172. [CrossRef] [PubMed]

18. Suda, T.; Masuyama, R.; Bouillon, R.; Carmeliet, G. Physiological functions of vitamin D: What we have learned from global and conditional VDR knockout mouse studies. *Curr. Opin. Pharmacol.* **2015**, *22*, 87–99. [CrossRef] [PubMed]
19. Oda, Y.; Uchida, Y.; Moradian, S.; Crumrine, D.; Elias, P.M.; Bikle, D.D. Vitamin D receptor and coactivators SRC2 and 3 regulate epidermis-specific sphingolipid production and permeability barrier formation. *J. Investig. Dermatol.* **2009**, *129*, 1367–1378. [CrossRef] [PubMed]
20. Kong, J.; Zhu, X.; Shi, Y.; Liu, T.; Chen, Y.; Bhan, I.; Zhao, Q.; Thadhani, R.; Li, Y.C. VDR attenuates acute lung injury by blocking Ang-2-Tie-2 pathway and renin-angiotensin system. *Mol. Endocrinol.* **2013**, *27*, 2116–2125. [CrossRef] [PubMed]
21. Girgis, C.M.; Cha, K.M.; Houweling, P.J.; Rao, R.; Mokbel, N.; Lin, M.; Clifton-Bligh, R.J.; Gunton, J.E. Vitamin D Receptor Ablation and Vitamin D Deficiency Result in Reduced Grip Strength, Altered Muscle Fibers, and Increased Myostatin in Mice. *Calcif. Tissue Int.* **2015**, *97*, 602–610. [CrossRef] [PubMed]
22. Ni, W.; Watts, S.W.; Ng, M.; Chen, S.; Glenn, D.J.; Gardner, D.G. Elimination of vitamin D receptor in vascular endothelial cells alters vascular function. *Hypertension* **2014**, *64*, 1290–1298. [CrossRef] [PubMed]
23. Wong, K.E.; Szeto, F.L.; Zhang, W.; Ye, H.; Kong, J.; Zhang, Z.; Sun, X.J.; Li, Y.C. Involvement of the vitamin D receptor in energy metabolism: Regulation of uncoupling proteins. *Am. J. Physiol. Endocrinol. Metab.* **2009**, *296*, E820–E828. [CrossRef] [PubMed]
24. Narvaez, C.J.; Matthews, D.; Broun, E.; Chan, M.; Welsh, J. Lean phenotype and resistance to diet-induced obesity in vitamin D receptor knockout mice correlates with induction of uncoupling protein-1 in white adipose tissue. *Endocrinology* **2009**, *150*, 651–661. [CrossRef] [PubMed]
25. Bhat, M.; Noolu, B.; Qadri, S.S.Y.H.; Ismail, A. Vitamin D deficiency decreases adiposity in rats and causes altered expression of uncoupling proteins and steroid receptor coactivator3. *J. Steroid Biochem. Mol. Biol.* **2014**, *144 Pt B*, 304–312. [CrossRef] [PubMed]
26. Alimirah, F.; Vaishnav, A.; McCormick, M.; Echchgadda, I.; Chatterjee, B.; Mehta, R.G.; Peng, X. Functionality of unliganded VDR in breast cancer cells: Repressive action on CYP24 basal transcription. *Mol. Cell. Biochem.* **2010**, *342*, 143–150. [CrossRef] [PubMed]
27. Huet, T.; Laverny, G.; Ciesielski, F.; Molnár, F.; Ramamoorthy, T.G.; Belorusova, A.Y.; Antony, P.; Potier, N.; Metzger, D.; Moras, D.; et al. A vitamin D receptor selectively activated by gemini analogs reveals ligand dependent and independent effects. *Cell Rep.* **2015**, *10*, 516–526. [CrossRef] [PubMed]
28. Menegaz, D.; Mizwicki, M.T.; Barrientos-Duran, A.; Chen, N.; Henry, H.L.; Norman, A.W. Vitamin D receptor (VDR) regulation of voltage-gated chloride channels by ligands preferring a VDR-alternative pocket (VDR-AP). *Mol. Endocrinol.* **2011**, *25*, 1289–1300. [CrossRef] [PubMed]
29. Batie, S.; Lee, J.H.; Jama, R.A.; Browder, D.O.; Montano, L.A.; Huynh, C.C.; Marcus, L.M.; Tsosie, D.G.; Mohammed, Z.; Trang, V.; et al. Synthesis and biological evaluation of halogenated curcumin analogs as potential nuclear receptor selective agonists. *Bioorg. Med. Chem.* **2013**, *21*, 693–702. [CrossRef] [PubMed]
30. Belorusova, A.Y.; Eberhardt, J.; Potier, N.; Stote, R.H.; Dejaegere, A.; Rochel, N. Structural insights into the molecular mechanism of vitamin D receptor activation by lithocholic acid involving a new mode of ligand recognition. *J. Med. Chem.* **2014**, *57*, 4710–4719. [CrossRef] [PubMed]
31. Dröse, S.; Brandt, U. Molecular mechanisms of superoxide production by the mitochondrial respiratory chain. In *Mitochondrial Oxidative Phosphorylation*; Springer: Berlin/Heidelberg, Germany, 2012; pp. 145–169.
32. Arnold, S. Cytochrome c oxidase and its role in neurodegeneration and neuroprotection. *Adv. Exp. Med. Biol.* **2012**, *748*, 305–339. [CrossRef] [PubMed]
33. Roy, D.; Felty, Q.; Narayan, S.; Jayakar, P. Signature of mitochondria of steroidal hormones-dependent normal and cancer cells: Potential molecular targets for cancer therapy. *Front. Biosci.* **2007**, *12*, 154–173. [CrossRef] [PubMed]
34. Galadari, S.; Rahman, A.; Pallichankandy, S.; Thayyullathil, F. Reactive oxygen species and cancer paradox: To promote or to suppress? *Free Radic. Biol. Med.* **2017**, *104*, 144–164. [CrossRef] [PubMed]
35. Hüttemann, M.; Pecina, P.; Rainbolt, M.; Sanderson, T.H.; Kagan, V.E.; Samavati, L.; Doan, J.W.; Lee, I. The multiple functions of cytochrome c and their regulation in life and death decisions of the mammalian cell: From respiration to apoptosis. *Mitochondrion* **2011**, *11*, 369–381. [CrossRef] [PubMed]
36. Piotrowska, A.; Wierzbicka, J.; Ślebioda, T.; Woźniak, M.; Tuckey, R.C.; Slominski, A.T.; Żmijewski, M.A. Vitamin D derivatives enhance cytotoxic effects of $H_2O_2$ or cisplatin on human keratinocytes. *Steroids* **2016**, *110*, 49–61. [CrossRef] [PubMed]

37. Shymanskyy, I.O.; Lisakovska, O.O.; Mazanova, A.O.; Labudzynskyi, D.O.; Khomenko, A.V.; Veliky, M.M. Prednisolone and vitamin $D_3$ modulate oxidative metabolism and cell death pathways in blood and bone marrow mononuclear cells. *Ukr. Biochem. J.* **2016**, *88*, 38–47. [CrossRef] [PubMed]
38. Zeeli, T.; Langberg, M.; Rotem, C.; David, M.; Koren, R.; Ravid, A. Vitamin D inhibits captopril-induced cell detachment and apoptosis in keratinocytes. *Br. J. Dermatol.* **2011**, *164*, 62–67. [CrossRef] [PubMed]
39. Uberti, F.; Lattuada, D.; Morsanuto, V.; Nava, U.; Bolis, G.; Vacca, G.; Squarzanti, D.F.; Cisari, C.; Molinari, C. Vitamin D protects human endothelial cells from oxidative stress through the autophagic and survival pathways. *J. Clin. Endocrinol. Metab.* **2014**, *99*, 1367–1374. [CrossRef] [PubMed]
40. Longoni, A.; Kolling, J.; Siebert, C.; Dos Santos, J.P.; da Silva, J.S.; Pettenuzzo, L.F.; Meira-Martins, L.A.; Gonçalves, C.-A.; de Assis, A.M.; Wyse, A.T.S. 1,25-Dihydroxyvitamin D3 prevents deleterious effects of homocysteine on mitochondrial function and redox status in heart slices. *Nutr. Res.* **2017**, *38*, 52–63. [CrossRef] [PubMed]
41. Uberti, F.; Bardelli, C.; Morsanuto, V.; Ghirlanda, S.; Molinari, C. Role of vitamin $D_3$ combined to alginates in preventing acid and oxidative injury in cultured gastric epithelial cells. *BMC Gastroenterol.* **2016**, *16*. [CrossRef] [PubMed]
42. Zheng, Y.; Trivedi, T.; Lin, R.C.; Fong-Yee, C.; Nolte, R.; Manibo, J.; Chen, Y.; Hossain, M.; Horas, K.; Dunstan, C.; et al. Loss of the vitamin D receptor in human breast and prostate cancers strongly induces cell apoptosis through downregulation of Wnt/β-catenin signaling. *Bone Res.* **2017**, *5*, 17023. [CrossRef] [PubMed]
43. Pilon, C.; Rebellato, A.; Urbanet, R.; Guzzardo, V.; Cappellesso, R.; Sasano, H.; Fassina, A.; Fallo, F. Methylation Status of Vitamin D Receptor Gene Promoter in Benign and Malignant Adrenal Tumors. *Int. J. Endocrinol.* **2015**, *2015*, 375349. [CrossRef] [PubMed]
44. Chandel, N.; Ayasolla, K.S.; Lan, X.; Sultana-Syed, M.; Chawla, A.; Lederman, R.; Vethantham, V.; Saleem, M.A.; Chander, P.N.; Malhotra, A.; et al. Epigenetic Modulation of Human Podocyte Vitamin D Receptor in HIV Milieu. *J. Mol. Biol.* **2015**, *427*, 3201–3215. [CrossRef] [PubMed]
45. Jurutka, P.W.; Remus, L.S.; Whitfield, G.K.; Thompson, P.D.; Hsieh, J.C.; Zitzer, H.; Tavakkoli, P.; Galligan, M.A.; Dang, H.T.; Haussler, C.A.; et al. The polymorphic N terminus in human vitamin D receptor isoforms influences transcriptional activity by modulating interaction with transcription factor IIB. *Mol. Endocrinol.* **2000**, *14*, 401–420. [CrossRef] [PubMed]
46. Pulito, C.; Terrenato, I.; Di Benedetto, A.; Korita, E.; Goeman, F.; Sacconi, A.; Biagioni, F.; Blandino, G.; Strano, S.; Muti, P.; et al. Cdx2 polymorphism affects the activities of vitamin D receptor in human breast cancer cell lines and human breast carcinomas. *PLoS ONE* **2015**, *10*, e0124894. [CrossRef] [PubMed]
47. Morán, M.; Moreno-Lastres, D.; Marín-Buera, L.; Arenas, J.; Martín, M.A.; Ugalde, C. Mitochondrial respiratory chain dysfunction: Implications in neurodegeneration. *Free Radic. Biol. Med.* **2012**, *53*, 595–609. [CrossRef] [PubMed]
48. Ma, J.; Zhang, Q.; Chen, S.; Fang, B.; Yang, Q.; Chen, C.; Miele, L.; Sarkar, F.H.; Xia, J.; Wang, Z. Mitochondrial dysfunction promotes breast cancer cell migration and invasion through HIF1α accumulation via increased production of reactive oxygen species. *PLoS ONE* **2013**, *8*, e69485. [CrossRef] [PubMed]
49. Bratic, A.; Larsson, N.-G. The role of mitochondria in aging. *J. Clin. Investig.* **2013**, *123*, 951–957. [CrossRef] [PubMed]
50. Sohn, J.H.; Han, K.-L.; Lee, S.-H.; Hwang, J.-K. Protective effects of panduratin A against oxidative damage of tert-butylhydroperoxide in human HepG2 cells. *Biol. Pharm. Bull.* **2005**, *28*, 1083–1086. [CrossRef] [PubMed]
51. Destefanis, M.; Viano, M.; Leo, C.; Gervino, G.; Ponzetto, A.; Silvagno, F. Extremely low frequency electromagnetic fields affect proliferation and mitochondrial activity of human cancer cell lines. *Int. J. Radiat. Biol.* **2015**, *91*, 964–972. [CrossRef] [PubMed]

International Journal of
*Molecular Sciences*

MDPI

*Article*

# Expression of TXNIP in Cancer Cells and Regulation by 1,25(OH)$_2$D$_3$: Is It Really the Vitamin D$_3$ Upregulated Protein?

**Mohamed A. Abu el Maaty, Fadi Almouhanna and Stefan Wölfl ***

Institute of Pharmacy and Molecular Biotechnology, Heidelberg University, Im Neuenheimer Feld 364, 69120 Heidelberg, Germany; Abdelgawad@stud.uni-heidelberg.de (M.A.A.e.M.); f.almouhanna@stud.uni-heidelberg.de (F.A.)

* Correspondence: wolfl@uni-hd.de; Tel.: +49-6221-544-880

Received: 21 February 2018; Accepted: 7 March 2018; Published: 10 March 2018

**Abstract:** Thioredoxin-interacting protein (TXNIP) was originally identified in HL-60 cells as the vitamin D$_3$ upregulated protein 1, and is now known to be involved in diverse cellular processes, such as maintenance of glucose homeostasis, redox balance, and apoptosis. Besides the initial characterization, little is known about if and how 1,25-dihydroxyvitamin D$_3$ [1,25(OH)$_2$D$_3$] induces TXNIP expression. We therefore screened multiple cancerous cell lines of different tissue origins, and observed induction, repression, or no change in TXNIP expression in response to 1,25(OH)$_2$D$_3$. In-depth analyses on HL-60 cells revealed a rapid and transient increase in TXNIP mRNA levels by 1,25(OH)$_2$D$_3$ (3–24 h), followed by a clear reduction at later time points. Furthermore, a strong induction in protein levels was observed only after 96 h of 1,25(OH)$_2$D$_3$ treatment. Induction of TXNIP expression by 1,25(OH)$_2$D$_3$ was found to be dependent on the availability of glucose in the culture medium, as well as the presence of a functional glucose transport system, indicating an inter-dependence of 1,25(OH)$_2$D$_3$ actions and glucose-sensing mechanisms. Moreover, the inhibition of de novo protein synthesis by cycloheximide reduced TXNIP half-life in 24 h, but not in 96 h-1,25(OH)$_2$D$_3$-treated HL-60 cells, demonstrating a possible influence of 1,25(OH)$_2$D$_3$ on TXNIP stability in long-term treatment.

**Keywords:** vitamin D; TXNIP; VDUP1; cancer

## 1. Introduction

1,25-dihydroxyvitamin D$_3$ [1,25(OH)$_2$D$_3$] is the biologically active form of vitamin D$_3$, and is the natural ligand of the nuclear vitamin D receptor (VDR) [1]. Upon binding to its receptor, 1,25(OH)$_2$D$_3$ induces immense changes in gene expression patterns in different cells, with several genes identified as either direct—harboring vitamin D response elements (VDRE)—or indirect targets of the molecule. Such genes include those involved in the regulation of proliferation, such as p21 and p27, apoptosis-related genes, like BAX, and tissue-specific differentiation genes, such as prostate specific antigen [1–3]. Among the putative vitamin D target genes is thioredoxin-interacting protein (TXNIP), which is located on chromosome 1q21.1.

TXNIP was originally reported by Chen and Deluca in 1994 as the VDUP1 (vitamin D$_3$ upregulated protein 1) [4]. Their search of 1,25(OH)$_2$D$_3$-responsive cDNAs in HL-60 (acute myeloid leukemia) cells led to the identification of this gene, whose mRNA levels were found to be induced by 1,25(OH)$_2$D$_3$ treatment in as early as 6 h, plateaued at 18 h, and was maintained till 24 h. In 1999, Nishiyama and coworkers reported that VDUP1 is the binding partner of reduced thioredoxin, and thus acts as its negative regulator through inhibiting its anti-oxidant function [5]. Since then, a plethora of publications from different groups has shown that, in addition to influencing redox homeostasis, TXNIP senses intracellular levels of glucose and glycolytic intermediates, as well as levels of adenosine-containing molecules [6–8]. In response to an increase in intracellular glucose levels, the heterodimer MondoA:MLX (Max-like protein X) translocates to the nucleus, and binds to carbohydrate response elements (ChoRE) on the promoter of the *TXNIP* gene, inducing its expression [6]. TXNIP in turn works to reduce intracellular glucose levels, by decreasing its uptake, possibly through limiting the membrane availability of glucose transporter 1 [9].

With regards to cancer, TXNIP is proposed to be a potential tumor suppressor due to its ability to induce oxidative stress-mediated apoptosis, as well as due to its reduced expression and silencing in tumor tissues and cancer cell lines [10,11]. Therefore, compounds that reactivate TXNIP expression are viewed as promising anti-tumor candidates, such as the histone deacetylase inhibitor suberoylanilide hydroxamic acid [12] and the histone methyltransferase inhibitor 3-Deazaneplanocin A [13]. Consequently, due to the "historical" association between 1,25$(OH)_2D_3$ and TXNIP as a vitamin D upregulated protein, it has been suggested that 1,25$(OH)_2D_3$ could be used to reactivate TXNIP in cancers [10]. To our knowledge, this association has only been shown in HL-60 cells [14], and thus extrapolation of this relationship to other cancers is highly speculative.

Recently, we showed that in the prostate cancer cell line LNCaP, 1,25$(OH)_2D_3$ treatment induces diverse metabolic changes that lead to a significant reduction of TXNIP levels [15]. This non-canonical regulation prompted us to investigate whether 1,25$(OH)_2D_3$ is capable of inducing TXNIP expression in different cancer types, including those with silenced TXNIP expression. An initial screen of different cancer cell lines that are treated with 1,25$(OH)_2D_3$ surprisingly showed all the possible options, induction, reduction, and no change in TXNIP levels in response to treatment. A detailed analysis of 1,25$(OH)_2D_3$ treatment in HL-60 cells, in which the up-regulation was initially discovered, demonstrated a transient increase in TXNIP mRNA levels in response to treatment that was diminished at later time points. Despite the reduced TXNIP mRNA levels upon longer term treatment, TXNIP protein levels were clearly higher after 1,25$(OH)_2D_3$ treatment compared to cells treated with dimethyl sulfoxide (DMSO). Inhibition of de novo protein synthesis in 1,25$(OH)_2D_3$-treated cells by cycloheximide (CHX) did not reduce TXNIP protein levels at a time point when mRNA levels were reduced (96 h), suggesting an increased TXNIP half-life in the presence of 1,25$(OH)_2D_3$. Additionally, we observed a lack of TXNIP induction on the mRNA and protein levels, as well as a lack of oxidative stress in response to 1,25$(OH)_2D_3$ treatment in the absence of glucose, hinting at a critical cross-talk between VDR mediated gene regulation and glucose-sensing transcriptional machinery like MondoA/MLX.

## 2. Results

### 2.1. 1,25$(OH)_2D_3$ Induces, Reduces and Has No Effect on TXNIP Levels in Cancer Cells of Different Tissue Origins

To investigate whether 1,25$(OH)_2D_3$ is capable of inducing TXNIP expression in different cancer models, cell lines of various tissue origins—hematological, prostate, pancreatic, liver, colorectal, and breast—were treated with either DMSO or 1,25$(OH)_2D_3$ (100 nM) for 72 h and TXNIP levels were analyzed using immunoblotting. The investigated cell lines exhibited disparate basal levels of TXNIP, where some cell lines had completely repressed TXNIP expression, with the colorectal cancer cell line HCT116 as an example, some cell lines had high basal expression of the protein, like the breast cancer cell line MCF-7, and finally, cell lines with lower TXNIP levels, like the pancreatic cancer cell line AsPC-1 (Figure 1).

In cell lines where TXNIP expression was silenced, 1,25$(OH)_2D_3$ did not appear to induce the expression of the protein, whereas in cell lines that expressed TXNIP at varying degrees, 1,25$(OH)_2D_3$ exerted either the expected "classical" induction, for example in U937 (histiocytic lymphoma), or non-canonical reduction, as observed in LNCaP, BxPC-3, and MCF-7 (prostate, pancreatic, and breast cancer cells, respectively), but also had no clear effect on TXNIP levels, like in HT-29 cells (colorectal cancer) (Figure 1).

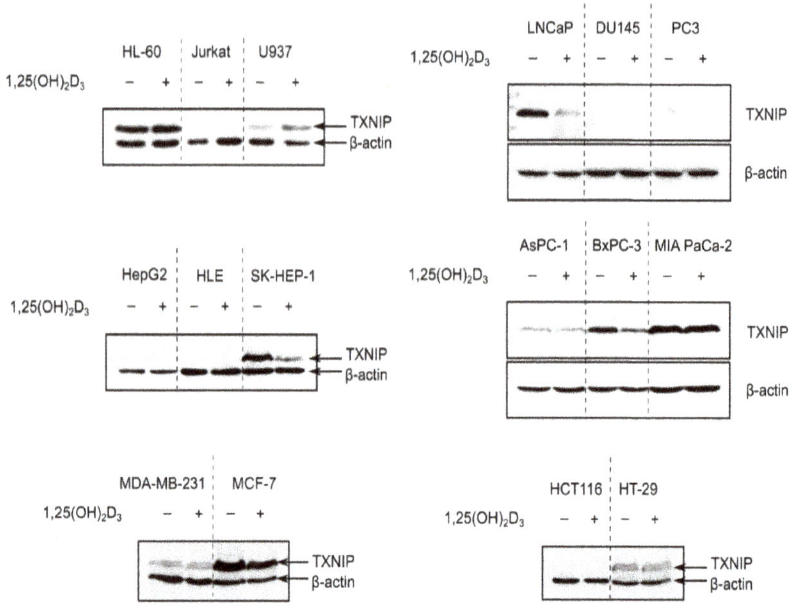


**Figure 1.** TXNIP is differentially regulated by 1,25(OH)$_2$D$_3$ in cancer cells. Immunoblots showing TXNIP protein levels in cancer cells of various tissue origins—hematological (HL-60, Jurkat, U937), prostate (LNCaP, DU145, PC3), liver (HepG2, HLE, SK-HEP-1), pancreatic (AsPC-1, BxPC-3, MIA PaCa-2), breast (MDA-MB-231, MCF-7), and colorectal (HCT116, HT-29)—after treatment with 100 nM 1,25(OH)$_2$D$_3$ for 72 h. In several cell lines, e.g., Jurkat cells, TXNIP is not detected and thus no influence of 1,25(OH)$_2$D$_3$ on its expression can be observed. In cases were basal levels are detected, 1,25(OH)$_2$D$_3$ either induced, reduced, or had no clear effect on TXNIP expression, e.g., U937, MCF-7, and HT-29, respectively. "+" and "−" denote the presence or absence of the indicated molecule/treatment, respectively.

Interestingly, we did not observe a clear induction in TXNIP levels in response to 1,25(OH)$_2$D$_3$ in HL-60 cells (Figure 1), and thus performed an extensive analysis into the temporal regulation of TXNIP by 1,25(OH)$_2$D$_3$, with the aim to identify possible modulatory mechanisms adjusting the metabolic response to 1,25(OH)$_2$D$_3$ in HL-60 cells, thereby triggering differential TXNIP regulation, as well as to investigate the possible dependence of this regulation on glucose-related signaling.

### 2.2. 1,25(OH)$_2$D$_3$ Transiently Induces TXNIP mRNA Levels but Increases Protein Levels in Long-Term Treatments

Since the original publication describing VDUP1 demonstrated the induction of TXNIP/VDUP1 by 1,25(OH)$_2$D$_3$ on the mRNA level at time points ranging from 6 to 24 h [4], we aimed to reproduce their findings, and add later time points to investigate whether their observed induction was transient or not. A time course spanning 1 to 96 h was employed and mRNA expression analysis was performed using RT-qPCR (Reverse Transcription-quantitative PCR). We observed a clear induction in TXNIP mRNA levels with 1,25(OH)$_2$D$_3$ treatment at time points 3, 6, and 24 h, and a surprising reduction at 72 and 96 h, demonstrating that the induction in TXNIP mRNA levels by 1,25(OH)$_2$D$_3$ is not sustained in the longer time treatments (Figure 2a).

Having analyzed mRNA levels we next analyzed TXNIP protein levels in HL-60 cells upon 1,25(OH)$_2$D$_3$ treatment at the same time points using immunoblotting (Figure 2b,c), and observed fluctuations in the basal levels independent of treatment, which could be attributed to changes in nutrient availability in the environment, or to the natural glucose-sensing, homeostatic mechanism that TXNIP is known to influence. We measured glucose levels in the culture medium of DMSO- and 1,25(OH)$_2$D$_3$-treated HL-60 cells and observed that while the glucose levels were clearly reduced in the medium of DMSO-treated cells, the levels were not diminished (Figure S1), demonstrating that the relative decrease in basal TXNIP level at the latest time point is not due to glucose depletion, but rather reflects the influence of glucose homeostasis on TXNIP expression. When compared to DMSO-treated cells, and independent of temporal fluctuations in basal expression, 1,25(OH)$_2$D$_3$ was found to mildly induce TXNIP levels in 24 h and strongly in 96 h (Figure 2c–g).

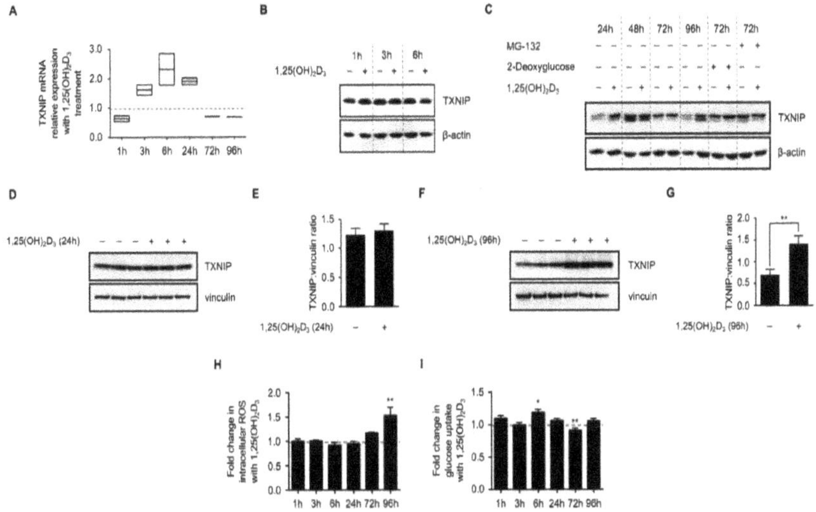

**Figure 2.** Time-dependent regulation of TXNIP expression and associated cellular processes in HL-60 cells by 1,25(OH)$_2$D$_3$. (**A**) 1,25(OH)$_2$D$_3$ (100 nM) induced a transient increase in TXNIP mRNA levels that was diminished at later time points. The dashed red line indicates the baseline TXNIP mRNA expression in DMSO-treated cells set to 1; (**B**,**C**) Immunoblot showing TXNIP protein levels in response to 1,25(OH)$_2$D$_3$; induction is observed after 24 h and also after 96 h of treatment. Basal TXNIP levels exhibit temporal fluctuations, independent of treatment. Treatment with 2-deoxyglucose (10 mM), a potent inducer of TXNIP expression, and MG-132 (5 μM), a proteasomal inhibitor to prevent degradation, were used as positive controls. The compounds were added into the conditioned medium of DMSO- and 1,25(OH)$_2$D$_3$-treated HL-60 cells either 24 h (2-deoxyglucose) or 6 h (MG-132) before the end of the initial treatment period. "+" and "−" denote the presence or absence of the indicated molecule/treatment, respectively; (**D**) Immunoblot and (**E**) densitometric quantification of TXNIP protein levels after 24 h of treatment of HL-60 cells with 1,25(OH)$_2$D$_3$ indicate a mild increase in TXNIP level, which lacks statistical significance; (**F**) Immunoblot and (**G**) densitometric quantification after 96 h of treatment of HL-60 cells with 1,25(OH)$_2$D$_3$ showing a statistically significant induction of TXNIP levels; (**H**) Intracellular ROS levels, presented as fold change of treated vs. control cells, were found to be significantly induced by 1,25(OH)$_2$D$_3$ only at the latest time point (96 h). The dashed red line indicates the baseline ROS level in DMSO-treated cells set to 1; (**I**) Glucose uptake, which is known to be inhibited by TXNIP, was found to be significantly induced by 1,25(OH)$_2$D$_3$ at an early time point (6 h), and reduced at a later one (72 h). Baseline glucose uptake level in DMSO-treated cells is indicated by the dashed red line set to 1. Statistical significance was calculated using a two-tailed Student's $t$-test. $p$-Values less than or equal to 0.05 and 0.01, are depicted by * and **, respectively. Error bars ± SD; $n$ = 3.

We then investigated changes in cellular parameters known to be influenced by TXNIP status, namely intracellular reactive oxygen species (ROS) levels and glucose uptake, in response to $1,25(OH)_2D_3$ treatment. Intracellular ROS levels were largely unaffected by the treatment across most of the investigated time points, except for the latest one (96 h), where $1,25(OH)_2D_3$ was found to strongly and significantly induce ROS levels (Figure 2h). On the other hand, an induction in glucose uptake was observed after 6 h of treatment with $1,25(OH)_2D_3$, an effect that was diminished in later time points (Figure 2i). In fact, after 72 h of treatment with $1,25(OH)_2D_3$, glucose uptake was found to be clearly reduced when compared to DMSO-treated cells (Figure 2i).

In view of the presented results, we hypothesized that: (i) glucose availability and subsequently recruitment of transcriptional machinery capable of regulating TXNIP expression, might be crucial in mediating $1,25(OH)_2D_3$'s effect, since basal expression levels exhibited profound temporal fluctuations, reflecting the time-dependent effects of cellular glucose homeostasis on TXNIP expression, and (ii) transcriptional induction of TXNIP by $1,25(OH)_2D_3$ may not be solely responsible for the observed upregulation on the protein level since at the latest time point, TXNIP mRNA levels were found to be reduced, whereas the protein expression induced by treatment, highlighting the possible involvement of protein stabilizing mechanisms at later time points, and finally (iii) $1,25(OH)_2D_3$ induces changes in glucose metabolism—glycolysis and/or mitochondrial respiration—that stimulate TXNIP expression.

### 2.3. Glucose Availability Is Crucial for $1,25(OH)_2D_3$-Mediated Regulation of TXNIP mRNA and Protein Levels as Well as Associated Oxidative Stress

As previously mentioned, tight regulation of glucose homeostasis by TXNIP is orchestrated through the nuclear translocation of the heterodimer MondoA/MLX, which regulates TXNIP expression by binding to ChoRE on the gene's promoter. Given the observed temporal fluctuations in TXNIP expression (Figure 2c), we postulated that glucose-sensing mechanisms might be involved in mediating $1,25(OH)_2D_3$'s effects on TXNIP mRNA and protein levels. We thus cultured HL-60 cells and performed treatments for various time points in glucose-free medium, and investigated TXNIP mRNA and protein expression. In the absence of glucose, TXNIP expression was found to be clearly diminished across all the time points, and addition of $1,25(OH)_2D_3$ was incapable of inducing its expression (Figure 3a). Similarly, in the presence of a single high dose of the glucose transporter inhibitor phloretin, TXNIP expression was not induced by $1,25(OH)_2D_3$ (Figure 3b). Additionally, the observed regulation of TXNIP mRNA expression by $1,25(OH)_2D_3$, whether induction at 24 h, or reduction at 96 h, was not observed in the absence of glucose (Figure 3c), unlike the clear induction in the mRNA levels of the $1,25(OH)_2D_3$ target gene CYP24A1, which was glucose-independent (Figure 3d). Interestingly, the strong increase in oxidative stress observed at the latest time point with $1,25(OH)_2D_3$ treatment, typically associated with elevated TXNIP levels, was not observed in the absence of glucose (Figure 3e). Altogether, we concluded that regulation of TXNIP expression and associated oxidative stress by $1,25(OH)_2D_3$ is glucose-dependent, indicating a possible cross-talk between VDR-signaling and glucose-sensing mechanisms.

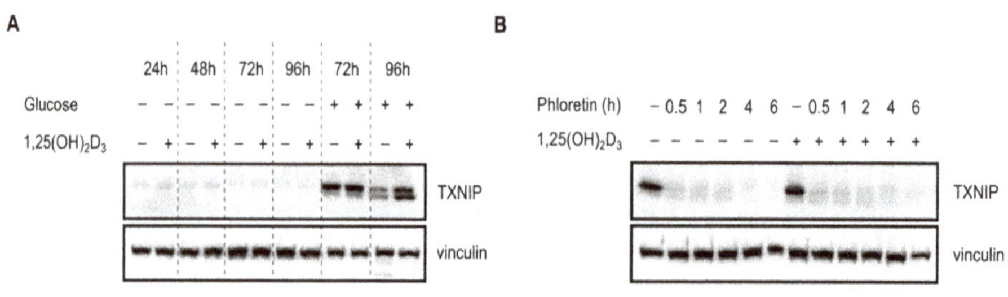


**Figure 3.** *Cont.*

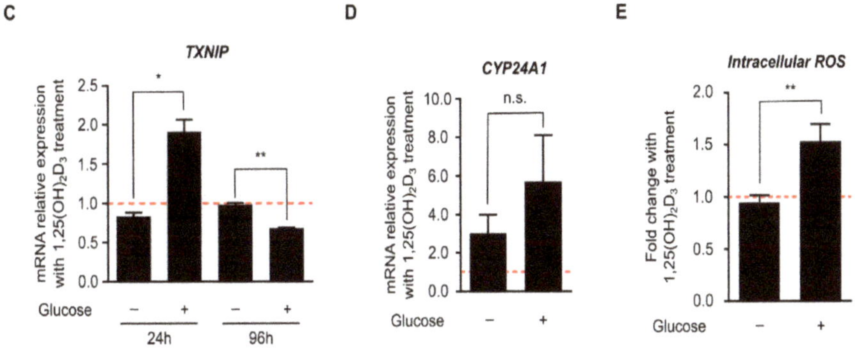


**Figure 3.** Regulation of TXNIP expression and redox balance in HL-60 cells by 1,25(OH)$_2$D$_3$ is glucose-dependent. (**A**,**B**) Immunoblots showing TXNIP protein levels. (**A**) In the absence of glucose, expression of TXNIP in HL-60 cells is significantly reduced, and appears to be unaffected by 1,25(OH)$_2$D$_3$ (100 nM) treatment. "+" and "−" denote the presence or absence of the indicated molecule/treatment, respectively; (**B**) In the presence of a single high dose of the glucose uptake inhibitor phloretin (200 μM), added at different time intervals prior to the end of the initial treatment period (24 h), 1,25(OH)$_2$D$_3$ is found to be incapable of inducing TXNIP expression; (**C**) Changes in TXNIP mRNA levels after 24 and 96 h of treatment with 1,25(OH)$_2$D$_3$ in HL-60 cells, cultured in the presence or absence of glucose, reveal that regulation by 1,25(OH)$_2$D$_3$ is glucose-dependent. The dashed red line indicates baseline mRNA expression in DMSO-treated cells set to 1. Error bars ± SD; $n$ = 2; (**D**) Unlike TXNIP, the 1,25(OH)$_2$D$_3$ target gene CYP24A1, is induced in response to 1,25(OH)$_2$D$_3$ treatment (96 h) independent of glucose availability. Error bars ± SD; $n$ = 2; (**E**) In absence of glucose, intracellular ROS levels are largely unaltered upon treatment of HL-60 cells for 96 h with 1,25(OH)$_2$D$_3$. The dashed red line indicates basal ROS levels in DMSO-treated cells set to 1. Statistical significance was calculated using a two-tailed Student's t-test, with p-values less than or equal to 0.05 and 0.01 depicted by * and **, respectively. Nonsignificant results are denoted as n.s. Error bars ± SD; $n$ = 3.

### *2.4. Induction in TXNIP Levels by 1,25(OH)$_2$D$_3$ Is Possibly Orchestrated through a Complex Interplay between Transcriptional Induction and Protein Stability*

We then aimed to address the second hypothesis on the possible involvement of protein stabilizing mechanisms. To confirm the lack of translational induction of TXNIP in HL-60 cells by 1,25(OH)$_2$D$_3$ at later time points, cycloheximide (CHX), which is an inhibitor of de novo protein synthesis, was added to 24 and 96 h DMSO- or 1,25(OH)$_2$D$_3$-treated HL-60 cells, for different periods (0.5, 1, 2, and 4 h before the end of the DMSO/1,25(OH)$_2$D$_3$ treatment period). In 24 h-treated cells, CHX treatment was found to reduce TXNIP protein half-life both in the presence and absence of 1,25(OH)$_2$D$_3$ (Figure 4a,b). On the other hand, in 96 h-treated cells, CHX did not hamper the induction in TXNIP levels in the presence of 1,25(OH)$_2$D$_3$, but moderately reduced its levels in the absence of 1,25(OH)$_2$D$_3$ (Figure 4c,d).

In view of this observation, we postulated that long-term treatment of HL-60 cells with 1,25(OH)$_2$D$_3$ induces TXNIP expression by reducing the levels of protein degradation machinery, thereby stabilizing TXNIP. It has been shown recently that the E3 ubiquitin ligase ITCH, targets TXNIP for proteasomal degradation [16]. We thus speculated that 1,25(OH)$_2$D$_3$ may reduce the expression of this protein. While ITCH mRNA levels were indeed found to be significantly reduced by 1,25(OH)$_2$D$_3$ after 24 and 96 h of treatment (Figure 4e), protein levels were insignificantly influenced by treatment across the same time points (Figure 4f–i). We therefore conclude that while ITCH regulation by 1,25(OH)$_2$D$_3$ may have mildly contributed to TXNIP stability, other uncharacterized factors may have also played a role, since 1,25(OH)$_2$D$_3$ treatment has been shown to reduce the expression of proteasome subunits in HL-60 cells, as well as modulate the expression of various genes of protein degradation machinery in other cell lines [17,18].

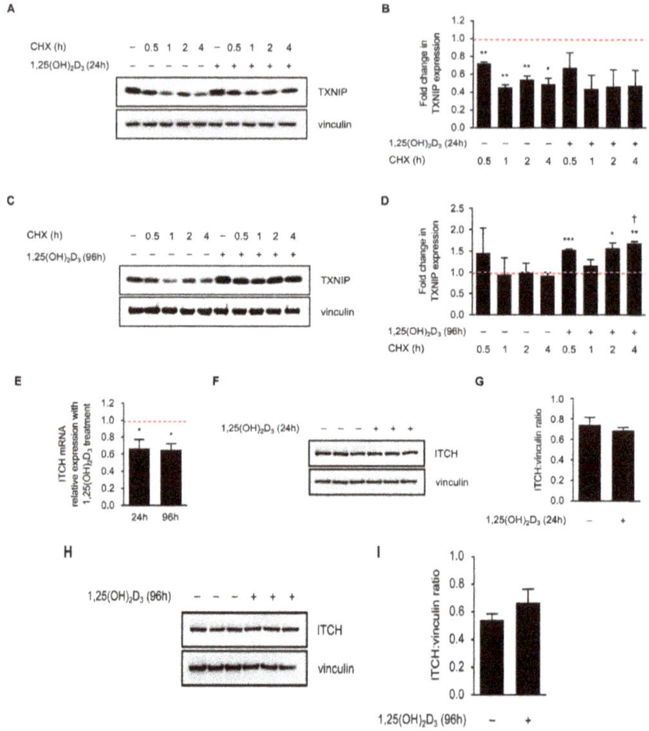


**Figure 4.** Cycloheximide (CHX) reduces TXNIP half-life in 24, but not in 96 h $1,25(OH)_2D_3$-treated HL-60 cells. (**A**) Inhibition of protein synthesis by CHX (10 μM) treatment, added into the conditioned medium of DMSO- and $1,25(OH)_2D_3$-treated HL-60 cells, at different time intervals prior to the end of the initial treatment period (24 h), led to reduced TXNIP protein levels in the presence and absence of $1,25(OH)_2D_3$. "+" and "−" denote the presence or absence of the indicated molecule/treatment, respectively; (**B**) Densitometric quantification of two similar biological replicates. The dashed red line indicates baseline TXNIP protein expression in DMSO-treated cells set to 1; (**C**) In 96 h-treated HL-60 cells, CHX reduced or had no effect on TXNIP protein levels in absence or presence of $1,25(OH)_2D_3$, respectively; (**D**) Densitometric quantification of 2 similar biological replicates. Statistical comparisons made between the different conditions and DMSO-treated cells were performed using a two-tailed Student's *t*-test. *p*-Values less than or equal to 0.05, 0.01, and 0.001, are depicted by *, **, and ***, respectively. A dagger indicates statistical significance compared to the corresponding mono-treatment. Error bars ± SEM; (**E**) ITCH mRNA expression analysis in response to 24 and 96 h of treatment with $1,25(OH)_2D_3$. The dashed red line indicates baseline ITCH mRNA level in DMSO-treated cells set to 1. Error bars ± SD; $n = 2$; (**F**,**G**,**H**,**I**) Analysis of ITCH protein levels after a (**F**,**G**) 24 h or (**H**,**I**) 96 h treatment with $1,25(OH)_2D_3$; (**G**,**I**) Densitometric quantifications illustrate only mild differences in response to treatment.

### *2.5. $1,25(OH)_2D_3$ Does Not Influence Glucose Metabolism in HL-60 Cells but Modulates Overall Intracellular Energy Levels*

Based on our previous findings in prostate cancer cells [15], we postulated that $1,25(OH)_2D_3$ could regulate TXNIP expression in HL-60 cells by inducing metabolic changes that stimulate its expression. Such metabolic alterations may include: (i) reduction of glycolytic rate leading to a relative accumulation of glycolytic intermediates capable of inducing TXNIP levels, and (ii) induction of mitochondrial activity and thus ATP production, which could drive ATP-requiring glycolytic reactions, e.g., that catalyzed by hexokinase, thereby increasing the levels of glycolytic intermediates and subsequently TXNIP expression.

To address this possibility, HL-60 cells were treated with either DMSO or 1,25(OH)$_2$D$_3$ and the pH as well as levels of dissolved oxygen in the culture medium were measured in real-time over the course of four days. No clear differences in the investigated parameters were observed after three days of treatment, however, at the beginning of the fourth day, both parameters were markedly reduced in DMSO-treated cells (Figure 5a). We attribute these differences to the strong reduction in cell number with 1,25(OH)$_2$D$_3$ treatment and not to actual metabolic reprogramming (Figure S2).

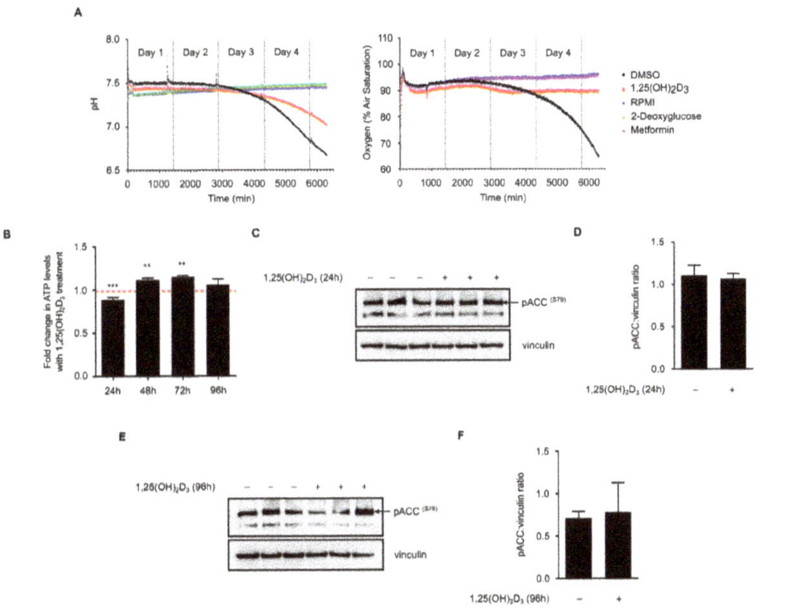


**Figure 5.** 1,25(OH)$_2$D$_3$ does not profoundly impact glucose metabolism but influences overall energy status. (**A**) On-line measurements of culture medium pH and amount of dissolved oxygen in response to different treatments over a time course of four days. Culture medium without cells (RPMI) was used as calibration reference for the measurement of both parameters. 2-deoxyglucose (10 mM) and metformin (2 mM) were used as additional controls for pH or oxygen measurements, respectively. 1,25(OH)$_2$D$_3$ treatment did not predominantly influence glucose metabolism in HL-60 cells. Differences between DMSO and 1,25(OH)$_2$D$_3$ treated cells observed during the fourth day of measurement could by large be attributed to the drastic inhibition of cell proliferation by 1,25(OH)$_2$D$_3$. Data presented are representative of two similar biological replicates; (**B**) ATP levels were significantly reduced with 1,25(OH)$_2$D$_3$ treatment in 24 h, but were elevated with treatment after 48 and 72 h. The dashed red line indicates baseline ATP level in DMSO-treated cells set to 1. Statistical comparisons are made between DMSO- and 1,25(OH)$_2$D$_3$-treated cells using a two-tailed Student's *t*-test. *p*-Values less than or equal to 0.01 and 0.001, are depicted by ** and ***, respectively. Error bars ± SD; $n$ = 3; (**C**,**D**) 24 h treatment of HL-60 cells with 1,25(OH)$_2$D$_3$ did not significantly influence phosphorylation of ACC $^{(S79)}$; immunoblot and densitometric analysis. "+" and "−" denote the presence or absence of the indicated molecule/treatment, respectively; (**E**,**F**) Similar to 24 h-treated cells, a 96 h treatment did not profoundly influence ACC $^{(S79)}$ phosphorylation. Error bars ± SEM; $n$ = 3.

On the other hand, intracellular ATP levels in 1,25(OH)$_2$D$_3$-treated HL-60 cells were found to be significantly reduced at 24 h and induced at the 48 and 72 h time points when compared to DMSO-treated cells (Figure 5b). In view of the lack of clear differences in glycolytic and oxygen consumption rates with 1,25(OH)$_2$D$_3$ treatment, we hypothesize that 1,25(OH)$_2$D$_3$ does not directly impact ATP production but rather utilization, possibly by inhibiting other ATP-consuming processes like fatty acid biosynthesis. We therefore postulate that preservation of ATP levels may constitute part of the mechanism through which 1,25(OH)$_2$D$_3$ regulates TXNIP expression.

Recent studies have demonstrated that the intracellular energy sensor AMP-activated protein kinase (AMPK) is activated by 1,25(OH)$_2$D$_3$ in cancer cells [15,19]. AMPK is known to be activated by different intracellular cues, namely increases in either intracellular calcium levels or in the AMP:ATP ratio [20]. In response to these cues, AMPK activates energy-producing pathways, such as fatty acid beta-oxidation and glucose uptake, and inhibits energy-consuming ones, including fatty acid and protein biosynthesis [20]. Wu et al. [9] have recently shown that the activation of AMPK leads to TXNIP degradation and increased glucose uptake. We thus speculated that the regulation of TXNIP expression by 1,25(OH)$_2$D$_3$ might be partly explained by modulation of this signaling pathway. We investigated the phosphorylation status of serine 79 of the AMPK substrate acetyl CoA carboxylase (ACC), as a biomarker of AMPK signaling activity, in response to 24 and 96 h of treatment with 1,25(OH)$_2$D$_3$. Treatment was not found to significantly influence this pathway at either time point (Figure 5c–f).

## 3. Discussion

Recent studies have shown that TXNIP plays pivotal roles in regulating glucose and redox homeostasis [6,10]. Additionally, it is currently being viewed as a putative tumor suppressor that is based on its ability to induce apoptosis in cancer cells on one hand, and its expression being down-regulated/silenced in tumors on the other [10,11]. Furthermore, it has been demonstrated that the loss of TXNIP increases the predisposition to hepatocellular carcinoma [21]; therefore, reactivating/inducing TXNIP expression is thought to be beneficial to anti-cancer therapy.

Although TXNIP was originally identified in HL-60 cells as the VDUP1 [4], reports on the ability of 1,25(OH)$_2$D$_3$ to induce its expression in cancer cells of diverse tissue origins are sparse, and have been largely limited to HL-60 cells. In our study presented here, we found no evidence for a direct link between vitamin D treatment and TXNIP levels. In fact, TXNIP levels are differentially regulated by 1,25(OH)$_2$D$_3$ in different cancer cell lines (Figure 1). Furthermore, induction of TXNIP in HL-60 cells on mRNA and protein levels by 1,25(OH)$_2$D$_3$ is glucose-dependent, unlike the regulation of the direct 1,25(OH)$_2$D$_3$ target gene *CYP24A1* (Figure 3). An overview of the described regulation of TXNIP by 1,25(OH)$_2$D$_3$ in HL-60 cells is presented in Figure 6.

Several findings of the current report elicit a number of questions pertaining to the nature of TXNIP regulation by 1,25(OH)$_2$D$_3$, such as whether the *TXNIP* gene is really a direct target of 1,25(OH)$_2$D$_3$, and, in instances where TXNIP expression is reduced by 1,25(OH)$_2$D$_3$, whether this is the result of direct VDR-mediated trans-repression or secondary/indirect effects. Stambolsky and coworkers [22] have demonstrated that mutant p53 alters the transcriptional activity of the VDR in response to 1,25(OH)$_2$D$_3$ treatment, converting the latter from an anti-cancer to a pro-survival agent. The authors demonstrated that in cell lines that are harboring mutant p53, such as MDA-MB-231 and SW480, the mRNA expression of TXNIP, a pro-apoptotic gene, was reduced in response to 1,25(OH)$_2$D$_3$ treatment [22]. While this may serve as an explanation for the non-canonical regulation of TXNIP by 1,25(OH)$_2$D$_3$ in certain scenarios, we posit that additional modes of regulation may exist since in cell lines with wild-type p53, such as LNCaP and MCF-7, we also observed a clear reduction in TXNIP expression in response to 1,25(OH)$_2$D$_3$, and in HL-60 cells, which lack p53 expression [23], 1,25(OH)$_2$D$_3$ induces TXNIP levels.

TXNIP appears to be at the crossroads of various signaling molecules implicated in tumorigenesis and anti-cancer treatment, such as Myc [24], AMPK [9], and mTOR [25], making its regulation subject to diverse cellular processes. Based on our previous findings in prostate cancer cells [15], as well as findings of the current study on HL-60 cells, we postulate that several pathways influenced by 1,25(OH)$_2$D$_3$ may contribute to the observed TXNIP regulation, such as regulation of metabolism-associated signaling molecules, namely AMPK, as well as the modulation of protein stability/degradation.

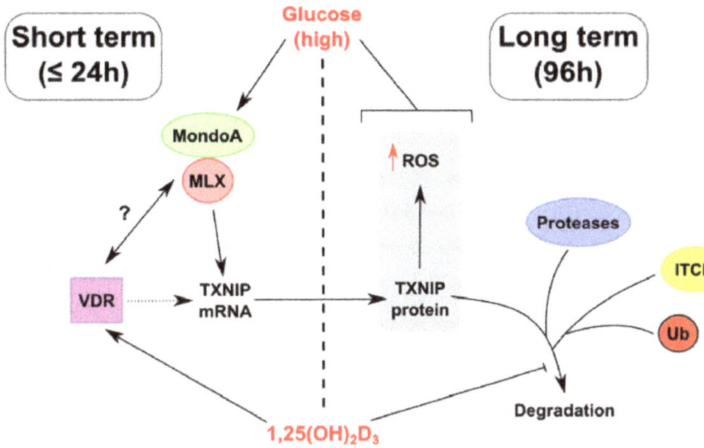


**Figure 6.** Proposed model of TXNIP regulation by 1,25(OH)$_2$D$_3$ in HL-60 cells. 1,25(OH)$_2$D$_3$ appears to regulate TXNIP expression through multiple mechanisms depending on the duration of treatment (long and short term treatment outcomes are separated in the figure by a dashed line). In short term treatments, 1,25(OH)$_2$D$_3$ induces TXNIP mRNA, but not protein levels, an effect that depends on the availability of glucose. This points toward the possibility that the VDR physically/functionally interacts with the transcriptional heterodimer MondoA/MLX, for example, via enhancing its binding to ChoRE on the TXNIP promoter, or via facilitating its recruitment/nuclear translocation. Such putative interactions are depicted in the figure by a question mark. On the other hand, in long term treatments, 1,25(OH)$_2$D$_3$ induces TXNIP protein expression as well as intracellular ROS levels, effects that are also glucose-dependent. Furthermore, long term treatment with 1,25(OH)$_2$D$_3$ influences TXNIP stability possibly through multi-modal regulation of TXNIP-degrading machinery, for example the E3 ligase ITCH, different proteases, or proteasomal subunits. Whether the TXNIP gene harbors VDRE is unclear, thus the vitamin D receptor (VDR) is linked to TXNIP regulation via a dotted line. A blunted arrow (t-bar) indicates inhibitory activity.

Studies have shown that the role of AMPK in cancer is contextual, with both beneficial and detrimental outcomes associated with its activation [26]. While the inhibition of mTOR activity as well as activation by the tumor suppressor LKB1 serve as a basis for AMPK's anti-tumor effects, induction of potentially pro-survival pathways like autophagy and glucose uptake have been proposed to challenge the molecule's anti-cancer role [26]. It is therefore possible that TXNIP degradation upon AMPK activation, described by Wu et al. [9], may contribute to the latter's pro-survival effects. Paradoxically, 1,25(OH)$_2$D$_3$ and its analogues have been shown to activate AMPK signaling in different cancer cell lines [15,19,27]. Assuming that 1,25(OH)$_2$D$_3$ is capable of inducing AMPK signaling in different tumors, it would be interesting to characterize whether the consequence of this activation is TXNIP degradation or induction (through increasing glucose uptake and subsequently glycolytic intermediates that are capable of driving TXNIP expression). In this study, we show that AMPK signaling is insignificantly affected by 1,25(OH)$_2$D$_3$ treatment (Figure 5c–f), suggesting no or only minimal involvement of this pathway in regulating TXNIP expression by 1,25(OH)$_2$D$_3$ in this cellular context. Furthermore, we did not observe clear differences in glucose metabolism of HL-60 cells in response to 1,25(OH)$_2$D$_3$

(Figure 5a). Despite this, our results demonstrate a clear induction in ATP levels with treatment at certain time points (Figure 5b), which, as previously mentioned, may drive TXNIP expression through increasing the availability of glycolytic intermediates.

Another interesting finding observed in this study is the lack of reduction in TXNIP levels upon addition of CHX to HL-60 cells treated with 1,25$(OH)_2D_3$ for 96 h (Figure 4c,d). This observation supports the possibility that 1,25$(OH)_2D_3$ influences TXNIP levels independent of direct transcriptional regulation. We explored the possibility that 1,25$(OH)_2D_3$ could lead to TXNIP stability through reducing the expression of the E3 ubiquitin ligase ITCH. Although ITCH mRNA levels were significantly reduced by 1,25$(OH)_2D_3$ after 24 and 96 h of treatment (Figure 4e), protein levels were found to be unaffected (Figure 4f–i). Results from others have illustrated the ability of 1,25$(OH)_2D_3$, and its analogues, to influence numerous players that are involved in regulating protein stability/degradation, including ubiquitin proteasome pathway (UPP) players, different proteases, as well as protease inhibitors [18]. For example, in colon cancer cells, 1,25$(OH)_2D_3$ treatment was found to mediate an overall repression of numerous genes encoding proteins belonging to the UPP [18]. Furthermore, in HL-60 cells, Shimbara et al. [17] showed that 1,25$(OH)_2D_3$ treatment reduced mRNA levels of different proteasome subunits. In view of this, we assume that in our setting, 1,25$(OH)_2D_3$ treatment reduced the expression of genes involved in TXNIP degradation besides ITCH. This could explain why 1,25$(OH)_2D_3$ treatment initially induced the expression of TXNIP at early time points (e.g., 24 h), but later stabilized its levels by preventing its degradation, and thus, a reduction in TXNIP levels by CHX treatment was not observed in 96 h 1,25$(OH)_2D_3$-treated HL-60 cells (Figure 4c,d). In support of this is the observed hastened and potentiated increase in TXNIP mRNA levels in HL-60 cells treated with 1,25$(OH)_2D_3$ and CHX, when compared to 1,25$(OH)_2D_3$ alone, described by Chen and DeLuca in their initial report [4]. Moreover, they also described an increase in TXNIP mRNA levels with CHX treatment alone [4], which altogether indicates that inhibition of de novo protein synthesis may reduce the expression of a TXNIP mRNA-degrading protein.

## 4. Materials and Methods

### 4.1. Cell Culture

The following cell lines were included in the study and were maintained in a standard tissue culture incubator set to 37 °C and 5% $CO_2$: HL-60, U937, and Jurkat (hematological cancers); LNCaP, DU145, and PC3 (prostate cancer); HepG2, HLE, and SK-HEP-1 (liver cancer); AsPC-1, BxPC-3, and MIA PaCa-2 (pancreatic cancer); MDA-MB-231 and MCF-7 (breast cancer); HCT116 and HT-29 (colorectal cancer). Cell lines representing hematological cancers were cultured in RPMI 1640-GlutaMAX™ medium (Gibco, Darmstadt, Germany) supplemented with 10% FCS (*v*/*v*) (Gibco), and 1% penicillin/streptomycin (*v*/*v*) (Gibco). Other cell lines were cultured in Dulbecco's Modified Eagle Medium (DMEM)-GlutaMAX™ (Gibco), supplemented with 10% FCS (*v*/*v*) (Gibco), 1% penicillin/streptomycin (*v*/*v*) (Gibco). Treatments with 1,25$(OH)_2D_3$ (Cayman Chemicals—Biomol GmbH, Hamburg, Germany) were performed in standard medium for different time points. 2-deoxyglucose (Fluka-Sigma-Aldrich, Steinheim, Germany), MG-132 (Sigma-Aldrich, Steinheim, Germany), CHX (Fluka-Sigma-Aldrich), metformin HCl (Sigma-Aldrich), and phloretin (Sigma-Aldrich) were used as indicated. For glucose deprivation experiments, RPMI 1640 without glucose medium (Gibco) was used.

### 4.2. RNA Isolation, cDNA Synthesis, and RT-qPCR

HL-60 cells were treated with 1,25$(OH)_2D_3$ for various time points, after which total RNA was extracted using QIAzol lysis reagent (Qiagen, Hilden, Germany). The purity and concentration of RNA samples were determined using a NanoDrop 2000 UV-Vis Spectrophotometer (Thermo Scientific, Darmstadt, Germany). 500 ng of total RNA were used to synthesize cDNA using ProtoScript® II first strand cDNA synthesis kit (New England Biolabs, Frankfurt am Main, Germany), following the manufacturer's instructions. Subsequently, qPCR was performed using the real-time thermal cycler qTower (Analytik Jena AG, Jena, Germany) to quantify mRNA levels. The following forward (for) and reverse (rev) primers (Eurofins Genomics, Ebersberg, Germany) were used: TXNIP for: 5′-CGCCTCCTGCTTGAAACTAAC-3′, rev: 5′-AATATACGCCGCTGGTTACACT-3′; CYP24A1 for: 5′-TGGGGCTGGGAGTAATACTGA-3′, rev: 5′-GAACGCAATTTCATGGGAGGC-3′; ITCH for: 5′-5TCTAGTAGCTGTGGTCGGGG-3′, rev: 5′-CACAAGGCCACCGTGAAATG-3′ and vinculin (as reference gene) for: 5′-CAGTCAGACCCTTACTCAGTG-3′, rev: 5′-CAGCCTCATCGAAGGTAAGGA-3′. Reactions were performed using ready to use master mix LightCycler® 480 SYBR Green I (Roche, Mannheim, Germany).

### 4.3. Intracellular ROS and Glucose Uptake Measurements Using Flow Cytometry (FACS)

HL-60 cells were seeded at a density of 200,000 cells/well in 12 well-plates and were immediately treated with 1,25(OH)$_2$D$_3$. For intracellular ROS determination at the time points indicated, cells were washed once with PBS (Gibco), incubated with 30 μM dihydroethidium (Biomol GmbH) for 15 min, harvested, and re-suspended in 500 μL PBS for FACS analysis.

For glucose uptake measurements, 50 μM of the fluorescently labeled glucose analog 2-[*N*-(7-nitrobenz-2-oxa-1,3-diazol-4-yl) amino]-2-deoxy-D-glucose (2-NBDG) (Cayman Chemicals—Biomol GmbH) was added to the culture medium 1 h before the treatment period was over, as previously described, with minor modifications [15,28]. Cells were subsequently harvested and re-suspended in 500 μL PBS for FACS analysis. FACS analysis was performed using the FACSCalibur instrument (Becton Dickinson, Franklin Lakes, NJ, USA) and the CellQuest™ software (Becton Dickinson).

### 4.4. Determination of Glucose in Medium Using the Glucose Oxidase (GOx) Assay

The assay was performed, as previously described [15]. The GOx enzyme mix was prepared using the following constituents: 50 mg/L GOx (Sigma-Aldrich), 250 mM Tris pH 8.0, 40 mg/L HRP (Sigma-Aldrich), and 100 mg/L *O*-dianisidine (Sigma-Aldrich). 10 μL of the medium supernatant of DMSO- and 1,25(OH)$_2$D$_3$-treated cells (initial seeding 20,000 cells/well in 500 μL medium in a 24-well plate) were collected at different time points, and were diluted in water 10 times. 240 μL of the enzyme mix were added to 5 μL of each sample or standard, and incubated for 1 h at room temperature. Absorbance at 450 nm was measured using a Tecan Ultra plate reader (Tecan, Crailsheim, Germany) and glucose concentration in samples was determined with a calibration curve obtained from reference absorbance values of six different glucose standards (concentrations 0.1–1.0 g/L).

### 4.5. On-Line Measurements of Cellular Bioenergetics

OxoDish and HydroDish 24-well plates (PreSens Precision Sensing GmbH, Regensburg, Germany) were used for on-line measurements of dissolved oxygen and the pH of the medium, respectively, as previously described [29]. Changes in dissolved oxygen reflect oxygen consumption by respiration and hence mitochondrial activity, whereas changes in medium pH indicate lactate production from glycolytic metabolism. The plates contain fluorescence based sensors that were embedded at the bottom of each well, that can be read out continuously using dedicated SensorDish Readers (SDRs) that were placed inside a standard cell culture incubator. For the measurements, 20,000 cells/well were seeded in 1 mL/well medium in either Hydrodish or Oxoplate, and placed on an SDR inside the incubator. Measurements were started immediately and signals were recorded every 3 min. When signals had stabilized—approx. 20 min after initiation—measurements were paused, and plates were taken from the incubator to start treatment adding either DMSO, 1,25(OH)$_2$D$_3$, and designated controls (metformin or 2-deoxyglucose for oxygen and pH measurements, respectively). Plates were then placed back inside the incubator and measurements were continued.

### 4.6. Determination of Cellular ATP

HL-60 cells were seeded at a density of 5000 cells/well in 100 μL medium in a black 96-well microplate with a clear bottom (Costar®, Corning Incorporated, New York NY, USA). Cells were subsequently treated with 1,25(OH)$_2$D$_3$ for different periods, after which 100 μL of the substrate solution provided with the ATPlite™ 1 step kit (Perkin Elmer, Rodgau, Germany) were added to each well. The Tecan Ultra plate reader (Tecan) was used to measure luminescence signals kinetically. The obtained values were then normalized to cell count.

### 4.7. Western Blotting

After treatment, cells were harvested, washed once with PBS, and lysed using 6 M urea buffer supplemented with a cocktail of protease and phosphatase inhibitors, namely aprotinin, leupeptin, pepstatin, PMSF, sodium orthovanadate, and sodium pyrophosphate. Protein content of samples was determined using Bradford reagent (Sigma-Aldrich). SDS-PAGE was subsequently performed to resolve the samples, and proteins were then transferred onto PVDF membranes (GE Healthcare, Munich, Germany). Membranes were washed once in TBS-Tween for 5 min, and then blocked for 1 h at room temperature using 5% non-fat dry milk in TBS/Tween. Membranes were then washed once in TBS/Tween for 5 min and were incubated overnight at 4 °C with the primary antibody. Anti-VDUP1 (TXNIP) antibody was purchased from MBL, whereas anti-β-actin and anti-vinculin antibodies were purchased from Santa Cruz Biotechnology. Anti-ITCH, anti-phospho-ACC $^{(S79)}$, as well as anti-mouse and rabbit IgG horseradish peroxidase (HRP)-linked antibodies were purchased from Cell signaling technologies. Western Lightning™ Plus-ECl (Perkin Elmer) was utilized as HRP substrate. Target proteins were detected using the Fujifilm LAS-3000 imaging system.

### 4.8. Statistical Analyses

GraphPad Prism and Microsoft Excel were used for statistical analyses. ImageJ software was used for densitometric analysis. Two-tailed Student's *t*-test was used to calculate significance in investigated paramters between 1,25(OH)$_2$D$_3$ and DMSO treatment. A *p*-value that was less than or equal to 0.05 was defined as statistically significant. In figures, *, ** and *** represent *p*-values less than or equal to 0.05, 0.01, and 0.001, respectively. Error bars ± SD unless otherwise stated.

## 5. Conclusions

This study describes the regulation of TXNIP expression by 1,25(OH)$_2$D$_3$ in different cancer models. The *TXNIP* gene is possibly not a primary target of 1,25(OH)$_2$D$_3$ since the canonical induction is not observed in all investigated cell lines, and, in HL-60 cells, the mRNA induction is transient and glucose-dependent. However, 1,25(OH)$_2$D$_3$ appears to induce a clear and sustained increase in TXNIP protein levels in the same cell line, possibly through regulating protein stabilizing mechanisms at late time points. Nonetheless, several questions remain unanswered with regards to regulation of the *TXNIP* gene by 1,25(OH)$_2$D$_3$, such as: (i) whether the VDR physically/functionally interacts with glucose-sensing transcriptional machinery, namely MondoA/MLX, and (ii) whether the *TXNIP* gene harbors VDRE and regulation by 1,25(OH)$_2$D$_3$ is subject to chromatin architecture. Additionally, the putative induction and stabilization of TXNIP levels by 1,25(OH)$_2$D$_3$ are potentially influenced by the myriad of cellular effects the molecule induces, such as regulation of glucose metabolism [15,30], protein degradation [18], and non-coding RNAs [31]. Moreover, the results question the role/importance of TXNIP in mediating 1,25(OH)$_2$D$_3$'s anti-cancer effects since in cell lines where TXNIP levels are reduced by 1,25(OH)$_2$D$_3$, such as LNCaP and MCF-7, treatment has been shown to induce profound anti-tumor actions [32–34]. On the other hand, the activation of TXNIP expression by 1,25(OH)$_2$D$_3$ in HL-60 cells described in this study coincides with a significant inhibition of cellular proliferation. Therefore, a better understanding of TXNIP's context-dependent roles would shed light on its importance to calcitriol's effects in tumor cells.

**Supplementary Materials:** Supplementary materials can be found at www.mdpi.com/1422-0067/19/3/796/s1.

**Acknowledgments:** Mohamed A. Abu el Maaty and Fadi Almouhanna are recipients of doctoral fellowships from the German Academic Exchange Service (DAAD). The authors acknowledge technical support from Nadine Pfaller. We acknowledge financial support by Deutsche Forschungsgemeinschaft within the funding programme Open Access Publishing, by the Baden-Württemberg Ministry of Science, Research and the Arts and by Ruprecht-Karls-Universität Heidelberg.

**Author Contributions:** Mohamed A. Abu el Maaty and Stefan Wölfl conceived and designed the study. Mohamed A. Abu el Maaty and Fadi Almouhanna performed experiments. Mohamed A. Abu el Maaty, Fadi Almouhanna and Stefan Wölfl analyzed the data. Mohamed A. Abu el Maaty and Stefan Wölfl wrote the manuscript. All authors approved the final version of the manuscript prior to submission.

**Conflicts of Interest:** The authors declare no conflict of interest.

## Abbreviations

| | |
|---|---|
| $1,25(OH)_2D_3$ | 1,25-dihydroxyvitamin $D_3$ |
| ACC | Acetyl CoA carboxylase |
| AMPK | AMP-activated protein kinase |
| CAMKK | Calcium/calmodulin-dependent protein kinase kinase |
| ChoRE | Carbohydrate-response element |
| CHX | Cycloheximide |
| DMSO | Dimethyl sulfoxide |
| GOx | Glucose oxidase |
| MLX | Max-like protein X |
| ROS | Reactive oxygen species |
| TXNIP | Thioredoxin-interacting protein |
| Ub | Ubiquitin |
| UPP | Ubiquitin proteasome pathway |
| VDR | Vitamin D receptor |
| VDRE | Vitamin D response elements |
| VDUP1 | Vitamin $D_3$ upregulated protein 1 |

## References

1. Feldman, D.; Krishnan, A.V.; Swami, S.; Giovannucci, E.; Feldman, B.J. The role of vitamin d in reducing cancer risk and progression. *Nat. Rev. Cancer* **2014**, *14*, 342–357. [CrossRef] [PubMed]
2. Abu El Maaty, M.A.; Wolfl, S. Effects of $1,25(OH)_2D_3$ on cancer cells and potential applications in combination with established and putative anti-cancer agents. *Nutrients* **2017**, *9*, 87. [CrossRef] [PubMed]
3. Abu El Maaty, M.A.; Wolfl, S. Vitamin D as a novel regulator of tumor metabolism: Insights on potential mechanisms and implications for anti-cancer therapy. *Int. J. Mol. Sci.* **2017**, *18*, 2184. [CrossRef] [PubMed]
4. Chen, K.S.; DeLuca, H.F. Isolation and characterization of a novel cDNA from HL-60 cells treated with 1,25-dihydroxyvitamin D-3. *Biochim. Biophys. Acta* **1994**, *1219*, 26–32. [CrossRef]
5. Nishiyama, A.; Matsui, M.; Iwata, S.; Hirota, K.; Masutani, H.; Nakamura, H.; Takagi, Y.; Sono, H.; Gon, Y.; Yodoi, J. Identification of thioredoxin-binding protein-2/vitamin $D_3$ up-regulated protein 1 as a negative regulator of thioredoxin function and expression. *J. Biol. Chem.* **1999**, *274*, 21645–21650. [CrossRef] [PubMed]
6. Stoltzman, C.A.; Peterson, C.W.; Breen, K.T.; Muoio, D.M.; Billin, A.N.; Ayer, D.E. Glucose sensing by mondoa:Mlx complexes: A role for hexokinases and direct regulation of thioredoxin-interacting protein expression. *Proc. Natl. Acad. Sci. USA* **2008**, *105*, 6912–6917. [CrossRef] [PubMed]
7. Yu, F.X.; Goh, S.R.; Dai, R.P.; Luo, Y. Adenosine-containing molecules amplify glucose signaling and enhance TXNIP expression. *Mol. Endocrinol.* **2009**, *23*, 932–942. [CrossRef] [PubMed]
8. Han, K.S.; Ayer, D.E. Mondoa senses adenine nucleotides: Transcriptional induction of thioredoxin-interacting protein. *Biochem. J.* **2013**, *453*, 209–218. [CrossRef] [PubMed]
9. Wu, N.; Zheng, B.; Shaywitz, A.; Dagon, Y.; Tower, C.; Bellinger, G.; Shen, C.H.; Wen, J.; Asara, J.; McGraw, T.E.; et al. AMPK-dependent degradation of TXNIP upon energy stress leads to enhanced glucose uptake via GLUT1. *Mol. Cell* **2013**, *49*, 1167–1175. [CrossRef] [PubMed]

10. Zhou, J.; Chng, W.J. Roles of thioredoxin binding protein (TXNIP) in oxidative stress, apoptosis and cancer. *Mitochondrion* **2013**, *13*, 163–169. [CrossRef] [PubMed]
11. Zhou, J.; Yu, Q.; Chng, W.J. TXNIP (VDUP-1, TBP-2): A major redox regulator commonly suppressed in cancer by epigenetic mechanisms. *Int. J. Biochem. Cell Biol.* **2011**, *43*, 1668–1673. [CrossRef] [PubMed]
12. Butler, L.M.; Zhou, X.; Xu, W.S.; Scher, H.I.; Rifkind, R.A.; Marks, P.A.; Richon, V.M. The histone deacetylase inhibitor saha arrests cancer cell growth, up-regulates thioredoxin-binding protein-2, and down-regulates thioredoxin. *Proc. Natl. Acad. Sci. USA* **2002**, *99*, 11700–11705. [CrossRef] [PubMed]
13. Zhou, J.; Bi, C.; Cheong, L.L.; Mahara, S.; Liu, S.C.; Tay, K.G.; Koh, T.L.; Yu, Q.; Chng, W.J. The histone methyltransferase inhibitor, DZNep, up-regulates TXNIP, increases ROS production, and targets leukemia cells in AML. *Blood* **2011**, *118*, 2830–2839. [CrossRef] [PubMed]
14. Shalev, A. Minireview: Thioredoxin-interacting protein: Regulation and function in the pancreatic beta-cell. *Mol. Endocrinol.* **2014**, *28*, 1211–1220. [CrossRef] [PubMed]
15. Abu El Maaty, M.A.; Alborzinia, H.; Khan, S.J.; Buttner, M.; Wolfl, S. 1,25(OH)$_2$D$_3$ disrupts glucose metabolism in prostate cancer cells leading to a truncation of the TCA cycle and inhibition of TXNIP expression. *Biochim. Biophys. Acta* **2017**, *1864*, 1618–1630. [CrossRef] [PubMed]
16. Zhang, P.; Wang, C.; Gao, K.; Wang, D.; Mao, J.; An, J.; Xu, C.; Wu, D.; Yu, H.; Liu, J.O.; et al. The ubiquitin ligase itch regulates apoptosis by targeting thioredoxin-interacting protein for ubiquitin-dependent degradation. *J. Biol. Chem.* **2010**, *285*, 8869–8879. [CrossRef] [PubMed]
17. Shimbara, N.; Orino, E.; Sone, S.; Ogura, T.; Takashina, M.; Shono, M.; Tamura, T.; Yasuda, H.; Tanaka, K.; Ichihara, A. Regulation of gene expression of proteasomes (multi-protease complexes) during growth and differentiation of human hematopoietic cells. *J. Biol. Chem.* **1992**, *267*, 18100–18109. [PubMed]
18. Alvarez-Diaz, S.; Larriba, M.J.; Lopez-Otin, C.; Munoz, A. Vitamin D: Proteases, protease inhibitors and cancer. *Cell Cycle* **2010**, *9*, 32–37. [CrossRef] [PubMed]
19. Abu El Maaty, M.A.; Strassburger, W.; Qaiser, T.; Dabiri, Y.; Wolfl, S. Differences in p53 status significantly influence the cellular response and cell survival to 1,25-dihydroxyvitamin D$_3$-metformin cotreatment in colorectal cancer cells. *Mol. Carcinog.* **2017**, *56*, 2486–2498. [CrossRef] [PubMed]
20. Hardie, D.G.; Ross, F.A.; Hawley, S.A. Amp-activated protein kinase: A target for drugs both ancient and modern. *Chem. Biol.* **2012**, *19*, 1222–1236. [CrossRef] [PubMed]
21. Sheth, S.S.; Bodnar, J.S.; Ghazalpour, A.; Thipphavong, C.K.; Tsutsumi, S.; Tward, A.D.; Demant, P.; Kodama, T.; Aburatani, H.; Lusis, A.J. Hepatocellular carcinoma in TXNIP-deficient mice. *Oncogene* **2006**, *25*, 3528–3536. [CrossRef] [PubMed]
22. Stambolsky, P.; Tabach, Y.; Fontemaggi, G.; Weisz, L.; Maor-Aloni, R.; Siegfried, Z.; Shiff, I.; Kogan, I.; Shay, M.; Kalo, E.; et al. Modulation of the vitamin D$_3$ response by cancer-associated mutant p53. *Cancer Cell* **2010**, *17*, 273–285. [CrossRef] [PubMed]
23. Wolf, D.; Rotter, V. Major deletions in the gene encoding the p53 tumor antigen cause lack of p53 expression in HL-60 cells. *Proc. Natl. Acad. Sci. USA* **1985**, *82*, 790–794. [CrossRef] [PubMed]
24. Wilde, B.R.; Ayer, D.E. Interactions between myc and mondoa transcription factors in metabolism and tumourigenesis. *Br. J. Cancer* **2015**, *113*, 1529–1533. [CrossRef] [PubMed]
25. Kaadige, M.R.; Yang, J.; Wilde, B.R.; Ayer, D.E. MondoA-Mlx transcriptional activity is limited by mtor-mondoa interaction. *Mol. Cell. Biol.* **2015**, *35*, 101–110. [CrossRef] [PubMed]
26. Liang, J.; Mills, G.B. AMPK: A contextual oncogene or tumor suppressor? *Cancer Res.* **2013**, *73*, 2929–2935. [CrossRef] [PubMed]
27. Hoyer-Hansen, M.; Bastholm, L.; Szyniarowski, P.; Campanella, M.; Szabadkai, G.; Farkas, T.; Bianchi, K.; Fehrenbacher, N.; Elling, F.; Rizzuto, R.; et al. Control of macroautophagy by calcium, calmodulin-dependent kinase kinase-beta, and BCL-2. *Mol. Cell* **2007**, *25*, 193–205. [CrossRef] [PubMed]
28. Cheng, X.; Kim, J.Y.; Ghafoory, S.; Duvaci, T.; Rafiee, R.; Theobald, J.; Alborzinia, H.; Holenya, P.; Fredebohm, J.; Merz, K.H.; et al. Methylisoindigo preferentially kills cancer stem cells by interfering cell metabolism via inhibition of LKB1 and activation of AMPK in PDACS. *Mol. Oncol.* **2016**, *10*, 806–824. [CrossRef] [PubMed]
29. Lochead, J.; Schessner, J.; Werner, T.; Wolfl, S. Time-resolved cell culture assay analyser (trecca analyser) for the analysis of on-line data: Data integration—Sensor correction—Time-resolved IC$_{50}$ determination. *PLoS ONE* **2015**, *10*, e0131233. [CrossRef] [PubMed]

30. Zheng, W.; Tayyari, F.; Gowda, G.A.; Raftery, D.; McLamore, E.S.; Shi, J.; Porterfield, D.M.; Donkin, S.S.; Bequette, B.; Teegarden, D. 1,25-dihydroxyvitamin D regulation of glucose metabolism in harvey-ras transformed MCF10A human breast epithelial cells. *J. Steroid Biochem. Mol. Biol.* **2013**, *138*, 81–89. [CrossRef] [PubMed]
31. Giangreco, A.A.; Nonn, L. The sum of many small changes: Micrornas are specifically and potentially globally altered by vitamin $D_3$ metabolites. *J. Steroid Biochem. Mol. Biol.* **2013**, *136*, 86–93. [CrossRef] [PubMed]
32. Swami, S.; Krishnan, A.V.; Wang, J.Y.; Jensen, K.; Horst, R.; Albertelli, M.A.; Feldman, D. Dietary vitamin $D_3$ and 1,25-dihydroxyvitamin $D_3$ (calcitriol) exhibit equivalent anticancer activity in mouse xenograft models of breast and prostate cancer. *Endocrinology* **2012**, *153*, 2576–2587. [CrossRef] [PubMed]
33. Narvaez, C.J.; Welsh, J. Role of mitochondria and caspases in vitamin D-mediated apoptosis of MCF-7 breast cancer cells. *J. Biol. Chem.* **2001**, *276*, 9101–9107. [CrossRef] [PubMed]
34. Skowronski, R.J.; Peehl, D.M.; Feldman, D. Vitamin D and prostate cancer: 1,25 dihydroxyvitamin $D_3$ receptors and actions in human prostate cancer cell lines. *Endocrinology* **1993**, *132*, 1952–1960. [CrossRef] [PubMed]

MDPI
St. Alban-Anlage 66
4052 Basel
Switzerland
Tel. +41 61 683 77 34
Fax +41 61 302 89 18
www.mdpi.com

*International Journal of Molecular Sciences* Editorial Office
E-mail: ijms@mdpi.com
www.mdpi.com/journal/ijms

www.ingramcontent.com/pod-product-compliance
Lightning Source LLC
Chambersburg PA
CBHW041752220926
51847CB00014B/179
*9783038975861*